Crop Management and Integrated Farming

Dr. S. C. Panda
Ph.D.
Emeritus Scientist, ICAR,
Former Dean, College of Agriculture,
Dean of Research, Dean, PGF-cum-DRI, OUAT,
Professor & Head, Dept. of Agronomy, ADR,
Chief Scientist (Water Management),
Regional Co-ordinator, NDP (Eastern Region of India),
Program Associate (Multiple Cropping Res.) Ford Foundation

AGROBIOS (INDIA)

Published by:
AGROBIOS (INDIA)
Agro House, Behind Nasrani Cinema
Chopasani Road, Jodhpur 342 003
Phone: +91-291-2642319, 2643993
E. mail: agrobiosindia@gmail.com

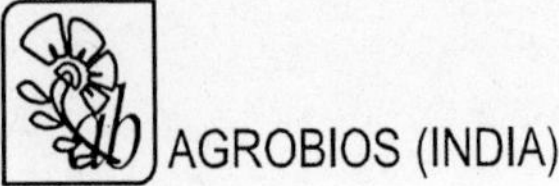

ISBN (13): 978-81-7754-264-6
***Price:* Rs. 2500.00 / US$ 90.00**

Published by: Dr. Updesh Purohit for Agrobios (India), Jodhpur
Lasertypeset at: Yashee Computers, Jodhpur
Cover Design by: Reena
Printed at: Agrobios Digitals, Jodhpur

AT THE FEET OF
LORD SHREE JAGANNATH,
PURI, ORISSA

Dedicated to My
Father, Shyam Sundar Panda
and Mother, Shnehalata Panda

Dr. B. Panda
B.Sc. (Vet.) Hons. (Mumbai),
M.S., Ph.D. (Maryland, USA)

Founder Director (Retd.),
Central Avian Research Institute (ICAR)
Former Member of the Board of Management,
OUAT, Bhubaneswar.
Fellow: International Poultry Hall of Fame
(World's Poultry Science Association)
National Academy of Veterinary Sciences (India)

Residence
C-6, Palaspalli,
Bhubaneswar – 751 020
Orissa (India)
Tel: 0674-590376,
Fax: 91-674-418142
(Attn. Dr. B. Panda)
E-mail: bpanda-29@hotmail.com

FOREWORD

The pressure of an ever increasing population and periodic famine due to unexpected flood and drought have forced and awakened the agricultural scientists of India to evolve new plant types and to develop a suitable production technology for such high input responsive crop plants. The last two decades had been an era of real revolution in agriculture, especially in the field of crop production which has brought a spectacular change in the existing cropping system in various agro-climatic zones of the country. The breakthrough in crop yields has made our nation self-sufficient in food grains and even some quantity of these grains is being exported to other needy countries. Thus, the modern technology has changed us from net importer to exporter of the food grains. The Indian farmers have shown a surprising dynamism, adaptability and progressiveness in accepting the recent crop production technology and predominant subsistence farming is being changed into an enthusiastic enterprise however, it lacks perfection. Moreover, even today the national average yield of various crops is miserably poor and incomparable to that obtained in other counties. Transfer of technology from research level to the farmers had been probably the major hurdle in achieving the targeted yield. Illiteracy and ignorance of our poor farmers are creating hindrance for new plant types to express their full yield potential when grown under faulty management in the field. Higher input levels, faulty planning coupled with poor marketing facilities often result in a marginal profit from farming due to which majority of marginal and sub-marginal farmers are obviously discouraged.

The technology of crop production is a vastly developing subject. In recent years, there has been such rapid progress in the research efforts in this branch in many of the developing countries of the world that has become difficult to keep pace with the subject. This is particularly true in the tropical regions of the world and in the Indian sub-continent. The students of Agronomy in the Agricultural Colleges need to be taught the latest technology so as to make them effective as extension workers as also to train them for becoming competent teachers and researchers. A comprehensive treatise on crop production would not only help the students but also the teachers in the Agricultural Colleges. In modern agricultural technology 'management' forms an essential part. If agriculture were to compete with other small and medium scale industries in rural areas, there should be proper management of the natural resources and the various inputs which go with it. Unfortunately, this aspect is mostly neglected in Indian agriculture.

Besides, the modern production technologies with high input requirement offers on tangible solution to the small and medium farmers' problem of low income and low productivity from cropping. To meet the growing demands, to solve small and medium farmers' problems and to achieve the targeted production levels with stability, ecological sustainability and equitability in the 21st century, a viable technology called 'Integrated Farming System Approach' suited to varied farm situations and agro-climatic conditions seem to be the answer. The farming system approach incorporated in this book has taken due importance through recycling of no cost/low cost material of one enterprise as input on the other enterprise thus achieving great reduction in the cost of cultivation. Agronomists have made efforts to develop low-cost farming systems based on the principles of productive utilization of farm wastes and available resources with restricted use of purchased inputs.

Farming Systems represent integration of farm enterprises as cropping systems, animal husbandry, fisheries, poultry farming, etc. for optimal utilization of resources bringing prosperity to the farmers. A judicious mix of cropping systems with associated enterprises like dairy, poultry, piggery, fishery, sericulture, etc. suited to the given agro-climatic conditions and socio-economic status of farmers would bring prosperity to the farmer.

The text-cum-reference book to meet precisely the felt need is an outcome of the author's active involvement in teaching, research and extension guidance in the field of agronomy for over thirty years. The author presented the book entitled, **"Crop Management and Integrated Farming"** in a scientific and systematic manner to understand the fundamentals clearly and easily which is the beauty of this book. I believe that this book will be very useful to the undergraduate and postgraduate students in agriculture, teachers in agricultural institutions and those who are interested in the subject. The book is divided into twenty-one chapters and covers comprehensively the content of the courses of crop management and integrated farming in agronomy for undergraduate and post graduate students of Agricultural Universities in the country. The chapters have been arranged in such a manner as to lead the students through the entire gamut of Agronomy.

I am confident that this book will serve as a text book for agronomy and veterinary students, a reference for research scientists and teachers in the areas of crop production, integrated farming systems, dry land agriculture, cropping systems, production technology management under different situations, soil fertility management, water management, weed management, nutrient management, and avian and animal sciences. This book will also serve as a guide to the extension officials of the department of agriculture. I congratulate Dr. S. C. Panda for his pains taking effort in bringing out this book covering comprehensively the content of all courses in agronomy offered for U.G. and P.G. students of the agricultural universities in the country and the latest technologies for crop production associated with integrated enterprises in farming systems to meet the growing interest in sustainable agriculture. I am confident that this book will be widely accepted among the students. I extend my best wishes to Dr. Sharat Chandra Panda for the success of this book.

Bhubaneswar **Dr. Bhagabat Panda**

Preface

Agriculture has been a mainstay of human being since time immemorial. Agriculture in any form has been practiced in those areas where men have been living permanently. They have been restrained from plundering and/or indiscriminate hunting rather domesticated a number of facultative organisms. They have been utilizing their diligence and intelligence to integrate natural germplasms of plants and animals with natural resources like land, light, air and other biotic and abiotic factors of production to produce usable products directly or after processing by their own muscle power or any aid found in nature or developed by them for ease, cheap and swiftness.

Agricultural development is multidirectional having galloping speed and rapid spread with respect to time and space. After the introduction of modern varieties of various durations with improved plant types along with improvement of input resources, farmers started using improved cultural practices in intensive cropping systems with labour-intensive programmes to improve production potential per unit land, time and input. A large number of alternate genotypes are available to the farmers. Agronomists provided suitable environment to these genotypes to foster them and to manifest their yield potential in newer areas and seasons.

Dealing with various types of soils and climatic conditions is a pre-requisite to grow crops. Understanding of relationships between crop plants and natural and manmade factors of production is a guideline to provide better environment to crop plants and to increase use-efficiency of these factors. A wide spectrum knowledge of soil, water, nutrients, weeds etc. helps to manage them efficiently in favour of crop production for years. A comprehensive idea on seeds and their handling, sowing and planting of different crop varieties under different agro-climatic conditions, cropping pattern and cropping systems helps to raise crops together or one after another considering prevailing conditions and available facilities.

Modern concept on improved cropping system is found to be successful in shifting the direction of agricultural practices from crop oriented one to system oriented approach. Modern agro-techniques on crop production, protection and preservation have opened the opportunities to absorb, adapt and apply new knowledge and technology to the farmers. Recently introduced methods of cropping also approach diversified farming systems with components complementary to each other. Non-traditional crops with respect to zone and season showed their promise. All these have declined the earlier traditional agriculture and established the scientific agribusiness.

India, with one of the biggest scientific manpower of the world, made a spectacular increase in its food grains production, after independence, especially after late sixties, when the country ushered into the world famous **'Green Revolution'.** The Green Revolution is manifested of the advances made in the agricultural sciences and can be summed up in three words, (1) seeds (High yielding Varieties), (2) water, and (3) fertilizers. The management of the last two factors of production, with a view to achieving the production potential of the first, i.e., High yielding varieties (seeds) is exclusively an agronomic domain.

The future of Indian agriculture depends on the development of appropriate farming system as applicable to resource poor farm families and as suited to different agro-ecological zones. Restricting the use of purchased inputs in farming could be achieved through multiple cropping and diversified farming including animal husbandry, forestry, dairy, duckery, fisheries, apiary, sericulture, etc. The demand must be met from limited resources such as land, water, energy and labour. Increasing food supplies with limited natural resources is a great challenge to the scientific communities. Under this situation, Integrated Farming System (IFS) seems to be the answer, considering the current scenario in agriculture in India. Besides, facilitating cash income and increased employment opportunities, Integrated Farming Systems minimize the quantum of purchased inputs in farming by effective recycling of products and by-products among the component enterprises, reduce the ill effects of inorganic fertilizers and chemicals (pesticides, herbicides etc.) or pollution hazards. Integrated farming system is not only a reliable way of obtaining fairly high productivity with substantial fertilizer economy, but also a concept of ecological soundness leading to sustainable agriculture.

Integrated farming system will generate appreciable employment potentialities providing more man-days in a year. Even for the educated person's particularly agricultural graduates, such farming systems are likely to be more profitable than white collar jobs thus partially reducing the job demands by the graduates. As a whole, such system will increase production of agriculture in the state, particularly, fruits, vegetables and produce from dairy, poultry, duckery and fishery which are of very much shortage in the state.

At present, there is no comprehensive text book on **'Crop Management and Integrated Farming'** and applied aspects suitable for farmers. This book will provide comprehensive information on the subject matter and fulfil the needs of students and other professionals. This is a book containing all sorts of valuable information in twenty-one chapters on crop management and integrated farming. Though this book primarily written to serve as a text book/reference for the students of agriculture in under graduate and post graduate levels and technologists in developing organizations, it is hoped that this book will be valuable for similar groups in the third world countries of Asia and Africa. This book also serves as a valuable reference for the candidates preparing Agricultural Research Services and other competitive examinations. Professional Institutions in Soil Conservation, Krishi Vigyna Kendras and Rural Institutions and similar other Institutions would find this book very much helpful. The farmers may refer this book to practice integrated farming and cropping systems as the considerable emphasis is placed for obtaining maximum, profitable production per unit area per unit time.

The author acknowledges his indebtedness to authors of books from which most of the material in the text has been drawn. In several cases, it has not been possible to obtain permission for reproduction for which the author and publishers offer their sincere apologies.

The author is deeply indebted to ICAR for its assistance provided at various levels for preparing the manuscript. Special mention is made for the valuable help received from Sri K.C. Sahoo, Research Fellow of the Emeritus Scientist Project, OUAT, Bhubaneswar.

The author would particularly like to thank Dr. Bhagabat Panda, Founder Director (Retd.), Central Avian Research Institute (ICAR) for his encouragement, invaluable help and providing excellent foreword. Not only has he checked the contents with meticulous care, but also he has made numerous suggestions for improvements and additions. His help and advice have added greatly to the value and reliability of a work such as this.

In preparing this book, I have received helps, suggestions and encouragements from Dr. B. C. Nayak, Dean, College of Agriculture; Dr. P. K. Mahapatra,, Professor and Head, Agronomy; Dr. L. M. Garnayak, Associate Professor, Agronomy and other staff members of the Department of Agronomy, College of Agriculture, OUAT, Bhubaneswar. I am grateful to all of them who communicated the encouraging comments regarding the text and constructing suggestions for the improvement of the book.

I express my profound love and affection to my wife Mrs Kalpana Panda and son Ar. Mrutyunjay Panda, Architect for their immense help in preparing this manuscript.

I also wish to thank the Agrobios (India), Jodhpur for the publication of this book.

Dr. S. C. Panda
2, Dharma Vihar, Kandagiri
Bhubaneswar - 751030 (Orissa),
Tel: (0674) 2351602/3102565

Contents

CHAPTER 1

Introduction

"If agriculture in this country fails, we fail,
The Government fails and the Nation fails,
There is no help for us but to succeed in Agriculture"

Nehru

Seventy per cent of the Indian population solely depends on agriculture those feed nearly 16% of the global population. About 26% of the gross national product comes from agricultural sector. The burgeoning population of India with a growth rate of 2.2% compounded with the effects of green revolution coming to a stand still has threatened our food security. Nearly 40% of the total Indian population live below the poverty line those cannot afford two squares of meals a day. If it is focused towards the middle of this century, the food and feed requirement will be nearly doubled that of the present situation. Out of 328.8 m ha of total geographical area, 18% is under forest, 13.5% is not available for cultivation. Only 144 m ha is available for cultivation which has to sustain the one billion of population.

Agriculture is practiced by 105 million farm families who live in 6 lakh villages. About 76.2% of the total farming community belongs to marginal and small farmers, who possess only 29% of total operated area (Agril Census 1985-86). On the other hand the per capita availability of land from the time of independence of the country has declined from 0.48 ha to 0.15 ha by 2000. The income from seasonal field crops alone from small and marginal farm is hardly sufficient now to sustain the farmers' family. With the gradual decline in farm size, it would be increasingly difficult to produce enough food for the farm family in coming years. Under such circumstances, the farmers will have to under take some land based enterprises which will compliment their existing farming activity to get more income and employment, leading to a better standard of living. Such enterprises include livestock, fishery, poultry, duckery, apiary, mushroom production and tree farming etc. A judicious combination of any one or more of such enterprises with cropping and effective recycling of residues/wastes would ensure better utilization of available resource and employment to the family labour during off season. Eastern India is characterized by the highest population density, low or stagnant growth in agriculture, per capita income and investment in infrastructure.

Agriculture in Orissa is largely rice based. Orissa has total geographical area of 155.7 lakh ha out of which 56.06 lakh ha is under forest, 4.45 lakh ha of land is cultivable waste, 60.48 lakh ha land is late sown area. Besides 8.38 lakh ha land is put to non agricultural uses (Orissa Agril. Statistics, 1998-99). This has only 30% of the net area sown under irrigation and fertilizer consumption of the state is only 36 kg per ha. Average rainfall during the cropping season is 1151.1 mm. Besides it has 480 km coast line, 0.49 m ha area under salinity, 0.34 m ha area under water logging and ill-drainage. Out of 33.5 million population of Orissa, 76% belong to small and marginal farmers and they occupy 62% of the total land holdings. Moreover, about 55% of the population mostly consisting of small and marginal farmers is below the poverty line. The average holding units for marginal farmers are less than one hectare. With the gradual decline in farm size, it would be increasingly difficult on the part of a marginal farmer to produce enough food for his family in coming years.

The benefits of modern technology have, however, been restricted to favourable farming situations. Only 25 to 20% of the modern technology have reached the farmers in the ecologically favourable areas but much less (< 5%) in the handicapped ecologies. This is often because the technology has not been consistent with the existing farming situations. In the horticulture and animal husbandry sectors, not even 1% of the technology has reached the small and marginal farmers. Only 44%, out of over 500 districts, are contributing to the half of total food grain basket of the country. This clearly suggests that the technology supposed to be scale and resource neutral has been confined to the districts with favourable farming conditions. Since, there is no further scope for horizontal expansion of land for cultivation and because of fragmentation and subdivision of land holdings and decline in the per capita availability of land day by day, there is no alternative left than to go in for vertical expansion and diversification of farm enterprises which have less demand on space and time particularly for small and marginal farmers with very limited resources, specially in rainfed areas (70% of total cultivated land). The new farming system research strategy, should therefore, necessarily concentrate on developing technology with participatory approach within the biophysical and socio-economic environments in which the farmers operate.

The most horrifying feature of many developing countries' demographic picture can be revealed from their burgeoning population over years. India with 2.2% (328 million ha) of the world geographical area supports about 16% of the global population increasing at a pace of 1.9% each year, which is equal to the population of Australia. The direct or/and indirect pressure has been exerted on the demand of food and land. There being no scope for horizontal expansion of cultivated land with dwindling man to land ratio, ever increasing problematic areas and with much touted "Green revolution" confined to a potentially endowed pockets benefiting well off farmers, the resource poor and socio-economically under privileged farmers' lot is becoming deplorable (Though there are 90 million small and marginal farmers out of 105 million farm families residing in 6 lakh villages). So, the only way out for maintaining sustainability in food production is through vertical expansion with diversification of farm enterprises particularly for resource poor small and marginal farmers by focusing on farmers' participatory research and generation of low cost location specific relevant and technically feasible technology and transfer of such technology through an effective and rejuvenated extension net work. Embracing all these purposes Farming Systems Research (FSR) has been developed for sustained increase in food production through vertical expansion.

World population has been increasing by leaps and bounds. The population of India is expected to reach 1370 and 1660 m in 2030 and 2050 A.D. respectively. The food production of the country has reached an all time high of 204 m t during 2000 A.D. A food production level of 289 and 349 m t is needed to satisfy the projected population in 2030 and 2050 A.D., respectively. The net cultivable area is 142.8 m ha. Unlike the population spurt and corresponding food need for 2050, there is every chance that the land area under cultivation will decrease due to diversion of some of the cultivable area to buildings and industrial purposes. It is anticipated that the land area available for cultivation in 2050 A.D. would be 130 m ha. So, it becomes necessary to increase the productivity almost to double of what we are producing today. This could be made possible by putting the land, both irrigated and rainfed, under intensive cultivation. Fortunately, most of our country lies in tropics and so is blessed with abundant solar energy, thus making cropping round the year possible. The only way to increase agricultural production is to increase the productivity per unit area per unit time. In the scientific cropping systems, genotypes, geometry of planting and management practices are designed to increase the productivity, simultaneously making efficient use of available resources and stabilizing yields (Palaniappan and Annadurai, 1999).

Modern agriculture emphasizes two more dimensions viz., **time** and **space**. Time concept relates to increasing crop intensification in a situation where there is no constraint for inputs including irrigation. In other words increasing the cropping intensity in areas where the production potential viz., land is underutilized even with full resource potential. It is a time bound programme where, for most of the field crops, it is considered for 365 days or 12 months. In the case of long duration and perennial crops, the duration of each rotation will vary from two or three years depending upon the duration of the constituent crops. The areas, where only one crop (100 per cent) two crops (200 per cent) three crops (300 per cent), are raised in a year, leaving the land fallow for two to eight months, the cropping intensity has to be increased to 200, 300 and 400 per cent, respectively within one year. This will ultimately help to enhance the productivity. In the case of rainfed areas where there is no possibility of increasing the intensity of cropping, the other concept viz., **'space concept'** can be applied. By way of building up of crops/other agricultural allied components in a vertical dimension (otherwise called multi-tier system), the land equivalent area can be increased. Thus by making use of these time and space dimensions either in irrigated or in rainfed areas within specified time span, say a year, and unit area of land, a hectare, productivity is sought to be increased by repeated and/or intensified cropping. This calls for urgent action on the part of Agronomists and Scientists of related disciplines to devote their attention on the design and testing of cropping systems for different regions. Income through arable farming alone is insufficient for bulk of the marginal farmers. Activities such as dairying, poultry, fish culture, sericulture, biogas production, edible mushroom cultivation, agro-forestry, agro-horticulture, etc., critical importance in supplementing their farm income. It fits well with farm level infrastructure and ensures fuller utilization of by-products.

Integrated Farming System (IFS) is the answer to the problem of increasing food production and farm income and for improving nutrition of the small-scale farmers with limited resources without any adverse effect on environment and agro-ecosystem. In a cropping system, the amount of by-products can be as high as or higher than marketable produce. This may go to waste if not utilized in an animal enterprise. Hence, integration of different agricultural and allied enterprises with crop activity as base will provide a way to recycle low cost produces at farm level/no cost waste materials of farm from one enterprises to another and thus reduce the cost of production of economic produce and finally to enhance the net income of the farm as a whole (Annadurai *et. al.*, 1994)

The future of Indian agriculture depends on the development of appropriate farming system as applicable to resource poor farm families and as suited to different agro ecological zones. Restricting the use of purchased inputs in farming could be achieved through multiple cropping and diversified farming including animal husbandry, forestry, sericulture etc., The demand must be met from limited resources--- land, water, energy and labour. Increasing food supplies with limited natural resources is a daunting challenge to the scientific community. Under this situation, Integrated Farming System (IFS) seems to be the answer, considering the current scenario in agriculture in India. Besides facilitating cash income and increased employment opportunities, IFS minimizes the quantum of purchased inputs in farming by effective recycling of products and by-products among the component enterprises; reduce the ill effects of inorganic fertilizers and chemicals (pesticides, herbicides etc.,)/or pollution hazards.

CHAPTER 2

Importance of Agriculture

The human society is dependent upon agriculture for its food, clothing and shelter. Agriculture plays a major role as the supplier of raw materials for industries and commerce apart from the basic role in supplying food, clothing and shelter. Therefore, the welfare of the state to a large extent is dependent upon the prosperity of its agriculture. Agriculture, therefore, is defined as the art, the science and the business of producing crops and livestock for economic purposes.

The word agriculture is derived from the Latin term ***Ager cultura*** meaning field cultivation. The term includes science of crop and animal husbandry, horticulture, dairy, forestry, fishery, etc. Agronomy deals with the principles and practices of field crops production. The term is derived from the Greek word ***Agros nomos*** meaning field to manage. For the management of field along with the soils and crops, the sciences of Biology, Chemistry, Physiology, Economics, etc. are made use of. Thus, knowledge of such allied science is found necessary for better management of crop production.

THE ART OF CROP PRODUCTION

As an art it embraces knowledge of the way to perform the operations of the farm in a skilful manner, but does not necessarily include an understanding of the principles underlying farm practices. The skill may be both physical and mental. Some farmers are able to do farm operations quickly, more efficiently than others and with great ease. They are said to be skilful in their art. The ability and capacity to plough well, to make a good hay stack, to handle the animals properly, to handle the farm implements and proper sowing of seeds, manures, fertilizers etc. are examples of physical skills which are obtained through experience. Similarly, deciding the proper time for agricultural operations like ploughing, selection of cropping pattern to suit the soil and climate, adopting improved agricultural practices, timing of production to get better prices are examples of metal skills involved in farming. Both the physical and mental skills are essential for successful farming.

THE SCIENCE CROP PRODUCTION

Science is generally, defined as a body of truths, systematized and arranged in such a way as to show the operation of general laws and principles. Such a body of knowledge is obtained through systematic study and research. Scientific principles are freely used for increasing production of quality crops and in their ultimate disposal to earn better price and income. Recent advances made in the fields of plant breeding and genetics, technology of production, use of fertilizers, pesticides, fungicides, herbicides and growth regulators, introduction of labour saving machineries, theories and concepts in the fields of mathematics, statistics and economics etc., have completely revolutionized agriculture for increased crop production.

In plant breeding and genetics, strains were evolved to suit different conditions of soils, climate and consumer tastes. Photo-insensitive, fertility responsive high yielding strains available in rice and wheat have induced green revolution in different parts of the world. Hybrid vigour in maize, pearl millet and sorghum have been properly harnessed in making available numerous strains and varieties doubling their yield. Collection of germplasm materials from countries around the world have helped in evolving strains resistant to pests and diseases. Research findings in isolating genes responsible for quality proteins have helped in screening varieties rich in lysine, methionine and tryptophane. Attempts are made to evolve strains to shorten their duration and for reducing straw-grain ratio so that crops can be cultivated in different agro-climatic conditions tailored to individual location.

The recent findings in the field of atomic science had greatly helped in the studies of soils, nutrition of crops, movement of water in soil and the related fields.

In the fertilizer technology both in their manufacture and application, the farmer is given the chance to utilize different types suited to different soil, climate and crop condition.

At present a series of new chemicals is marketed to help greatly to the farmers in reducing the loss of crops due to pests. Similarly, a number of new chemicals are used for the control of diseases and weeds.

THE BUSINESS OF CROP PRODUCTION

Agriculture is considered as a way of life in most of the under-developed countries. In these countries agriculture is carried on at subsistence levels. Traditionally farming is handed over from father to son and no effort is taken to increase agricultural production on business lines. In the developed countries, the farmer raises the crop and livestock with the object of getting more income. An industry run on business motive will work more efficiently and diligently with the object of getting more profit from less investment. Similarly, in agriculture the land is the workshop from where greater returns are expected. To obtain these returns, the factors like land, labour and capital are utilized in such a way that greater profits are obtained. Agriculture has been mechanized and commercialized with the main objective of promoting business to greater profit. The factors like land, labour and capital are utilized in such a way that greater profits are obtained. Just like any industrial business, the farming operations also comprise such problems as production, consumption, barter, trade, employer and employees, marketing, transportation, international trade, economic interdependence, taxation, custom and tariffs in their relation between the farmer and other operations such as merchants, manufacturers and to earn greater profits, agriculture is mechanized and commercialized to run a business

Starting from the selection as an enterprise for one's livelihood, agriculture is treated as any other business in the developed countries. The availability of employment opportunities in those countries induces thinking on the pros and cons of selecting agriculture for their livelihood. By making inventories on the availability of land, labour and capital, the selection of different branches of farming such as field crop production, live stock rearing, growing of horticultural craps, dairying etc. are decided. If the land available is small, additional land is acquired and organizational abilities are properly made use of. In the absence of sufficient land not available one may go in for part time employment in other business.

After the decision has been taken for farming, the choice of growing a particular crop to a particular area is decided on the possibility of earning the maximum profit. Similarly, the exact amount of inputs such as fertilizers, pesticides, fungicides, herbicides, etc., is decided on the profit motive. Due to the consumer's market with a variety of agricultural inputs available, the farmer is making his choice of purchases with an eye on profit.

The farmer is supplied with innumerable alternate plans for efficient farm management within the resource restrictions of land, labour and capital. The farmer who has taken up agriculture on these business lines is well educated and knows the trick of his trade and enjoys the most modern amenities. For making the correct decision in all aspects of farming, the new tool of linear programming is made use of.

AGRICULTURE AND AGRONOMY

Advances in basic sciences such as chemistry, physics, biology and genetics have helped in understanding the factors affecting growth of plants and to increase the production processes in agriculture. Fertilizers, pesticides, new varieties and hybrids of crops and methods of cultivation are largely the offspring of science. Scientific agriculture played an important role in the process of enrichment of developed countries. If development is to occur in the developing countries, agriculture must be able to produce surplus food and exportable commercial crops on cost-effective basis. Any country seeking to develop its economy has to give significant priority to agriculture, particularly scientific agriculture. Whenever scientific methods are applied, experience world over shows they got agriculture revolutionized. The term **agriculture** is derived from the Latin words ***'ager'*** or ***'agri'*** meaning **'soil'** and ***'cultura'*** meaning **'cultivation'.** Agriculture is a very broad term encompassing all aspects of crop production, livestock farming, fisheries, forestry etc. The scenario in food production is changing fast in the country with the advances made in all branches of Agricultural Sciences. However, the science of **Agronomy**, a specialized subject dealing with all aspects of field crop production, has accelerated the pace of food production, aided by the progress made in understanding the intricate relationships between crop growth and yield, and between crop and its environment of climate, soil, biotic factors and management practices.

The science of **Agronomy** has made it possible to generate suitable technology for varied agro-climatic regions and integrate the results of other allied sciences, so as to deliver these appropriate innovative and tested practices to the farmers at large. Agronomy is a division of Agricultural Science embracing preparation of soil in accordance with crop demands, enriching the soil with organic matter and plant nutrients, choice of crops and varieties to fit the climate, crop rotations and intercropping, suitable dates of sowing, moisture regulation, drainage, and weed management, harvesting and processing to an extent. It deals essentially with all aspects of soil, crop and water management to increase productivity of crops. It is, therefore, essential that the students of agriculture have a good grasp of the **Principles of Agronomy**.

Agronomy is a branch of agricultural science which deals with principles and practices of soil, water and crop management. This term is derived from Greek words ***'agros'*** meaning **'field'** and ***'nomos'*** meaning **'to manage'**.

In recent times, Agronomy has assumed newer dimensions and can be defined as a branch of agricultural science that deals with methods which provide favourable environment to the crop for higher productivity.

Environment is defined as the aggregate of all the external conditions and influences affecting the life and development of an organism. Norman (1980) has defined agronomy as the science of manipulating the crop environment complex with dual aims of improving agricultural productivity and gaining a degree of understanding of the process involved. The environment of crop constitutes both soil and aerial, as part of the plant is inside the soil and the other part is exposed to the aerial atmosphere. Environment influences growth and development of crops. Soil environment is amenable for manipulation through tillage, irrigation, fertilizer application etc. The soil physical environment comprises of soil air, soil water and soil temperature. Soil solution pH, electrical conductivity, nutrient concentration constitutes the soil chemical environment. The soil biological environment constitutes the living organisms present in the soil and the resultant reactions like mineralization, immobilization etc. The soil management part of agronomy deals with the manipulation of soil environment for better crop growth.

Aerial environment includes solar radiation, rainfall, temperature, relative humidity and wind velocity. Crop production is basically conversion of environmental inputs like solar energy, carbon dioxide, water and soil nutrients into economic products in the form of human and animal food or industrial raw materials. The aerial environment can not be altered easily. Season for raising each crop has to be selected to attain the highest productivity from the available climatic resource.

"A man without food for three days will quarrel,
for a week will fight and for a month or so will die"

Agriculture is a branch of applied science. The term agriculture has been derived from the Latin word 'ager' meaning land or field and 'cultura' meaning cultivation *i.e.*, the science and art of farming including the work of cultivating the soil, producing crops and raising livestock. The Greek 'geoponic' (cultivation in earth), 'hydroponic' (cultivation in water) and 'aeroponic' (cultivation in air) refer to the three main spheres of agriculture.

Agriculture is the most important enterprise in the world. It is true sense it is a productive unit where the free gifts of nature namely, land, light, air, temperature, rain water, humidity, etc., are integrated into a single primary unit (crop plants or their usable parts) indispensable for human beings. Secondary productive units namely, animals including livestock, birds, and insects, feed on these primary units and provide concentrated products such as meat, milk, wool, hide, eggs, honey, silk, and lac.

The important cultural energies utilized for the production and protection of agricultural commodities are irrigation and drainage, organic, biological and mineral fertilizers, chemicals, farm equipment and draft power. These are used to maximize productivity per unit time, water, land, labour and rupee investment.

Agriculture provides as food, feed, fibre, fuel, furniture, raw materials and feed-back materials for and from factories, funds, flood control, a free, fair and fresh environment, abundant food driving out famine and friendship eliminating fights. It also considers employment generation, economics, education, ecology, energy consumption, use of equipment and earning for production, protection, processing, consumption, preservation and war against waste, transport and trade. Even though agricultural commodities are mostly seasonal, bulky and perishable in nature, they help the nation to earn and conserve a greater amount of foreign exchange and to build up the national economy. Satisfactory agricultural production brings peace, prosperity, harmony, health and wealth to individuals of a nation by driving away distrust, discord and anarchy. It helps to elevate the community consisting of different casts and communities to a better social, cultural, political and economic life. Agriculture consists of growing plants and rearing animals in order to yield produce and thus, it helps to maintain a biological equilibrium in nature. It congregates the interaction of all environmental factors namely, water, heat, light, air and soil distributed in the different spheres such as the lithosphere, pedosphere, atmosphere, hydrosphere and photosphere. Agriculture helps to meet the basic needs of human beings and their civilization by providing food, clothing, shelter, medicine and recreation. Their eyes and minds are soothed by dynamic changes from grey (bare soil) to green (growing crop) to golden (mature crop) and gay harvests.

In wide sense agriculture implies the effective use of land, water, light and other resources of environment through the production of field crops, forage crops, fruit crops, farm animals, fisheries and forestry.

The word agriculture, thus be expanded as Activities on the Ground for Raising Intended Crops for Uplifting Livelihood. Agriculture deals with enterprises as stated below:

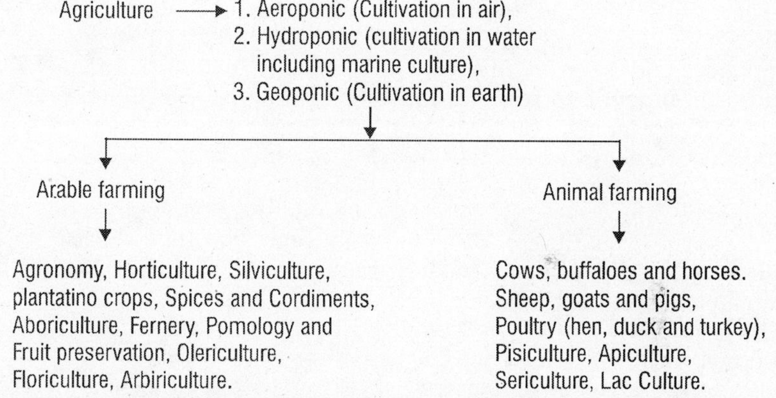

Advances in basic sciences such as chemistry, physics, biology and genetics have helped in understanding the factors affecting growth of plants and to increase the production processes in agriculture. Fertilizers, pesticides, new varieties and hybrids of crops and methods of cultivation are largely the offspring of science. Scientific agriculture played an important role in the process of enrichment of developed countries. If development is to occur in the developing countries, agriculture must be able to produce surplus food and exportable commercial crops on cost-effective basis. Any country seeking to develop its economy has to give significant priority of Agriculture, particularly scientific agriculture. Wherever scientific methods are applied, experience world over shows they got agriculture revolutionized.

The scenario in food production is changing fast in the country with the advances made in all branches of Agricultural sciences. However, the science of Agronomy, a specialized subject dealing with all aspects of field crop production, has accelerated the pace of food production, aided by the progress made in understanding the intricate relationships between crop growth and yield, and between crop and its environment of climate, soil, biotic factors and management practices.

Agriculture encompasses the applied aspects of the following basic sciences.

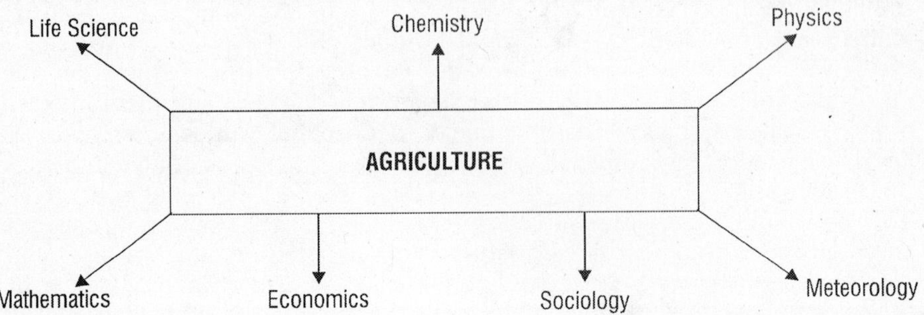

Basic Concepts of Agronomy

The word agronomy has been derived from two Greek words- 'agros' meaning field and 'nomos' meaning to manage. Literally it means the art of managing fields and technically it means the science and economics of crop production by management of farm land. In other words it is the art and underlying science in production and improvement of field crops with the efficient use of soil fertility, water, labour and other factors related to crop production. Agronomy is the field of study and practice of ways and means of production of food, feed and fibre crops. Thus, agronomy as a branch of agricultural science deals with principles and practices of field management for the production of field crops. Among all the branches of agriculture, agronomy occupies a pivotal position and is regarded as the mother or primary branch. Like agriculture, it is nothing but an integrated and applied aspect of different disciplines of pure sciences. It has three clear branches: (1) Crop science (mainly field crops); (2) Soil science; and (3) Environmental science, that deals only with applied aspects.

The central theme of agronomy is of soil-crop-environment relationships. Field crops without soil can not be considered and soil without crops is barren. The core of agronomy is the field of crop plants with the theme of controlling the environment (microclimate) and the nature of agronomy is based on soil-plant-environment relationships. The study of the magnitude of variation in yield, cause and effect relationships, internal and external factors and their interrelationships, techniques of increasing use-efficiency of inputs, evolving technologies for better management practices of soil, water, plant nutrition, weeds and crop plants are the major aspects of agronomy to boost production and its usable products per unit land, time and input. Scientific crop production includes crop environment improved agro-techniques, the ameliorating agro-climate of the locality and other aspects of the surrounding area for the entire field duration of the crop concerned.

Agronomy thus denotes Activities on the Ground to Raise Outspread and Noble crops to Obtain Massive Yields.

Basic Principles of Agronomy

Science deals with five w's – they are what, when, where, why and how? As an applied science agronomy deals a little more in detail in most cases, *i.e.*, which, whether, who, whom and how many, how much and how long.

A principle means a scientific law that explains natural action and agronomic principles are the ways and means for the better management of soil, plants and environment for economically maximum returns per unit area fir years.

Some of the factors of crop production are controlled by the environment and rest is man-made factors. The environmentally controlling factors are modifiable to a limited extent, for instance drought is mitigable by irrigation or gales are retrievable with shelterbelts.

The principles of management depend largely on the type of farming namely, specialized, diversified, mixed and integrated and also on the physical and technological facilities available – irrigated farming, dry farmig and rainfed (*barani*) farming.

The fundamental principles of agronomy may be listed as:

1. Planning, programming and executing measures for maximum utilization of land, labour, capital, sunshine, rain-water, temperature, humidity, transport and marketing facilities.
2. Choice of crop varieties adaptable to the particular agro-climate, land situation, sol fertility, season and method of cultivation and befitting to the cropping system.
3. Proper field management by tillage, preparing field channels and bunds for irrigation and drainage, checking soil erosion, leveling and adopting other suitable land improvement practices.
4. Adoption of multiple cropping and also mixed or intercropping to ensure harvest even under adverse environmental conditions.
5. Timely application of proper and balanced nutrients to the crop or crops in sequence and improvement of soil fertility and productivity. Correction of ill-effects of soil reactions and conditions and increasing soil organic matter through the application of green manure, farm yard manure, organic wastes, bio-fertilizers and profitable recycling of organic wastes.
6. Choice of quality seed or seed material and maintenance of requisite plant density per unit area with healthy and uniform seedlings.
7. Proper water management with respect to crops, soil and environment through conservation and utilization of soil moisture as well as by water for life saving or protective irrigation, scheduling irrigation at critical stages of crop growth and also growing crops in nontraditional seasons or belts with irrigation or rain or residual moisture.
8. Adoption of adequate, need-based, timely and exacting plant protection measures against weeds, insect-pests, pathogens, as well as climatic hazards and correction of deficiencies and disorders.
9. Adoption of suitable and exacting management practices including intercultural operations to get maximum benefit from dearer and difficult to get, low-monetary and non-monetary inputs. Adequate input supply should not be a substitute for deficiencies in management practices.
10. Adoption of suitable method and time of harvesting of crop to reduce field-damage and to release land for succeeding crop(s) and efficient utilization of residual moisture, plant nutrients and other management practices. Adoption of suitable post-harvest technologies.

LAND UTILIZATION STATISTICS OF INDIA

India has a subcontinent has the following land utilization statistics:

1.	Total geographical area	328.848 million hectare
2.	Total reporting area	304.300 million hectare
3.	Area under cultivation	143.000 million hectare
4.	Total cropped area	179.750 million hectare
5.	Area sown more than once	36.750 million hectare
6.	Area not available for cultivation	161.300 million hectare
7.	Forest	66.400 million hectare

The Past and Present forms of Agriculture

It appears from excavation, legends and remote sensing tests that about ten thousand years ago man invented agriculture. It is believed that women were the pioneers for cultivating useful plants from the wide flora, while men went out for hunting and gathering food. Some nomadic aborigines used fire and weapons. They did not have tools except for digging sticks and moll use shells. During the drier season they wandered in bands. The wetter and colder seasons restricted their movements. They would depend on their food gathering. They would dig out edible roots and rhizomes having buried the small ones therefore, subsequent harvests. They would hunt wild animals and use their flesh as food and skin as clothing. Thereafter, they entered upon a more secure stage of living. Probably woman by her intrinsic insight first observed that plants come up from seeds. Man might have consciously planted seeds and nurtured the sprouts to harvest. They used fire and digging sticks to prepare garden plots in which they would grow small grains. Once the seeds were planted the band would move on and return later for the harvest.

A primitive form of agriculture was shifting cultivation, in which people working with the crudest of tools, cut down a part of the forest, burnt the undergrowth and started new garden sites. After a few years when these plots lost fertility, choked with grass or bush or became heavily infested with soil-borne pests, cultivators would shift to a new site. This system would support a sparse population where abundant land was available.

Man invented the use of five animals as beasts of burden. They started domesticating wild animals facultative in nature. People with more members in a group selected somewhat better areas (near a stream or river) and settled down in permanent village sites and started cultivating the same land more continuously. However, the tools, crops and cropping methods might have been primitive. This system was probably the rudimentary system f settled farming. Since then agricultural activity became subsidiary to gathering and hunting.

Thereafter, an advanced form of primitive agriculture was the subsistence form of agriculture. This was based on the principle “grow it and eat it” instead of on a commercial basis. In some cases it was based on the crop production and in other cases on a more intimate blend of crops and livestock. Hunting and gathering were subsidiary occupations. Men in meaningful members began to settle down to sedentary life.

A sparse population of nomadic tribesmen followed their flocks and herds in the seasonal pursuit of water, forage and shelter. Commercial ranching gradually blended with subsistence farming. Under shifting cultivation land abandoned became degraded. Subsequent subsistence farming followed naked fallow rotation. Thereafter, legume rotation was adopted. Under mixed farming comprising field crops and animals, field grass husbandry was adopted and the same fields were used for cropping and grazing.

The concept of property in land evolved and the individuals held the privilege of autonomous action. The area over which a band wandered and carried on their food-gathering was considered to be for the joint and mutual exploitation of the group. They held a territory assuming they could defend it from other bands by the right of prior 0ccupance. Any individual property in land was right to a certain garden plot and domestic site, earned by virtue of having cleaned, improved and used it. During their movement from one location to another they would come in contact with a large number of plant and animal types. They selected the most useful ones for domestication and improved them.

Battles, fights on rivalries among different groups and individuals for property or passion helped to blend newer ideas for the components, commodities and comfort of life. Mutual exchanges, gifts, offerings etc., also blended the style of living, food habits and nature of earning. This right to use the land with the progress of time was recognized on a continuing basis year after year. With still further evolution land was held to be inheritable and the ownership title was granted and defended by group authority. These progressive stages of security in mobile and immobile properties, contributed proportionately to the greater stability and well being of the society. They concentrated on the welfare of their individual families, and the produce of their land and animals. They devoted themselves more intensively to utilize their land to the best of their ability. However, they held together to defend any unwanted enemy of their society. This was still the transitional stage from food gathering to found growing.

The self sufficient agricultural village was the least common denominator of a permanent human society. It developed and perpetuated itself without many tools or other capital and without dependence on the outside world. Several crafts also developed for and around agricultural production. Agriculture then experienced extreme adversity as long as the minimum of food, feed and fibre were raised. The standard of living was low. With the span of time there was the periodic and geographic rise of different civilizations. Agricultural production expanded and towns and cities developed, trade and arts flourished.

Downtrodden people, slaves and bonded labour were the principal sources of human labour. Draft animals were the plodding oxen. Farming equipment was animal drawn and consisted largely of hand tools. It was organized on a commercial basis and management was developed to a fine art. The skills developed were inherited through the generations of cultivators who became a major section of the then civilized society. Advanced farming practices included seed selection, green manuring with legumes, rotation of crops, use of animal and crop refuse as manure, irrigation, pasture management and rearing of birds.

In subsequent stages the feudal manor emerged and was essentially an agricultural village. It consisted of a nucleated village built around the manor-house and surrounded by the village land. The best lands were laid out in large arable fields, the medium lands were used as grasslands for mowing and grazing and the least productive areas were used for grazing, wood lands, funeral grounds and other common activities. Being a self-contained the village included a worship alter and a meeting place.

Arable lands were divided into different plots according to topography and use. During the favourable season crops were grown in the fields and livestock was kept away from the crops but after the harvest cattle would graze in these areas under the supervision of herdsmen. During the cropping season livestock were allowed to graze in common grazing lands and in adjoining woodlands and forests. By-products of crops were used to feed the animals during stall feeding. Quack remedies based on locally available plants with therapeutic values and minerals were the major method of treatment for common ailments of animals. Infestation of insect-pests was also combated by indigenous methods.

There were ponds and tanks in the village. Some of these water reservoirs were reserved for drinking water, washing purposes, fisheries, or small-scale irrigation. The banks of water systems were used for arboriculture and fruit growing.

Emperors, monarchs, kings, zamindars, governors, heads of states, etc., would claim authority over their respective geographical land areas and rule over them. Cultivable areas would be distributed to lease holders, tenants or share croppers directly or through intermediaries. Such cultivators or tillers would enjoy little benefit except for mere subsistence.

The enclosure movement began to be consolidated in some countries during the 14th century. Open and scattered fields of a tenant were compacted into a farm unit under an individual owner. These consolidated, fenced and individually controlled farm units were managed by the farmers with commercial intent. Land as an inheritable and transferable property was subdivided and fragmented to an indefinite extent without the consideration of its economic viability to the tenants who depended on subsidiary occupations to supplement their earnings subsequently. In long and medium term cycles, natural calamities affected agricultural production and the farming community and the kingdom or nation as a whole would fall in distress. Facing such famine was a futile effort except by large scale emigration. The toll of human and animal lives at such times was considerable. No other occupation could compensate for these disasters.

Land, though an immobile property was capable of producing an enormous quaintly of mobile short to long –term crops meant for the livelihood of a large number of people. The helping hands of the peasantry and ploughmen were usually family members. There were also different types of human labourers and work animals. The then tools and implements were a slightly improved form of primitive ones. Bullock or horse drawn tools were limited to the iron banded wooden plough, bullock cart, etc. Therefore, wooden and iron implements and equipment were developed. All of them were manually operated or drawn by animal power.

There were many ups and downs, shortage and disasters in this expended (both spatial and temporal) development of agriculture in site of the diligent endeavours of human for a greater dependence on agricultural production. Modern agriculture started during the early eighteenth century. Crops were grown in sequence without allowing the field to remain fallow. Crop rotation was introduced. Organic recycling was intensified. A number of exotic crop plants and animals were introduced. The crops selected were clean, stable in yield and well adapted to the local agro-ecological systems. Land improvement was introduced. Farm implements and equipment were improved. The care and management of crops were advanced and specialized, crop protection practices were introduced and preservation and processing improved. Gradually livestock and flock breeding and rearing and crop farming became diversified. Sheep meant for wool were bred for the production of better and more efficient meat animals. Dairying became a profession. Fish rearing and fishing became an occupation. Greater emphasis was laid on feeding, breeding, caring for and weeding out of different livestock and birds meant to meet various requirements more efficiently and economically. Crop plants also drew due attention and they were grown in different categories as field crops, plantation crops and garden crops.

In the nineteenth century the real science of agriculture started. Nutrition, physiology, ecology, etc., developed. Research and development in fundamental and basic sciences were brought under applied aspects in agriculture. It looks the shape of a teaching science. The most neglected occupation of a downtrodden and dull community drew the attention of aristocrats, scientists, researchers and others in almost all walks of life. Research on specialized branches of several disciplines was developed and scientific findings were translated into production practices. Laboratories, farms, stations, centres and institutes for research, teaching, training and demonstration were developed. Books, journals and scientific and popular literature were introduced. News media and audio-visual aids were developed to disseminate new findings and information to the rural masses engaged in agriculture.

At present a lot of development on hydrological, mechanical, chemical, genetical and technological aspects of agriculture is in progress. Government is apportioning a greater share of development for agricultural improvement. Central and State Departments of Agriculture, Agricultural Universities and Institutions are flourishing. International Research Institutes are being set up. Farmers specially small and marginal ones of the underdeveloped and developing countries are supplied with improved seeds, fertilizers, irrigation, equipment, electrical energy, pesticides, improved genetic stocks of cows, chicks, kids, pigs, fingerlings, silk worms and their improved feed. Even funds for both recurring and non-recurring expenditure with reasonable cost or rate of interest, or free of cost, as well as technology services are being provided freely. Preserving, processing, pricing, marketing, insuring, distributing, consuming and exporting importing policies are strengthening. Agro-based industries and crafts are developing. Need based planning; programming and executing are in progress. Improved use-efficiency of non-monetary, less-monetary inputs is in the process of adoption. From specialized farming, diversified and integrated farming is emerging. Land reforms, development and diversion from degradation are going on. As the size of operational holdings is rapidly deceasing to marginal and sub-marginal levels, farming by individuals is going to be recast to co-oprative, community or contractual farming.

Agriculture now is not merely production oriented but is becoming an agribusiness consisting of enterprises such as livestock, poultry, fishery, silviculture, sericulture, apiculture and lac culture along with field and forage crops and plantation crops. These are the stepping stones of the green, white and yellow revolutions of today and simultaneously of the multicolour revolution with manifold quantitative, qualitative and generative yields.

It is surprising that even in the era of nuclear energy in industrial development and expeditions to space; man retains his animality and primitiveness in food habits and is a competitor with his animals for food grains.

Even though agriculture was, is and will have tremendous potential for life and longevity in this luminous world, yet it is far behind the technologies developed in other sectors such as industry, space expeditions and defense.

The science of Agronomy has made it possible to generate suitable technology fir varied agro-climatic regions and integrate the results of other allied sciences, so as to deliver these appropriate innovative and tested practices to the farmers at large.

Relation of Agronomy to other Sciences

Agronomy is a synthesis of several disciplines like soil science, agricultural chemistry, crop physiology, plant ecology, biochemistry and economics. The soil physical, chemical and biological properties have to be understood thoroughly to effect modification of the soil environment. Similarly, it is necessary to understand the physiology of crops to meet their requirements. Developments in these subjects help in developing new practices which are simpler and economical to provide favourable environment to the crop. The herbicidal properties of 2,4-D have made revolutionary changes in weed control. Advances in economic analysis helped in production of crops economically. For efficient utilization of resources, systems approach is better than individualistic approach. Instead of taking individual crop into consideration for taking decision, cropping system as a whole is considered. Fertilizer recommendation for an entire cropping system saves considerable amount of fertilizer nutrients as compared to recommending fertilizers for individual crops.

SCOPE OF AGRONOMY

Agronomy is a dynamic discipline. With the advancement of knowledge and better understanding of plant and environment, agricultural practices are modified or new practices developed for high productivity. For example, availability of chemical fertilizers has necessitated the generation of knowledge on the method, quality and time of application of fertilizers. Similarly, availability of herbicides for the control of weeds has led to development of a vast knowledge about selectivity, time and method of application of herbicides. Gigantic irrigation projects are constructed to provide irrigation facilities. However, these projects created side effects like water-logging and salinity. To overcome these problems, appropriate water management practices are developed. Population pressure is increasing but area under cultivation is static. More number of crops has therefore, to be grown on the same piece of land in an year. As a result, intensive cropping has come to vogue. Similarly, no tillage practices have come in place of clean cultivation as a result of increase in cost of energy. Likewise, new technology has to be developed to overcome the effect of moisture stress under dry land conditions.

As new varieties of crops with high yield potential become available, package of practices has to be developed to exploit their full potential. The factors restricting increased agricultural production are low soil fertility, crop varieties of low genetic yield potential, poor agronomic practices, inadequate or non-existent control of diseases and insects, non-availability of production inputs, government economic policies affecting agriculture and weak research and extension programmes. Restoration of soil fertility, preparation of good seed beds use of proper seed rate, correct date of sowing for each improved variety, proper conservation and management of soil moisture and proper control of weeds are agronomic practices to make the land water resources more productive.

ROLE OF AGRONOMIST

Agronomist aims at obtaining maximum production at minimum cost. He exploits the knowledge developed by basic and allied applied sciences for higher crop production. In a large sense, agronomist is concerned with the production of food and fibre to meet the needs of the growing population. Agronomist carries out research on scientific cultivation of crops taking into account the effects of factors like soil, climate. Variety of crop and adjusts production techniques suitably depending on the situation. He is a key person with working knowledge of all agricultural disciples and coordinator of different subject matter specialists.

CHAPTER 3

Factors Affecting Crop Production

The life of a plant is so intimately related with the environmental factors of a place that every species of plant seems to be fitted to a particular set of environmental conditions within the limits of which it is able to grow and reproduce its own kind. The preferred environment of a plant is called its **habitat.**

Though the environmental factors have to be studied individually, they do not act in isolation from one another. All these factors interact with one another to influence the plant (Fig. 3.1).

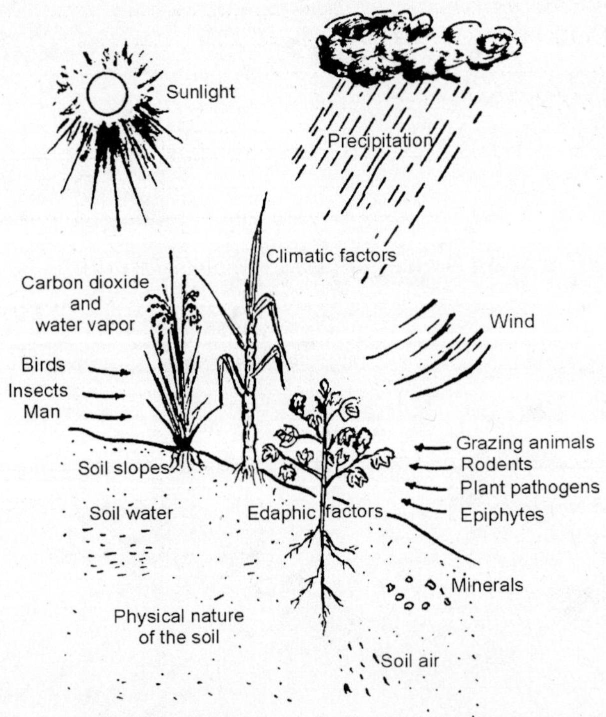

FIG. 3.1: Environmental factors affecting a plant.

Every living organism is a product of its biological heredity and its environment. To survive and reproduce with which it regularly comes in contact. The crop may not attain the maximum efficiency in any one of the environments occupied; but to achieve the maximum efficiency the environmental conditions should be ideal. The factors can be broadly classified under the following types:

CROP GROWTH

1. Internal Factors (Genetic or Hereditary)
2. External Factors (Environmental)

a) Climatic
b) Edaphic
c) Biotic
d) Physiographic
e) Man (Anthropic)

GENETIC FACTORS

The importance of genetics in the growth of agricultural crops is clearly shown by the increase in yields that have been obtained by the use of improved seeds. Plant Breeders have incorporated many desirable characteristics into their hybrids and varieties. The following outline gives a list of useful characters which are considered as desirable breeding aims although it may not be possible to combine all these useful properties in a single variety.

1. High yields under given environmental conditions.
2. Early maturity (in some cases, later maturity).
3. Better resistance to lodging.
4. Drought, flood and salinity tolerance.
5. Greater tolerance to insects and diseases.
6. Chemical composition of grains (high percentage of oil, increase in protein quality or quality, etc.).
7. Quality of grains (fineness, coarseness, etc.).
8. Quality of straw (sweetness, juiciness, etc.).

The increased yields and other desirable characters are related to the genetics make up of the plant. These characters are inherent in each individual and are transmitted from one generation to another by genes.

ENVIRONMENTAL FACTORS

Climate

Variations in the climate can take place in relatively small areas due to the topographical or other factors, though the climate of a particular region is fairly uniform over a wide area. These local variations are limited to small areas in which they can be arranged both vertically and horizontally. Some of the examples are the microclimates of rice fields in garden lands, groves of coconut and areca-nut and sugarcane plantations. The atmospheric factors which affect the crop plant are called climatic factors.

1. Precipitation,
2. Temperature,
3. Atmospheric humidity,
4. Solar radiation
5. Wind velocity, and
6. Atmospheric gases.

Precipitation

World's supply of fresh water is obtained almost entirely as precipitation resulting from evaporation of sea water and land surfaces. The processes involved in the transfer of moisture from the sea to the land and back to the sea again form what is known as the **hydrologic cycle**. The first stage in the hydrologic cycle is the evaporation of water from the oceans. This vapour is carried over the continents by moving air masses. If the vapour is cooled to its dew point, it condenses into visible water droplets which form cloud or fog. Under favourable meteorological condition, the tiny droplets grow large enough to fall to earth as precipitation.

Precipitation includes all water which falls from the atmosphere to the earth's surface. Precipitation occurs in a variety of forms such as rainfall, snow, hail, etc. Besides, there may be fog drip and dew that may contribute to moisture. Fog consists of water droplets so small that their fall velocities are negligible. Fog particles which contact vegetation may adhere; coalesce with other droplets and eventually from a drop large enough to fall to the ground. On clear nights, the loss of heat by radiation from the soil causes cooling of the ground surface and of the air immediately above it. Condensation of the water vapour present in the air results in a deposit of dew. Crops grown in winter such as wheat and cabbage are efficient dew collectors and can be grown under dry condition with little irrigation. The yearly precipitation, both in total amount and seasonal distribution greatly affects the choice of cultivated crops of a place. Low and ill-distributed rainfall are common features of dry land farming wherein drought resistant crops like sorghum, pearl millet (*Pennisetum typhoides*), Italian millet (*Setaria italica*) and other minor millets are commonly grown. On the other hand in places of heavy and regular rainfall areas such as in the Western Ghats of India crops like rice are grown in flat areas while tea, coffee, rubber, etc., are grown in the slopes.

Rainfall is one of the most important factors influencing the vegetation of a place. Most of the crops receive their water supply from rain water. Rain water the source of soil moisture so essential for the life of a plant. The interaction of rainfall with temperature has a very profound effect on the vegetation and soil of a place. Very heavy rainfall and high temperature in the equatorial and tropical zones cause the formation of most highly developed vegetation of the world. The kind of vegetation and soil formed due to climatic conditions are indicated below.

Relationship between Climate, Natural Vegetation and Soil

Dry cold	**Wet cold**
Perpetual snow and ice Tundra Taiga	Perpetual snow and ice Tundra soils Podozol soils

Arid Desert Grasses & Shrubs	**Semi arid steppe**	**Subhumid grass land**	**Humid rain forests**
Sierozems and desert soils Dry hot	Chestunut and brown soils	Prairee soils nad chernozems	Podzol Grey brown podzol soils Red yellow podzol soils Laterite soils Wet hot

It may thus be seen that while desert grasses and shrubs are found in desert soils of dry hot climate, rain forests are seen in laterite soils of wet hot climate.

Temperature

Temperature may be considered as a measure of intensity of heat energy. The range of maximum growth for most agricultural plants is between 15 and 40°C (59 to 140°F). The temperature of a place is largely determined by its distance from the equator (latitude) and the altitude. Based on the above, the vegetation are classified as tropical, temperate, taiga, tundra and polar as indicated in Fig. 3.2.

Some investigators have classified the vegetation of the world into four classes on the basis of prevailing temperature conditions as stated in Table 3.1.

TABLE 3.1: Classification of vegetation based on temperature

Class	Region	Temperature conditions	Type of vegetation	Common crops cultivated
Megatherms	Equatorial and tropical	High temperature throughout the year	Tropical rain forest	Tropical crops like rubber, cassava, varieties of rice, etc.
Mesotherm	Tropical and Sub-tropical	High temperature alternating with low temperature of winter	Tropical deciduous forests	Sub-tropical crps like maize, sorghum, etc.
Microtherms	Temperate and high altitude plants (upto 4000 m of tropical and sub-tropical)	Low temperature	Mixed coniferous forests	Temperate crops like wheat, oat, potato, etc.
Hekistotherms	Arctic and alpine regions (above 5000 m in tropics and 4000 m in temperature regions)	Very low temperature	Alpine vegetation	Pines, spruce, etc.

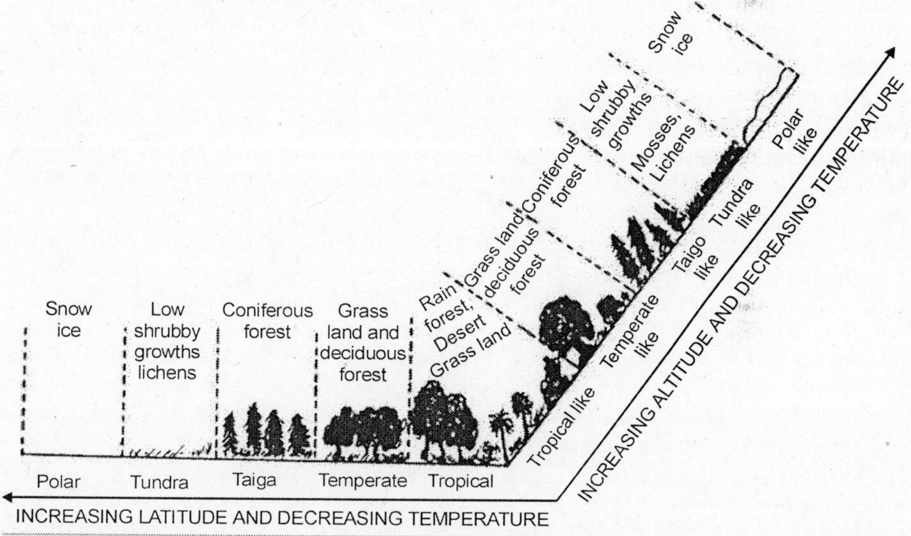

FIG. 3.2: Relationship of latitude and altitude to vegetation.

Every plant community has its own minimum, optimum and maximum temperature known as their cardinal points. The cardinal points of some of the common crops are indicated in Table 3.2.

TABLE 3.2: Cardinal temperature of certain crops

Crops	Minimum (°F)	Optimum (°F)	Maximum (°F)
Wheat	39-40	77	86-90
Barley	39-40	68	82-86
Oats	39-41	77	86
Maize	46-50	89	104-111

Crops	Minimum (°F)	Optimum (°F)	Maximum (°F)
Sorghum	46-50	89	104
Rice	50-53	90	97-101
Tobacco	55-57	82	95

The effect of temperature in affecting the yield of a crop is indicated below for potatoes (Table 3.3.).

TABLE 3.3: Yield of potatoes at varying temperature

Place	Mean temperature in growing period in °F	Average yield in q/ha
Nilgiris (Tamil Nadu)	60	125
Delhi	64	118
Patna (Bihar)	66	83
Bangalore (Karnataka)	70	52

A part from the reduction in yield many injuries on the plants and adverse effect on soil conditions occur under both extremes of temperature.

***Cold injury*:** Plants are injured due to cold temperature in the following ways:

1. **Chilling injury:** Plants growing in hot climate, if exposed to low temperature (which is above the freezing point) for some time are found to be killed or injured severely. Some effects of chilling are chlorotic condition or bands on leaves of sugarcane, sorghum and maize when exposed for 60 hours at 2 to 4°C.
2. **Freezing injury:** This is generally, caused in plants growing in temperate region. When the plants are exposed to very low temperature, water is frozen into ice crystals in the intercellular spaces. With further fall in temperature, water is withdrawn from the cells, resulting in increase in the size of ice crystals in the intercellular spaces. The protoplasm of the cell is dehydrated and mechanical distoration takes place resulting in killing of the cells. Frost damage to potatoes, tea etc., in winter in the hilly areas like Nilgiris is a typical example of the freezing injury.
3. **Suffocation:** During winter the ice or snow form a thick cover over the ground and the crop suffers for want of oxygen. Ice in contact with roots inhibits diffusion of carbon dioxide and the respiratory products may become harmful to plants.
4. **Heaving:** Injury to plants is caused by a lifting upward of the plant along with the soil from its normal position in temperate regions where snowfall is a common phenomenon.
5. ***Heat injury*:** Very high temperature often stops growth. The plant faces incipient starvation due to high respiration rates. The plant is stunted and if such a condition persists for a long period the plant is killed. Direct temperature effects are noticeable in young seedlings and transplanted crops. High temperature causes sterility in flowers. The general effects of excessive heat are defoliation, pre-mature dropping of fruits and in extreme cases death of plants.

Effect of temperature on crop production: Plants can grow only within certain limits of temperature. For each species and variety there are not only optimal temperature limits, but also optimal temperatures for different growth stages and functions, as well as lower and upper lethal limits.

While light is involved in only one aspect of photosynthetic process namely as a source of energy for the conversion of carbon dioxide and water to carbohydrates, there are certain biochemical processes preceding and following the reduction of carbon dioxide, which are affected mainly by temperature. As long as light is limiting, temperature has little effect on the rate of photosynthesis. However, when light is not limiting, the biochemical processes associated with photosynthesis become the limiting factor, so that the effect of favourable temperatures on the rate of photosynthesis is measured.

In general, high temperature associates growth processes. Rarely are high temperatures per se the direct cause of death of plants, provided the water supply is adequate. However, beyond a certain limit, which depends on the crop, the stage of development and physiological processes involved, high temperatures may have detrimental effects on crop production. The photosynthetic process appears to become heat inactivated at excessively high temperatures which however, do not inhibit respiration, so that apparent photosynthesis declines rapidly under these conditions.

Reduction of growth and difficulties in fertilization, even in heat loving crops such as sorghum, occur at temperatures that are often well below the lethal limit. The harmful effects of excessive temperatures are usually aggravated by lack of available moisture. Hot, dry winds will further increase the damage.

In the Great Plains of U.S.A., it is found high summer temperatures may even override the effects of summer rainfall on the yield of crops. Thus, the variability in yields of spring wheat was clearly associated with variability in average maximum temperatures in June, and variability in maize yields was most closely associated with variability in average maximum temperatures in July. With increasing temperatures, evapo-transpiration increases. Hence, a moisture supply that is adequate in a cooler climate may be deficient in a warmer region – a situation which may be aggravated by hot winds. The daily alternations of high and low temperature may affect grain production.

Atmospheric Humidity

Water is always present in the atmosphere in the form of invisible water vapour. This is normally known as **humidity** of the air. When the atmosphere contains the maximum possible amount of water vapour it is said to be saturated at the particular temperature and pressure. Any increase in temperature, water remaining constant, will make the air unsaturated. In unsaturated condition, the water vapour content of air is usually expressed as **relative humidity** which is the ratio between the actual humidity per cent and the saturated humidity possible at that temperature.

The relative humidity of a place, besides being affected by temperature and pressure, is also affected by wind, exposure, vegetation and water content of soil. The evaporation of water from plants or a body of water is directly dependent on the relative humidity of the atmosphere. To be more precise, the rate of transpiration is directly proportional to the difference between the amount of water vapour actually present and the amount that would be needed to saturate the atmosphere. The difference is termed **vapour pressure deficit** and as compared to relative humidity is much more inactive of the potential rate of evaporation. Evapo-transpiration of crop plants increases with temperature but decreases with high relative humidity affecting the quantity of irrigation water. Moist air favours the growth of many fungi and bacteria and these affect seriously the crop. The blight diseases of potato and tea are common examples of diseases spread under moist weather. Similarly, many kinds of insect parasites such as aphids and jassids thrive well under moist conditions.

Solar Radiation

Solar energy provides two essential needs of plants – (i) **light**, required for photosynthesis and for many other functions of the plant – including seed germination, leaf expansion, growth of stem and shoot, flowering, fruiting and even dormancy and (ii) **thermal conditions** required for the normal physiological functions of the plant.

Light: Light is one of the most important climatic factors for many vital processes of the plant. It is indispensable for the synthesis of the most important pigment of the plant, *i.e.*, the green chlorophyll. The chlorophyll is capable of absorbing radiant energy and converting it into potential chemical energy of carbohydrates. The carbohydrates manufactured by the plants during photosynthesis are the only link between the solar energy and the living world. It regulates the rate of transpiration by controlling the opening and closing of stomata.

Light affects the plants in four ways:

1. Light Intensity,
2. Quality (wave length) of Light,
3. Duration (photoperiod) of Light, and
4. Direction of Light.

Light intensity: The variations in light intensity are always accompanied with change in temperature and relative humidity and therefore, it is difficult to evaluate light effects alone. Generally speaking light intensity falling at a particular place is normally enough for the plants and their physiological phenomena viz. photosynthesis.

In photosynthesis about one per cent of the light energy is converted into potential chemical energy. Very low light intensity reduces the rate of photosynthesis and may even result in the closing of the stomata. This results in reduced vegetative growth of the plants. Very high light intensities are detrimental to plants in many ways. It increases the rate of respiration and thus disturbs the photosynthesis-respiration balance. It causes rapid loss of water resulting in the closure of stomata. The most harmful effect of light intensity is the phenomenon of solarization in which all the cell contents are oxidized by atmospheric oxygen. This oxidation is different from respiration and is termed photo-oxidation.

Light intensity varies greatly under natural conditions. Most of the crops show marked response to changes of light intensities. During cloudy weather the production of dry matter is limited. Depending upon the nature of the crops such as *heliophytes* (sun loving) and *sciophytes* (shade loving), the dry matter production is affected. Many species produce maximum dry matter under high light intensity if water is available in plenty. However, crops like betel, sunflower, buckwheat and tobacco grown during summer produce greater dry matter if slightly shaded. Except under glass house or shade conditions, intensity of light can not be controlled in the production of most agricultural crops.

Light intensity and crop yield: Light is the most important factor influencing the photosynthetic activity of a crop, as it is the energy source for the whole process. Light utilization by the crop has two limitations: (a) The maximum quantum yield at low light intensities and (b) Light saturation at high light intensities. The variation in yield of crops particularly the photo insensitive varieties such as dwarf rice in tropics under same input and management conditions is attributed to the effect of sunlight and the connected thermal conditions.

Quality of light: When white light is passed through a prism it is dispersed into wavelengths of different colours; violet 400-435 millimicrons (mm); blue 435-490; green 490-574; yellow 574-595; orange 595-626 and red 626-750. The principal wavelengths absorbed in photosynthesis are in the violet-blue and the orange-red regions. Among these short rays beyond the violet such as X-rays and gamma rays and longer rays beyond red such as infrared are detrimental to growth. Red light seems to be the most favourable light for growth followed by violet-blue.

Ultraviolet and shorter wavelengths of the visible light are scattered and infrared and longer wavelengths are absorbed by moisture of the atmosphere. On clear days such scattering and absorption mat be 10-15 per cent of the total light, whereas on cloudy days it may be upto 100 per cent. Such a light is called **diffuse light** or **sky light.** Ultraviolet and shorter wavelengths kill bacteria and many fungi.

Duration of light: This is considerable importance from the farmers' point of view in selecting the variety of a crop. The length of the day has greater influence than the intensity. The response of plants to the relative length of day and night is known as '**photo-periodism**'. Plants which develop and produce normally when the photo-periodism is greater than a critical minimum (more than 12 hours of illumination) are called '**long day plants**' and those develop normally when the photoperiod is less than a critical maximum (less than 12 hours illumination) are called '**short day plants**'. Some plants are found to be unaffected by photoperiod and are called '**day neutral**' plants.

Plant characters like floral development, floral initiation, bulb formation, rhizome production etc. are all influenced by photo-periodism. If a long day plant is subjected to short day periods the internodes may be shortened to give a rosette appearance and flowering will not take place. In the same way when a short day plant is subjected to long day period, the growth of the plant becomes abnormal and there is no floral initiation.

Among crop plants, soybean, maize and millets are examples of short day flowering plants while sugar beet, wheat, barley are long day plants and others are intermediate (day neutral) in that they flower at any day lengths. Among the horticultural crops while straw berries, chrysanthemum, etc., are short day plants, crop like lettuce, spinach and cruciferous plants are long day plants. Tomato, asparagus, etc. are not sensitive to day length.

Direction of light: Shoots roots and leaves show different orientation to the direction of light. In temperate regions, the southern slopes show better growth of crops than the northern slopes due to the direction of light contributing more sunlight towards the southern side.

Wind Velocity

The velocity of wind at a place depends on various factors such as geographical situation, topography, altitude, distance from sea shore, flat plains, vegetation etc. Its action on plants may be direct or indirect. As a direct effect, wind causes mechanical lacerations and bruises on the tissues of crop plants. High winds tear the leaves and the plants are in danger of being uprooted and blown away. Lodging of crop plants in wheat, maize, sugarcane, rice, etc. are caused by violent winds.

A very high velocity of wind (gale, blizzard, hurricane, cyclone etc.) breaks dead and living branches of trees and even uproots them completely. The effect is however, less when there is a windbreak. In bare deserts high velocity of wind causes constant soil erosion and this makes it difficult for plants to grow. Shelter belts of vegetation are raised to arrest the movement of soil by erosion. Wind is also responsible for causing rainfall to a very large extent. In India, the monsoon type of rainfall is largely determined by particular patterns of wind movement. Wind has a powerful effect on the humidity of atmosphere.

An important function of wind is to help in pollution of flowers and dispersa of seeds, fruits and micro-organisms. Indirectly the velocity of wind affects the rate of transpiration in plants. Much damage is caused to crops by hot dry winds aqt the time of flowering. The internal water balance of plants is affected resulting in poor seed setting. Another for of damage is blossom injury caused by evaporation of secretions from the stigmas.

Effects of wind on crop production: Winds affect growth mechanically and physiologically. The sand and dust particles carried by the wind may damage plant tissues. Emerging seedlings may be completely covered or alternately the roots of young plants may be exposed by strong winds. Winds may also cause considerable losses by inducing lodging, the breaking of stalks and shedding of grain.

The physiological effects of winds consist mainly in increasing transpiration as well as evaporation from the soil. Hot dry winds, may however, adversely affect photosynthesis, and hence productivity, by causing closure of the stomata even when soil moisture is adequate.

The hot dry winds may also occasionally have beneficial effects. The main reason for the low incidence of the dangerous yellow rust of wheat (*Puccina glumarum*) is the susceptibility of the pathogen to high temperature that occurs when seasonal winds blow. In rare years when no winds blow in the spring season, the rust may cause devastating reductions in yield.

Moderate winds have a beneficial effect on photosynthesis by continuously replacing the carbon dioxide absorbed by the leaf surfaces. The favourable effect of wind on photosynthesis is most marked under low turbulence of wind at the lower leaf-layers, increasing wind strength permits even deeper leaf layers to approach the maximum rate of photosynthesis under the amounts of radiation received.

Atmospheric Gases

The atmosphere surrounding the earth contains a mixture of gases viz., carbon dioxide (0.03%), oxygen (20.95%), nitrogen (78.09%), argon (0.93%) and miscellaneous gases (0.02%) in a constant proportion. Variations from this may be found over industrial plants which release large quantities of gaseous fumes such as SO_2, CO_2 and CO as well as water vapour and mineral particles.

1. ***Carbon dioxide***: The CO_2 is the main source of carbon for the various types of organic compounds in the body of the plants. It is the main raw material for the manufacture of carbohydrates by the photosynthetic process of the green plants. Photosynthesis is approximately proportional to the concentration of CO_2 in the air surrounding the foliage of the crop. The CO_2 which gets

incorporated into the organic compounds of the plants returns to the atmosphere by respiratory break down of these compounds and by death, decay and combustion of plants.

a) Increased growth and greater yield of vegetables are possible under green house conditions by increasing the CO_2 content. Certain gases like SO_2, CO and HF when released into the air in sufficient quantities are toxic to plants. Injuries to crops have also been reported in the vicinity of factories producing alumina and phosphates. In both cases fluorine is released and this causes the damage. Near about cement factories, the fall out of cement particles also causes damage to the crops.

b) The CO_2 content dissolved in water is fairly high. So the aquatic plants get CO_2 from water. In soil, excess CO_2 causes poor root respiration. This causes adverse effect on the metabolic activities of roots in plants.

2. ***Oxygen*:** Every living organism whether plant or animal, sustains its life because of oxygen. Atmospheric oxygen is vital to its life and normal growth. The amount of oxygen is normally constant in the air because plants give off oxygen during photosynthesis.
3. ***Nitrogen*:** The nitrogen of the atmosphere is made available to the soil by lighting, rainfall and nitrogen fixing micro-organisms. Free living bacteria like *azotobacter*, symbiotic bacteria like *rhizobium*, blue green algae, etc. fix a good amount of nitrogen in the soil. The decomposition of the protein content of the dead plants and animals by micro-organisms also adds to the nitrogen pool of the soil. The nitrogen in soil is made available to the crops by the activity of nitrifying bacteria. The plant tissues are largely consumed by animals as food. Part of the nitrogen in the tissues is again returned to the soil in the form of animal excreta or dead animal tissues.

Edaphic Factors

The land plants are completely dependent on the soils in which they grow for anchorage, water and mineral nutrients. The nature of the soil is normally so complex and dynamic and its effects on plants so pronounced that an agronomist has to attach great importance to the study of soil and its management. The factors which affect the crop growth through the soil are (1) soil moisture, (2) soil air, (3) soil temperature, (4) soil mineral matter, (5) soil organic matter, (6) soil organisms and (7) soil reaction.

Soil Moisture

Plant tissues constitute about 90 per cent of water. The moisture lost through transpiration is made up only by absorbing water from the soil. The important functions of water in the plant system are:

1. It gives turgidity to plants and keeps them erect.
2. It regulates the temperature in the plant body.
3. It acts as a solvent and nutrient carrier.

Indirectly water present in the soil helps the plants in many ways:

1. It supplies the essential raw materials for production of carbohydrates by photosynthesis.
2. It promotes physical, chemical and biological activities in the soil.
3. The physical properties of soil such as proper structure formation, friability, penetrability plasticity, shrinkage, cohesion etc. Are altered by the soil moisture content. Besides the thermal properties of soil such as thermal

conductivity, absorptivity etc. is affected by moisture content.

4. The forms and contents of different elements in the minerals, the different chemicals changes such as hydration, hydrolysis etc. and the concentration of different salts affecting plant growth are affected by moisture content.
5. The forms and types of different microbial organisms which may affect plant growth both beneficially and adversely are controlled by the moisture content of soil.
6. Gaseous diffusion in soil for proper aeration depends on the moisture content.

Soil moisture is in a highly dynamic condition and varies from time to time. The moisture is held within the soil by the attractive forces of the soil particles.

Soil Air

Aeration of the soil is absolutely essential for the absorption of water by roots. Absorption of water by the roots takes place at a rapid rate in well aerated soils while at a greatly retarded rate in soils deficient in oxygen supply.

Oxygen is needed for the respiration of roots. In well aerated soils, carbon dioxide released in respiration of roots and micro-organisms is freely exchanged with the air above the soil. In poorly aerated soil the carbon dioxide gets accumulated and is detrimental for the water absorption process of the plant.

Sol air is also useful in increasing the nutrient availability of the soil, (1) by breaking down the insoluble minerals into soluble salts, (2) by decomposing plant and animal remains and (3) by bringing about nitrifying and nitrogen fixing processes of bacteria.

Soil Temperature

The prevailing temperature of the soil besides affecting the physical and chemical processes going on the soil also influences to a great extent the rate of absorption of water and solutes, the germination of seeds and the rate of growth of the underground portions of the plant body. The maximum metabolic activities of the plants and the maximum absorption of water by the roots take place generally between 20 and 30°C. Lowering of temperature below 20°C causes an appreciable reduction in the rate of absorption of water. Cold soils are therefore, not conducive for rapid growth of most agricultural crops.

Soil temperature is one of the factors that control the micro-biological activity and the process involved in the availability of nutrients to plants. Nitrification does not begin until the soil reaches a temperature of about 5°C (40°F).

Soil Mineral Matter

The mineral content of the soil is derived from the weathering of the rocks and minerals and consists of particles of different sizes. The principal minerals occurring in the earth's crust are indicated in Table 3.4.

TABLE 3.4: Principal minerals in earth's crust

Minerals	Per cent
Feldspar	48
Quartz	36
Mica	10
Limestone and magnesium limestone	2
Hornblend and Augite	1

Minerals	Per cent
Olivine and Serpentine	1
Clays	1
Other minerals	1

1. Feldspars are aluminium silicates with varying amounts of silicates of K, Na and Ca.
2. Quartz is silicon dioxide and is found in all crystalline rocks; also occurs as sand and is the principal component of granite and sand stones.
3. Mica is silicate of Al and K with silicates of Fe, Mg and Na.
4. Limestone contains carbonate of Ca. With magnesium carbonate it occurs in dolomite.
5. Hornblend is composed of silicate of Ca, Mg and Fe with sodium silicate.
6. Augite is made up of silicates of Ca, Mg, Fe and Na in varying proportions.
7. Olivine is a ferro-magnesium silicate, and serpentine is hydrated magnesium silicate.
8. Clays are hydrated aluminium silicates and may contain varying amounts of Ca, Mg, Na and Fe. They usually contain iron oxide and carbonaceous matter together with varying amounts of quartz, feldspar and mica.
9. Other minerals occurring in the soil include Tourmaline (borosilicate of Al with alkali metals and Fe or Mg), Rutile (Titanium oxide), Zircon (Zirconium silicate), Glauconite (hydrated silicate of Fe and K), Magnetite and Hematite (oxides of Fe).

In course of weathering of rocks, the readily soluble products are carried away in drainage waters whereas comparatively insoluble products accumulate to form soil. This is particularly so in humid climates and in consequence, silica, alumina and iron oxide, as such or in silicate combinations form the major portions of soils in heavy rainfall areas. Silicon in silicate form also dissolves, leaving laterites, which consist chiefly of aluminium iron oxides. In arid regions, leaching plays an insignificant part and therefore, the chemical composition of soils is very similar to that of the rocks from which they originate.

Inorganic Compounds

Compounds of Si, Ca, Mg, Fe, K, Na and Al form the principal chemical constituents of soils. Besides these, the sol contains small quantities of a large number of other mineral elements like B, Mn, Mo. Zn, Cu, Co, I and F which are known to be essential for the well being of plants in very small quantities. Such elements are called **minor-** or **trace-elements**.

The total amounts of elements contained in soils depend partly on the nature of the rocks from which they are formed and partly on their age and the extent to which soluble products have been leached away. The chemical composition of different horizons of a soil also shows a good deal of variation. Generally, speaking the A horizon is rich in insoluble compounds as compared with the B horizon. Usually the elements that are commonly leached out are the ones required in largest quantities by plants.

TABLE 3.5: Percentage inorganic composition of representative Indian soils

Type	State	SiO_2	Al_2O_3	Fe_2O_3	MnO	CaO	MgO	K_2O	Na_2O	P_2O_5	SO_3
Sandy	Rajasthan	91.72	2.92	2.36	-	0.35	0.78	0.33	0.08	0.08	0.04
Sandy loam	Uttar Pradesh	84.84	5.30	4.52	-	0.91	0.52	0.16	0.03	0.01	-
Sandy loam	Punjab	-	10.89	-	-	0.06	0.29	0.71	-	0.34	-
Loam	Tamil Nadu	82.11	5.19	5.70	2.12	1.10	1.27	0.24	0.86	0.06	3.04
Loam	Uttar Pradesh	86.98	4.10	3.52	-	0.42	1.29	0.56	-	0.06	-
Silt loam	Assam	86.49	9.16	7.00	0.21	0.14	0.90	1.85	-	0.24	-
Clay	Tamil Nadu	65.16	13.76	9.27	0.25	2.18	2.47	0.14	0.01	Tr	-
Clay	Maharashtra	63.80	Not available		-	1.09	1.60	1.60	-	0.23	-

Source: Hand Book of Indian Agriculture, I.C.A.R., New Delhi.

TABLE 3.6: Percentage chemical composition of different horizons of red Lateritic soil from Nilgiris, Tamil Nadu

Horizons	Depth in cm	SiO_2 (soluble)	$Fe_2O_3 + Al_3O_3$	CaO	MgO	K_2O	P_2O_5	N
1	0-22.5	0.53	32.60	0.140	0.03	0.656	0.138	0.154
2	22.5-75	0.58	30.22	0.084	0.02	0.667	0.138	0.109
3	75-112.5	0.57	37.15	0.056	0.03	0.568	0.149	0.050
4	112.5-117.5	0.55	26.15	0.098	0.04	0.480	0.245	0.030
5	117.5-162.5	0.50	35.73	0.084	0.02	0.580	0.222	0.030

Source: Hand book of Indian Agriculture, ICAR, New Delhi.

TABLE 3.7: Composition of soil

Content	Per cent
Mineral matter	30
Soil water	30
Soil air	30
Soil organic matter	5 – 10

TABLE 3.8: Textural classification of soil

Name of particles	Particle size (mm)
Clay	0.0001-0.005
Silt	0.005-0.05
Very fine sand	0.05-0.10
Fine sand	0.10-0.25
Medium sand	0.25-0.50
Coarse sand	0.50-2.00

TABLE 3.9: Mechanical composition and textural class of some typical Indian soils (surface soils)

Soil group	Per cent composition of different size fraction				Textural classes
	Coarse sand	Fine sand	Silt	Clay	
Desert soil (Ajmer)	73.6	18.5	3.8	4.4	Coarse sand
	79.4	13.4	3.7	4.8	Coarse sand

Soil group	Per cent composition of different size fraction				Textural classes
	Coarse sand	Fine sand	Silt	Clay	
	89.4	8.7	1.4	3.4	Coarse sand
Alluvial soil (Kanpur)	0.2	57.8	25.7	14.1	Sandy loam
	1.2	36.3	27.9	27.8	Loam
	2.0	72.1	14.5	8.4	Loamy sand
Black cotton soil (South India)	0.7	11.4	21.8	42.2	Clay
	18.9	13.5	16.1	47.4	Clay loam
	41.2	18.2	11.4	23.9	Coarse loam
Laterites and Red soil (Bengal)	10.0	43.0	20.0	27.2	Sandy loam
	32.0	19.3	18.0	32.3	Loam
	24.0	29.0	31.2	18.0	Sandy loam

Source: Hand book of Indian Agriculture, ICAR, New Delhi.

Table 3.5 give the percentage inorganic composition of eight representative Indian soils and Table 3.6 indicate the successive horizons of a lateritic soil. The different components of the soil may be classified as in Table 3.7.

The mineral matter of soil may be classified texturally based on the size of particles (Table 3.8).

Based on the proportion of different size groups of soil particles the soil is classified for better understanding on their characters (Table 3.9).

Sands tend to menace the size of pore spaces in soils and this aid in the movement of air and water. In general, fine textured soils hold much of water and make them available to crop growth over longer periods. Root penetration is retarded by high silt and clay content of the soil.

Loam soils are the best for the growth of plants since they contain coarse sand as well as fine silt and clay particles. Loam soil has good aeration, good water infiltration and water movement, easy root penetration and also good water holding capacity and fertility.

Soil Organic Matter

Besides inorganic substances, soils contain organic matter in amounts ranging from less than 1 per cent in arid sandy soils to as much as 90 per cent in peaty soils. The presence of organic matter distinguishes soil from disintegrated rock material.

The organic matter is added to the mineral matter of the soil every year. Though the percentage of organic matter is less than 5% of dry weight of well developed soil, it has a marked influence on the soil properties and growth of crops. The organic matter of the soil is largely derived from (1) dead and decaying roots of plants and living organisms present in the soil and (2) dry leaves, twigs, dead plants and animals.

The organic matter of the soil is a source of mineral nutrients essential for the crop growth. During decomposition, the humus is mineralized. It has increased water holding capacity due to its content of organic colloids. It can hold by adsorption a great amount of minerals particularly the cations in the ionic form, thus increasing the exchange capacity of soils. It is a source of food for most of the soil organisms. As a source of plant nutrient it contains 95% of the total nitrogen, 50 to 60% of total phosphorus and 10 to 20% of total sulphur.

Soil Organisms

Raw organic matter in the soil is not directly used by plants as food. It must be broken down first into humus and then into simpler products before it can be utilized. This work is done by different kinds of micro-organisms which inhibit the soil by the billions. Decomposition of organic matter primarily forms part of the feeding and growth processes of these microscopic plants and animals. Sugars, starches and proteins are broken down first, then cellulose and fatty substances (lipids), and lastly lignin or woody substances.

The principal kinds of living organisms found in the soil are as follows:

Plants	Animals
Bacteria Actinomycetes Fungi Algae Roots, rhizoids and rhizomes	Protozoa Nematodes Mites Insects (especially ants and beetles) earthworms burrowing vertebrates (Moles, gophers, rats etc.)

A large number of bacteria and fungi bring about the decay of the organic substances. Their mineralization process releases a variety of nutrients to be made available to the plants. The ammonifying bacteria convert proteins into ammonia. The nitrifying bacteria oxidize ammonia to nitrites and nitrates. A number of bacteria and blue green algae fix atmospheric nitrogen in the soil under anaerobic conditions of water logged soils; certain bacteria cause denitrification, releasing free nitrogen which gets lost into the air. Many bacteria and fungi are pathogenic causing diseases of plants. Some of the fungi form mycorrhizal association with root system of higher plants for helping them in the absorption of water and minerals.

The larger organisms like earthworms; rodents etc. improve the aeration of the soil due to their burrowing habit. They play a key role in the weathering of rock and minerals and in the formation of soil.

Soil Reaction

Soils may be neutral, acidic or alkaline depending upon their content of the basic salts and acidic components. Neutral soils are the best for the growth of most crops.

Soil acidity beyond a particular limit is injurious to plant growth due to the following reasons:

1. High acidity (particularly high alumina content) is toxic to plants.
2. High acidity interferes with the absorption of several nutrients particularly the cations such as K, Ca and Mg due to their low content in soil. Nutrients like P are fixed in acidic soils.
3. Decomposition of organic matter by micro-organisms may be reduced.
4. The activities of nitrifying and nitrogen fixin bacteria may be checked.
5. Certain types of fungoid diseases like potato scab may be favoured.

Similarly, high alkalinity of the soil also adversely affects crop growth. The soil alkalinity is due to the excess of cations like Na, K, Ca and Mg found in the sol. Among these the presence of high amount of sodium interferes with nutrient and water absorption. This may interferes with water movement and infiltration in the soil as well as the microbial activity.

Biotic Factors

Beneficial or harmful effects are caused by other plants and animals on the crop plants. These are called as biotic factors.

Plants

1. ***Competitive and complimentary nature among field crops***: Competition between plants occur when there is demand for nutrients, moisture and sunlight. Different crops do have optimum leaf area indices for the maximum utilization of the available resources of sunlight and nutrients for their maximum production. While closer spacing reduces the yield of individual plants, wider spacing will reduce the total yield because of lesser population of plants per unit area. Hence, optimum spacing of crops is an important agronomical practice. But in the case of different crops when grown together as in the case of mixed cropping, there is mutual benefit resulting in better yield. Ceral-legume mixture is a common practice throughout the world.
2. ***Competition between weed and crop***: Weeds are plants growing where they are not wanted. In nature, certain plants, the seeds of which are not sown come up at all times under various conditions of soil, climate and environment. Weeds reduce crop yields on account of their competition with crops for water, soil nutrients and light. They increase the cost of labor and equipment, render harvesting difficult, reduce the quality and marketability of agricultural produce, harbour insect, fungal, virus and bacterial organisms and some are poisonous to human beings and livestock.

   In non-irrigated areas the competition between weeds and crops is largely for water. The transpiration coefficient for Bermunda grass (*Cynondon dactylon*) is 813, while it is only 430 for sorghum. A saving of 750 to 1250 tons of water in a hectare of soil, 180 cm deep, is possible by keeping it free of weeds. In irrigated tracts, the competition is severe for nutrients. The removal of nutrients by the weeds in the unweeded control is much higher as compared to the weeded plots.

   Weeds in fallow land deplete the soil of both moisture and nutrients. Weeds block drainage and impede the flow of water in rivers and canals. Control of weeds is a must for the well being of crops, soil and water.
3. ***Plants as parasites***: A plant parasite is dependent on its host plant for its existence. Under favourable conditions these parasites exert a profound influence on the plant community. Common examples of this are the parasitic fungi, bacteria, virus etc. These cause different kinds of diseases on agricultural crops. These micro-organisms obtain their food either by breaking down dead plant and animal remains (saprophytes) or by attacking living plants and animals (parasites). In order to obtain their nutrients, the parasitic organisms kill the tissues and cells of the host plant as a result of which either the whole plant or a part of it is damaged and killed, or considerable disturbance takes place in the normal metabolic processes of the living plant. In some cases they also produce toxic substance

   Controlling the diseases caused by these parasitic organisms has become a routine practice in crop production. They may be controlled by evolving disease resistant varieties, use of chemicals and cultural practices and field sanitation.

Symbiosis

The apparent independent existence of individual plants, on close scrutiny, is found to be not so. Certain organisms are normally found associated with others to form a community. In the community, the different organisms have manual relationship with each other and with the prevailing environment of the place. This biological relationship among the organisms is termed as '**symbiosis**'.

The symbiotic relationship between legumes and rhizobia which results in nitrogen fixation is of great significance to crop production. Plants by themselves are unable to utilize elemental nitrogen present in the atmosphere. Two groups of bacteria participate in the capture of gaseous nitrogen and help plants in utilizing it, the nodule organisms (*Rhizobium sp.*) of the legumes and the free living bacteria (*Azotobacter*) of several kinds.

The legume bacteria use the carbohydrates of their host as energy and fixes up atmospheric nitrogen. A part of it is passed to the infected host. The free living bacteria acquire their energy from the soil organic matter, fix the free nitrogen and make it a part of their own tissue. When they die, the nitrogen available by their body tissues is used by the crop plants.

Animals

1. ***Soil animals***: The soil fauna including protozoa, nematodes, rotifers, snails and insects constitute a highly important part of the environment of plant roots. All these organisms contribute to organic decomposition, when using the organic matter for their living.
   Among these, insects and nematodes cause considerable damage as crop pests. Crops are subjected to insect attack throughout their period of growth. Even after harvest, the grains may be damaged by insects. The average loss due to insects is reported at about 20% throughout the world.
2. ***Beneficial organisms***: Many crop plants are pollinated primarily by insects. Bees and wasps are probably the pollinators of the greatest importance. Moths, butterflies and beetles do considerable pollination. Earthworms are influential in changing soil conditions. Burrowing facilities aeration and drainage and the ingestion of organic matter and mineral matter result in a constant mixing of these materials in the soil. This probably tends to make better plant growth.
3. ***Small animals***: Many small animals like rabbits, squirrels and field rats constitute an important factor. They effect changes by consuming green plants, burrowing and soil mixing. These cause extensive damage to field and garden crops.
4. ***Large animals***: Domesticated animals like cattle, goat, sheep and horses as well as wild animals like deer, elephants, pigs and porcupines do considerable damage to crop plants by their grazing and browsing habits. Extensive areas nearer their habitat suffer greater damage if no protection is made.

Physiographic Factors

The geological strata of a place and topography affect the crop considerably.

Geological strata: The nature of the geological strata accounts not only for the kind of parent material utilized in soil formation, but also on the nature of crops grown in these soils for proper utilization.

Topography: The nature of the surface of the earth is known as topographic factors. They include:

1. Altitude of the place.
2. Steepness of slope.
3. Exposure of the slope to light and wind.
4. Direction of mountain chains.

The topographic factors affect the crops indirectly by modifying climatic and edaphic factors of a place.

Altitude of the Place

The continuous increase in altitude and its effect on the vegetation can be best observed on the mountains. Increase in altitude is usually associated with:

1. Decrease in temperature.
2. Increase in precipitation,
3. Increase in wind velocity.

It has been found that the temperature decreases at the rate of 6 to 7°C per 1000 metres except in the valleys and lowlands due to the downward descent of cold air. Due to increased rate of precipitation and strong winds, the type and nature of soil and vegetation are affected. The organic matter of the soil increases and this in turn increases the nitrogen content and the acidity of the soil.

As we ascend the mountains, changes in temperature etc. give a definite pattern of vegetational zonation, similar to the sequence of vegetation met with the moving from equator to the poles.

Steepness of Slope

Steep slope causes swift run-off of water after a rainfall. This decreases the moisture content of soil. Besides the soil becomes unstable and may even be completely eroded. The humus is also not allowed to accumulate. The bare rocks are exposed. These areas may not be useful for raising field crops. To counteract the harmful effect of steep slopes, soil conservation measures are employed to reduce the flow of water and to conserve soil.

Exposure of the Slope to Light and Wind

A mountain slope exposed to weak intensity of light and strong dry winds as the case of northern slopes of temperate regions and the Himalayas may have poor crops due to want of moisture and sunlight. Similarly, the slopes exposed to high velocity of wind as in the case of western slopes in the hilly areas of Tamil Nadu have poor crops due to the damage caused by winds.

Direction of Mountain Chains

The distribution of rainfall throughout the country during the monsoon seasons is largely governed by the direction of mountain chains. This pattern of rainfall largely governs the type of agricultural crops grown under the dry conditions under different soil conditions.

Anthropic Factor

Man or woman has produced profound changes in plant environments. Improvements effected by plant breeders towards increasing crop yields, introduction of exotic plants etc. have greatly influenced crop growth.

Successful farming does not depend merely on a good knowledge of the physical, chemical and biological properties of a soil. It is a matter of both soil and its management. The most important consideration in the latter is the correct application of the relationship between the soil and the crop to be grown on it.

Although problems of soil management differ widely with the nature of the soil, the climatic conditions and the types of crops to be grown, yet, there are certain fundamental factors that should govern the choice of soil management practices everywhere. Good soil management must be based on the following guiding principles.

1. Selection of the right crops for a given soil or the right soil for a given crop.
2. Maintenance of soil so that it provides suitable conditions for the successful growing of selected crops.
3. Improvement of the productive capacity of soil.
4. Recommendation of only economically profitable husbandry methods.

In carrying out these, one has to collect and carefully weigh accurate information on the different points mentioned above. Field surveys, enquiries from experienced farmers, actual experiments in farms, and the results of investigations in laboratory and glass house must all be pressed into service.

In doing so, the scientific soil and crop management practices have greatly influenced the crops by increasing their yield manifold.

CHAPTER 4

Systems of Crop Production

The relationship between man and his environment is hardly ever a static or 'closed' system and if so only for a limited time, because sooner or later human ingenuity or mistakes, natural disasters or chance discoveries such as rice as a food crop will bring about changes from within the ecosystem. It may be altered by external influences, by new ideas, techniques and implements or by exotic plants such as potato and groundnut.

Food is the basic requirement of man. The successful domestication of plants, animals, soil and water by settled human communities has been the starting point of culture. Tropical Asia provides a typical example. Here the predominant pattern has been one of almost uninhabitable uplands side by side with heavily populated lowlands where the cultivation of rice as a staple food crop has developed over the past five millennia supported by irrigation under monsoonal, seasonally dry conditions.

HUMAN SETTLEMENT AND LAND USE

The basic spatial pattern of population and land use as described in early colonial times in tropical Asia must have existed for a long time over five thousand years depending on the geography and climate of a location. For example in peninsular India, three main spatial units had evolved whose general features have persisted until date.

The most densely populated and cultivated regions were and still are the coastal and riverine alluvial plains, mainly on account of their high potential of continuous irrigated rice cultivation. In the Western Ghats, abundant rainfall during the major part of the year and unfailing water supply along numerous mountain streams even during the drought period seemed to intensify multiple cropping of rice, the staple food crop. In the adjoining latertitic high lands, luxuriant growth of coconut and areca nut, palms, tapioca, banana, spices and condiments are seen. In the East, where the rainfall is low, thousands of tanks have been constructed to store the rain water for extensive stretches of rice fields. Whenever irrigation is not possible as in the case of uplands, a variety of millets, pulses, oilseeds and fibre crops are grown under rainfed conditions.

In the hilly areas, the widespread land use of shifting cultivation of the past has been given place to permanent intensive agriculture of plantation crops such as tea, coffee, rubber, cardamom, etc., and the growing of profitable vegetable crops like potato, cabbage, cauliflower, carrot, pea and beans apart from fruit trees, like pears and plums. At present different cropping systems are intensively practiced in many parts of the world.

SYSTEMS OF CROPPING

Sun light, soil, water, plants, animals and human beings constitute the basic resources of agriculture. In tropics, due to sun light available throughout the year, more than two crops can be easily raised if other resources are available. Growing of more than two crops in the same field in sequence in a year is called **multiple**

cropping.

In many instances, the seeds/plants of second crop are sown /planted before the harvest of the first crop. Similarly, the seeds/plants of third crop are sown/planted before the harvest of the second crop. Growing of more than two such crops in the same field in a year is called **relay cropping**.

It is a common practice, to raise two or more crops in the same field in the same season particularly under rainfed agriculture. The seeds may be mixed sown or sown separately either in paired rows or alternate rows or in different proportions of rows depending on the need of the farmers and successful growth of the mixed crops. This practice is called **mixed cropping** if seeds are sown mixed and **inter-cropping** if raised in rows.

A combination of these different cropping methods is called **crop-cafeteria** wherein every available space is used for growing crops in the same field over time.

In case of plantation crops, three or more crops are raised simultaneously in a mixed pattern with different heights and canopy to take advantage of the sunlight. Growing together of coconut, areca nut, pepper, banana, coco, pine-apple, fodder grasses etc. are a common practice. This is called **multi-storied cropping** or **multi-tier cropping.**

Multiple and Relay Cropping

The cultivation of three or four crops during a year has become possible in irrigated areas of tropics. The development of high yielding varieties in tuber crops namely potato, sweet potato and tapioca has added a further dimension to yield possibilities. Hundreds of different cropping patterns to suit different soils, climate and locations have been tried in many parts of the country particularly in India. The general approach has been the standardization of various alternative rotation patterns all of which are scientifically sound from the view point of maintaining the long term productivity of the soil but from which the farmer can choose according to his capacity to mobilize inputs, market conditions and seasonal behaviour.

New systems of cropping are rapidly coming into existence in many parts of irrigated areas. In the states of Punjab, Haryana and parts of Uttar Pradesh, rice during *kharif* and wheat during *rabi* are now popular. In Bengal, wheat is popular as a *rabi* crop in rice fields. In the rice fallows of South India, sorghum, bajra (*Pennisetum*) and cotton are raised during the off-season. The following data (Table 4.1) indicate the possibilities.

TABLE 4.1: Yields and total production of three crop annual sequence at various centres during 1970-71

State	Centre	Soil type	Crop sequence			Yield in kg/ha			Grain produced per day Kg/ha/day
			Kharif	Rabi	Summer	Kharif	Rabi	Summer	
Andhra Pradesh	Tirupathi	Red sandy	Rice (IR 8)	Rice (IR 8)	Rice (Hamsa)	4438	6410	3090	42.9
	Maruteru	Medium block	Rice (Pankaj)	Rice (Padma)	Mung (Kaveri)	4666	5583	910	34.1
U.P.	Benars	Alluvial	Rice	Wheat	Rice	3732	5042	4309	39.4
Karnataka	Mangalore	Laterite	Rice (MTU	Rice (Padma)	Rice (Padma)	2450	1770	4570	34.3

State	Centre	Soil type	Crop sequence			Yield in kg/ha			Grain produced per day Kg/ha/day
			Kharif	Rabi	Summer	Kharif	Rabi	Summer	
			20)						
Tamil Nadu	Bhabanisagar	Red loam	Rice (Adt 27)	Rice (IR 8)	Rice (Co 29)	3643	3908	5106	35.6
Orissa	Bhubaneswar	Laterite	Rice (Padma)	Rice (T 141)	Rice (Padma)	2648	1970	3127	28.8
			Jute (JRO 212)	Rice (T 141)	Rice (Padma)	1738	2055	4485	33.7

Source: Annual Progress Report of All India Co-ordinated Agronomy Experiment (1969-71).

Potentials

Experiments conducted in various research stations have indicated that more than 20 tonnes per hectare of grain per year can be produced. The continuous rice cropping at Tamil Nadu Agricultural University, Coimbatore had produced 22 tonnes per hectare with rice IR 8 over a period of one year with three crops. Similar results have been obtained at Inter National Rice Research Institute, Manila. The principles used are the improvement of harvest index (proportion of economic yield to the total biological yield) as well as the ability of crops to respond to good management through restructuring of plant hybrid vigour and the enhancement of per day productivity through the introduction of photo-insensitive varieties thereby facilitating multiple cropping.

Multiple cropping provides an effective agricultural method to harvest solar energy. Green plants can be made to harness sunlight throughout the year, wherever there is water to sustain crop growth. There is vast potential for using solar energy in arid and semi-arid areas for household purposes thereby saving valuable organic matter to soil.

Advantages

Multiple and relay cropping is a very potent tool for improving the economic position of a farmer with a small holding and for banishing unemployment and underemployment. Some of the beneficial effects of the technology are the following:

1. Increase of yield and income.
2. The employment potential of agriculture can be greatly enhanced.
3. There is scope for better utilization of resources such as animals and machineries, irrigation water etc.
4. Better advantage of market particularly in case of horticultural crops.
5. There is scope for self-sufficiency in meeting the needs of animals for feeds and fodders.
6. There is appreciable increase in real income among all classes of farmers big, medium and small as well as landless labour.
7. If properly handled, this may become a powerful method of minimizing the incidence of pests and diseases.

Scope of Multiple Cropping

When crops are grown continuously in sequence and with relay and mixed crop patterns, there is greater scope for reducing the cost of cultivation of crops without adversely affecting yield.

1. **Tillage:** When single or double crops are grown in the same field in a year, the fallow period available in between the crops help in soil hardening, weed growth and loss of nutrients due to erosion. This necessitates more number of tillage operations than needed. Experiments on tillage have indicated that the concepts of minimum and zero tillage have reduced the cost of production without affecting the yield. The case of relay cropping such as sowing cotton in the stubbles of finger millet, has clearly demonstrated that there is no need for preparatory cultivation for cotton for successful growth. Similarly in the paired row system of mixed crops, the inter-space is covered by the second crop and helps in controlling weeds and reducing the time needed for after cultural operation. Thus, this technology helps in reducing the cost if tillage operations considerably.
2. **Manures and fertilizers:** In the package of practices formulated for any crop, the cost on manures and fertilizers is considerable. Many experiments conducted on cropping system, instead of individual crop, has shown considerable saving on the manorial budget particularly with P and K without affecting the yield. The usage of a leguminous crop in the sequence helps in the reduction of N requirement for the following next crops.
3. **Water:** In the irrigated garden land mixed crops, the water loss by evaporation is considerably reduced. The design of irrigation method is changed due to the paired row system and considerable water saved.
4. **Plant Protection:** The opportunity for greater weed growth in the multiple, relay and mixed cropping system is considerably reduced due to lesser area and time for its proliferation. Again due to the mixing of crops, the spread of pests and diseases are reduced. These help in the reduction of chemicals needed for their control.

Some Ground Rules for Introduction

1. No two crops sharing in common, the same pests and diseases should be grown either mixed or succession.
2. One crop should be deep rooted and another with more shallow roots so that different layers also differ in rooting pattern and if a deep rooted crop such as cotton is grown during *kharif*, a wheat variety such as Sonalika or Sharbathi sonara may suffer less from micronutrient deficiency than Kalyan sona which also taps nutrients from the lower layers of the soil.
3. For the restoration of soil fertility, at least one leguminous crop should find a place in the sequence. If the stalks of the soil would help to improve soil structure.
4. The requirement of different nutrients very among the crop. So also their capacity to absorb them. Hence, it is better to grow crops of different families in sequence.

Multiple Cropping in Farmers Fields

In India up to mid 1960s wheat was grown in the *rabi* season while rice was confined to *kharif* in the northern regions as a single crop in their respective areas depending on soil and rainfall pattern. Even the irrigated wheat was cultivated as a mono-crop after *kharif* fallow. But with the development of high yielding, photo-insensitive and short duration varieties and hybrids of maize, sorghum, pearl millet, wheat and rice, an entirely new era in crop management has been ushered. It was established that there is no need to keep the land fallow. Double and triple cropping has become common in most of the irrigated area where water supply is assured.

Major states growing irrigated wheat in India are Punjab, Haryana, Uttar Pradesh, Bihar and Rajasthan. The new varieties of rice fitted very well in rotations with wheat in Punjab and Haryana replacing maize, pearl millet or pulses, the normal *kharif* season crops (Table 4.2).

TABLE 4.2: Statistics in the growing importance of rice in the traditional wheat growing states

	Haryana		Punjab	
	1971-72	**1978-79**	**1971-72**	**1978-79**
Area (000 ha)	291	463	450	1053
Production (000 t)	536	1241	920	3073
Yield (kg/ha)	1842	2680	2044	2918

Research in multiple cropping has resulted in the development of several crop rotations of different cropping intensity as shown in Table 4.3.

TABLE 4.3: Crop rotations developed

Location	**Crop sequence**			**Yield in q/ha**			**Total yield per year q/ha**
Ludhiana (Punjab)	Rice	Wheat	Mung	67.8	36.3	7.4	111.5
	Maize	Wheat	Mung	34.1	38.6	9.4	82.1
Bichpuri (U.P.)	Pearl millet	Wheat	Mung	32.4	57.6	10.2	100.2
Pantnagar (U.P.)	Maize	Wheat	Mung	43.5	40.7	5.9	90.1
Delhi	Rice	Wheat	-	50.0	38.0	-	88.0

Among these 'rice-wheat' double cropping and rice–wheat-green gram triple cropping to a limited extent have become popular with the farmers.

Constraints for Adoption

Although the experiments conducted at research stations have indicated the high profitability of cropping pattern of 300 per cent and above, these are not much popular with the farmers due to the inherent constraints for adoption.

1. The land, water and other farm resources are decidedly inferior to those of Experimental Stations and farms in which experiments are conducted.
2. The farms, being small ranging between 0.5 to 3.0 hectares, needs fencing and protection unless the adjoining farmers also practice the same pattern.
3. Many farmers are share-tenants with limited assets. Hence, high profit may not be an inducement.
4. The family labour and cash flow are limited and as such the farmer may not be in a position to tackle the needs of multiple cropping.
5. The non-farm and other farm activities competed for labour at the critical times.
6. In most of the villages, the contract able labour pools and credit sources are limited.

7. Inputs that can not be purchased locally at farm level costs may not be applied as suggested.
8. Crops that local markets can not absorb at profitable prices may not be adopted.
9. In areas depending on onset of rainfall and water resources, the field operation may not be started on time. In these areas, the risks are very high because of likely water shortage.
10. While quantitative and qualitative resource deficiencies account for part of the gap, socio-economic factors hinder adoption of modern techniques and efficient management.

These constraints indicate the need for systematic cropping systems research both in the experimental stations and the farmers holding as on site research.

Cropping Systems Research

Although the evolution of different cropping patterns is concerned with the agronomists, the adoption by farmers needs a co-ordinated approach involving crop management research and farm management economics research. Besides there is the need for farmer involvement and scrutiny so as to evoke an alternate cropping pattern with economic viability. There is need to consider the following aspects in the research activities.

1. The resource base of farm family and community that governs adoption potential.
2. The research findings should have greater variability in physical and socio-economic environment for wide scale adoption.
3. The indigenous technical knowledge evolved in a location on practical experiences over centuries should be given its due importance when new methodologies are tried.
4. As the research needs in all these activities require a large number of human resources, the limited supply of qualified research personnel have to be taken into account.

Need for On-site Research

The data generated in the experimental farms may not take into account the variability found in the physical and socio-economic environment of the large number of farmers,. Soil, physiographic and climate variability are more pronounced in tropics. Homogenous, agro-ecological zones are smaller. Hence, there is greater need for land evaluation to assess variability. Similarly, in the socio-economic environment, the farming community can be differentiated into small, marginal, medium and large farmers, illiterate and semi-literate, owner and tenant cultivators etc. To cater to needs of such a clientele in the physical environment, there is need for on-site research, the results of which may provide a basis for judging the following.

1. Technology that is within the management and resource limitation of small and marginal farmer and can be adopted without Government programme.
2. Technology that faces specific resource constraint such as weak market, managerial deficiency, etc. have to be modified to meet the needs of the farmers of the locations.
3. Technology that encounters problems which can not be solved or that can not be over come by a programme of resource augmentation, market development or support or extension activities may not be any use to the

farmers.

4. The technology suggested should be such that it should be flexible and provide alternate plans in carrying the farm operations.
5. The technology suggested should have taken into account all the environmental and socio-economic resources of the farmer and due importance given to the indigenous technical knowledge developed in the locality.

Cropping systems research methods that evaluate limitations by the environment and farm resource base is a continuous process and the farmers should be given alternate plans indicating their benefits and drawbacks. Where components of pattern are economically viable; the farmers adopt those components rapidly. When components are not compatible with physical, biotic or socio-economic factors, cooperative research with farmers in their field helps recognize field level constraints. For large scale following by farmers, the studies on the cropping pattern may help in suggesting to Government for its input wherever necessary and for to the researchers to correct defects on the technology proposed.

CROPPING SYSTEM

A system consists of several components which are closely related and interacting among themselves. In agriculture, management practices are usually formulated for individual crops. However, farmers are cultivating different crops in different seasons based on their adaptability to a particular season, domestic needs and profitability. Therefore, production technology or management practices should be developed keeping in view all the crops grown in a year or more than one year if any sequence or rotation extends beyond one year. Such a package of management practices for all the crops leads to efficient use of costly inputs, besides reduction in production cost. For instance, residual effect of manures and fertilizers applied and nitrogen fixed by legumes can considerably bring down the production cost if all the crops are considered than individual crops. In this context, cropping systems approach is gaining importance.

System

System means an arrangement of components which process inputs into outputs. Each system consists of boundaries, components, interactions between components, inputs and outputs.

Crop System

Crop system is an arrangement of crop populations that transform solar energy, nutrients, water and other inputs into useful biomass. The crop can be of different species and variety, but they only constitute one crop system if they are managed as a single unit. The crop system is a sub-system of a cropping system. For example, in the maize crop system, maize is the dominant crop which is grown in association with other crops.

Cropping System

Cropping system is a land-use unit comprising soils, crop, weed, pathogen and insect subsystems that transform solar energy, water nutrients, labour and other inputs into food, feed, fuel and fibre. The cropping system is a sub-system of the farming system.

The cropping patterns used on a farm and their interaction with farm resources, other farm enterprises, available technology and environment (physical, biological and socio-economic) which determine their make up, constitute the cropping system.

Livestock System

Livestock system is a land-use unit comprising of animals and auxiliary feed sources transforming plant biomass into animal products.

Farming System

Farming system is a decision making and land-use unit comprising the farm household, cropping and livestock systems that produces crop and animal products for consumption and sale. The farming system is a sub-system of a higher level land-use system, such as village or water-shed that, in turn, forms a component of the agricultural sector of the regional system.

Entire complex of development, management and allocation of resources as well as decisions and activities which, within an operation form unit or combination of units, results in agricultural production and the processing and marketing of the products.

A farming system is a complicated interwoven mesh of soils, plants, animals, implements, workers, inputs and environmental, physical, biological and social influence with the strands held and manipulated by a person called the farmer who, given his preferences and aspiration, attempts to produce outputs from the inputs and technology available to him.

Farming system is a complex inter-related matrix of soil, plants, animals, implements, power, labour, capital and other inputs controlled in part by families and influenced to varying degrees by political, economic, institutional and social forces that operate at many levels.

Farming system is the scientific integration of different interdependent and interacting farm enterprises for the efficient use of land, labour and other resources of a farm family which provides year round income to the farmers specially located in the handicapped zone.

Integration of various farm enterprises in a farm ensures growth and stability in overall productivity and profitability. It also ensures recycling of residues, optimization of resources, and minimization of risk and generation of employment. Various enterprises that could be included in farming system are crops, vegetables, fruits, flower cultivation, dairy, poultry, fish, goat, pig, sericulture, mushroom cultivation, agro-forestry, bee keeping, silviculture, agro-based industries and food processing. A judicious mix of enterprises complementary to cropping and suited to the given farm situation and farmer's preference would bring overall prosperity.

Under farming system, the farm is viewed in a holistic manner. Farmers are subjected to many socio-economic, bio-physical, institutional, administrative and technological constraints. Farming system conceptually is a set of elements or components that are inter-related which interact among themselves. At the centre of the interaction is the farmer exercising control and choice regarding the type and results of interaction.

Farming system is a resource management strategy to achieve economic and sustained production to meet diverse requirement of farm household while preserving resource base and maintaining a high level environmental quality (Lal and Millar, 1990).

Farm Household System

Farm household system is a group of usually related people who, individually or jointly, provide management, labour, capital, land and other inputs for the production of crops and livestock, and who consume at least part of the farm produce.

Agro-Ecosystem

Agro-ecosystems are ecological systems modified by human beings to produce food, fibre or other agricultural products. Like the ecological system they replace, agro-ecosystems are often structurally and dynamically complex, but their complexity arises primarily from the interaction between socio-economic and ecological processes.

Agro-Ecological Zone

Agro-ecological zone is an area of similar soil, vegetation, and population density characteristics, resulting in a similar (sub) type of cropping system. Because temperature and rainfall do not vary significantly through out the region, climate is not included as a variable in the definition of agro-ecological zones.

Efficient Cropping Zone

The efficient cropping zone is the zone where both yield and spread indices are maximum.

$$\text{Yield index of a crop} = \frac{\text{Average yield of state/zone}}{\text{Average yield of country}} \times 100$$

$$\text{Spread index of a crop} = \frac{\text{Area in state/zone}}{\text{Area in country}} \times 100$$

The most efficient zone of a crop is where both the indices are above 100. in case the indices are below 100 that zone is the least efficient zone of a crop.

Cropping System and Patterns

Cropping system varies widely from the simplest system of two crops a year in sequence to complex intercropping with many crops. Multiple cropped lands can be broadly grouped into lowlands, irrigated uplands, and rainfed uplands.

The cropping system and patterns are the yearly sequence and spatial arrangement of crops or of crops and fallow on a given area.

Shifting Cultivation/Land Rotation/Jhumming

1. Forest land is cleared and cultivated. Due to cultivation of the same crop on the same cleared forest land year after year; soil productivity is lost and the crop is shifted to other slashed and burnt land.
2. Here same crop is grown year after year. In this case land is rotated but crop is fixed. Therefore it may also be called land rotation.
3. Shifting of land hence called shifting cultivation. Also called *Jhum* cultivation.
4. It causes soil erosion.
5. Practised in north-eastern states of India, Chhotanagpur Plateau of Jharkhand, M.P. and in hilly areas.

Crop Rotation

1. Crop rotation is the reverse of land rotation. Here land is fixed but crop is rotated year after year.
2. On a certain land, repeated cultivation of crop or crop fallow land in a certain sequence is called crop rotation.
   a) Rice – Wheat – Mung (Green gram)
   b) Rice – Mustard – Chilli
   Here wheat is rotated by mustard and mung by chilli in the next agriculture year.
3. Maintains and even improves soil fertility and stabilises income.
4. It checks the soil erosion and conserves moisture.
5. Ensures a balanced programme of work throughout the year.
6. Suitable crop rotation is the key of modern scientific agriculture which aims to produce maximum yield by maintaining soil productivity.

Ley Farming

To define, "leys" are temporary, short term---two to five years specially sown pastures consisting of grasses and legumes. These are used for grazing in succession with cultivated crop (Sample, 1951). The leys are established with intention of plough the land again after a predetermined period (Davies, 1960). Ahlgren (1956) defined it as a system of farming in which grasses and legumes are included in proper rotation for hay, silage and pasture to care for maximum livestock needs and to improve and conserve the soil fertility.

Thus, ley farming is a dynamic and integral system of agriculture which encompasses soil, plants and livestock together. It denotes a system of farming whereby emphasis is placed on the value of grass-legume mixtures to get fodder for livestock and build up soil fertility for obtaining higher yields of subsequent arable crops.

The production of a grass-legume mixture for fodder to provide adequate amounts of proteins and carbohydrates in cattle feed would also be the basis of an improved livestock industry to provide more milk for better nutrition of our primarily vegetarian people and better draught power for the cultivation of land. A cut and carried ley crop would, therefore, be helpful in producing more food for animals and enhancing soil fertility.

The practice of ley farming can be employed to introduce a measure of stability in our agriculture. It would contribute enormously for increasing productivity or agriculture under conditions of traditional farming where farm inputs are low. Ley farming has proved to be a beneficial practice in western countries. In Europe, major part of land is under grasses *e.g.* acreage under grasslands in United Kingdom alone amounted to 73.8% in 1949 which increased to 76.7% in 1958 most of this increase being in temporary leys (Davis, 1960).

Sustainable Agriculture/Eco-Farming

The concept of sustainable agriculture has come up because yield from modem farming technique (modern commercial agriculture) reaching a plateau and the environmental problems due to excessive use of chemicals, fertilizers and pesticides in food chain.

Types of Agriculture	Objective	Quantity of Input	Output
Subsistence Agriculture *i.e.* low level equilibrium	To sustain life and family need	Low	Low
Commercial Farming *i.e.* high level equilibrium	To obtain high Income	High	High
Sustainable Agriculture *i.e.* Natural equilibrium or ecological equilibrium	Ecological balance	Low	High

Sustainable agriculture is that form of farming which produces sufficient food to meet the needs of the present generation without eroding the ecological assets and productivity of life supporting systems of future generations. Natural farming is an excellent illustration of sustainable agriculture. It is also known as ecological farming/ eco-farming or organic farming or perm culture. It is called eco farming because ecological balance is given importance and organic farming because organic matter is the main source for nutrient management. It is a system of cultivation with use of manures, crop rotation and minimal tillage. Sustainable agriculture also involves agro-forestry and multi-level cultivation and integrated animal husbandry. The term sustainability denotes the characteristic of a process that can be maintained indefinitely and sustainable use of the eco system refers to making use of the system without impairing its capacity for renewal or regeneration.

Cropping Systems

System means a set of elements/components that are inter-related and interacting among themselves. The objective of any cropping system is efficient utilization of all resources viz. land, water and solar radiation, maintaining stability in production and obtaining higher net returns. The efficiency is measured by the quantity of produce obtained per unit resource in a unit time.

The objective of traditional agriculture was to increase the production by two means:

a) by increasing the area of cultivation.
b) by increasing the productivity per unit area of the crop. But two more dimensions are added to modern agriculture.
c) to increase the production per unit time.
d) to increase the production per unit space.

In the traditional cropping systems, mixtures and rotation were developed by the farmers over years of experience by trial and error to suit specific ecological and socioeconomic conditions to attain yield stability, whereas modern scientific cropping system has three pillars: (i) Genotype, (ii) Geometry of planting and (iii) Management practices.

1. Genotype means genetic make up of seed.
2. Geometry of planting means:
   a) Shape of planting pattern on the land surface.
   b) Space of the area available for the individual plant. Geometry of planting may be circular, rectangular, square type or cubical. It is indirectly related to plant population. Cubical pattern of planting has maximum plant population. Plant population may be defined as
      i) Size of area available to the individual plant.

ii) Number of plants per unit area.

3. Management practices: include all the practices of crop production. For the cropping systems, management includes.
   a) Type and arrangement of crops in time and space *i.e.* cropping pattern.
   b) Choice of variety.
   c) Method of stand establishment.
   d) Pest management and harvest.

Difference between Cropping Pattern and Cropping System

Cropping Pattern means	Cropping System means
Crop rotation practised by a majority of farmers in a given area or locality.	Cropping patterns and its management to derive benefits from a given resource base under specific environmental condition.
Type and arrangement of crops in time and space.	The cropping patterns used on a farm and their interaction with farm resources, other farm enterprises and available technology which determine their make up.
Yearly sequence and spatial arrangement of crops or of crops and fallow on a given area.	Pattern of crops taken up for a given piece of land, or order in which the crops are cultivated on a piece of land over a fixed period, associated with soil management practices such as tillaqe, manuring and irrigation.
The proportion of area under various crops at a point of time in a unit area.	

Farming Systems

It represents an appropriate combination of farm enterprises viz. cropping system, livestock, poultry, fisheries, forestry and the means available to the farmer to raise them for increasing profitability. The term ‘farming systems’ and’ mixed farming’ are used interchangeably but have some subtle differences.

Mixed Farming/Integrated Farming System

Defined as a system of farming on a particular farm which includes crop production, raising livestock, poultry, fisheries, bee keeping etc. to sustain and satisfy as many needs of the farmer as possible. The objective of mixed farming is subsistence while higher profitability without altering ecological balance is important in farming systems. Cropping system is an important component of a farming system.

Sole Cropping/Solid Planting

It is defined as the cultivation of one crop variety alone in pure stand at normal density in a certain time and place.

CROPPING SCHEME

The plan according to which, crops are raised on individual plots of a farm with an object of getting the maximum return from each crop without impairing the soil fertility. Soil fertility and crop yields are important factors in determining which cropping programme to follow, but at the same time, the amount of labour and capital required and prices in relation to the resources influence the adoption of a particular cropping scheme.

Monoculture

Monoculture is the repetitive growing of the same crop on the same land.

Or

Monoculture is also called single cropping or mono-cropping. A system of growing the same crop on the same land year after year is called mono-cropping or monoculture.

Cropping Intensity/Cropping Index

CI = No of crops grown in a year in a piece land x 100

Cropping intensity in India is about 127%

TYPES OF CROPPING SYSTEMS

1. **Mono Cropping:** C I is always 100%
2. **Multiple cropping:** Growing of two or more crops on same field in a year. It is the intensification of cropping in **time** and **space** dimensions *i.e.* more number of crops within a year and more no. of crops on the same piece of land at any given period. The principal limiting factors in adoption of multiple cropping are low intensity of irrigation and preponderance of long duration varieties.

Recently one more dimension *i.e.* to maintain the soil fertility, is added in the definition of multiple cropping. Thus multiple cropping is defined as the cultivation of two or more crops on the same field in a year without deteriorating the soil fertility.

Intensive Cropping

Cropping systems based on climate, soil and water availability have to be evolved for realizing the potential production levels through efficient use of available resources. The cropping system should provide enough food for the family, fodder to the cattle and generate sufficient cash income for domestic and cultivation expenses. These objectives could be achieved by adopting intensive cropping. Methods of intensive cropping include multiple cropping and intercropping. Intensive cropping may pose some practical difficulties such as shorter turn-around time lapse for land preparation before the succeeding crop and labour shortage at periods of agricultural activities. These practical handicaps can easily be overcome by making modifications in the cropping techniques. Alteration of crop geometry may help to accommodate intercrops without loosing the base crop population.

Multiple Cropping

The intensification of cropping in time and space dimensions is growing two or more crops on the same field in a year.

a) Inter Cropping
b) Mixed Cropping
c) Sequential or sequence cropping or non-overlapping Cropping.
d) Relay cropping or overlapping Cropping.
e) Double, Triple and Quadruple Cropping
f) Ratoon Cropping
g) Multi-storey / Multilevel/Multi-tier Cropping

h) Paira / Utera Cropping

Inter-cropping

It refers to growing of two or more crops simultaneously, on the same piece of land, base crop necessarily in distinct row arrangement. Crop intensification is in both time and space dimensions. There is intercrop competition during all or parts of crop growth.

Or

Intercropping is the cultivation of two or more crops simultaneously on the same field, with or without a row arrangement (row intercropping or mixed cropping). Relay intercropping is the growing of tow or more crops on the same field with planting of the second crop after the first one has already flowered. The crops grown in intercropping are called a crop association, and are denoted with a + sign. The crops involved can be annuals only, a combination of annual and perennials or perennials only. Important intercropping systems for various regions are given in Table 4.4 and 4.5.

TABLE 4.4: Some recommended intercropping systems for Indo Gangetic alluvial plains

S No.	System	Ratio
1.	Wheat + Raya	8 : 1 or 12 : 4
2.	Winter maize + Peas (green pods)	1 : 1
3.	Toria + Gobhi Sarson	1 : 1
4.	Potato + Mustard	3 : 1
5.	Potato + Wheat + Summer moong	1 : 2 : 2
6.	Mustard + Gram	4 : 1
7.	Autumn cane + Wheat	1 : 2
8.	Autumn cane + Potato	1 : 1
9.	Autumn cane + Raya	1 : 2
10.	Autumn cane + Toria	1 : 2
11.	Autumn cane + Winter Maize	1 : 1
12.	Maize + Green/Black Gram	1 : 1
13.	Maize + Soybean	1 : 1
14.	Pigeon pea + Green/Black gram	1 : 1

TABLE 4.5: Intercropping systems for region other than the Indo-Gangetic alluvial plains

S No.	System		Ratio
1.	Semi-arid Alfisol of Southern Telengana	(i) Sorghum + Pigeonpea	2 : 2
2.	Vidharbha Region of Maharshtra	(i) Sorghum + Mungbean	1 : 1
		(ii) Sorghum + Pigeonpea	2 : 1
3.	Malwa plateau of Madhya Pradesh	(i) Sorghum + Pigeonpea	2 : 1
4.	Vetisols of south cast Rajasthan	(i) Sorghum + Mungbean	1 : 1
		(ii) Sorghum + Pigeonpea	2 : 1
5.	Dccan region of Maharashtra	(i) Pearl millet+ Pigeonpea	2 : 1
6.	Deccan region of Karnataka	(i) Pearl millet+ Pigeonpea	2 : 1
7.	Vertic inceptisols of Rajkot region	(i) Pearl millet+ Pigeonpea	4 : 1
		(ii) Peanut + Pigeonpea	6 : 1

S No.	System		Ratio
8.	Sub-humid Alfisols region of Orissa	(i) Rice + Pigeonpea	5 : 2
9.	Sub-humid Alfisol region of Chotanagpur	(i) Rice + Pigeonpea	4 : 1
10.	Semi- arid Alfisols Ralayaseema region	(i) Peanut+ Pigeonpea	5 : 1

Here cropping intensity in space dimension is achieved. Such crops are usually grown along with those crops which have larger row interval. Intercropping includes alley cropping, strip cropping, contour cropping, paired row cropping, skip cropping, parallel cropping, companion cropping, multi-storied cropping, and synergetic cropping. But the major intercropping systems are

Parallel Cropping

Cultivation of such crops which have different natural habit and zero competition *e.g.* Black gram/Green gram + Maize

The peak nutrient demand period for black gram or green gram is around 30-35 days after sowing while it is 50 days after sowing for Maize.

Companion Cropping

Such intercropping where the production of both intercrops is equal to that of its solid planting. *e.g.* Mustard/Potato/Onion + Sugarcane.

By growing of the above three intercrops individually with sugarcane, the production of none is hardly affected.

Multi-storey/Multi tiered/Multi level

Cultivation of two or more then two crops of different heights simultaneously on a certain piece of land in any certain period. *e.g.* Sugarcane + Mustard + Onion/Potato

Synergetic Cropping

The yields of both crops are higher than of their pure crop on unit area basis *e.g.* Sugarcane + Potato. On the basis of per cent of plant population used for each crop in intercropping system; intercropping is of two types viz, additive series and replacement series.

a) **Additive series intercropping:** In such type of intercropping, one crop is main crop or base crop and another crop is intercrop. Intercrop is introduced into the base crop by adjusting or changing crop geometry *i.e.* addition of intercrop to the base crop. Here plant population of base crop is same to what recommended population in pure stand whereas that of intercrop is less. Cropping husbandry is according to the base crop. This type of intercropping is prevalent in India and its main objective is to get additional income and to cover risk. *e.g.* sowing of potato in the field of sugarcane in between the rows of cane at a 90 cm.

b) **Replacement series intercropping:** Both the crops are component crops. Neither is the base crop nor is the intercrop. It means the plant population of both component crops is less than their recommended population in pure stand. Such type is widely practiced in western countries. Her 10th row of wheat crop is replaced by mustard and row to row distance of wheat crop is somewhat reduced to adjust the mustard crop.

Mixed Intercropping

Growing two or more crops simultaneously with no distinct row arrangement is also referred to as mixed cropping.

Row Intercropping

Growing two or more crops simultaneously where one or more crops are planted in rows is often simply referred to as intercropping.

Patch Cropping

The component crops are planted in patches.

Strip Intercropping

Growing two or more crops simultaneously in strips wide enough to permit independent cultivation but narrow enough for the crops to interact agronomically.

Relay Intercropping

Growing two or more crops simultaneously, during the part of life cycle of each is called relay intercropping. A second crop is planted after the first crop has reached its reproductive stage of growth but before it is ready for harvest. Often it is simply referred to as relay cropping.

Main Crop/ Base Crop

This is the one which is planted at its optimum sole crop population in an intercrop situation and the second crop is planted in between rows of the main or base crop with a view to obtain some extra intercrop yield without sacrificing the main or base crop yield.

Component Crop

The component crop is used to refer to either of the individual crops making up the inter-cropping situation.

Pre-requisites of Successful Intercropping

The main objective of intercropping is higher productivity per unit area in addition to stability in production. Intercropping utilizes resources efficiently. There are certain important pre-requisites of intercropping:

1. The time of peak nutrient demands of component crops should not overlap.
   a) Maize + Urd/Mung.
   b) The peak nutrient demand is 30-35 days after sowing in case of urd/mung where as it is 50-55 days after sowing of maize.
2. Competition for light should be minimum among the component crops.
3. Complimentarily should exist between the component crops.
4. The differences in maturity of component crops should be at least 30 days.
5. Competition for CO_2 and water should also be minimum among component crops.

Advantage and Disadvantage of Intercropping

The advantage and disadvantage of intercropping are presented below: Advantage and disadvantage of intercropping.

Advantages of Intercropping	Disadvantages of intercropping
Better use of growth resources including light, nutrients and water.	Labour intensive
Suppression of weeds reduced plant and disease incidence yield stability	Control of pests and diseases or chemical weed control may be difficult
Ecological stability *i.e.* improvement of soil health and agro-ecosystem	Mechanisation is difficult to realize
Other *e.g.* spreading of labour, physical support of one crop to another and home gardening leading to a more pluriform food supply is a good example of realization of intercropping advantages.	Disadvantages caused by adverse competitive effects or by allelopathy (can of course be avoided).

Mixed Cropping

Mixed cropping is the cultivation of two or more than two crops simultaneously on the same land without definite row pattern or fixed ratio. Sowing of seeds is generally by broadcasting method commonly practised in dryland area of India. It is subsistence farming. Main objective is to lessen the risk of total crop failures, and to satisfy the farmers in food and fodder. Intercropping is the new concept of mixed cropping. Scientific study of mixed cropping was firstly done by La-Flitze (1928)

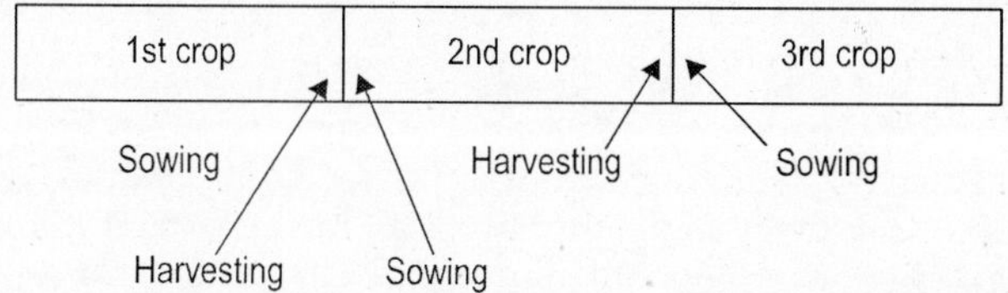

Sequence/Sequential Cropping or Non-overlapping Cropping

Growing of two or more crops in quick succession, on the same piece of land in a farming year is called sequential cropping. The sowing of the succeeding crop and harvesting of the preceding crop may be done simultaneously or in a quick succession. *e.g.* just after the harvesting of maize, potato is sown and just after digging of potato, chilli is sown. Sequential cropping is also called non-over lapping cropping because there is no overlap between the two or more cropping. Ratooning or ratoon cropping may also be classified under sequential cropping.

Relay Cropping or Overlapping Cropping

Its concept has been derived form relay race in which four runners run in the field having in their hand. The first runner passes on its flag to the succeeding partner and 2nd to 3rd and 3rd to the 4th runner. Relay cropping can be defined as growing two or more crops simultaneously during the part of the life cycle of each. Succeeding crops are planted before harvesting the preceding crop. Generally 2nd crop is planted after the first crop has reached its reproductive stage of growth but before it is ready for harvest *e.g.* Potato is planted before the harvesting of Maize and Radish is sown before harvesting of potato. Paira (Bihar & W.B.) or Utera (M.P) cropping may be example of relay cropping. Paira or utera cropping means sowing of lathyrus or lentil (generally pulses) before the harvesting of rice in lowland area. Planting of nursery crop may also be an example. Dr. Sher Singh Bains has suggested the following relay cropping pattern for north-west part of India

Mung – Maize – Potato – Wheat

Ist crop	IInd crop		IIIrd Crop		IV Crop	
Sowing of 1st crop	Sowing of 2nd crop	Harvesting of 1st crop	Sowing of 3rd crop	Harvesting of 2nd crop	Sowing of 4th crop	Harvesting of 4th crop

Double, Triple and Quadruple Cropping

- **Double cropping** Growing two crops a year in sequence.
- **Triple cropping:** Growing three crops a year in sequence.
- **Quadruple cropping:** Growing four crops a year in sequence.

Ratoon Cropping

The cultivation of crop regrowth after harvest, although not necessarily for grain.

Multistorey/Multilevel/Multi-tier

Intercropping is mostly prevalent in plantation crops like coconut and arecanut. There is scope for intercropping in coconut garden upto the age of 8 years and after 25 years. During this period, there is adequate light transmission to the ground which permits cultivation of intercrops. The practice of growing different crops of varying height, rooting pattern and duration is called 'multi-tier cropping'. The objective of this system of cropping is to utilize the vertical space more effectively. In this system, the leaf canopies of intercrop components occupy different vertical layers. The tallest components have foliage tolerant of strong light and high evaporative demand and the shorter component(s) with foliage requiring shade and or relatively high humidity. *e.g.*, Integrated Farming System

Coconut + black pepper + cocoa + pineapple. The schematic representation of planting of these crops is shown in Fig 4.1

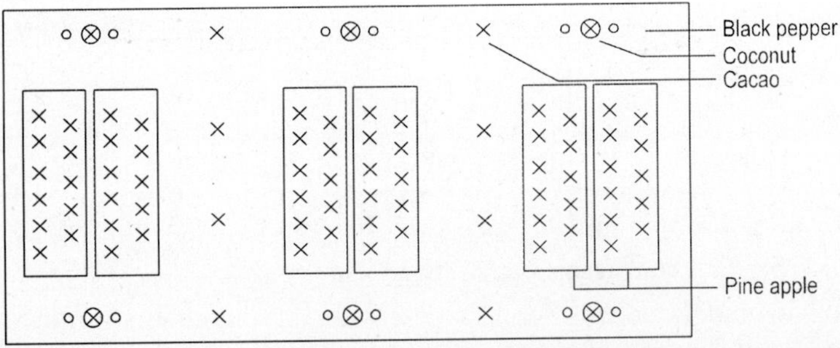

FIG. 4.1: Multistoried cropping

In this system, coconut is planted with a spacing of 7.5 m. Rooted cuttings of black pepper are planted on either side of coconut about 75 cm away from the base. On the coconut trunk at a height of about one metre from the ground level, the vines of pepper are trailed. A single row of cacao is planted at the centre of space between coconut rows. Pine apple is planted in the interspace.

Coconut growing to a height of more than 10 m occupies the top floor. Black pepper is growing to about 6-8 m height forms the second floor. Cacao with its pruned canopy of about 2.5 m height and pine apple growing to about 1 m height form the first and ground floors respectively.

In another multi-tier system in coconut, ginger or turmeric and partial shade loving vegetables form the first tier, banana in second tier, pepper in third tier and coconut or arecanut in the final tier.

In the arecanut plantation, tuber crops are predominantly intercropped. Elephant yam, tapioca, greater yam and sweet potato are common intercrops in humid tropics. Banana and pine apple are also cultivated as intercrops in arecanut gardens.

In coffee based multi-tier cropping system, first tier is with pine apple, second tier with coffee, third tier with cacao/mandarin orange and final tier with fast growing shade trees necessary for coffee plantation. (*e.g.*, dadaps and silver oak).

Paira / Utera Cropping

Growing of such crops sown a few days or weeks before harvesting of standing mature crops is called paira/utera cropping and the sown crop is called paira/utera crop *e.g.* Lathyrus in rice. Paira cropping in succession may constitute relay cropping.

Pulses like mung, black gram, gram, lentil and lathyrus come up well as paira cropping in the *rabi* season. Where the land after harvest of paddy remains slaucy making it unsuitable for tillage, these crops may be broadcast atleast 15 - 20 days before harvest using 1.50 times the seeds required for normal sowing.

CHOICE OF CROPS FOR MULTIPLE INTENSIVE CROPPING

1. The crops should be arranged in such a way that Allelopathic effect, temporary immobilisation of nutrients and depletion of nutrients from the same layer of soil do not occur.
2. Legumes have a definite place in any cropping system because of their limited demand on resources like water, nutrients and light, their adaptability to varying environmental conditions and capacity to fix atmospheric N in root nodules.
3. Vegetables deserve their due place in intensive cropping because of
   a) high cash and nutritional values,
   b) most vegetables can be raised as seedlings and planted,
   c) thus reducing main field duration and competition with the associated crop.
   d) they can be grown in interspaces in orchards and plantations,
   e) they can be inter or relay cropped in cereal base crops.
4. Short duration, photoperiod insensitive genotypes are suited for intensive cropping system.

RELATED TERMINOLOGY USED IN MULTIPLE CROPPING SYSTEM

1. **Cropping index:** The number of crops grown per annum on a given area of land multiplied by hundred.
2. **Cropping pattern:** The yearly sequence and spatial arrangement of crops or crops and fallow on a given area.
3. **Cropping system:** The cropping patterns used on a farm and their interaction with farm resources, other farm enterprises and available technology which determine their makeup.
4. **Monoculture:** Repetitive growing of the same sole crop in the same land.
5. **Sole cropping:** One crop variety grown alone in pure stand at normal density.

6. **Rotation:** Repetitive cultivation of an ordered succession of crops (or crops and fallow) on the same land. One cycle often takes several years to complete.
7. **Mixed farming:** Cropping systems which involve the raising of crops, animals and/or trees.
8. **Land equivalent ratio (LER):** Ratio of the area needed under sole cropping to one of intercropping at the same management level to give an equal amount of yield. **LER** is the sum of the fractions of the yield of the intercrops, relative to their sole crop yields.
9. **Income equivalent ratio(IER):** The ratio of the area needed under sole cropping to produce the same gross income as one ha of intercropping at the same management level. **IER** is the conversion of **LER** into economic terms.

ADDITIONAL TERMS USED TO DESCRIBE VARIOUS FORMS OF MULTIPLE CROPPING

1. **Agrisilviculture:** Growing of trees for timber, but with cultivated crops beneath.
2. **Area-time equivalency ratio (ATER):** The ratio of number of hectare-days required in monoculture to the number of hectare-days used in intercropping to produce identical quantities of each of the components.
3. **Calorie index:** A single productivity index which incorporates caloric equivalents by all crops in a system.
4. **Competition effect:** Competition of intercropped species for light, nutrients, water, carbon dioxide and other growth factors.
5. **Complementary effect:** Effect of one component on another which enhances growth and productivity, as compared to competition above.
6. **Component crops:** Individual crop species which are a part of the multiple cropping system.
7. **Component technology:** The procedure for growing each component crop.
8. **Farm enterprise:** An individual crop or animal production function within a farming system which is the smallest unit for which resource use and cost-return analysis is normally carried out.
9. **Farming systems research (FSR):** The study of whole farm system, which includes all the enterprises on the farm, their biological, economic and cultural operations and usually implies some involvement of the farmer in the research process.
10. **Interplanting:** All types of seeding or planting a crop into a growing stand. It is used especially for annual crops grown under stands of perennial crops.
11. **Maximum cropping:** Attainment of highest possible production per unit area per unit time without regard to cost or net return.
12. **Monetary index:** A single productivity index which incorporates monetary value of all crops in a system.
13. **Over yielding:** Production of component crop in an intercrop which is higher than the sum of appropriate monoculture crops. This is indicated by an **LER** greater than unity.
14. **Residual effect:** Effect of the previous crop in a sequential cropping pattern on the productivity of the current crop.
15. **Simultaneous poly culture:** Simultaneous growth of two or more useful plants on the same plot. It includes mixed cropping, intercropping, interculture, interplanting and relay planting.

16. **Spatial arrangement:** The physical or spatial organisation of component crops in a multiple cropping system.

PLANT INTERACTIONS

Interaction Between Component Crops

In intensive cropping, crops are grown in association (intercropping) or in sequence (sequential cropping). In such situations there is possibility of interaction between the component crops. The interaction is mainly due to response of one species to the environment as modified by the presence of another species. Interaction may be competitive or non-competitive or complementary. If the crops are grown in association and the growth of either of the concerned species is not affected, such type of interaction is called as non-competitive interaction or interference. If one species is able to help the other it is known as complementary interaction. *e.g.* supply of N fixed by the legumes to the associated non-legumes. (This is otherwise known as annidation). One species may have greater ability to use the limiting factor and will gain at the expense of the other. This is called as competitive interaction or interference. Interaction may also occur in some other manner by way of producing toxic substances and affecting the establishment and growth of the associated species. This is called allelopathy.

Interactions in Intercropping

Factors such as light, water, nutrient, oxygen and CO_2 are required for plant growth. In mixed or intercropping situations, the component species compete for the growth factors. The close proximity of species causes sub-optimal utilization of the growth factors and hence there is inequitable distribution of resources among the plants. Generally competition will develop between two components or within the components. The nature of such competition may vary depending on the density and proportion of each component and planting pattern.

Light

Photo-synthetically active radiation (light) is directly related to photosynthetic rates of component crops. Intercropping can increase light interception by as much as 30-40%. When one component is taller than the other in an intercropping system, the taller components intercept most of the solar radiation. The pattern of light interception also varies according to the age of the crop. Light transmission by coconut is only about 20% at the age of 8-10 years and this remains almost constant till about 25 years of age. Subsequently the% light transmission increases progressively as the canopy coverage of ground decreases. When the palms are at about 40 years of age, light transmission increases to about 50%. In an intercropping situation where the component crops have different growth durations, the peak demand for light would occur at different times. For example, in maize + green gram intercropping, the peak light demand for maize occurs at 60 days when green gram is ready for harvest. In such combinations, competition for light is less among the component crops and there is greater light use in intercropping than in pure stands. In general, the component crops under intercropping situations are grown in such a way that competition for light is minimized. Proper choice of crops and varieties, adjustment of planting density and pattern are the techniques to reduce competition and increase the light use efficiency.

Moisture and Nutrients

Competition for moisture and nutrients may result in two types of effects on the less successful or suppressed component. First, the roots to this component may grow less on the sides towards plants of aggressive component. Secondly, plants affected by competition for soil factors may have increased root/shoot ratio. The aggressive component generally absorbs greater quantity of nutrients and soil moisture. In legume and non-legume combination, the latter takes up large amount of P, K and S. As a result, the legume may show deficiency of these nutrients. Such effects, however, may be mitigated by appropriate fertilizer application. Among intercrops, sorghum and pearl millet are more competitive in extracting nutrients. Generally the intercropped stands remove greater amount of nutrients than sole crop stands.

Allelopathy

Allelopathy is any direct or indirect harmful effect that one plant has on another through the production of chemical substances that escape into the environment. Some crops may be unsuitable to be grown as intercrops because they may produce and excrete toxins into the soil which are harmful to the associated components. Allelochemicals produced from the leaves of Eucalyptus globules drastically reduce the germination of mustard sown underneath.

Annidation

Annidation refers to complementary interaction which occurs both in space and time.

- **Annidation in space:** The canopies of component crops may occupy different vertical layers with taller component tolerant to strong light and high evaporative demand and the shorter component favouring shade and high relative humidity Multi-storeyed cropping in coconut and planting shade trees in cocoa and tea plantations use this principle. Similarly, root systems of component crops may exploit the nutrients in different layers of soil and hence utilize the resources in a better way with much less competition.
- **Annidation in time:** when component crops of widely varying duration are planted, their peak demand for light and nutrients are likely to occur at different periods, thus reducing competition. In a combination having early and late maturing crops (sorghum + redgram), when early maturing crops are harvested, conditions become favourable for the late maturing crop (redgram) to putforth its full vigour.

Other Complementary Effects

In an intercropping system involving a legume and a non-legume, part of the N fixed in the root nodule of the legume may become available to the non-legume component. With the presence of rhizosphere microflora and mycorrhiza, one species may lead to mobilization and greater availability of nutrients not only to the species concerned but also to the associated species. Provision of physical support by one species to the intercropped climbing species may improve the yield of the climber. Examples are coconut + pepper and maize + beans. The taller component acts as wind barrier protecting the shorter components from lodging.

Interactions in Sequential Cropping

In sequential cropping, sole crops are grown in succession. Hence, competition for light, water and nutrients as found in mixed cropping does not occur. In relay cropping, a short span of overlapping occurs between two crops in sequence. Hence the relay sown crop at the seedling stage may have competition for light from the standing crop. Such type of competition may be minimized by proper choice of crops and varieties and adjustment of time and method of planting.

In intensive multiple cropping involving two or more crops in sequence, the main adjective is to harvest as much solar energy per unit area per unit time as possible. In rice based cropping system, the solar energy use efficiency ranged from 1.58 to 2.02% of PAR in Uttar Pradesh. In Coimbatore, the efficiency ranged from 0.82 to 1.77% of PAR for the whole year in cotton based system. Inclusion of a C_4 plant in the summer increased the efficiency.

In sequential cropping, the preceding crop has considerable influence on the succeeding crop. This includes the complementary effects such as release of N from the residues of the previous crop, particularly legume, to the following crop and carry-over effects of fertilizer applied to the preceding crops. The adverse effects include allelopathy, temporary immobilization of N due to wide C/N ratio of the residues and carry over effect of pest and diseases. Introduction of pulses as summer crops results in improved yield of rice in *kharif* season.

SEQUENTIAL CROPPING SYSTEMS

In India, food crop is predominantly grown in most suitable seasons and thus particular food crop is basic to the cropping system followed by the farmers. Accordingly the cropping systems are usually referred to as:

1. Rice-based cropping system
2. Sorghum-based cropping system
3. Pearl millet-based cropping system
4. Wheat and gram-based cropping system

Some of the cropping systems based on commercial crops are

1. cotton-based,
2. groundnut-based
3. sugarcane-based
4. plantation crop-based, and
5. vegetable-based cropping system.

The grain production potential in different regions of the country ranges from 11 to 18 t/ha. In maize-potato or toria -wheat- mung system followed at IARI, New Delhi, it was possible to produce 14-15 tonnes of food per ha per annum without impairing the soil health. The results of multiple cropping demonstrations under irrigated conditions showed that production potential can be as high as 19.8 t/ha in cereal-based cropping system of rice – rice – rice. The yield potential of multiple cropping varies from region to region depending up to the physical and socio-economic conditions.

RICE BASED CROPPING SYSTEM

Sequential cropping is the most suitable cropping pattern of rice-rice in the rice areas. Traditionally single crop of rice with long duration cultivars is planted during the rainy period. However, due to the advent of short duration, photo-insensitive varieties, multiple cropping is being now practiced. The cropping system adopted under lowland condition can be grouped into two broad categories such as all rice and mixed rice and upland crop sequential cropping systems. These can be further grouped as balow.

All rice cropping pattern

1. Rice- Rice (R-R)
2. Rice- Rice – Rice (R –R –R)

Mixed rice and upland crop sequence.

1. Rice –upland crop (R-UC)
2. Upland crop – Rice –upland crop (UC-R-UC)

ALL RICE SEQUENTIAL CROPPING SYSTEM

Rice crop is repeatedly grown wherein two or three crops are raised. With three crops of rice, the land is cropped at least ten months in year. This is possible only in irrigated areas where water is adequate even during the dry periods. Crop failure due to water stress normally does not exist. Among the three crops, productive rice crop is the one which coincides with the dry months. During this period, solar radiation is high and the occurrence of there calamities such as cyclone or pest infestation is very low.

Two rice crops are mostly cultivated in many parts of the rice growing areas in India. The first crop is a short duration crop of rice which is cultivated during the *Kharif* season. This is followed by the second crop of rice during *rabi* season with medium duration varieties. Single crop of rice with the medium or long duration varieties is most prevalent in the command areas. Cultivation is restricted to the period of water availability in the canal. All rice sequence is followed in some pockets of the river command areas of Orissa, Andhra Pradesh, Karnataka and Tamil Nadu.

MIXED RICE AND UPLAND CROPS

Only one rice crop is grown during the peak rainy period. In such situations, there is enough time left to grow a fast- maturing annual crop between the first monsoon rain and the time of transplanting. Such crops must mature early. The crops uniquely suited for these conditions are vegetables, such as green corn, cowpea, green soybean or a green manure. The crop can easily be harvested within 80 days since only cobs are harvested.

In most occasions immediately after the harvest of the main rice crops, it is possible to plant another crop (rice fallow crop). Because rice crop is usually harvested at the end of the rainy season when the soil is still wet, there is usually enough moisture in the soil to support the succeeding crop. Some of the crops that are suited to this condition are green gram, black gram, cowpea, soybean, etc. Rice–upland crop is the commonly used system. Green gram is the popular choice as upland crop after rice.

SYSTEMS WITH MORE THAN THREE CROPS

With ratooning or relay cropping and special cultural practices that avoid puddling in lowland rice, up to five crops can be harvested in the same rice field over 12 month period (Bradfield, 1970). In rice fields with adequate irrigation and good water control, Bradfield experimented at IRRI, Philippines, with various cropping patterns that could maximize productivity per unit area. The most successful pattern was a sequence of five crops consisting of rice, sweet potato, soybean, sweet corn and green soybeans. The tillage pattern varied for each crop and number of days to harvest were 102, 100, 85, 66 and 60 respectively. This pattern gave three times higher gross income than normally obtained in a rice-rice pattern.

PROMISING RICE–BASED CROPPING SYSTEMS IN INDIA

Promising rice-based cropping systems with high productivity and intensive land use for diverse agro-climatic situations of India have been identified in multi-locational experiments under Cropping Systems Research Project (Table 4.6).

TABLE 4.6: Promising rice based cropping systems in India

System	Region
Rice – Indian mustard	Kashmir valley
Rice– wheat – sorghum (Fodder)	Jammu region
Rice – wheat – green gram	Punjab, Haryana, U.P.
Rice - Mustard	North Konkan
Rice – field bean	South Gujarat
Rice – groundnut	South Karnataka
Rice – black gram/sesame/cotton	Thanjavur, Tamil Nadu (Old Cauvery delta)
Rice – groundnut	Thanjavur, Tamil Nadu (New Cauvery delta)
Rice – mustard	Assam
Rice – wheat – jute	West Bengal
Rice – groundnut	Coastal Andhra Pradesh and Orissa
Rice – green gran	Eastern Madhya Pradesh.

CROP SEQUENCES UNDER IRRIGATED UPLAND CONDITIONS

Under irrigated uplands (garden lands) depending upon the water availability or irrigation potential, more than two crops can be grown in a year. Crop sequences and their production potential have been tested under different agro climatic conditions. At Delhi, four crop sequence of green gram – maize – potato- wheat gave the highest production of 13.6 t/ha. However, three crop sequence of green gram – maize – wheat was found to be the most ideal for the small and marginal farmers (Table 4.7).

TABLE 4.7: Crop sequences and their production potential, New Delhi, India

Crop sequences	Total production (q/ha)
Two crops	
Maize – wheat	90.1
Three crops	
Green gram – maize – wheat	97.7
Green gram –pigeon pea – wheat	68.8
Cowpea – pearl millet –wheat	92.9
Four crops	
Green gram – maize – potato –wheat	135.9
Green gram –maize – toria –wheat	112.8

Under All India Co-ordinated Agronomic Experiments Project, Production potentials of the crop sequences under irrigated condition were studied. Some of the important sequences and their yield potential are given in Table.4.8.

TABLE 4.8: Crop sequences, duration and yield levels in different states of India

State	Crop sequences	Duration (days)	Yield (q/ha)		
			Grain	Fodder	Vegetable
Himachal Pradesh	Soybean - wheat	323	77	--	--
	Soybean – potato	283	34	--	72
	Maize – wheat	299	65	--	--
Haryana	Pearl millet - wheat – cowpea (f)	323	83	63	--
	Maize – wheat - cluster bean (f)	323	82	103	--
	Pearl millet - wheat – maize (f)	232	81	126	--
Uttar Pradesh	Maize – wheat - pearl – millet (f)	301	108	316	--
	Sorghum – wheat - green gram	306	63	--	--
	Sorghum – wheat - cowpea	313	59	--	--
Rajasthan	Pearl millet - wheat – green gram	339	35	--	--
	Green gram - wheat – green gram	295	22	--	--
Madhya Pradesh	Maize – wheat - cowpea	323	70	--	--
	Maize – wheat - green gram	325	70	--	--

f-fodder

In multi – location experiments, under Cropping Systems Research Project, promising cropping systems with high productivity as well as intensive land use for diverse agro climatic situations have been identified (Table 4.9).

TABLE 4.9: Potential cropping systems for different agro climatic situations

Region	System
Western Haryana	Cotton – wheat
Rajasthan	Groundnut – wheat
Madhya Pradesh	Soybean – wheat
North Gujarat	Pearl millet – mustard
Sourashtra region – Gujarat	Cotton – groundnut
Western Maharashtre	Sorghum – wheat
Western Maharashtra	Groundnut – wheat
Marathwada region of Vidarba region	Cotton – groundnut
North Karnataka	Sorghum–bengalgram
South Karnataka	Cotton – groundnut

CROPPING SYSTEMS IN DRY LANDS

Normally only one crop is grown under dryland condition and cultivation is restricted during the rainy season. However, the intensity of cropping can be increased through sequential or intercropping, depending on the rainfall and moisture storage capacity of the soils (Table 4.10).

TABLE 4.10: Cropping pattern with varying rainfall and soil moisture storage capacity

Rainfall (mm)	Storage capacity of soil (mm)	Cropping pattern
350 – 625	100	Single crop in *kharif*
650 – 750	100	Intercropping can be attempted
780 – 900	150	Sequential cropping is possible
900 – above	200	Sequential cropping is assured

Source: Sharma, 1980

Choice of crops: Selection of a suitable crop and genotype is important to get higher yield in dry farming conditions. Deep rooted crops extract moisture from the deeper layers and they also have slow rate of transpiration. Leguminous crops are very well adopted for rainfed conditions because of their root system. Among the cereal, barley has better ability to harvest moisture from deeper layers than wheat. Moisture loss from the plant surface is less in barley. Similarly linseed has greater ability to trap soil moisture under water scarce situation.

CROPPING SYSTEMS FOR DRY LANDS

Important factors such as total rainfall, soil type and water availability have to be considered while practicing a particular cropping pattern in a place/ region. Cropping system for different regions as suggested from the results of AICARP (1970) are given in Table 4.11.

TABLE 4.11: Cropping systems for different region of India under dry lands

Regions/soil type	Crops	
	Kharif	**Rabi**
Northern Region		
Samba (jammu)	Maize	Wheat
	Green gram	Barley
	Cowpea	Wheat
	Sunflower	Wheat / barley
	Groundnut	Barley
	Pearl millet	Bengal gram
Punjab	Sorghum	Wheat
	Maize	Wheat / Potato / Barley
Hisar (Haryana)	Pearl millet	Cluster bean / Chickpea
Arid soils (sierozemic soils)	(Mono-cropping is the rule)	
Central Region		
Dehra Dun (Uttar Pradesh)	Maize	Wheat
	Rice (upland)	Wheat / Bengal gram
Agra (Uttar Pradesh)	Gram	Mustard
Entisols (Alluvial soils)	Pearl millet	Chickpea / Barley
Varanasi (Uttar Pradesh	Sesame	Chickpea / Mustard
Entisols (Alluvial soils)	Bengal gram	Barley / Mustard
	Rice	Chickpea / Mustard
	Pearlmillet	Chickpea / Mustard

Regions/soil type	Crops	
	Kharif	Rabi
Eastern Region		
Ranchi (Bihar)	Rice	Chickpea/Linseed/Barley
Alfisols and related red soils	Maize	Rapeseed / Bengal gram
Bhubaneswar (Orissa)	Rice	Linseed / Mustard
Alfisols and related red soils	Finger millet	Horse gram
	Maize	Horse gram
Rewa region (M.P)	Rice	Chickpea / Lentil
Vertisols and related black soils	Sorghum	Bengal gram / Wheat
	Green gram	Wheat / Chickpea
Jhansi region (M.P)	Sorghum	Bengal gram
Vertisols and related black soil	Pearl millet	Bengal gram
	Black gram	Safflower / Chickpea
	Sesame	Bengal gram
Indore (M.P)	Maize	Chickpea
Vertisols and related black soils	Soybean	Safflower
	Soybean	Wheat
Western Region		
Udaipur (Rajasthan)	Green gram	Safflower
(Vertisols and related black soils	Sorghum	Bengal gram
	Pearl millet	Bengal gram
	Maize	Wheat
	Black gram	Wheat
	Sunflower	--
Anand (Gujiarat)	Pearl millet	Wheat
	Pigeon pea	--
	Castor	--
	Sunflower	--
Akola (Maharashtra)	Green gram	Safflower
Vertisols and related black soil	Sorghum	Safflower
Sholapur (Maharashtra)	Green gram	Safflower
Vertisols	Pearl millet	Bengal gram
	Black gram	Sorghum
Southern regions		
Anantapur (AP)	Sorghum	Safflower / Horse gram
	Pearl millet	Cowpea / Black gram
	Black gram	Sorghum
	Green gram	Pearl millet
	Cowpea	Pearl millet / Sorghum
	Mono-cropping is generally adopted	
Hyderabad	Pearl millet	Cowpea
Alfisol	Sorghum	Horse gram
Bijapur	Sorghum	Green gram / Black gram

Regions/soil type	Crops	
	Kharif	**Rabi**
Vertisols	Green gram	Sorghum / Safflower
Bangalore (Karnataka)	Cowpea	Finger millet
Alfisols	Cowpea	Chilli
Mysore (Karnataka)	Groundnut	Sorghum
	Setaria	Safflower
	Hybrid	Cotton
	Pearl millet	
Bellary (Karnataka)	Cotton	Sorghum
	Black gram	Bengal gram
		Safflower
Kovilpatti (TN)	-	Cotton, Sorghum Pearl millet, Cowpea Horse gram, Chilli

Intercropping System

Multiple cropping in the form of intercropping is predominant in the regions of dry, humid and semi-arid tropics. Farmers normally grow two or more crops together on the same piece of land. Different crops are cultivated in mixed cropping: simple mixture of only two or more crops is the most common.

The objectives of this practice are:

1. Insurance against total crop failure under aberrant weather conditions or pest epidemics.
2. Increase in total productivity per unit land area.
3. Judicious utilization of resources such as land, labour and inputs.

When two crops are to be grown together, they are chosen in such a way that there is variation in their growth duration. The peak periods of growth of the two crop species should not coincide. In such arrangements, a quick maturing crop completes its life cycle before the other crop starts. Willey (1979) described the concept as temporal complimentarity. Greater the differences in maturity and growth demands of the crop components, more opportunity is provided for greater exploitation of growth factors and over yielding. This will be achieved either by genetic difference in crop species or manipulation of planting dates. Normally short and long duration crops are grown together. Food crops are usually mixed with cash crops to ensure both sustenance and cash income.

Combination of species under intercropping depends on climate, location preferences and other site specific factors. Some of the important intercropping systems are described below:

Intercropping in Cereals

Intercropping with cereal is an excellent way of improving the resource utilization because the cereal utilizes the rainy season resources while late maturing crop exploits the post-rainy season resources such as residual moisture. Sorghum is most commonly inter cropped with pigeon pea on a variety of soils. Sorghum is harvested after 3½ to 4½ months and pigeon pea matures in about 6-9 months depending on the genotype. Grain legumes such as cowpea, green gram, black gram and groundnut are the other recommended intercrops with sorghum.

Maize + bean (*Phaseolus vulgaris* L.) is a predominant intercropping system in Central and South America and parts of East Africa, where rainfall ranges from 700 to 1500 mm spreading over 4 - 5 months. Soybean is also a good compatible companion crop with maize.

Pearl millet is a quick tillering and fast growing crop that attains full canopy development within 20 - 30 days of seedling establishment. It can be intercropped with groundnut, black gram or castor.

Finger millet is widely intercropped with field beans (*Dolichos lab- lab*) in Karnataka State, India, where 4 – 12 rows of the millet are alternated with one row of beans.

Intercropping in Pulses

For pigeon pea, short duration grain legumes such as black gram and soybean are the best companion crops in peninsular India. Groundnut is also a suitable intercrop. Sorghum is commonly intercropped with pigeon pea.

Intercropping in Cotton

Cotton is initially a slow growing crop. Any short duration and fast growing crops such as green gram, black gram, groundnut or cluster bean are the compatible companion crops. Onion is also a preferred intercrop in cotton.

Intercropping in Sugarcane

Sugarcane is slow growing upto 80 – 90 days. Since the crop is planted in rows in 0.8 – 1.0 m apart, considerable space is available for intercropping. Short duration crops maturing in 80 – 90 days can be advocated as intercrops. Black gram and soybean are found suitable. The green manure, dhaincha *(Sesbania aculeata)* can be sown in the inters pace and incorporated at about two months of its growth.

Intercropping for Dry Lands

Intercropping recommended for different regions of India under drylands conditions are given in Table 4.12.

TABLE 4.12: Intercropping for different regions in India

Region	Intercropping
Samba (Jammu)	Castor + black gram / sesame / cowpea / sunflower
	Maize + cowpea (fodder)
	Maize + green gram
Hissar (Haryana)	Pearl millet + cluster been / green gram
Dehra Dun (U.P.)	Pigeon pea + soybean
	Maize + pigeon pea
Agra (U.P.)	Pearl millet + black gram / green gram
	Pigeon pea + green gram / pearl millet
Varanasi (U.P.)	Pigeon pea + maize / black gram / millet
Ranchi (Bihar)	Barley + rapeseed
	Wheat + Bengal gram / rapeseed / lentil
Bhubaneswar (Orissa)	Pigeon pea + black gram / sorghum / groundnut
Rewa region (M.P.)	Pigeon pea + sorghum + finger millet black gram / sunflower
Jhansi (M.P.)	Sorghum + black gram / green gram / soybean
	Wheat + Bengal gram / mustard
	Barley + Bengal gram / mustard

Region	Intercropping
	Bengal gram + linseed / mustard.
Indore (M.P.)	Pigeon pea + sorghum
	Cotton + soybean
Undaipur	Sorghum + black gram
(Rajasthan)	Pearl millet + black gram
	Sunflower + black gram
Anand (Gujarat)	Pearl millet + red gram / castor / sunflower
Akola	Pigeon pea + black gram / green gram / sorghum / groundnut
(Maharashtra)	Cotton + sunflower / black gram / groundnut
Sholapur	Pigeon pea + pearl millet / sorghum / black gram
(Maharashtra)	Cotton + sunflower / groundnut
Anantapur (A.P)	Pigeon pea + green gram / pearl mill / sorghum / foxtail millet/ groundnut / sunflower
Bijapur (A.P)	Cotton + groundnut / sorghum / pearl millet / green gram
Mysore region	Pigeon pea + pearl millet
(Karnataka)	Cotton + pearl millet
	Groundnut + cotton
Kovilpatti (T.N)	Cotton +black gram / green gram
	Sorghum + black gram / cowpea / Dolichos lab lab
	Pearl millet + green gram / cluster bean / cowpea / Dolichos lab lab

CROP ROTATION

Crop rotation is a process of growing different crops in succession on a piece of land in a specific period of time, with an object to get maximum profit from least investment without impairing the soil fertility. One cycle may take one or more farming years to complete. Farming year is 12 months for irrigated areas and is limited to the period of adequate soil-water availability for crop growth in rain fed areas.

PRINCIPLES OF CROP ROTATION

- The crops with tap roots should be followed by those which have a fibrous root system. This helps in proper and uniform use of nutrients from the soil.
- The leguminous crops should be grown after non-leguminous crops because legumes fix atmospheric N into the soil and add more organic matter to the soil.
- More exhaustive crops should be followed by less exhaustive crops because crops like potato, sugarcane, maize etc. need more inputs such as better tillage, more fertilizers, greater number of irrigations etc.
- Selection of the crop should be demand based.
- The crop of the same family should not be grown in succession because they act like alternate hosts for insects, pests and diseases.
- An ideal crop rotation is one which provides maximum employment to the family and far in labour, the machines and equipments are efficiently used and all the agricultural operations are done timely, simultaneously maintaining soil productivity.

- The selection of the crops should be problem based
- On sloppy lands which are prone to erosion an alternate cropping of erosion promoting and erosion resisting crops like legumes should be adopted.
- Under dry faming the selection of crops should be such which can tolerate the drought.
- In low-lying and flood prone areas the crops should be such which can tolerate water stagnation.
- The selection of crops should suit farmer's financial conditions.
- The crop selected should also suit the soil and climatic conditions.

ADVANTAGES OF CROP ROTATION

- Agricultural operations can be done timely for all the crops because of less competition.
- Soil fertility is restored by fixing atmospheric nitrogen, encountering microbial activity, avoiding accumulation of toxins (HCN etc.) and maintaining physical properties of soil.
- An ideal crop rotation helps in controlling insects, pests and diseases. It also controls the weeds in the fields.
- Proper utilization of all the resources and inputs could be made by following crop rotation.
- The farmer gets a better price for his produce because of its higher demands in the locality or market.

ROTATIONAL INTENSITY

This is calculated by counting the number of crops grown in a rotation and is multiplied by 100 and then divided by the duration of the rotation.

$$\text{Rotation intensity (RI)} = \frac{\text{No. of crops grown in a rotation}}{\text{Duration of rotation}} \times 100$$

e.g., Maize	Barley	Late potato	Green gram	First year
Paddy	Potato	Wheat	Black gram	Second year

$$\text{R1} = \frac{8}{2} \times 100 = 400\%$$

CROPPING SCHEME

The cropping scheme is a plan according to which crops are grown on individual plots of a farm during a given period of time with the object of obtaining maximum return from each crop without impairing the soil fertility. Thus a cropping scheme is related to the most profitable use of resources, land, labour, capital and management.

PRINCIPLES OF CROPPING SCHEME

- *Area under building and layout.* A proportionate area is allocated for buildings and layout. If the farm area is 50 hectares or less, then the area under buildings and layout allocated is 8 -10 per cent of the total area, if the farm area is more then 50 hectares then the area allocated is only 5 per cent.
- *Number of plots.* The number of plots should be either equal to the duration

of rotation or a multiple of that. *e.g.*, if the total duration of rotations in a cropping scheme is 5 years, then the number of plots must be either 5, 10, 15, or 20 etc

- *Selection of the crops.* Selection of crops depends upon the situation of the farm as discussed below.
- *Near a city.* If the farm is near a city, the farmer should grow vegetables on at least 60% of his net cultivated area.
- *Near a sugar factory.* If the farm is situated near a sugar factory then the farmer should put at least 60% of his net cultivated area under sugarcane or sugar beet.
- *Near a canning factory.* In such a locality, the farmer should grow vegetables, *e.g.*, peas, tomato etc on at least 60% of his net cultivated area.
- *Near a vanaspati factory.* The farmer should grow oil seed crops, e. g., groundnut, sunflower etc. in such a locality.
- *Near a dairy farm.* If the farm is located near a dairy farm, the owner should go for fodder cultivation. Apart from fodder he can raise oil seeds and pulses on 10 – 15% of his net cultivated area.
- *Near a cold storage.* In such areas, the farmer should grow potatoes, onions etc.
- *Near a highway or railway station.* Then farmer can go for perishable crops like vegetables etc. In such cases, the farm is considered to be within the proximity of a city area.
- *Facilities available on the farm.* Selection of crops also depends upon the facilities for power, irrigation, inputs, labour, transport etc.
- *Area of individual plots.* The area of individual plots should be equal unless topography, soil conditions or any other person compels the farmer to layout.
- *Area allotted to crops for calculation of cropping intensity.* If the farmer takes crop mixtures then the relative area under different crops has to be considered so that none of them are adversely affected due to their mutual competition.
- *A crop planning* which can utilize inputs and other resources available on the farm without wastage should be adopted by the farmer.

CONCEPTUAL BASES OF MULTIPLE CROPPING

A cropping system is the kind and sequence of crops grown on a given area of soil over a period of time. It may be a regular rotation of different crops in which the crops follow definite order of appearance on the land or it may consist of only one crop grown year after year on the same area (monoculture). Other cropping systems may include different crops but lack definite or planned order in which crops follow one another or growing of two or several crops mixed together. Multiple cropping is thus a one year cropping system in which two (double cropping) or more crops are grown in succession within a year. If all these crops are cereals, multiple cropping will resemble monoculture in its advantages and requirements. If legumes are also included it will have certain features of traditional rotations.

A soil management system includes the cropping system plus the other associated practices in raising the crops, such as frequency and method of tillage, use of fertilizers, irrigation, drainage etc. thus effect of cropping system on soil productivity depends on other practices used with it in making up the systems of soil management.

Recent developments in the field of soil management (frequency and method of tillage, residue management and use of fertilizer, herbicides, insecticides and fungicides) have enabled a multiple cropping system to be feasible, desirable and commendable in tarai where the temperature and the sun shine are such that one or the other crop could be grown throughout the year provided water is available in plenty for irrigation.

If the system of soil management is attuned to the long term maintenance of soil productivity with the newer tools then the maximization of yield over a unit area in a year will involve in bio-energetic terms (Energy is the only common scientific denominator of the many sciences with which agriculture is concerned) the conversion of optimum amount of electromagnetic light energy received from the sun into useful yield of chemical energy through the photosynthetic mechanism of crop plants. This chemical energy has to be digestable, palatable, movable and storable. The wastage of the converted energy is to be minimized through the control of pests and diseases which convert chemical energy in the form which is not usable by man. The output of a crop will thus depend on the rate of working (net assimilation rate), size (leaf area) and the duration of the photosynthetic system. Recently we have varieties and hybrids of cereals available in the country which have a morphological frame to withstand high fertilizer doses. The disposition of their leaf area in space and time (dating, size and geometry of leaf area in its duration) can make full use of the growing condition and in particular of the available light energy. Fortunately these varieties have shorter duration than the traditional ones an attribute which fit them very well in a multiple cropping system. In paddy the newer varieties are not only short duration but insensitive to photoperiod and weekly thermo sensitive. These attributes enable their cropping throughout the year in some parts of the country.

Vegetable growers of our country have a long record of multiple cropping. They have developed soil management systems which maintain the productivity of their soil at a higher level than those of their neighbours who have been growing grain crops. The system of their soil management included, heavy dressing of organic manures, more tillage and inter culture, manual removal of weeds and following a rotation of vegetable crops which kept the diseases at minimum. With the availability of the short duration high yielding varieties of cereals the multiple cropping system is now to be extended to cereals but with a quite different soil management system. Users of this system will be mostly dependent on fertilizer use, chemical herbicides and insecticides, disease resistant varieties of plants, soil fumigation etc.

Two sets of forces determine the most profitable cropping system for a farm. One set involves the effect on an individual field and is measured in long term crop yield possibilities. The other involves effect on the whole farm and therefore, includes not only yield but also the economic and management factors of investment in implements and machinery, seasonal distribution of labour and prices and prices and of the produce:

Let us examine the first set in retrospect and prospect. Cropping systems followed in major parts of the country are:

1. Continuous cropping in which on conscientious effort is made to maintain productivity. This is one year cropping system in which the main *kharif* crop alternates with *rabi* fallow or *kharif* fallow alternates with *rabi* crop;

2. Rotation of two or more than two years involving occasional fallowing, legumes for fodder, seed or green manure;
3. Mixed cropping of one year either during *kharif* or *rabi* or during both the seasons;
4. Cropping systems with a lack of definite planned order, and
5. In certain pockets shifting cultivation in which soil is cleared, cropped for three or four years and then abandoned.

This classification is, of course, too generalised and is not exhaustive. There is considerable and necessary overlap between these systems. Some of these will continue in the dry farming area but wherever water has been made available either in dry farming area or humid areas all will give way to multiple cropping systems. During the recent years a renewed emphasis has been placed on minor irrigation programmes in almost all states. All these minor irrigation sources are destined to become nucleus around which a programme of multiple cropping will develop.

For these traditional cropping systems soil management programmes were based on occasional fallowing, green manuring, frequent and cultivation during fallow period to remove weeds and hot weather cultivation to speed up the burning of organic matter for releasing nitrogen. Experimentally, continuous wheat with *kharif* fallow has given steady yield of nearly 900 kg/ha. In North India, mono-culture of paddy over centuries has given similar yield without manuring. Thus wheat and paddy yields have stabilized on lower level and maintained through the operation of natural nitrogen fixing process. The year to year variation in yield around these averages is due to seasonal variation. Green manuring-wheat rotations have shown no long term effect on soil productivity. In short term they have given response equal to the application of 20-30 kg N/ha. Old rotational trials have shown the value and utility of rotations under the management followed. Long term trials at Pusa, Coimbatore, Cuttack have shown that plots receiving only balanced inorganic fertilizers have maintained yield and organic matter content of the soil. Since nitrogen was almost universally deficient for maximum yields of non-legumes there was an urgent need of including a legume in rotation and rotations like sorghum-wheat were utter failure without manuring. The whole picture has changed since nitrogen has become more and more available. Often the farmers and some agronomists raised the question 'How would the soil physical factors be affected under intensive cereal cropping in multiple cropping systems?'

The importance of good tilth for high yield has been generally accepted but seldom found in research results to be as important as it appears. On sandy soils the tilth is seldom a problem. On medium textured soil (silt loams and loams) an acceptable level of soil tilth can be maintained in the plough layer without bringing legumes in the cropping system with minimum tillage. On fine textured soil (silty clays, clay and clayey loams) favourable tilth is more necessary than in coarse and medium textured soils. On such soils problem of structure may become more severe when one grows the crops which residues to the soil. Farmers, who have such soils, should watch carefully the water retention in their soil. If seed bed is cloddy or field dries slowly after rains or water stands between the rows in low places then the soil management system used is not maintaining soil structure. More fertilizer to grow more residues or less tillage or both is needed. If paddy is included in such a cropping system to alternate with upland crop, it should be grown like upland crop without puddling. One agronomic fact has always to be remembered *i.e.* any damage from over cropping is temporary and can be corrected by changing the soil management system which provide greater return of organic residues, less frequent tillage and perhaps more deep rooted crops.

As far as the economic factors are concerned, the cost benefit ratios are more favourabie now for following multiple cropping. Improved fertilizer technology, higher analysis of fertilizers to reduce handling and transportation costs, more efficient application techniques including bulk handling of dry mixture etc. are expected further to reduce cost. The peak labour requirement for harvesting of crops and sowing of *rabi* crops is generating a demand for labour saving machines. Thus the multiple cropping programme will not only accelerate the chemicalization of the India agriculture, but it would also lead to mechanization. Double cropping with no or inadequate manuring or with inclusion of a legume has been practiced for long in the country. What is needed is to provide more purchased inputs. The economics is in favour.

Although written in another context the basis of multiple cropping according to Melsted (1954) is, it is a philosophy of maximum crop production per acre of land with a minimum of soil deterioration. The philosophy is based on the concept that high production is good for soil that minimum tillage promotes soil tilth and conserves soil organic matter, that high fertility promotes high yield, and lessens the losses of soil humus, that large amount of decomposable organic matter in soil is essential to good soil tilth and good soil physical conditions, and cover in the form of living mulch is good protection against all forms of erosion.

Many of the concepts on which multiple cropping is based are only tentative. The preliminary results which have been achieved in national demonstrations, by progressive farmers and in experimental plots are encouraging. They seem to indicate that the concepts are sound and future will support them. When multiple cropping will be adopted on large scale many soil management problems are likely to come up. For their solution integrated research done by plant breeders, agronomists, soil scientists and agricultural engineers will be absolutely necessary.

PRE-REQUISITES FOR MULTIPLE CROPPING

There are a number of essential features for developing a successful multiple cropping programme.

Even in less intensive cropping, success is the end result of the timeliness of the following operations: Choosing the proper crop, selecting the correct variety, providing high quality vigorous seed, treating the seed before planting, preparing the soil properly, planting the seed and correct setting of the plants, cultivation for stimulating plant growth, controlling weed and insect pests, preventing and controlling diseases and proper and timely harvesting. The timeliness of cultural operations becomes much more critical factor in multiple and relay cropping. A few of the conditions considered to be pre-requisite to successful multiple cropping are enumerated here.

1. There must be adequate supply of available water and it must be available at the time it is needed. Since intensively conducted multiple cropping will involve irrigated farming and because the timeliness of the operations is extremely critical, it is absolutely essential that a dependable source of water be available. Fortunately tube wells and pumping sets are becoming available

in greater and greater numbers year after year. As three to four crops will be taken from every acre the area that will be commanded by a well will be only 5 – 10 acres especially during the summer period from April to June because the consumptive use of water will be very high due to the higher evaporative demand during this period. With the available supply of under ground water it is possible to follow relay cropping over large areas. It is also equally important that the water available is well controlled in its distribution. For this purpose land levelling and provision of adequate drainage is absolutely necessary.

2. The second pre-requisite is an adequate and timely supply of seed, fertilizer and plant protection materials. These basic inputs have to be available when required and in sufficient quantities if multiple cropping is to be successful. The time available between two crops is very short and this period will be the peak period of farm activities, hence all these inputs should be available to the farmer without wasting much of his time.
3. The use of short duration varieties and hybrids. Crops of short duration are generally more useful taking into consideration the yield per unit area per unit time rather than the longer duration ones. Fortunately, now such varieties of paddy and wheat, jowar and bajra and maize are available and the more are in the nursery of the breeders for release. For ensuring continuous increase in the yield of crops extensive research on various aspects of crop production is necessary because new problems of soil management always crop up.
4. A fourth pre-requisite, is the development of new market, processing and new uses of the crops which are not traditionally grown in the locality but have a comparative advantage during some period. For example, crops like soybean, cheena and zaid legumes fit very will in the multiple cropping. New uses of these crops have to be developed for generating a higher demand for them.
5. Another important aspect of relay cropping is a high man-hour reguirement which can be reduced by the availability of machinery for tilling the land, harvesting of crop, and pumping irrigation water. In a multiple cropping system, the time available between two crops is extremely short and time in this system literally means money. Time available for many of the farming operations under the multiple cropping system will be reduced. This implies higher degree of mechanization. When relay cropping becomes more and more widespread, there will likely be increasing demand for machinery to work the land and harvest the crops.
6. A sixth pre-requisite is a better under standing of scientific agriculture, or in other words, a need for better farm managers. The farmers must learn to be efficient in irrigation water application; efficient in harvesting techniques; and he must learn to be judicious in the selection of crops and varieties which he grows. There are many skilled farmers capable of undertaking such scientific agriculture, but they are dependent on research information on which to base decisions. There is a need for production-oriented research, including this area of intensive planning, to lead the way so that the farming community can invest and expand along an already explored and proven path.

CHAPTER 5

Classification of Crops

Crop production and management is rapidly converting itself as a management science, wherein the aim is to obtain maximum income with not only increased production but also by reduced cost of inputs, with the resources available to the farmer. They various systems of multiple and relay cropping have not only become popular in India, but also being adopted in the developed world, wherein single cropping was common so far. The potentials and constraints for adoption as well as the need for on-site research information are dealt in details in this chapter.

Field crops are classified into different groups based on various criteria. The most useful classification is from the agronomic standpoint based on their usage. Field crops are classified as cereals, pulses, oilseeds, fibre crops, sugars, forage crops, narcotics, plantation crops, spices, etc. Roots and tubers, plantation crops, flowers spices, fruit trees and vegetables are generally referred to as horticultural crops. Among the field crops, cereals and pulses are called as food crops; the others as cash or commercial crops. At present only 200 species are used as food and 80% of the food supply come from eight species.

Before going to any agronomic field operations some basic knowledge about the crop plants must be acquired. Some basic information about the crop plants and how wisely they can be included as components in a particular farming system are discussed below.

PRINCIPLES OF CLASSIFICATION OF CROPS

Classification is an acceptable product of scientific studies and is obviously based on well demarcated characteristics of the things or concepts so classified.

A crop is an organism grown or harvested for obtaining yield. Crop plants do not belong to any particular portion of the plant kingdom. Several crop plants are alike with respect to ontology (development of an organism), morphology, anatomy, physiology and requirement of particular type of ecology at different stages of their growth and development. All these help to generalize similar crop plants as a class for attaining a better understanding of them. Of course, while considering some crops as a group, some exceptions must always be expected.

Crops are grouped in several ways namely, according to the range of cultivation, place of origin and distribution, different characters, uses, cultivation requirements and other common behaviour.

Crop plants are grouped into three main classes according to the range of cultivation. They are:

a) **Garden crop:** Crop plants that are grown on a small scale in gardens such as in kitchen gardens, flower gardens and backyard gardens for instance onion, brinjal.
b) **Plantation crop:** Crop plants that are grown on a larger scale on estates. They are permanent (need not be replanted after each harvest) in nature *i.e.*, harvesting continues for a prolonged period from a single planting, for instance tea, coffee, cocoa.

c) **Field crop:** Crop plants that are grown on a vast scale. They are mostly seasonal – such as rice, wheat, cotton.

Among these three classes agronomy deals with field crops only. Field crops are classified in the following ways:

1. According to the place of origin;
2. Botanical or Taxonomic classification;
3. Special classification based on special purpose crops;
4. Commercial classification;
5. Economic classification;
6. Seasonal classification;
7. Classification according to ontogeny;
8. Agronomic classification; and
9. Classification based on leaf morphology.

CLASSIFICATION ACCORDING TO THE PLACE OF ORIGIN

A number of crops may be grown elsewhere but each crop has its native place where there are a large number of wild relatives even today. The agro-climatic situation of the place of origin may provide relevant information about the requisite or ideal environment for its adaptability. The wild relatives may be used as a gene pool for any further improvement of the crop concerned.

According to the place of origin crops are grouped into:

i) **Native:** Which are grown within the geographical limits of their origin. Crops of Indian origin are rice, barley, black gram, green gram, castor, sarson, sugarcane and cotton.
ii) **Exotic or alien or introduced:** Which are grown even beyond their site of origin. Some of the crops which are now grown in India but introduced from other countries, are tobacco, potato, jute, and maize.

BOTANICAL OR TAXONOMIC CLASSIFICATION OF CROPS

Crop plants are dealt with under the natural orders or the families under which they come, in a systemic arrangement. The classification has certain advantages in the understanding of the morphological character of any particular family of crop plants with different agro-botanical or morpho-agronomical characteristics or peculiarities. Crop plants are grouped into families:

1. Poaceae (Graminae) or Grass family: **Cereals**, **Sugarcane, Napier, Paragrass.**
2. Papilionaceae (Leguminosae) or pea family: Pulses, **Legumes, Groundnut, Sunnhemp.**
3. Cruciferae/Mustard famity: **Mustard, Radish, Cabbage, Cauliflower, Knolkhol.**
4. Cucurbitaceae/Gourd family: All gourds (*e.g.* bottle gourd, bitter gourd), Cucumber, Pumpkin.
5. Mulvaceae/Cotton family: **Cotton, Bhindi, Roselle.**
6. Solanaceae/Brinjal Family: **Brinjal, Potato, tomato, tobacco, Chillies.**
7. Tiliaceae: **Jute, Phalsa**
8. Asteraceae (Compositae): **Sunflower, Safflower, Niger.**
9. Chenopodiaceae: **Spinach, Beet, Sugarbeet.**
10. Pedaliaceae: **Sesame (**Sesamum).

11. Euphorbiaceae: **Castor, Tapioca**
12. Convolvulaceae: **Sweetpotao**
13. Umbelliferae: **Coriander, Cumin, Carrot.**
14. Aliaceae: **Onion, Garlic.**
15. Zingiraceae: **Ginger, Turmeric.**

SPECIAL CLASSIFICATION BASED ON SPECIAL PURPOSE CROPS

Crop plants which are grown with the intention of serving purposes in addition to obtaining yield or to meet a particular purpose are grouped into special purpose crops.

1. **Arable Crops:** Which require preparatory tillage *e.g.* potato, tobacco, rice, sugarcane, maize, etc.
2. **Alley Crops:** Such arable crops which are grown in alleys/passages formed by trees or shrubs, established mainly to hasten soil fertility restoration, enhance soil productivity and reduce soil erosion slight shade tolerance and non-trailing habit are the prerequisites of Alley crops *e.g.* Arable crops like sweet potato, urd, turmeric and ginger are grown in the passages formed by the rows of Eucalyptus, subabool and cassia.
3. **Augmenting Crops:** Such crops are sown to supplement the yield of the main crops *e.g.* Japanese mustard with Berseem, Chinese Cabbage with mustard. Here Japanese mustard and Chinese cabbage help in getting higher yield in the first cutting.
4. **Avenue Crops:** Such crops are grown along farm roads and fences *e.g.* Arhar, Glyricidia, Sisal, etc.
5. **Border Crops/Barrier/Guard Crops:** Such crops help to protect another crops from trespassing of animals or restrict the speed of weed and are mainly grown as border *e.g.* safflower(thorny oilseed crop) is planted around the field of chick pea.
6. **Braker Crops:**
   a) To break the continuity of agro-ecological situation of the field under multiple cropping systems.
   b) To reduce the inoculum of soil born harmful biotic agents such as weeds, pest and improves soil condition for crop growth *e.g.* Legume in rice-wheat system.
   c) Are also used to designate guard crops – particularly those which help to break (retard) the wind speed and project crops from wind hazards.
7. **Cash Crops:** Such crops are grown for sale to earn hard cash *e.g.* Jute, Cotton, Tobacco, Sugarcane.
8. **Catch/Contingent/Emergency Crops:** Such crops are cultivated to catch the forthcoming season when main crop is failed. They are of very short duration, quick growing, fast bulking, harvestable or usable at any time *e.g.* green gram, urd, cowpea, onion, radish etc.
9. **Cleaning Crop:** Whose agronomical practices makes the field clean *e.g.* potato, maize etc.
10. **Cole Crops:** 'Cole' is derived from colewart. Colewart is the ancestor of wild cabbage. Therefore, cole crops are essentially cold weather crops belonging to the conciferae capable for withstanding considerable frost. *E.g.* cabbage, cauliflower and brussels sprouts.
11. **Contour Crops:** Grown on or along the contour lines to protect the land

from erosion *e.g.* marvel grass etc.

12. **Commercial Crops:** Such crops are grown to earn money *e.g.* Jute, Cotton, Tobacco, Sugarcane, etc. Cash crops are certainly commercial crops.
13. **Companion crops:** In intercrop situations when subsidiary crops are usually of shorter duration in a long duration main crop *i.e.*, the main crop gets the company of a short duration crop for a certain period at the early stages of its growth, for instance potato, onion and spinach in autumn planted sugarcane, basella, lady's finger and amaranths in spring planted sugarcane; radish in potato; cowpea, rice bean, green gram in napier.
14. **Cover Crops:** Able to protect the soil surface from erosion through their ground covering foliage and/or rootmats *e.g.* lobia, groundnut, urid, paragrass, sweet potato.
15. **Complementary Crops:** Each other crop is benefited in intercropping *e.g.* Jowar + Lobia. Jowar receives nitrogen from lobia and lobia requires support from jowar.
16. **Competitive Crops:** Such crops compete to each other and are not suitable for intercropping *e.g.* two cereals.
17. **Supplementary Crops:** Such crops are neither competitive nor complementary *e.g.* maize + cucurbits.
18. **Exhaustive Crops:** Such crops leave the field exhaustive after growing *e.g.* cereals.
19. **Energy Crops:** To obtain liquid energy such as ethanol and alcohol *e.g.* sugarcane, potato, maize, tapioca.
20. **Fouling Crops:** Whose cultural practices allow the infestation of weeds intensively *e.g.* direct seeded upland rice.
21. **Fuel crops:** These crop plants are grown to obatain fire wood or solid fuel as a by-product along with their economic yield, for instance jute, sugarcane, pigeonpea, cotton, mustard and sesame.
22. **Green manuring crops:** These crop plants are grown to be incorporated into the soil fresh to increase the fertility of the soil. Green manuring may be (a) green leaf manuring or (b) green manuring *in situ*, such as *dhaincha*, sunnhemp, *Glyricidia*, Subabool.
23. **Leaf crops:** These crop plants are grown to harvest leaves for economic yield, for instance tobacco, sisal, spinach, basella and greens.
24. **Ley Crops:** Any crop or combination of crops is grown for grazing or harvesting for immediate or future feeding to livestock *e.g.* Berseem + Mustard. Such cropping is called ley cropping.
25. **Mixed crops:** These consist of two or more crops that are grown simultaneously in the same field without preserving their identy with respect to field area. Seeds of these crops may be mixed together before sowing and broadcasted irregularly or drilled in rows or may be sown at the same time and grown with the same management practices. They may be harvested together or separately, for instance peas+rai, wheat+mustard, gram+safflower, maize+cowpea, Jowar+cowpea, leenseed+toria, berseem+mustard.
26. **Mono or whole or sole or pure crops:** These crop plants are grown as pure or solid stands, for instance transplanted rice, jute, tobacco, oats, and rice bean.
27. **Mulch Crops:** To conserve soil moisture, such crops are grown *e.g.* cowpea.
28. **Nurse Crops:** Such crop helps in the nourishment of other crops by

providing shade and acting as climbing sticks *e.g.* rai in peas, jowar in cowpea.

29. **Paira/Utera Crops:** Growing of such crops sown a few days or weeks before harvesting of standing mature crops is called paira/utera cropping and the sown crop is called paira/utera crop *e.g.* Lathyrus in rice. Paira cropping in succession may constitute relay cropping.
30. **Paired row Cropping:** Each third row is removed or growing of crops in paired row is called Paired row cropping. It is suitable for dry land and objective is to conserve soil moisture.
31. **Parallel crops:** These consist of two or more crops which are grown simultaneously but do not have inter-competitive effects *i.e.*, they parallel with respect to competition, for instance maize+cowpea, sorghum+pigeon pea.
32. **Plant crop or first harvest crop or stub crop or first cycle crop:** In perennial or multicut crop plants this refers to the first harvest after sowing or planting the crop, for instance in sugarcane, napier, cotton, peogeon pea, berseem, oats, paragrass, Guinea grass and marvel grass.
33. **Ratoon crop or stubble crop or coppice or providence or regenerated or second or third** –cycle crop: In perennial or multicut crop plants this refers to the subsequent harvests taken from the regrowth of the root stocks, stubble and stumps, after the first harvest such as for sugarcane, napier, berseem, rice, jowar, bajra, oats, guinea grass, and rice bean. Such crops reduce the cost of cultivation, do not need and sowing, reduce the crop cycle period, produce a higher yield per unit time and often require less input.
34. **Restorative Crops:** Such crops provide a good harvest along with enrichment or amelioration of soil *e.g.* Legumes.
35. **Riparian Crops:** Grown along irrigation and drainage channels or waterbodies *e.g.* waterbind weed (kalmi sak), para grass. They help to protect the soil from erosion.
36. **Seed crops:** These crop plants are grown for seed production (generative yield), for instance the seed crop of jute, cabbage, cauliflower, potato, tobacco, oats, berseem, cowpea and maize. Such crops require special care and management to maintain the satisfactory genetic identity and purity of the crop variety (agro-ecotype) and healthy crop.
37. **Skip Cropping:** A line is left unsown in the regular row series of sowing is called skip cropping.
38. **Silage Crops:** Such crops are grown to preserve in pits in a succulent condition by a process of natural fermentation or acidification for feeding livestock during months or offseason *e.g.* cowpea, jowar, etc.
39. **Smother Crops:** Able to smother (suppress) the population and growth of weeds by providing dense foliage and quick growing ability *e.g.* cowpea, mustard.
40. **Sod or turf crops:** These crop plants belong to the grass family and have sod type tillers with matted foliage and roots close to the soil surface. They are grown to conserve soil from erosion particularly in non-arable areas, for instance marvel grass, *Digitaria sanguinalis*, *Cynodon dactylon*.
41. **Soiling Crops:** Grown to harvest while they are still green and fed fresh to livestock in stalls *e.g.* barseem, napier etc.
42. **Strip crops:** They are of six types:
    a) **Erosion permitting crops:** These crops are grown to permit soil erosion

in large sloping (where vertical distance is more as compared to horticultural distance, say, more than ten per cent) lands which require land shaping for successful arable farming, Such crops have lower foliage cover and holding capacity of roots to surface soil, for instance jowar, bajra and maize. They are sown along the slope in strips.

b) **Erosion restricting or resisting crops:** These crop plants are grown to restrict soil erosion. They are grown across the slope in strips alternately with erosion permitting strip crops to introduce land shaping by cultural means, for instance marvel grass, groundnut, *Atylosia*, *kudzu*, *shyama*, horse gram and kidney bean.

c) **Contour strip crops:** These crop plants are grown in strips of suitable width across the slopes (bi- or multi- directional) on the level or contour lines alternating with erosion restricting crops particularly where land shaping is not needed *i.e.*, in low gradient grounds, for instance jowar, pigeonpea, groundnut and marvel grass.

d) **Field strip crops:** The crop plants that are grown on farm lands in more or less parallel strips across fairly uniform slopes but not on exact contours, for instance maize, pigeonpea, guar, jowar, bajra, cowpea, sun hemp, mesta, green gram, sesame.

e) **Wind strip crops:** The crop plants (mainly tall crops such as jowar, bajra, maize, pigeonpea, safflower and mustardthat are grown with low crops such as potato, peas, groundnut, black gram and green gram in alternately arranged straight and long but relatively narrow, parallel strips laid right across the direction of the prevailing wind regardless of the land contour. Some times bi-directional rows of wind strip cropsare also sown to combat wind from any direction.

f) **Permanent buffer strip crops:** These are crop plants such as permanent legume, grass or brush or shrubs that are grown on a permanent or temporary basis on the strips that are established to take care of critically steep or highly eroded slopes in fields under contour strip cropping, for instance subabool, babul, *assia*, *Prosopis*, *Glyricidia*, *Tephrosia*, *sar (Saccharum munjo)*, Guinea grass, *Panicum repens*, lemon grass, *Cymbopogon naraus*, *C. maritini*, *Andropogon squarrosus*, *Imperata cylindrical*(thatch grass).

43. **Trap Crops:** Grown to trap insect-pests and soil-borne harmful biotic agents such as parasitic weeds *e.g.* cotton red bug is trapped by growing Bhindi around the cotton and **orobache**(weed) is trapped by **solanaceous plants** *i.e.* striga by sorghum.
44. **Truck Crops:** Grown to market fresh *e.g.* Bhindi, Spinach.
45. **Ware Crops:** Such crops are grown for temporary storing as intact in warehouse for future use or sale *e.g.* potato.

COMMERCIAL CLASSIFICATION

The plant products (Usable products) which come into the commercial field go under popular grouping such as:

a) **Food crops:** Rice, wheat, green gram, soybean, groundnut,
b) **Feed crops:** Oats, napier, sorghum, maize, stylo, ricebean, berseem, Lucerne.
c) **Industrial or commercial crops:** Cotton, sugarcane, sugar beet, tobacco, jute.

d) **Food adjuncts:** Turmeric, cumin, garlic.

ECONOMIC OR AGRARIAN OR AGRICULTURAL CLASSIFICATION OR CLASSIFICATION ACCORDING TO USE OF CROP PLANTS AND THEIR PRODUCTS

Crops and Botanical Name

Crops	Botanical Name
Cereals	**Graminae**
Rice	*Oryza sativa*
Wheat	*Triticum aestivum*
Maize	*Zea mays*
Sorghum/Jowar	*Sorghum bicolor*
Bajra	*Pennisetum typhoides* (Pearl millet)
Barley	*Hordeum vulgare*
Madua/Ragi/Finger Millet	*Eleusine coracana* (good for diabetes Patients)
Cheena/Proso millet	*Panicum miliaceaum*
Sawan/Barnyard millet	*Echinochloa frumentacea*
Kodo	*Paspulum scrobiculatum* (Coarsest of all food grain) Kakun/Italian/German
Millet/Foxtail	*Setaria italica*
Pulse Crops	***Leguminosae***
Gram/Chickpea/Bengal gram	*Cicer arietinum*
Lentil	*Lens esculenta*
Pea	*Pisum* spp.
Arhar/Red gram/Pigeon Pea	*Cajnus cajan*
Green gram/Mung	*Vigna radiate*
Black gram/Urd	*Vigna mungo*
Cowpea/Lobia	*Vigna sinensis*
Soybean	*Glycine max*
Oilseed Crops	
Groundnut/Peanut/Earthnut/Monkeynut/Manillanut	*Arachis hypogea* (Leguminosae)
Sesame/Til/Sesamum	*Sesamum indicum* (Pedaliaceae)
Castor	*Ricinus communis* (Euphorbiaceae)
Papeseed and Mustaerd	*Brassica* spp. (Crucifereae)
Linseed/Flax	*Linum usitatissimum* (Linaceae)
Safflower	*Carthamus inctorius* (Compositae)
Sunflower	*Helianthus annus* (Compositae)
Fibre Crops	
Cotton	*Gossypium* spp. (Malvaceae)
Jute	*Corchorus capsularis* (White jute) (Tiliaceae)
Corchorus olitorius (tossa) (Tiliaceae)	
Sunhemp/banaras Hemp/Sann-Hemp/ Bombay Hemp/Indian Hemp/WarangalHemp	*Crotalaria juncea* (Leguminoseae)
Mesta (Bimli in Andhra)	Andhra Pradesh is the leading state in both area and production.

Crops	Botanical Name
	Hibiscus spp. --- *Sabdarifa canabinus*
Forage	
Oat	*Avena sativa* (Graminae)
Napier/Elephant grass	*Pennisetum purpureum* (Graminae)
Berseem/Egyptian clover	*Trifolium alexandrinum* (Leguminoseae)
Lucerne/Alfalfa	*Medicago sativa* (Leguminoseae)
Guar	*Cyamopsis tetragonoloba* (Leguminoseae)
Sugar Crops	
Sugarcane	*Saccharum officinarum* (Graminae)
Sugarbeet	*Beta vulgaris* (Chenopodiaceae)
Miscellaneous	
Potato	*Solanum tuberosum* (Solanaceae)
Tobacco	*Nicotiana tabacum* (Solanaceae)
Root and Tuber Crops (Starchy)	
Five crops on global basis	
Potato	*Solanum tuberosum*
Sweet potato	*Ipomea batatas*
Cassava/Tapioca	*Manihot utilissima*
Yams	*Dioscorea* spp.
Cocoyam Taro	*Colocasia esculenta*
Tannia	*Xanthosoma sagittifolia*
In addition to above five crops, three more starchy crops are grown in India:	
Elephant foot yam/Zimikand in Hindi	*Amorphophallus campanulatus*
Chinese Potato (Coorka)	*Coleus parviflorus*
Arrow root/Koova/Ararut	*Marantha arundinaceae*

SEASONAL CLASSIFICATION

Different annual crops are grown in different seasons of the year. Such seasons are most prominent in the plains in contrast to coastal and mountainous ranges and in areas receiving monsoonal climate. Crops are grouped under the seasons in which are their major field duration:

1. ***Kharif*** **or South-West monsoon crops:** Crop plants grown during June-July to September-October such as rice, maize, jowar, bajra, castor, groundnut, sesame, sun hemp, green gram, soybean. They require warm-wet weather during their major period of growth and shorter day-length for flowering.
2. ***Rabi*** **crops:** Crop plants grown during October-November to January-February such as wheat, mustard, barley, oats, bengal gram, peas, linseed, lentil, berseem, cabbage and cauliflower. They require cold-dry weather for their major growth period and longer day-length for flowering.
3. ***Zaid*** **or summer crops:** Crop plants grown during February-March to May-June such as black gram, green gram, sesame and cowpea. They require warm-dry weather for their major growth period and longer day-length for flowering.

With respect to rice crops grown during different season's *aus*, *aman* and *boro* seasons are prevalent. The word *aus* has been derived from the Sanskrit *ashu* meaning quick or early. It is designated to the rice varieties grown during the pre-monsoon period *i.e.*, harvested in August-September. It is also knownas autumn rice with respect to the time of harvest. The word *aman* has been derived from Arabic meaning safely which indicates stability of the crop. It is synonymous to *kharif* or winter crop of rice. *Boro* refers to the rice crop grown as submerged land lower in elevation. It is grown during January-February to April-May. It is also known as summer rice.

This seasonal classification is not a universal one. It only indicates the period of the year when a particular crop has been grown, for instance *kharif* rice, *kharif* maize, *rabi* maize, *boro* rice, *rabi* pulse, summer pulse.

Some crops overlap the seasons such as Jute which may be sown in February-March and harvested in July-August. Sesame is an adaptable crop grown during summer in the southern plains of West Bengal whereas it is a suitable crop during the monsoon period in the north and both in the *kharif* and *rabi* seasons in south India. Some photo neutral or weakly photo- and thermo-sensitive crop varieties may be grown throughout the year provided irrigation is available, for instance rice, maize, groundnut. Again some long duration temporary crops such as sugarcane, cotton, pigeon pea, etc., do not follow the regular pattern except that low temperature in winter provide a triggering factor contributing towards the maturity of the crops.

CLASSIFICATION ACCORDING TO ONTOGENY

Each crop plant competes its life cycle after passing through progressive stages of development. Crop plants are grouped according to their life cycle:

1. **Annual crops:** Crop plants that complete their life cycles within a season or year such as rice, wheat, maize, mustard and tobacco. They produce a crop seed and die. Some of these crop plants may produce tillers. If such rooted tillers are separated from the main shoot and planted, each tiller will survive that season as a new plant, but will not live another season.
2. **Biennial crops:** Crop plants having a life span of two consecutive years. The first year's growth is purely vegetativewith top growth usually confined to rosette of leaves. The tap root is often fleshy and serves as a food storage organ. During the second year it produces a flower stalk from the crown. After producing seed the plant dies, for instance sugar beet, cabbage, radish, carrot. Due to unfavourable weather conditions, a floral stalk may come out (bolting) even in first year.
3. **Perennial crops:** Crop plants that live for three or more years. They may be seed bearing (supermophytes) or non-seed-bearing. They may propagate by seeds and /or by vegetativepropagules that be developed in aerial or subaerial or in both parts, such as sugarcane, napier, ginger, sweet potato and garlic. In general, perennial crops occupy land for more than 30 months.

AGRONOMIC CLASSIFICATION OR CLASSIFICATION ACCORDING TO CULTURAL REQUIREMENTS

A group of crop plants are alike in cultural requirements due to their similar agro-botanical or morpho-agronomical character:

According to Suitability of Toposequence

1. Crops grown on high land such as arhar, groundnut, sunnhemp, maize, jowar, bajra, cotton, sesame, napier and Dinanath grass. They can not tolerate water stagnation.
2. Crops grown on medium land such as jute, potato, sugarcane, upland rice, ragi, wheat Bengal gram, black gram, barley, oats and mustard. They require sufficient soil moisture but can not tolerate water stagnation.
3. Crops grown on low land such as rice, dhaincha, paragrass and jute (*capsularis*). They require an abundant supply of water and can stand prolonged water-logged conditions.

According to the Suitability of the Textural Groups of Soils

1. Crops of sandy to sandy loam (light) soils, potato, sweet potato, sugarbeet, onion, carrot, turnip, green garam, black gram, sunflower, jowar, bajra.
2. Crops of silty to silty loam (medium) soils: jute, sugarcane, maize, cotton, mustard, tobacco, Bengal gram, peas, pigeon pea, berseem, Lucerne, cowpea.
3. Crops of clay to clay loam (heavy) soils: rice, wheat, barlet, linseed, lentil, chickling-pea, para grass, Guinea grass, marvel grass.

According to the Tolerance to the Problem Soils

1. Crops tolerant to acidic soils: wet-rice, potato, mustard.
2. Crops tolerant to saline soils: chillies, cucurbits, wheat, raya, bajra, jowar, sesame, guar, barley.
3. Crops tolerant to alkine soils: barley, cotton, chick pea, peas, beans, berseem, groundnut, maize.
4. Crops tolerant to waterlogged soils: wet-rice, dhaincha, paragrass, napier, Guinea grass.
5. Crops tolerant to soil erosion: marvel grass, groundnut, black gram, rice bean, moth bean.

According to Tillage Requirement

1. Arable crops: Crops requiring preparatory tillage: potato, radish, berseem, sugarcane, rice, tobacco, maize.
2. Non-arable crops: Crops that may not be requiring preparatory tillage: black gram, para grass, cow pea.

According to the Depth of Root System

1. Shallow-rooted crops:rice, potato, toria, onion.
2. Moderately deep-rooted crops: wheat, groundnut, castor, tobacco, barlet, chick pea, berseem.
3. Deep-rooted crops: maize, cotton, jowar, bajra, sugar beet, sweet potato.
4. Very deep-rooted crops: sugarcane, safflower, Lucerne, pigeon pea.

According to the Tolerance to Hazardous Weather Conditions

1. Frost tolerant crops: Sugar beet, beet.
2. Cold tolerant crops: cabbage, mustard.
3. Drought tolerant crops: bajra, jowar, barley, safflower, castor, cotton, chillies, sesame.

According to Water Supply

1. Irrigated crops: potato, *boro* rice, cabbage, berseem, wheat.
2. Rainfed upland crops: jute, maize, ragi, upland rice, pigeon pea, cotton, castor, sesame, cowpea, Guinea grass, Dinanath grass.
3. Rainfed but partial irrigated crops: Bengal gram, wheat, jowar, bajra, mustard.
4. Residual or conserved soil moisture crops: toria, bajra, barley, safflower, lentil, linseed.
5. Rainfed crops with supplemental irrigation: *kharif* rice, sugarcane, black gram, green gram.
6. Rainfed plus flooded crops: deep water rice, tidal wet land rice, jute (capsularis), sugarcane, dhaincha.

According to Method of Sowing or Planting

1. Direct seeded crops: upland rice, wheat, mustard, maize, jowar, bajra, groundnut, peas, beans, berseem, Dinanath grass, rice bean, stylo.
2. Planted crops: sugarcane, potato, sweet potato, napier, Guinea grass, para grass.
3. Transplanted crops (after raising seedlings in the nursery: transplanted rice, tobacco, onion, ragi, brinjal, cabbage, cauliflower, bajra.

According to Requirements of Intertillage Specially Earthing up

1. Intertilled crops: potato, sweet potato, groundnut, maize, napier, sugarcane, turmeric, arum, ginger.
2. Non-intertilled crops: jowar(fodder), Dinanath grass, para grass, oats, lentil, linseed, chickling pea

According to Length of Field Duration

1. Very short (up to 75 days) duration crops: green gram, black gram, toria, maize, radish, spinach, beet.
2. Short (75 to 100 days) duration crops: upland rice, potato (early), lady's finger, cauliflower, sunnhemp, sunflower.
3. Medium (100 to 125 days) duration crops: rape, wheat, jowar, bajra, barley, potato (main), groundnut, Bengal gram, soybean, sesame, cabbage, tomato, guar, tobacco (*rustica*), onion, *boro* rice, jute.
4. Long (120 to 150 days) duration crops: mustard, coriander, tobacco (*tabacum*), anise, cumin, *kharif* rice, mesta, roselle, cotton, rice bean, berseem.
5. Very long (150 days and above) duration crops: Sugarcane, sugar beet, castor, pigeon pea, napier, para grass, Guinea grass.

According to the Method of Harvesting

Different crops are harvested by different methods: by

1. Reaping: rice, wheat, mustard, suagercane.
2. Uprooting by pulling: toria, lentil, Bengal gram, khesari, black gram, radish, cabbage.
3. Uprooting by digging: potato, onion, garlic, ginger, sugar beet, sweet potato, colocasia, turmeric, groundnut, carrot.
4. Picking: cotton, brinjal, lady's finger, gourds, maize, cowpea, green gram, chillies.

5. Priming: tobacco.
6. Cutting: oats, napier, berseem, spinach, amaranth.
7. Grazing: para grass, pasture grasses.

According to Post-Harvest but Pre-Processing Requirements

Some crops require special care before they are used or sold, for instance:

1. Curing: tobacco, mustard, sesame.
2. Stripping: jute, mesta, roselle, sunnhemp, flax.
3. Shelling: groundnut, maize, sunflower.
4. Ginning: Cotton.
5. Seasoning: turmeric, chillies.
6. Grading and sorting: potato, rice, wheat, fibre crops. Tobacco.

CLASSIFICATION BASED ON LEAF MORPHOLOGY

Crop plants have two types of foliage leaves:

1. **Broad leaves:** mustard, peas, potato, tobacco, cotton, jute, berseem, cowpea.
2. **Narrow leaves:** rice, wheat, sugarcane, napier, para grass, onion, garlic, ginger, turmeric, asparagus.

ORIGIN OF THE CROP

Crop		Origin
Cereals		
1.	Rice	Indo-Burma (Indo-Myanmar), Jaypur (Orissa)-Secondary
2.	Wheat	South West Asia
3.	Maize	Mexico
4.	Jowar	Ethiopia and Sudan (Africa)
5.	Bajra	Africa
6.	Barley	Abyssinia (Ethiopia)
Pulses		
1.	Gram	South West Asia
2.	Field Pea	Mediterranean region
3.	Arhar	Africa (Nile river and Angola)
4.	Mung/Urd	India (Central Asia)
5.	Cowpea	Central Africa
6.	Soybean	China
Oilseeds		
1.	Groundnut	Brazil
2.	Sesame	South West Africa
3.	Castor	Ethopia
4.	Rai	China
5.	Rapeseed	Afganistan, Pakistan and India
6.	Sunflower	Mexico and South USA(Central America)
Other Crops		
1.	Cotton	India
2.	Corchorus olitorius	Africa
	Corchorus capsularis	Indo-Burma (Indo-Myanmar)

Crop		Origin
3.	Tropical Cane	Oceania (New Guinea)
	Indian Cane (Saccharum Spontaneum) (Kans)	North Eastern India
4.	Sugarbeet	Mediterranean region
5.	Potato	South Africa (Peru)
6.	Tobacco	Mexico and Central America

PROTEIN CONTENT IN CEREALS

Carbohydrate........Around 70% in all cereals
Rice........6-7%
Wheat........11-12%
Maize........10%
Sorghum........10-12%
Bajra........11-12%
Barley........11.5% (albuminoids)

PROTEIN CONTENT IN PULSES

Carbohydrate........Around 60%
Gram........21.1%
Lentil........25%
Pea........22.5%
Arhar........21-25%
Mung........25%
Urd........24%
Cowpea (Vegetable Meat)........23.4%

PROTEIN CONTENT OF OILSEEDS

Soybean........42%
Groundnut........26%
Sesame........18-20%
Linseed........36%
Safflower........40-45% (Oilcakes)

AREA AND PRODUCTION

1.	Rice	West Bengal (both area and production)
2.	Wheat	Punjab (both area and production)
3.	Jowar	Maharashtra (both area and production)
4.	Maize	U.P. (both area and production)
	Summer maize	Bihar
5.	Total Pulses	M.P. (both area and production)
6.	Total oilseeds	M.P. (both area and production)
7.	Gram	M.P. (both area and production)
8.	Arhar	Maharashtra (both area and production)
9.	Groundnut	Andhra Pradesh (both area and production)
10.	Rape and Mustard	Rajasthan (both area and production)
11.	Sugarcane	U.P (both area and production)

12.	Cotton -Area Production	Maharashtra, Punjab
13.	Tobacco	Gujarat (70% of total area and 80% of total production)

Irrigated wheat: Haryana (Maximum area)

Total Consumption of Fertilizers

1. Maximum : U.P. (N P K)
2. Total N + Total P: U.P.
3. Total K: Maharashtra
4. Total Fertilizer per hectare: (1) Pandicheri (2) Chandigarh
5. Food grains : (Cereal + Pulses) only.

Production of Cereals in Descending Order in India

(1) Rice (2) Wheat (3) Jowar (4) Maize (5) Bajra.

Area of Cereals in India

(1) Rice (2) Wheat (3) Jowar (4) Bajra (5) Maize.

Production of Cereals (World)

(1) Wheat (2) Maize (3) Rice (dehusked) (4) Jowar.

Oilseeds in India production wise

(1) Groundnut (2) Rape and Mustard (3) Soybean (4) Sesame (5) Sunflowr (6) Niger (Edible) (7) Castor (non-edible) (8) Linseed (9) Safflower (Maharashtra).

– ***Kharif*** **Crops:** Generally short day plant.

– ***Rabi*** **Crops:** Generally long day plant.

CHAPTER 6

Production Technology of Crop : *Cereals*

(Ecological Requirements, Cultural Practices and Crop Production Management)

RICE (*ORYZA SATIVA* L.)

Rice is the premier food crop of India. It belongs to family *gramineae*. According to D. Chatterjee (1948), there are altogether 23 species of genus *Oryza* of which 21 are wild and 2 (two) viz. *Oryza sativa* and *Oryza glaberrima* are cultivated. *Oryza sativa* is grown in all rice growing areas, but *Oryza glaberrima* is confined to the West Africa only. Thus it indicates that there might have been two centres of origin of our cultivated rice; South-Eastern Asia (India, Burma and Thailand) and West Africa. It is also identified that Jaypur of Orissa State is the secondary centre of origin.

Rice is the most widely cultivated crop in the world occupying an area of 150.8 m ha. Total rice output is 562.3 million tones and the average yield is 3.7 tonnes/ha (FAO, 1996). During the last 25 years, area under rice has increased only to a small extent from 135.4 m ha in 1970 to 149.2 m ha in 1995. However, rice production increased gradually from 307 m t to 550 m t. World demand for rice in 2025 is estimated to be about 765 million tones. Rice is the crop of tropical climate. However, it is grown successfully in humid regions of sub-tropics and temperate climate. Rice growing countries include those in Asia, West and North Africa, South and Central Latin America, Australia and the United States of America. More than 90% of the world's rice area is in Asia, which is the home for more than half of the world's poor and more than half of world's rice cultivators. Out of 2.4 billion rice consuming people in the world, more than 95% live in Asia, China and India are the two major countries producing rice in Asia. These two countries together have 49.1% of the world rice area (74.1 m ha) with the 55.2% of world production (310.1 m t). Though China has less area than India, its production is high. Other important rice growing Asian countries are Indonesia, Bangladesh, Thailand, Vietnam, Burma, Philippines, Japan, and Pakistan etc.

Distribution

Rice is distributed geographically over a wide range of conditions between attitudes 45°N and 40°S, embracing Central America, Southern half of the USA, the West Indies, most of South America, whole of Africa, Australia, India, China, Japan and South-East Asia. However, 90 per cent of the total area under rice is situated in wet tropical South and South-East Asia. Rice area, production and productivity level of some important rice growing countries of the World are presented in Table 6.1 and 6.2. Rice is also flourishes in temperate counties like Spain, Italy, France, Hungary and Portugal. In fact, the highest yields are recorded between latitudes 30° and 45° north and south of the equator.

TABLE 6.1: Rice production in different continents of the World, 1996

Countries/Continent	Area (m ha)	Production (m t)	Yield (kg/ha)
World	150.8	562.3	3730
Asia	135.1	513.7	3802
Africa	7.3	16.0	2208
N.C. America	1.7	9.9	5829
South America	5.9	18.3	3108
Europe	0.7	3.4	5171
Oceania	0.1	1.0	6585

Source: FAO year book- production, 1996.

TABLE 6.2: Rice production in the Asian countries, 1996

Country	Area ('000 ha)	Production ('000 t)	Productivity (kg/ha)
Bangladesh	10030	28008	2792
Burma	6475	20865	3222
China	31360	190100	6062
India	42700	120012	2811
Indonesia	11332	51165	4515
Japan	2100	13000	6191
Nepal	1497	3579	2391
Pakistan	2264	5551	2451
Philippines	3951	11284	2856
Republic of Korea	680	2800	4118
Sri Lanka	800	2241	2801
Thailand	9220	21800	2364
Vietnam	7300	26300	3603
World	150758	562259	3730

Source: FAO Yearbook- Production, 1996.

In India, the area under rice has increased from 30.8- m ha in 1950-51 to 42.9 m ha in 1995-96. There was also a steady increase in the production of rice. Production doubled from 20.6 m t in 1950-51 and to 40 m t in 1970. It has again doubled to 80 m t in 1995-96. The yield level also increased from 668 kg/ha in 1950 to 1855 kg/ha in 1996. However, comparing the world average (3730 kg/ha), there is enormous scope to further push the yield level. Recent studies have shown that the demand for rice would reach about 82 m t by 2000 A.D. and would touch 117 m t by 2020. Rice is cultivated in almost all states of India. However, the states having large area under rice are West Bengal, Uttar Pradesh, Madhya Pradesh, Bihar, Orissa, Andhra Pradesh, Assam, Tamil Nadu, Punjab, Maharashtra and Karnataka. These states contribute 92% of the area and production (Table 6.3).

India has the potential to become the number one rice producer in the world. India has reached a plateau in its irrigated rice yield, yet in the coming years it is expected to play a crucial role in the international rice production and technology. The focus is to be on tapping the production potential in the North Eastern region. New varieties will have to be bred to enhance the yield per hectare. Area under hybrid rice should be extended beyond the frontiers of South India, where it is been adapted more favourably than in the North.

TABLE 6.3: Rice production in different states in India, 1995-96

States	Area ('000 ha)	Production ('000 t)	Yield (kg/ha)
East			
Arunachal Pradesh	122.2	140.0	1146
Assam	2503.4	3309.0	1354
Bihar	5031.1	6910.9	1374
Manipur	134.9	338.1	2506
Meghalays	102.3	118.9	1162
Mizoram	65.7	101.5	1545
Nagaland	140.0	85.0	1321
Orissa	4529.0	6026.2	1375
Sikkim	15.9	21.9	1377
Tripura	231.5	465.5	2011
West Bangal	5953.4	11887.0	1997
North			
Delhi	0.5	2.9	1450
Haryana	837.0	1860.0	2222
Himachal Pradesh	82.6	111.2	1346
Jammu and Kashmir	273.0	508.0	1863
Punjab	2161.0	6768.0	3132
Uttar Pradesh	5575.7	10408.1	1867
South			
Andaaman & Nicobar Island	12.6	33.0	2619
Andhra Pradesh	3678.8	9194.8	2499
Karnataka	1268.6	3018.7	2380
Keral	472.0	931.7	1974
Pondicherry	26.0	67.1	2523
Tamil Nadu	2228.5	7562.6	3394
West			
Dadra & Nagar Haveli	13.5	24.1	1785
Daman and Diu	2.0	1.8	3600
Goa	53.5	128.1	2394
Gujarat	570.0	826.6	1450
Madhya Pradesh	5168.2	5705.1	1104
Maharashtra	1517.1	2562.8	1689
Rajasthan	139.5	117.6	834
All India	**42910.0**	**79618.1**	**1855**

Source: Fertilizer statistics, 1996-97.

Economic Importance

Rough rice (Paddy) consists of about 20% hull and 80% grain (hulled or brown rice). There is a loss about 10% of brown rice while milling as bran and germ. Some parts of the endosperm is also removed at polishing. Rice has a high calorific and nutritional value and rich in easily digestible starch and carbohydrates. Whole rice contains high amount of vitamin B (thiamin, riboflavin and niacin). Most of the vitamins are removed while polishing in the modern rice mills. The grains contain appreciable amount of K, P Mg and I.

All parts of the rice are used economically. The husk is used as fuel and rice grains are used for making starch and vinegar. In many rice growing Asian countries, alcohol beverages are also brewed from rice.

Situation in India

In India, rice is grown to varying extents in practically all the states. Bihar and West Bengal lead in the area under rice while West Bengal and Tamil Nadu have the highest rice production. The yield per hectare is the highest in Punjab.

The rice areas in the country can be broadly grouped into the following five regions according to the rice cropping season and other agricultural conditions:

1. The North-Eastern region comprising Assam, West Bengal, South Bihar and Orissa. Rice is grown in the basins of the Brahmaputra, Ganga and Mahanadi rivers and the region has the highest intensity of rice cultivation in the country. This region enjoys heavy rainfall and rice is grown mainly under rainfed conditions.
2. The Southern region comprises deltaic tracts of the Godavari, Krishna, Cauveri and Tambrabarni rivers and the non-deltaic rainfed areas of Tamil Nadu and Andhra Pradesh. Rice is grown under irrigated conditions in the deltaic tracts.
3. The Central region comprises Madhya Pradesh, the Telengana area of Andhra Pradesh and parts of Karnataka. Except in Andhra Pradesh, rice is sown broadcast as a rainfed crop in this region. One crop of rice per year is the normal practice in this tract.
4. The West Coast Region comprises Keral and the coastal districts of Maharashtra and Karnataka. Rice is grown largely under rainfed conditions during the period May-June to November-December. There is heavy rainfall during the monsoon period in this region. In Keral, two to three crops of rice are grown in a year.
5. The North Region comprising Jammu and Kashmir, Punjab, Haryana, Himachal Pradesh, Uttar Pradesh and North Bihar, has low winter temperatures and only a single crop of rice is raised from May-June to September-October.

Rice as crop shows very wide adaptability to different soil and climatic conditions and can, therefore, be cultivated over a vast area in the tropical and subtropical regions and yet, like other crops, it too has its own soil and climatic requirements under which it performs best.

Climate

1. Rice is grown under a damp warm climate. It is the best suited to the regions which have high temperatures, high humidity, prolonged sunshine and an assured supply of water. A temperature range of 20 to 37.7°C (68 to 100°F) is required for optimum growth of rice. At the time of tillering, the crop requires a higher temperature than for growth but the temperature requirement for blooming is practically constant for all the varieties in the range of 26.5° to 29.5°C. Humidity requirements, however, vary according to the variety.
2. Rice, however, provides its own micro-climate, raising the relative humidity

within the crop area much above that of adjacent unirrigated areas.

3. Photo-periodically, rice is a short day plant but there are varieties which are non-sensitive to photo-period.
4. In India, rice is grown under widely varying conditions of rainfall, altitude and climate. In parts of eastern region and Peninsular India, the mean temperature throughout the year are favourable for rice cultivation and two or even three crops of rice are taken in a year. In the northern and western parts of the country, where winter temperatures are fairly low and the rainfall is not high, only one crop of rice is taken during the months of May to November.

Season

There are three main seasons for growing rice and these are named according to the seasons of harvest of the crop. The most important is the *kharif* (winter rice) sown in June-July and harvested in November-December. A little over 51 per cent of India's rice crop is grown in this season and it usually consists of medium and long duration varieties. The autumn rice is grown in May-June and harvested in September-October. About 44 per cent of the rice crop is grown in this season. This comprises mostly varieties of early and medium duration ranging from 90 to 110 days. The summer rice is sown in December-January and harvested April-May. The area under summer rice is limited (4.4 per cent) and early maturing varieties are mostly grown in this season.

These three (3) types of season for rice cultivation in India are otherwise called:

1. **Aus/Autumn:** Word 'aus' derived from Sanskrit 'ashu' meaning quick or early. Rice varieties grown during pre-monsoon period is called aus rice and harvested in August-September hence called 'Autumn Rice' due to harvest time.
2. **Aman/Kharif/Winter:** 'Aman' derived from Arabic words means safety which indicates stability of crop (Aman-chain). Aman rice is also called winter rice due to harvesting time in winter.
3. **Boro/Summer:** Boro refers to rice grown in submerged land lower in elevation, grown dueing January-February to April-May, also called summer crop because Boro rice is harvested in summer season.

Soil

As with climate, rice is adaptable to all kinds of soil and practically all soil types are found in the world's rice growing areas. The suitability for rice cultivation depends more on the conditions under which the plants are grown than upon the nature of the soil itself. The semi-aquatic nature of the crop necessitates a heavy soil through which the irrigation or rain water will not be easily drained away, because for rice the requirement of water is more crucial than soil conditions. Provided that sufficient water is available either through assured rainfall or through irrigation, rice can be grown on a wide range of soils from sandy to heavy soils.

Rice is able to tolerate a wide range of soil reaction. It grows well in soils having a pH range between 5.5 and 8.0.

Rice, being a semi-aquatic crop, grows the best under submerged or water-logged conditions. A major part of rice crop in India is grown under 'low land' conditions, that is, under 'flooded' or 'submerged' conditions. To retain water in rice fields, the fields are enclosed by bunding or with dikes. Five or more centimeters of water is retained in the field throughout the greater part of the rice growing season. The origin of water may be impounded rainfall or irrigation.

On level areas without bunds or dikes or areas that are not terraced, rice may be grown during the rainy season without impounding the water in the fiels. Rice grown under such conditions is known as 'upland' rice.

Soils under 'lowland' or 'submerged' or 'water-logged' conditions develop fundamental characteristics different from those of soils under 'upland' conditions. Water-logging causes physical, biological and chemical changes in the properties of the soil because of the physical reactions between the soil and water and also because of the biological and chemical processes set in motion as a result of the excess water. These changes have a profound bearing on nutritional and fertilization aspects of rice cultivation.

Physical Changes

Upon flooding, the pore spaces in soil become saturated with water. The soil swells and hard clods soften and break down into small aggregates. The structure of the soil, which is considered of great importance for air-water movement and root penetration in the soil under **upland** conditions and in well drained soils, is destroyed by a minimum manipulation due to the weakening effect of water on the bonds holding the soil particles together as stable aggregates.

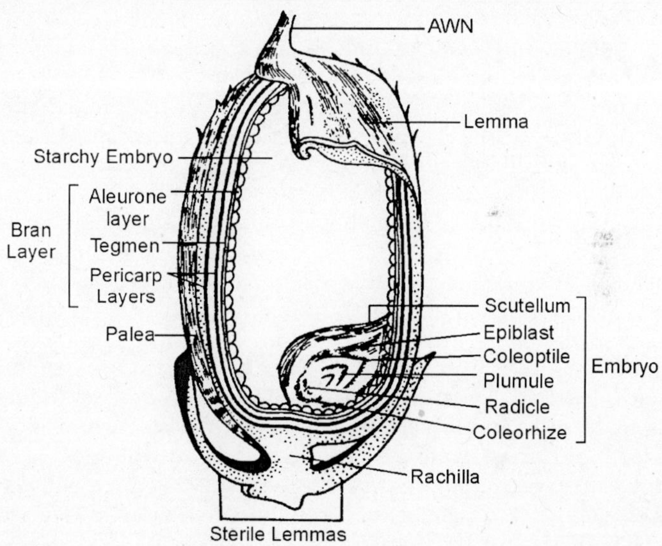

FIG. 6.1: Differentiation of a water-logged soil into a surface oxidized layer and an underlying reduced layer.

Puddling, *i.e.* working of soil under water-logged or submerged conditions, which is considered as a necessary operation in land preparation for **lowland** rice, renders the soil less pervious, drastically reduces downward movement of water due to percolation and completely destroys the structural aggregates in the soil. The diffusion of air (exchange of air) between the atmosphere and the puddle soils is impeded.

The most important change in the soil as a result of water-logging is the conversion of the root zone of the soil from an aerobic environment to as anaerobic or near aerobic environment where oxygen is limiting. The flooded or water-logged puddled soil develops two zones (Fig 6.1). The upper zone (oxidized zone), a thin 1 to 10 mm thick zone, absorbs oxygen from the water, turns brown in colour and reacts to nitrogen like an unflooded upland soil. The lower zone, the remaining lower portion of the puddle soil, turns to a dark or blue grey colour (these colour changes vary with the chemical composition and organic matter content of the soil). The colour change takes place as iron compounds in the flooded soils lose their oxygen. This soil zone is said to be in a reduced state.

The effects of these two zones in flooded soil have to be considered for efficient nitrogen fertilization of **lowland** rice.

Biological Changes

The absence of soil air (oxygen) in flooded and puddle soil results in drastic changes in the kinds of micro-organisms that decompose organic matter and release plant nutrients. The rate of decomposition of organic matter is considerably reduced in water-logged soils. The products of organic matter decomposition under submerged conditions differ from those formed under aerated upland conditions. When large amounts of organic matter are added or soils are rich in organic matter, some temporary toxic end-products of decomposition such as different organic acids are formed under water-logged conditions. Research has shown that to save the seedlings from toxic effects of these end-products, 2 to 3 weeks time should be allowed to elapse after rice fields are flooded before rice seedlings are transplanted.

Other biological changes of importance due to flooding relate to the fate of native or applied nitrogen. The nitrogen in the organic matter (incorporated plant residues or native soil organic matter) which is present in organic form (*e.g.* proteins, amino acids, etc.) is changed to ammonium form (NH_4^+) during decomposition. The ammoniacal form of nitrogen is stable under flooded conditions (reduced conditions). It is absorbed on the soil particles and thus retained against leaching losses and subsequently used by the growing rice plants. If the soil is not flooded immediately after ploughing, a second microbiological change takes place during which the ammoniacal form of nitrogen is oxidized to nitrate form (NO_3^-) as in **upland soils.** The nitrate form of nitrogen is not held by the soil particles and remains very mobile. If the soil is flooded, the nitrate nitrogen is leached to the lower layers (reducing zone) where it gets reduced to elemental form of nitrogen and gets lost into the atmosphere (Fig.6.2). Only a part of the nitrate nitrogen may be used by rice plants through root system present at the surface. The loss of nitrogen through conversion of nitrate to elemental nitrogen is called **denitrification.** Since nitrate denitrifies easily under submerged or flooded conditions, nitrogenous fertilizers containing nitrate form of nitrogen are less efficient for **lowland** rice.

If ammoniacal nitrogen fertilizers are applied at the surface and not incorporated into the soil, they are nitrified to nitrate form which leaches down and gets denitrified to gaseous nitrogen under flooded conditions. This loss of nitrogen can be prevented by incorporating ammoniacal nitrogen into the lower zone (reduced zone) of the soil where it is held.

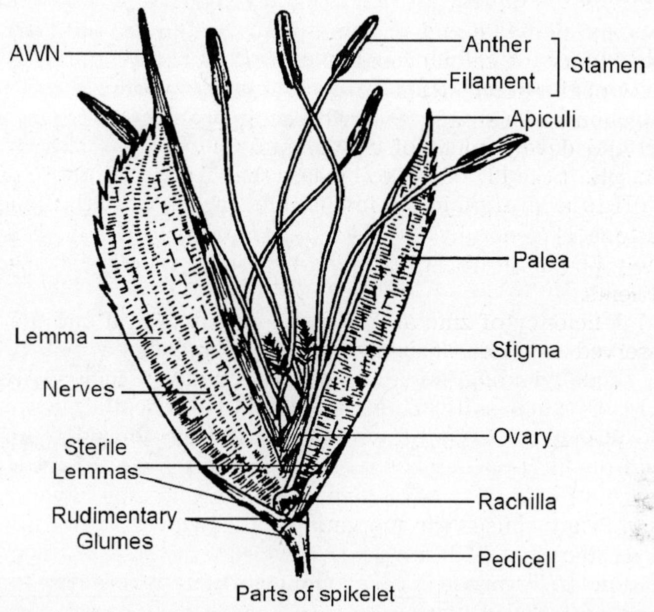

FIG. 6.2: Fate of nitrogen in flooded soils.(Source: Villegas, L. M. and Feuer, R., "The lowland or flooded rice" in "Rice Production Manual", Univ. Philippines, 1970).

Chemical Changes

Among the important chemical changes that take place when a soil is kept flooded are: (1) increase in the solubility of phosphorus and (2) reduction of sulphate form of sulphur to injurious sulphide form. Flooding the soil increases the availability of native soil phosphorus. The increased availability of phosphorus takes place because under flooded conditions iron, aluminium, manganese and calcium become more soluble and chemically release the phosphorus which under aerated **upland** conditions is chemically bound with these elements. The rise in pH of acid soils and lowering of pH of alkaline soils upon flooding likewise increase phosphorus availability.

There are a few cases of high sulphate soils and/coarse textured soils which release toxic amounts of hydrogen sulphide(H_2S) upon flooding. Hydrogen sulphide is toxic to roots or rice plants. Some cases of sulphide toxicity have been reported from Andhra Pradesh in India and widely from Japan. Hydrogen sulphide inhibits the iorn and copper containing enzymes, retards water and nutrient uptake and at times proves lethal. When such toxicity is observed through the characteristics smell of the hydrogen sulphide (H_2S) gas, materials containing available iron should be applied. In such soils, fertilizers which do not contain sulphur such as urea, ammonium chloride, diammonium phosphate, etc. should be used and also application of organic matter should be avoided. By excluding any of the factors which promote reduction such as presence of readily decomposable organic matter, low iron content, flooding. etc., the risk of reduction of sulphur to hydrogen sulphide can be reduced.

In India, rice is grown in almost all types of soils ranging from the coarse laterite soils of east and south India to black clay, clay loam and heavy clay soils of west and central India. In the north, it is mainly grown in alluvial soils most of which are loams and clay loams. It grows equally well in the red loams and red earths of south and central India. Low lying clay soils are preferred for long duration varieties and loams and lighter soils for quick maturing types.

Though almost all types of soils in the country grow rice, successful cultivation depends on soil and water management, adequate availability of water and more particularly on sound fertilizer practices.

Rice Genotypes

Vavillov considered that India and Burma would be the centres of origin of cultivated rice. Rice belongs to the family ***Gramineae*** and the genus ***Oryza***. Two cultivated species of rice are *Oryza sativa* (L) and *Oryza glaberrima* (steud). *Oryza sativa* is a diploid species having 24 chromosomes *e.g.* **2n – 24.** *Oryza sativa* is widely grown in Asia, parts of Europe and America. Cultivation of *Oryza glaberrima* is restricted to Africa. The cultivated species of *Oryza sativa* are further divided into three varietal types/sub-species *i.e.* ***indica***, ***japonica*** and ***javanica***, based on the morphological and physiological characteristics. These sub-species also differ in their geographical distribution.

1. **Indica Rice:** It is tropical rice, grown in India, awnless or short awn, late in maturity, long stem.
2. **Japonica Rice:** Temperate or sub-tropical rice, grown in Japan, early maturity, photosynthetically very active, short stem, very responsive to nitrogenous fertilizer. No lodging.
3. **Javanica Rice:** Wild form of rice, grown in Indonesis.

India rices are indigenous to the humid regions of tropics and sub-tropics of Asia. The japonica types are limited to temperate zones and sub-tropics. Javanicas are mainly grown in parts of Indonesia.

Varietal Improvement

Rice Breeding programme in India was started as early as 1911. The Imperial Council of Agricultural Research established the Central Rice Research (CRRI) at Cuttack in 1946 primarily for developing high yielding rice varieties and technologies for different parts of the country (Ramaiah, 1953). In 1965, the All India Coordinated Rice Improvement Project (AICRIP) was established to accelerate and co-ordinate rice research at national level for rapid generation and identification of varieties and management techniques appropriate to different agro-ecological situations in partnership with State Agricultural Universities and departments of Agriculture and Central Agricultural Institutes. A major break-through was made in rice breeding with the release of IR-8 in 1966 from the introduced material of the International Rice Research Institute (IRRI). This variety bestowed practical expression to the concept of ideotypes. A number of such high yielding varieties have been released subsequently through selection and hybridization. Varieties are released for different purposes such as earliness, deep water and flood resistance, lodging resistance, drought resistance, non-shedding of grains, disease resistance and high response to heavy manuring. Varieties are also available with good quality and aromatic characters. Hybrids have also been released by for higher productivity.

Early Maturing Varieties

Early maturing rice varieties play greater role in intensive cropping systems.

Farmers could fit them in varied cropping systems and increase productivity. Early maturing high yielding dwarf cultivars of rice replaced large area under millets and oil seeds in 1970's. Some of the quick maturing varieties are Pusa 2-21, Pusa 33, Saket-4 and Annapurna. These varieties are capable of yielding of yielding as high as 90% of medium duration varieties under irrigated conditions.

Resistance to Pests and Diseases

In rice about 25-30% yield loss occurs due to pests and diseases incidence. The problem became serious after the introduction of high yielding semi-dwarf varieties. These varieties provide a favourable host environment due to their genetic homogeneity. Large areas are planted with a few high yielding varieties. Changes in the cropping pattern, crop intensification and high input management also added to the problem. Of late, varieties have been released for specific or multiple resistance tp biotic and abiotic stresses. The impact of host-plant resistance is visible through reduced crop losses and less frequent reoccurrence of pest epidermics. Their resistance could also be assessed with stable yield in pest endemic areas.

Insect Resistance

Varieties resistant to some predominant insect pest:

1. Gall midge: Phalguna, Surekha, Pothana, Abhaya, Suraksha, Triguna.
2. Brown Plant Hopper (BPH): Krishnaveni, Chaitanya, Jyothi, Sonasali, Kanakam
3. Stem Borer: TKM 6, Sasyasri, Vikas, Ratna

Disease Resistance

Important diseases in rice are bacterial leaf blight (BLB), sheath blight, rice tungro virus (RTV). The estimated yield loss is about 10%. Varieties resistant to the diseases are:

1. BLB: Ajaya, PR 110, PR 111, Panidhan 12
2. RTV: Vikramarya, Nidhi, Bharani, Vasundhara
3. Blast: CO 25, CO 29, CO 30, CO 38, MTU 17:
4. Helminthosporium: Ch 13, Ch 45, IET 1328

Special Situations

Drought Tolerance

Tolerance to drought is generally associated with

1. High germination in polyethylene glycol,
2. Less starch disintegration in root and low destruction of chrophyll,
3. Tillering in quick succession before the onset of drought.
4. Longer root length (> 25 cm),
5. High leaf moisture content and greater accumulation of reducing sugars.
6. Varieties: CO 31, MTU 17, Ch 45, Bala, Kiran, CRM 13-3241.

Tolerance to Water Logging

Tolerance to water logged situations is associated with (i) fast, early tillering, (ii) liberation of more oxygen from the root, (iii) moderate pectin-methyl esterase activity in the culm and (iv) greater chlorophyll b and higher photosynthetic rate of the top leaves above water level.

Varieties: CR 1014, Mashsuri, Pankaj, Jagannath, Gauthami, Visista.

Deep Water Condition

The standing water in the field for a significant period of time is more than 50 cm. The Huntra area of Thailand, water usually remains on the fields for four months or longer. Deep water areas are found in deltas, estuaries and river valleys of India, Bangladesh, Burma, Thailand, Vietnam, Kampuchea and Indonesia. In India, there is a large deep water rice area in West Bengal, Assam, Orissa and Kerala. Drought of the early phase, weed competition, poor stand establishment and submergence of rice at various growth stages are some of the problems encounted in deep water rice. Varieties should have good seedlings vigour, ability to elongate if the water level increases, intermediate plant height (about 100 cm), submergence tolerance and with kneeing ability. The duration of the varieties varies from 150 to 240 days.

Varieties: Jaladhi 1, Jaladhi 2, PAL-2, ARC-353-146, Jaisuria, Jalamagna, Ptb 15, Ptb 16, TNR-1, TNR-2.

Saline and Alkaline Tolerance

CST 7-1, Lunishree, CSR 10, Panvel 1, Panvel 2, Vytilla 2, Vytilla 3, Vytilla 4 and CSR 5 tolerate saline conditions. For alkali areas, Vikas, CSR 5, TRY 1 are suitable.

Scented Rice Varieties

Scented rice varieties are known for their characteristic fragrance when cooked. There are several scented varieties of rice which emit aroma and form a group by themselves. Genetic studies showed that single dominant gene controls fragrance of flowering (Jodon, 1944). The aroma inheritance reportedly depends on one to three complementary factors. However, little is known about the influence of environment of the intensity of aroma. Studies on flavour component in scented rice varieties showed their higher content of carboryl component than ordinary varieties.

Basmati rices are an elite group among rice varieties characterized by aroma and specific cooking and eating quality. They fetch the highest premium since they are priced about three times than that of high quality non-Basmati type. The traditional scented varieties are Krishnabhog, Prasadbhog, Bhadshabhog, Rajbhog, etc. All these traditional scented are tall, most of them are weak strawed, lodge heavily and are poor yielders. The most popular scented varieties in the North India are Basmati-370 and Basmati-217. Dwarf aromatic varieties (Pusa Basmati 1, Kasturi) are dwarf, high yielding and comparable with highly prized Basmati varieties like **Basmati-370.**

In Punjab, Hartana and Western U.P., Basmati is cultivated in about 7-8 lakh ha. About 60% of the Basmati rice produced are exported. Major markets for this scented rice are the middle East and the Gulf (Saudi Arabia, U.A.E. and Kuwait). Scented rice varieties are normally grown in smaller areas due to low yield (1.5 to 2.0 t/ha).

Timely harvest is important for obtaining high milling percentage. Harvest is to be taken up at 25 days from general flowering when the last few grains at the base of the panicle are still green but in their hard dough stage. Further, drying in shade will increase the head rice recovery by avoiding sun-cracks which will increase the percentage of broken rice.

Hybrid Rice

Hybrid rice is a land mark technology development in China. Professor Yuvan

Long Ping, the father of hybrid rice revolution in China, developed the first hybrid rice. The work on hybrid rice was started by him with the idea of exploiting heterosis in rice. Since 1976, hybrid rice is being cultivated on a commercial scale in China. Hybrid rice cultivation has helped China to increase rice production by nearly 200 million tones frm 1976 to 1991. Hybrid rice technology has enable has enabled China to improve rice yields by 15-20% over the commercial rice varieties. China's success with hybrid rice technology has generated great interest in the other Asian countries.

International Rice Research Institute, Philippines, started work on hybrid rice in 1980's. The other countries such as India, Indonesia, Korea, Thailand and Vietnam have also joined in this pursuit. Having recognized heterosis breeding of rice as one of the most feasible methods is to increase the productivity and production of rice in the country in 1989. Government of India initiated a goal-oriented, time-bound project entitled, 'The Development and use of Hybrid Rice Technology'. This ICAR project has been further strengthened by financial support from the United Nations Development Programme (UNDP) and by technical support from FAO since September 1991. As a result of extensive evaluation, hybrids have been released in the states of Andhra Pradesh, Tamil Nadu, Karnataka and West Bengal (Table 6.4).

TABLE 6.4: Hybrids released in India and their characteristics

Hybrids	Duration (days)	Yield (t/ha)	Adaptability
APHR 1	130-135	7.10	Telengana and Rayalseema region of A.P.
APHR 2	129-125	7.5	Telengana and Rayalseema region of A.P.
DRRH 1	125-130	8.0	AP, TN. Karnataka
CORH 1	110-115	6.0	*Kuruvai* (June-Sept.) in Tamil Nadu
CORH 2	120-125	7.0	Late *Thaladi*(OCT-Dec) in Tamil Nadu.
ADTRH 1	115-120	7.1	Cauvery delta zone of Tamil Nadu
KRH 1	120-125	6.0	Irrigated areas of Karnataka
KRH 2	130-135	6.0	Irrigated areas of Karnataka
KRH 4	130-135	8.2	Irrigated areas of Karnataka
CNHR 3	125-130	8.1	*Boro* season in West Bengal

In the last decade 0f 1970, hybrid rice was first developed by China by using cytoplasmic male sterility-Fertility restorer gene system. After China, in India, in 1994, first set of five hybrid rice varieties have been released for commercial cultivation. Out of these five hybrids, four varieties APRH 1, APRH 2(Andhra), KRH 1(Karnataka), and MGR 1 (Tamil Nadu) have been released by A.P., Karnataka and Tamil Nadu. After two years, another three hybrids – CNRH 1, KRH 3 and DRRH 1 have been released by West Bengal and Andhra Pradesh respectively. PHB 71 is the only hybrid rice variety which is released by private organization.

The hybrids released are capable of yielding 1-1.5 tonnes more yield per hectare than the best variety. During 1996, about 1300 tonnes of hybrid rice seeds were produced and hybrid rice was cultivated on an area of about 50,000 ha. It is estimated that the coverage will increase to two million hectares by 2000 A.D.

The latest hybrids have both high yielding property and better grain quality. They have the yield advantage of 20-30% over the best high yielding varieties of comparable maturing period. They can withstand adverse growing conditions better than the conventional semi-dwarf varieties. Most of the rice hybrids developed in and outside China are based on cytoplasmic male sterility (CMS) and fertility restoration system.

The basic problem encounted with hybrid rice production is developing male sterile lines adapted to tropical countries. A CMS line must possess stable pollen sterility, adaptability to target environment, good crossing potential and good combining ability. Seed production is the major constraint in commercializing rice hybrids.

Introduction and Release of High Yielding Varieties of Rice

- **Taichung Native 1 (TN-1)** was developed after IInd world war in Taiwan,

$$\text{TN-1} = \frac{\text{Dee geo woo gen}}{\text{(dwarf \& N responsive)}} \times \frac{\text{Tasai Yung Chung}}{\text{(Tall drought resistance)}}$$

  Dr. T. T. Chang brought TN-1, Dee-geo-woo-gen and I-geo-tze from Taiwan to IRRI, Manila.
- **IR-8 was** envolved by breeder Henery M. Beachell at IRRI, Manila.
  One of the parents of IR-8 was Dee-geo-woo-gen.
  IR-8 - Dee-geo-woo-gen X Peta (from Indonesia)
- First time in India, TN-1 (2 kg in 1964-65) was introduced by G. V. Chalam (G.M. of NSC) from IRRI.
- IR-8 was introduced in India in 1966 and IR-8 outyielded TN-1.
- Literally meaning of Dee-geo-woo-gen: Brown tipped short legged.
- Jaya = TN-1 (from Taiwan) x T 141 (Indian Variety) evolved by Dr. Shastry, first rice variety is developed under India's rice programme. It is outyielded TN-1, IR-8 both, hence called 'Miracle Rice' in India.
- Padma is the reverse cross product of the parent of Jaya. Padma – T 141 x TN-1.
- Jagannath: Mutant variety of T 141.
- CR-1014 = T 90 (Indica) x Urang Urangan (Javonica)
  Released in 1988, is popular in Orissa, Andhra Pradesh and West Bengal. It is super fine grain variety capable of yielding 3-4 tonnes/ha under submerged condition in semi-deep water regime.
- **Pusa Basmati-1:** World's first high yielding dwarf variety under quality rice has been developed by IARI through convergent breeding.
- Deep Water Rice Varieties: Chakia-59, Madhukar, Jalamagan, Jaisuria, Jaladhi-1, Jaladhi-2, Panidhan-II, Jalapriya.
- Aromatic (Scented) Rice: Sabaramati, Basmati-370, Kamal Local, Dehradun Basmati (T3), Pusa Basmati-1, Kasturi, TaraoriBasmati, Basmati-385, Haryana Basumati-1
- Non-Basmati Suprfine rice: IR-64, PR 106, Gaurav and Punjab No. 1
- Max. Rice exporter: 1. Thailand (4 mt) (1990-91) 2, US (2.4 m t).

India at the 5th position Half of the world market is now captured by Thailand which is managing rice on the principles of the corporate sector.

As a result of the ICAR Japonica-Indica Hybridization Projrct, the variety ADT 27 was developed in 1964 and became the ruling variety for *Kuruvai* season in Tamil Nadu. In 1966-67, a number of exotic high yielding varieties of rice like Taichung Native 1, Tainan-3, IR-8, Taichung 65 were introduced on a large scale in

India. Simultaneously, the breeding programme to develop indegeneous high yielding varieties suitable for specific agro-climatic zones of the country was taken up in the All India Coordinated Rice Improvent Project under the aegis of the Indian Council of Agricultural Research. As a result of this coordinated effort, two varieties Jaya and Padma were released for general cultivation in 1968. Subsequently, more varieties have been released. Some State Governments have also released high yielding varieties evolved in their States to meet the requirements of the specific agro-climatic and soil zones.

The special features of these high yielding varieties as compared to the indigenous tall improved varieties are:

1. They are dwarf and posses a stiff straw which reduces lodging.
2. Some of these varieties are of short duration and early maturing, thus making it possible to adopt in multiple cropping.
3. The dwarf plant type rice varieties are generally just as suited for rainfed direct seeded conditions (**upland rice**) as for irrigated transplanted conditions (**lowland rice**). Rainfed rice is subjected to moisture stress at one time or the other during the growing period. Since drought is a non –specific stress, high yield potential in the variety chosen is even more relevant under rainfed conditions. In the choice of variety, particularly under rainfed upland conditions, the main consideration should be duration. Duration of these varieties is reduced 7 to 10 days under direct seeded upland conditions.
4. While a large number of varieties has been released since 1965 both by the Central Varietal Release Committee as well as the State Governments, with the exception of Masuri, IR-20, Jaya, IR-8, RP 4-14, Ratna, ADT 31, Sita, Pankaj, Jagannath, Rasi, Tellahamsa, Vani, Phalguna and a few others, the coverage under the rest is limited. With the development and release of such varieties, it is hoped that the coverage under HYV will go up significantly.

Concept of New Plant Types in Rice

1. Early maturity (110-125 days).
2. Insensivity to photoperiod to give flexibility in planting date.
3. Moderate tillering to minimize mutual shading and over elongation of internodes.
4. Short, study culm to minimize lodging.
5. Small, thick, erect, dark green leaves to maximize light utilization.
6. Resistant to the prevailing races of blast.
7. Seed dormancy to avoid germination in the panicle.
8. Moderately firm thresh ability to reduce shattering losses.
9. Glabrous leaves and husks to facilitate mechanical harvesting and processing.
10. High milling yield, and
11. Acceptable shape, size and cooking quality of grains.

High Yielding Varieties Released for Cultivation

Varieties Released by the Central Varietal Release Committee.

IR 8, Jaya, Padma, Pankaj, Jagannath, Bala, Cauvery, Jamuna, Kanchi, Krishna, Sabarmathi, Ratna, IR 20, Vijaya, Jayanthi, Sona, Kannagi (Pusa 2-21), Vani, Akashi, Rasi,

Varieties Released by States and Agricultural Universities

Andhra Pradesh

Hamsa, Tella Hamsa, Kakatiya, Vasistha, Gautamai, Surekha, Prabhat, Rajendra, Phalguna, NLR 9672, NLR 9674,

Bihar

Deepa, Archana, Sita, Panidhan 1, Panidhan 2, Prahalad, Narasingh, Vishnu, Kanchan, Kiran, Rajendradhan 201, Rajendradhan 202, 64-117.

Gujarat

Gaur 1, Gaur 10, Gaur 100, Gaur 2, Gaur 11, Gaur 3.

Himachal Pradesh

IR 579, Himdhan.

Jammu and Kashmir

K 84, Giza 14, Tawi (PC 19), K 78-13.

Karnataka

Suma, Kusuma, Manila paddy (Introduction from Philippines), Madhu, Vikram, Mangala, Pragati, Prakash, Pushpa, Intan, Red Annapoorna.

Kerala

Annapoorna, Triveni, Rohini, Aswathi, Bharathi, Jyothi, Sabari, Suvarani Modan (ARC 11775).

Madhya Pradesh

Anupama, Garima, Kranti, Jagriti, Pragati.

Maharashtra

Ratnagiri 24, Palgarh 60, Kanjat 184, Tujapur 1, Satya, Surya, Suhasini 6, Radhnagari 185-2, Karjat 14-7, Ratnagiri 8-1-1, Improved Ambemohar, Ratnagiri 711-4-1, Jalgaon 5.

Meghalaya

Ngoba, Konoruln.

Orissa

Shakti, Kalina 1, Kaling 2, Supriya, Kumar, Rajeshwari, Parijat, Suphala, Hema, Pathara, Khandagiri, Lalat, Kanark, Kharavel, Surendra, Gajapati, Mahanadi, Kanchan, Indravati, Ramchandi, Prachi.

Punjab

Hybrid Mutant 95, Palman 579, PR 103, PR 106.

Rajasthan

Chambal.

Tamil Nadu

Adt 28, TKM 9, Adt 22, PVR 1, Pennai (ASD 14), Karuna, Ponni(Mahsuri), Karikalam, Co 36, Adt 30, Vaigai, Adt 31, Bhagawathi, Amaravathi, Pusa 33, Bhavani, Pusa 4-1-11, AU 1 (Annamalai Uvar Nel), CR 146-7027-224, CR 149-3244-198, C 7306, Raja Rajan (Co 40), Adt 32, Co 41, Co 42, IET 4786, ADT 33, ADT 34, ADT 35, ADT 15, IR 34, ADT 36, MDU 1.

Uttar Pradesh

Sudha, Majhera, Madhukar, Jalamagna, Saket 4, Sarjoo 49, Sarjoo 50, Prasad.

Growth and Development Phases in Rice

The life cycle of rice may be physiologically divided into three groups viz. Vegetative, reproductive and ripening phases.

Vegetative Phase

1. **Seedling Stage:** The seed sprouts into young seedling s having seminal and lateral roots and green leaves. These seedlings depend upon carbohydrate reserve of the grain upto 15th day or until first two leaves have come out. The phase ends after fifth leaf has developed and during this period the seedlings exhaust almost entire endosperm.
2. **Transplanting Stage:** This stage covers the period from uprooting, transplanting of seedlings till full recovery following transplanting. The secondary adventitious roots develop within 4-10 days after transplanting and they start absorbing nutrients from soil and sustain the growth. The root injury due to uprooting of seedlings can be reduced if they are uprooted carefully in presence of water. In case of direct seeded rice there is no recovery phase and the vegetative growth phase is correspondingly reduced by 4-10 days.
3. **Tillering Stage:** The period is of 10 days after seedling establishment to maximum tillering stage which comes between 35-42 days after transplanting. Usually, plant stops producing tillers after tertiary tillers have been produced. The tiller number declines after the plants have reached the maximum tillering stage.
4. **Vegetative Lag Phase:** This is the period between maximum tillering and panicle initiation in which some of the tillers die because of tiller mortality. This phase is absent in early maturing varieties but it occurs in long duration ones. Lag is a sort of rest period. The vegetative lag phase is also known as (P.S.P.). This is the end of the basic vegetative phase (bvp).

Reproductive Phase

1. **Panicle initiation stage:** This is the beginning of the productive phase which succeeds the maximum tillering stage in long duration varieties but in early maturing varieties it may proceed or even overlap the maximum tillering stage. This stage occurs approximately from 21 to 25 days before heading but flowering primordium can be recognized only a week after initiation stage has occurred.
2. **Internode elongation and booting stage:** After panicle initiation stage the internal auxin concentration in the plans accelerated the elongation of the internodes. This elongation takes place soon after panicle initiation stage. In early maturing varieties the elongation of upper internodes may begin much earlier than the panicle initiation stage. This shows the beginning of the booting stage when the developing panicle causes bulging in the leaf sheath.
3. **Heading stage:** The emergence of panicle tip or heading takes place after the booting stage is over. This comes about 15 days after the booting in all varieties.
4. **Flowering stage:** Flowering occurs about 20-25 days after booting. It continues successively until all the spikelets in the panicle bloom. This is

followed by pollination and fertilization.

Ripening Phase

The rice ripening is completed between 25 to 35 days after the crop has passed through the following stages:

1. **Milky Stage:** Between 7 to 12 days after anthesis the watery content of the grain turn milky in consistency of the stage is said to be milky stage.
2. **Dough Stage:** The milky grains turn into soft dough stage and later turn into hard dough stage. The period required from milky to dough stage is about 2 to 3 weeks.
3. **Maturity Stage:** This is the stage at which the grains turn hard, clear and free from greenish tint.

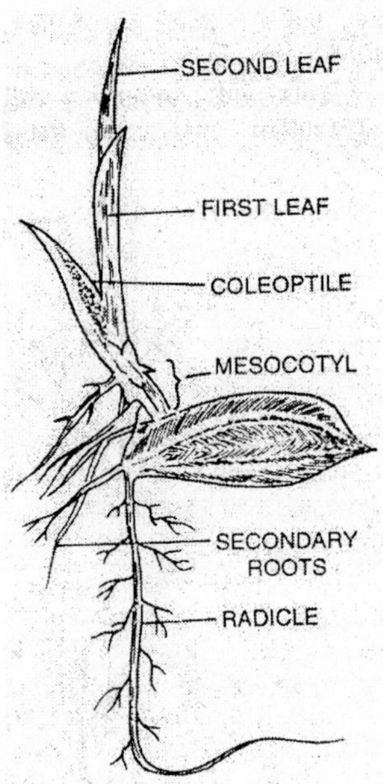

FIG. 6.3: Parts of rice germinating seedling (Chang & Bardenas, 1965).

Production Technology

Different Systems of Rice Cultivation

Rice growing conditions prevailing in different rice regions of India largely determine the system of rice cultivation. The principle systems in India are **Dry, Semi-dry and Wet.**

The dry and semi-dry systems of cultivation are mainly confined to tracts which depend on rains and do not have adequate irrigation facilities. The wet system is practiced in areas with assured and adequate supply of water either by way of rainfall or through irrigation. In some areas like Assam, Bihar, U.P. and West Bengal, flooded conditions occur in the monsoon.

Dry and Semi-dry Systems

The basic system of cultivation for dry and semi-dry areas is the same. The land is repeatedly ploughed and harrowed in the summer season for obtaining the necessary tilth. Farm yard manure is evenly spread one to one and a half month before sowing. About 5 to 6 ploughings at intervals of about seven days are given. The seed is directly sown in the field in these systems after the onset of rains by either (1) broadcasting the seeds and then working the soil to cover it or (2) sowing the seeds behind the plough furrow, dibbling two or three seeds per hill or (3) drilling it in lines. Of these, line sowing by dibbling or drilling method is preferable as it ensure uniform stand, requires a lower seed rate and facilitates inter-cultural operation. The main difference between lines should be 15 and 20 cm. The main difference between the dry and the semi-dry systems is that in the latter the rain water is impounded in the field when the crop is 5 to 6 weeks old and the field is ploughed crosswise with a light plough. This practice known as ***bushening*** is **prevalent in Orissa and Madhya Pradesh.** Ploughing of field with the standing crops serves three purposes, namely, **weeding, thinning and inter-culturing the crop.**

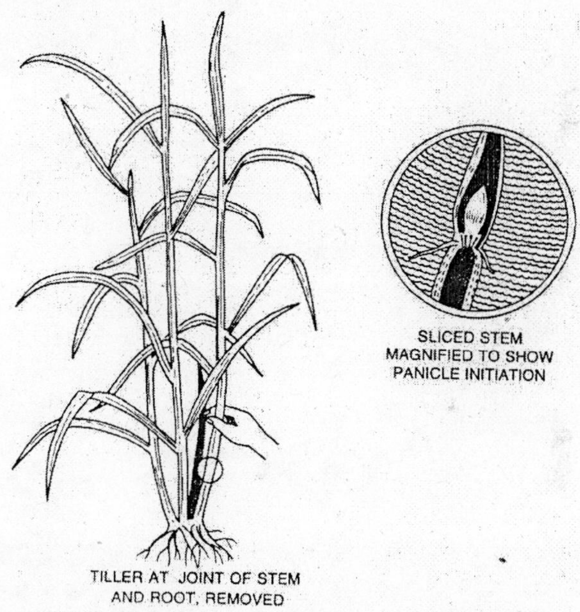

FIG. 6.4: Dissecting a rice stem to show panicle initiation.

Wet System

Under this system, the land is thoroughly ploughed and puddle with 5 to 6 cm of standing water in the field. The aim of puddling is to get a soft sol so that the transplanted seedlings can establish themselves quickly. Tractors/power tillers can also be used for puddling the soil. The wet field is repeatedly ploughed 4 to 6 times at intervals of 4 to 5 days and then leveled up by planking.

In this system, sprouted seeds are directly sown in a puddle and leveled field or alternatively seedlings grown in a nursery are transplanted. Transplanting in a puddle field has the following advantages:

1. Transplanting enables the cultivator to prepare the field thoroughly,
2. The puddling operation buries all weeds which would otherwise compete with rice plants for nutrients in the soil,

3. It is easy to achieve proper stand establishment and plant population becomes more certain,
4. Availability of most plant nutrient elements, specifically of iron and phosphorus, is increased and N is conserved,
5. The seedlings on being transplanted in soft puddle soil are able to put forth fresh roots in all directions. The root system is thus extended enabling the seedling to utilize food materials from the area all round,
6. Since nurseries occupy only 1/10th of the area of the field crop, the cost of irrigation and plant protection is reduced as compared to a broadcast wet crop of the same age. Nurseries raised in advance enable the crop to be planted in the field at the right time, and
7. The young crop is often subjected to insect attack which can be easily and effectively controlled in a seed-bed whereas any control measures in the case of a broadcast wet crop could be more expensive.

For the above reasons, transplanted crop generally gives more yield than direct-seeded crop. However, good yields from direct-seeded crop can also be obtained by effective water management, use of suitable variety and high quality seed and control of weeds, rats and other predators.

Deep Water Rice

In areas such as the Gangetic alluvium of Assam, Bihar, Eastern U.P. and West Bengal where prolonged flooding occurs during the monsoon period, rice is cultivated under deep submerged conditions, often with 2 to 3 metres of standing water in low lying areas. The varieties grown under such conditions are photo-sensitive, resistant to submergence and of long duration. They are usually broadcast by the end of March to April so that the seedlings attain the required age before flood water starts entering the fields. The crop is normally harvested from end of October to November. While the basic field preparation is similar to that in other areas, special efforts have to be made to control weeds particularly grassy ones and also wild rice which sometimes seriously hampers the crop growth.

Seed Rate

To provide seedlings for one hectare, the nursery area should be 1/10th of a hectare and the seed rate should be 30 to 40 kg. In case of a directly sown crop, about 100 to 125 kg of seeds per hectare are used for the broadcast method and 60 to 80 kg and 80 to 100 kg for a line-sown dibbled and drilled crop respectively.

Seed Selection and Treatment

Only well filled viable seeds should be used for sowing. Light seeds floating in a solution of common salt should be rejected. It is desirable to obtain treated seeds from a recognized source. Untreated seeds should be treated with the chemical Agrosan G. N. at the rate of 100 g per 50 kg of seeds. The untreated seeds of high yielding varieties should be soaked for 12 hours in a solution of wettable Ceresan 0.1 per cent (1 kg in 1 litre of water) and Streptocycline (0.6 g in 22.5 litres of water). The seeds should then be dried in shade and used for sowing.

Soaking and Incubating the Seeds

In order to ensure uniform and rapid germination, selected seeds should be soaked for 24 hours in clean water, drained and incubated in a warm, moist place for another 36 to 48 hours. Sprouted seeds should be used for sowing.

Nursery

Early raising of nursery between 10th to 25th December is highly desirable so that transplanting is over before the end of January. Early planting reduces stem borer attack and the crop is harvested early which leaves enough time for land preparation of *kharif* crops for timely sowing of jute in the coastal areas. Transplanting should be done between 10th to 25th January.

Treat seeds with Monosan or Emesan or Thiram (0.3%) as a protection against seed borne diseases namely blast, bacterial leaf blight and *Helminthosporium* leaf spot. Fifty to seventy kg seeds will be required per hectare depending on the grain weight of the variety. Always sow pre-germinated seed on a well puddle nursery bed. Soak seeds for 24 hours in water and incubate for 48 hours for sprouting.

The seedlings should always be raised on well puddle beds. Draw furrows at 1.5 metre in the puddle field and sow sprouted seeds on these strips. A total nursry area of about 1000 sq. metres will be required for transplanting one hectare of land.

To hasten early growth in cold weather apply 2 baskets of FYM per 40 sq. metres of nursery area along with application of N, P and K @ 10:15:20 g/m^2. Thoroughly mix the manure and fertilizer in the soil before drawing the furrows and sowing the seed. Apply granular insecticide in the nursery at 3 weeks after sowing.

Raising Nursery

Nurseries may be raised under dry and wet conditions. **Dapog** method of raising seedlings is also practiced.

1. **Dry Nursery:** The field should be brought to a fine tilth by repeated ploughing, about 4 to 4 times at interval of 5 days. Fifteen to 20 tonnes of farm yard manure should be evenly spread per hectare 4 to 6 weeks before sowing of nursery and incorporated into soil. For a nursery of 10 metres length and 1.5 metres width 30 kg of manure should be applied.

   Raised beds 150 cm wide, 15 cm high and any convenient length should be prepared with a channel half a metre wide all round them to facilitate drainage. The treated seeds or treated and sprouted seeds may be sown evenly in the bed or planted in close spaced rows. The sown seeds should be covered with a layer of fine compost and earth. Dry nursery should be irrigated by sprinkling water through a can once in 2 or 3 days depending on the nature of the soil. In case the soil is light, more frequent irrigation is necessary,
2. **Wet Nursery:** The nursery area should be ploughed twice in the dry conditions and then puddle by giving 3 or 4 ploughings in standing water, the interval between ploughings being 5 or 6 days. Green manures, if available may be applied at the rate of about 8,000 kg of leaf per hectare after the second ploughing about 3 to 4 weeks before sowing the nursery and turned into the soil. The field may be leveled properly and beds of one metre width and of any convenient length formed. Sprouted seeds may be broadcast evenly on the soft mud, keeping only a thin film of water. It is necessary to regulate irrigation so as to maintain the field in saturated condition with a thin film of standing water till germination is completed and coleoptiles start turning green. Then water level may be gradually raised and maintained at 2 to 3 cm. In the *Rabi* season (in low temperature areas), it is desirable to apply

P fertilizers at the rate of 0.5 kg P_2O_5/100 sq. metres.

3. **Dapog method of Nursery: Dapog** method of raising nursery consists of growing seedlings on a concrete floor or a raised soil bed covered with polyethelene sheets. The method is used especially in places where supply of water in the seed-bed is dependable.

The main merit of this method is that less area is needed to raise seedlings. Twenty five to thirty five sq. metres of area are enough to raise seedlings sufficient for planting one hectare. The seedlings are raised faster by this method and get ready within 9 to 14 days of sowing regardless of the duration of the variety used. Hence this method should be used in emergency cases after the onset of rains and when puddling of the main field has begun.

Land preparation is done similarly as in the wet bed method. After the last ploughing, raised beds should be prepared. These beds should then be packed, leveled and covered with polyethelene sheets. Banana leaves with the midribs removed can be used. Cement or fertilizer bags can also be used for this purpose; however, care should be taken so that no roots stick to the gunny bags. Covering is necessary to prevent roots of the seedlings from coming in contact with the soil, thus facilitating the separation of seedlings during transplanting. Pregreminated seeds should be sown on top of the cover at the rate of one kg per sq. metre of nursery. The germinating seeds should be sprinkled with water and pressed down lightly by hand or with a wooden flat board twice a day for 3 to 6 days. This operation keeps the roots of the seedlings is contact with water retained on the surface and prevents their drying. After this period, the seed bed should be irrigated continuously to a depth of 1 to 2 cm of water.

Dapog seedlings should not be pulled. While transplanting, the bed should be divided into convenient sizes and rolled like a mat with roots outward. Water level in the main field where dapog seedlings are planted must be carefully controlled during early stages of seedling growth.

Fertilization of Nursery

When the seedlings in the nursery are well established, *i.e.*, about 10 to 15 days after sowing, 10 kg N per ha should be applied. For a seed bed of 10 x 15 metres, 0.2 kg of N would be sufficient. A dilute solution in water may be applied evenly on the seed bed. In the dapog method, fertilization of nursery is not necessary.

For high yielding varieties, adequate fertilizer is essential, *i.e.*, for each 100 square metres of nursery bed, one kg of N, 0.5 kg each of P_2O_5 and K_2O may be applied just before sowing the seeds in the nursery. The phosphate dose may be doubled in locations where cool temperatures retard seedling growth. The fertilizers should be broadcasted on seedbeds and lightly worked in.

Care of Seedlings

Seedlings are susceptible to diseases and insect attack. Therefore, adequate care must be taken to protect them by spraying insecticides and fungicides.

Transplanting

In the case of indigenous varieties, the age at which the seedlings should be transplanted from nursery to the main field depends upon the total growth period of the variety. The following schedule may be followed.

Total growth period (days) of the variety (including growth in nursery)	Days after which seedlings should be transplanted
25	32
150	35
180	42

In case of high yielding varieties, however, the seedlings should be transplanted at the 4 to 5 leaf stage of the seedlings. Delayed transplanting of these varieties leads to poor tillering, early flowering of the main tillers and reduction in yield.

Shoot Pruning

If the seedlings are beat over grown a certain amount of top foliage is removed at the time of transplanting. By so doing the strains of transplantation on the plant which has to produce new roots and shoots is little minimized. Further, some of the insects are also found in the tip of the leaves and by so doing, they are also destroyed.

Age of Seedling

Optimum seedling age at the time of transplanting is about 3 to 3½ weeks for short duration varieties (*e.g.* Cauvery and Ratna), about 3½ to 4 weeks for a medium duration variety Like Jaya and about 5 to 5½ weeks for a long duration variety like Jagannath. During *rabi* season, when cooler temperature conditions prevail, it may take longer for the seedlings to reach the 4 to 5 leaf stage. Under stress conditions of salinity or impeded drainage, however, slightly over-aged seedlings perform better.

Transplanting of healthy seedlings may be done at the 4 to 5 leaf stage or when they are about 20 to 25 cm tall as soon as possible after uprooting. Two to three seedlings may be used per hill. Seedlings should not be planted deeper than 2 to 4 cm, as deeper planting delays and inhibits tillering. When dapog seedlings are planted, a larger number of seedlings (4 to 6) per hill may be used. In low lying situations also it is advisable to plant more number of seedlings per hill.

Spacing

Planting should be done in lines to ensure a uniform stand of the crop and to facilitate intercultural operations. Early varieties should be planted at 15 x 15 cm or 15 x 10 cm distance, medium and late varieties at 20 x 10 cm or 20 x 15 cm and high yielding varieties at 20 x 10 cm in *Karif* and 15 x 10 cm in *Rabi*. With low tillering varieties or over-aged seedlings spacing should be 15 x 10 cm or 10 x 10 cm and more number of seedlings should be planted per hill

All the missing hills should be replanted within 7 to 10 days after transplanting and this practice should be repeated, if necessary, one week later.

Classification of Rice Culture

Rice is grown in following three ways:

1. Direct seeding on dry soils,
2. Direct seeding on puddle soil, and
3. Transplanting.

But some others classify rice based on cultural practices as:

1. Upland Rice,

2. Transplanted Rice, and
3. Deep Water Rice

Rice Culture

1. A wide array of high yielding and improved varieties of rice in different maturity groups, grain quality and resistance to pests and diseases are available. Selection of variety for any situation should be made depending on topography, texture of the soil, availability of irrigation water and crops to follow in the cropping sequence.
2. Choose relatively shorter duration rice varieties than grown at present for high, medium lands from which water can be drained at harvest. This will enable growing a second crop on rainfed lands with residual moisture. On irrigated land, it will provide adequate time for land preparation for wheat or potato and will enable to grow three crops in sequence.
3. High yielding varieties: High yielding varieties have greater yield potential. They are short, erect, stiff strawed and responsive to high dose of nitrogen. Under comparable conditions they have consistently given higher yields than the improved or local varieties. Cultivation of these varieties especially on high and medium lands both under irrigated and rainfed conditions may be extensively popularized.

Field Preparation

Summer ploughing should be done following summer showers. It (1) reduces weed population, (2) helps in fixation of atmospheric nitrogen, (3) reduces pests and diseases incidence, (4) makes land preparation easier before sowing and (5) makes soil nutrients available to the crop. Final land preparation is done with pre-monsoon showers for sowing.

Stale seed bed (Paga Bhangiba) method be followed by allowing 5-7 days time after a pre-monsoon shower to have the first batch of weeds come up and destroyed by final ploughing.

Summer rice should be transplanted or pre-germinated seeds can be sown in lines on a well puddle and leveled field. The land should be flooded two to three weeks before transplanting and ploughed and harrowed at intervals to obtain a good puddle. The stubbles of the previous crop and organic manure should be thoroughly incorporated into the soil at puddling. A well leveled field is important for uniform fertilizers distribution, water control and weed control

Interculture

In line sown crop the inter-space can be worked with a narrow plough or rotary weeder or rake weeder at 3 to 4 weeks from germination. In broadcast crop, beusan is the common practice for killing weeds. This operation is to be followed by proper ***Khelua*** to maintain adequate plant population as per the transplanted crop. To fill up the gaps properly about $1/10^{th}$ of the area of the field be thickly sown as nursery to supply additional seedlings for the purpose.

It is to be noted that if monsoon falls to set in time it is convenient to take up weeding with hoes if the crop is sown in lines.

Timely weed control of the direct seeded crop is very important. Weeds compete with young rice plant for space, food, water and light. Weeding should be done at two weeks of germination. In the line sown crop it will be economical to work the inter space with a rake weeder or any hoe in broadcast crop ploughing of the land on the third day of sowing (mendha) and working with a tooth harrow (bida) two weeks after germination in a sunny weather reduces weed population.

Chemical weed control is to be done by post emergence application of propanil at the rate of 2.5 L a.i. per hectare (1000 litre of water) 12-15 days after sowing synchronizing with 3-4 leaf stage of the weeds. Spraying be done in a bright sunny weather to have the best effect of the chemical. This will reduce the cost of manual weeding.

Weeding should be done three weeks after transplanting. For the direct seeded crop, weed at 4-5 leaf stage (when plants attain 15-20 cm height) should be removed from the field by using a weeder between lines.

Weed Control

In broadcast crop, 3 or 4 weedings should be done by hand at intervals of 20 to 25 days. In transplanted crop, intercultural operations should be done before the first topdressing of nitrogen. These operations also serve to remove the weeds. Timely weeding is an absolute necessity. Available information indicates that weed control during first 3-4 weeks of sowing or planting is critical for rice. Timely weed control is necessary as weeds deprive the crop of nutrients, reduce tillering and ultimately the crop yields are reduced. The extent of yield reduction in rice due to weeds alone is estimated at 15-20 per cent in transplanted rice, 30-35 per cent under upland conditions.

The following cultural practices should help in effective weed control:

1. Thorough soil preparation,
2. Maintaining water at a depth of at least 5 cm, and
3. Hand weeding and rotary weeding.

For direct sown rice, stam F-34 (Propanil, chemically called DPA, 3,4-Dichloropropionanalide) may be sprayed at the rate of three litres active ingredient (a.i.) per hectare in 600 litres of water, three weeks after sowing the crop. Propanil should not be applied within a week before or after the application of organo phosphorus compounds as it causes severe phyto-toxicity under such conditions. In case there are too many broad leaved weeds, MCPA (2 methyl, 4-chlorophenoxy acetic acid) may be sprayed at the rate of two kg acid equivalent (a.e.) in 600 litres of water per hectare, five weeks after sowing.

For control broad-leaved weeds in transplanted rice, MCPA should be sprayed at the rate of 2 kg acid equivalent per hectare in 600 litres of water or 2,4-D (2,4-Dichlorophenoxy acetic acid) at the rate of one kg of active ingredient (a.i.) in 150 to 250 litres of water per hectare, 2 to 3 weeks after transplanting. This spraying will no kill grass weeds but only broad-leaved weeds and one hand weeding or interculture by paddy weeder is, therefore, necessary and should be done a week after spraying. The field should be drained, sprayed and reflooded after 1 to 2 days.

Where available, some new granular weedicides like Butachlor (Machete) applied to the soil surface 4 to 5 days after transplanting gives excellent control when used at the recommended rate of 1.5 kg a.i./ha 2,4-D granules are also available now-a-days which when used as pre-emergence application 5-6 days after sowing or planting rice at the rate of 0.8 kg a.i./ha is found useful. Basalin is yet another granular herbicide which has been found promising for transplanted rice. Granular herbicides are easy to transport, store and apply. Another advantage is that it is not necessary to drain the field for applying the granular herbicides.

Whenever liquid formulations are sprayed they should be sprayed uniformly all over the field. The crop will not be affected by these spraying. However, leaf-tips may sometimes get burnt or leaves may become yellow. In such cases the crop will, however, recover within one week.

Manuring and Fertilization of Rice

Role of Plant Nutrients

An adequate and balanced supply of plant nutrients is a prerequisite to maximize crop production. Rice is no exception. It needs adequate nutrition for growing well and producing high yields. Maximum utilization of plant nutrients (either derived from inherent soil fertility or supplied through fertilizers) for optimum growth and high yields can be achieved only when other factors are optimum.

The major nutrients required by rice are nitrogen, phosphorus and potassium. Throughout the rice growing areas in the country, nitrogen is the most limiting nutrient. Nitrogen is usually deficient where rice is grown because the same conditions of climate which favour rice production cause rapid turnover and loss of nitrogen from the soil. The requirement of phosphorus by rice is not as high as that of nitrogen. However, widespread phosphate deficiency in rice soils and response of rice to applied phosphorus are observed in the country. Responses of rice to the application of potassic fertilizers are not as marked as those to nitrogen. Yet potash deficiency and responses to potash applications are observed in some rice growing tracts of the country. This holds particularly true where fertilizer responsive high yielding varieties are grown under improved cultural practices. In fact, efficient utilization of applied nitrogen can be achieved only when it is applied in balanced proportions with phosphate and potash.

Increased yields arising from the application of lime and secondary elements like calcium and magnesium have been reported from acid rice soils of Bihar, Karnataka, Orissa, West Bengal, etc. Of the micro-nutrients, deficiency of zinc is fairly widespread in the rice growing areas of the country. To some extent deficiencies of iron and manganese are also reported. Spectacular response to application of these micro-nutrients is reported from such areas.

Nitrogen

Rice plants require large amounts of nitrogen at the early and mid-tillering stages to maximize the number of tillers. It also requires considerable amount of nitrogen even in the ripening stage. Application of nitrogen in rice results in:

1. A dark green colour to plant,
2. A rapid vegetative growth (increases height and number of tillers, the latter being particularly important for increased grain yield),
3. Increased size of the leaves and grains, and
4. Increased protein content of the grain and thus improved quality.

Rice plants well fed with nitrogen possess better stems and larger grains. In nitrogen deficient plants, growth is stunted, development of the stem is poor, the plants have a yellowish green appearance and the grain yield is considerably reduced. Rice can utilize nitrogen in both ammoniacal and nitrate forms efficiently. However, it has been observed that the ammoniacal form is more effective and is preferred during the early stages of growth. This is particularly so under water-logged conditions where ammoniacal form of nitrogen is more stable and not lost through leaching and denitrification as is the case with the nitrate form of nitrogen.

Phosphorus

Rice like any other cereal crop requires considerable quantities of phosphorus for vigorous growth and high grain yields. Besides contributing greatly to high yields, it

1. Stimulates growth and development of roots, making plants more resistant to drought and facilitates better nutrient absorption,
2. Promotes early flowering and ripening by means of which the unfavourable influence of late transplanting is eventually reduced or neutralized,
3. Encourages active tillering which enables rice plants to recover more rapidly after any adverse conditions encountered in the early stages of growth, and
4. Improves the quality of rice owing to higher P content of the grains.

Rice plants suffering from phosphate deficiency are small in size, have fewer tillers and show poorer root growth. Panicle development and ripening are delayed.

In general, response to phosphorus in lowland rice is somewhat less in magnitude than the response to added nitrogen because flooded rice is able to draw upon sources of soil P, which are relatively less easily available to upland crops. The soil conditions created by water-logging influence the transformation and availability of both native phosphate and applied fertilizer phosphate. There is relatively little difference in phosphorus sources for flooded rice except in extremely acid or saline-sodic soils. In extremely acid soils, water soluble phosphate gets fixed as insoluble iron and aluminium phosphates. In such cases liming the soil 4 to 6 weeks prior to phosphate application corrects soil acidity and improves crop yields. Liming should therefore, be practiced in such areas. Application of finely ground rock phosphate is effective in acid soils. In saline-sodic soils, the soil conditions have to be corrected through soil amendments (gypsum or sulphur application with farm yard manure or compost followed by irrigation available to leach down the harmful salts) without which fertilizer application is not efficient.

Potassium

Potassium is one of the three important plant nutrient elements and is needed in large quantities by rice plants. It

1. Exerts a favoourable influence in tillering and on the size and weight of the grains,
2. Stimulates building up and translocation of carbohydrates and grain development, and
3. Renders the crop resistant to diseases and adverse effects of unfavourable climatic conditions by strengthening and stiffening plant calls.

Generally, however, response of rice to potassium application is comparatively less marked than that to nitrogen and phosphorus. Neverthless, responses to potassium application are obtained in coarse textured soils where leaching causes considerable loss of bases. Acute potassium deficiency in rice is characterized by brownish yellow discourisation of older leaves spreading inwards from the tip, irregular necrotic spots on the leaves and panicles, formation of long thin ears and tendency of the plant to lodge. Lack of potash at the time of tillering and grain formation reduces the yield of the crop.

There is not much controvercy regarding source of potassium for rice. Muriate of potash is commonly used as a potassic fertilizer, either as a single fertilizer or through mixtures or complex fertilizers.

Secondary Elements and Micronutrients

Of the 10 secondary elements and micronutrients, which are essential for plant growth, the importance of calcium, sulphur, zinc, iron and manganese is now being felt in view of the deficiencies of these elements observed in some of the Indian soils. Calcium and sulphur, the two secondary elements, play important roles in growth and development of rice plants. Calcium has also a corrective influence on soil pH. It the pH of the soil is less than 5.5, addition of lime helps in increasing the pH to a level which is favourable for rice growth. Lime stone or dolomitic limestone is generally used for neutralizing acid soil. Basic slag, if available in finely powdered form, can also be used for liming. It also supplies some micronutrients.

Widespread deficiency of zinc and to some extent of iron and manganese is now being observed in India. This has come about as a result of intensive cropping with high yielding varieties which make a heavy demand on micronutrient reserves of soils. Zinc application is essential in areas of its deficiency. Appearance of rusty brown spots and discolouration of the older leaves starting from 2 to 3 weeks after transplanting are symptoms of zinc deficiency. The field may look rusty brown in colour and may have uneven crop stand. Acute deficiency causes the margins of older leaves to dry up giving the appearance of K deficiency. Tillering and growth are adversely affected. **Khaira** disease observed in the Tarai area of U.P. and elsewhere is reported to be caused by zinc deficiency. Wherever zinc or manganese deficiency is suspected, it must be confirmed and corrective measures viz., application of zinc sulphate or manganese sulphate in appropriate quantities through soil or in the form of soliar spray as advised by the expert.

A large number of investigations, particularly in Japan, has shown that silicon is responsible for improved growth of the rice plant. The way in which silicon affects rice growth is still obscure. However, silicon is reported to increase the resistance of the rice plant to blast disease and insects. Silicon nutrition is also reported to be related to decreased transpiration, increased oxidizing power of roots and detoxification of some heavy metals like iron, manganese, etc. It also enhances the development of rice fro heading to ripening. Silicon improves rice growth by maintaining more erect leaves, hence increasing photosynthesis. One of the usual characteristics of the rice plant grown under silicon-deficiency conditions is drooping leaves.

Nutrient Removal

Rice crop removes large quantities of nutrients from the soil. For example, a crop of early variety, Ptb-10, yielding 24.2 quintals of grain per hectare removes 45 kg nitrogen, 12 kg of P_2O_5 and 54 kg of K_2O. In the cases of a high yielding variety, the drain of nutrients from the soil is of a much higher order because of high yield levels. The quantities of nutrients removed by different high yielding rice varieties grown in the *Kharif* or *Rabi* seasons are given in Table 6.5.

It is seen that even a single crop of hig yielding variety of rice removes large quantity of nutrients from the soil. When two or three crops of rice are taken from the same field in one year, as is often the case in peninsular and eastern India, the quantities of nutrients removed from the field would be correspondingly higher. Even the most fertile soils can not supply these nutrients on a sustained basis unless they are replenished by application of manures and fertilizers.

TABLE 6.5: Nutrient removed by some of the released high yielding and local varieties

Variety	Season	Grain yield (kg/ha)	Nutrient removed (kg/ha)		
			N	P	K
HYV					
IR 8	Kharif	6,308	95	25	149
	Rabi	6,739	99	23	124
	Total	**13,047**	**194**	**48**	**273**
Jaya	Kharif	6,136	84	25	147
	Rabi	6,635	105	23	129
	Total	**12,771**	**189**	**48**	**276**
Padma	Kharif	4,954	87	22	140
	Rabi	5,435	95	21	115
	Total	**10,389**	**182**	**43**	**255**
Bala	Kharif	4,262	59	19	108
	Rabi	5,713	102	20	103
	Total	**9,975**	**161**	**39**	**211**
Krishna	Kharif	5,399	94	22	101
	Rabi	6,574	96	23	109
	Total	**11,973**	**190**	**45**	**210**
Cauvery	Kharif	4,457	85	17	96
	Rabi	5,690	88	21	116
	Total	**10,147**	**173**	**38**	**212**
Kanchi	Kharif	4,470	76	20	115
	Rabi	5,785	103	24	141
	Total	**10,255**	**179**	**44**	**256**
Local					
Ptb-10	Kharif	2,417	45	12	54
	Rabi	1,800	22	8	29
Total	4,217	67	20	83	
MTU	Kharif	1,942	43	13	49
	Rabi	3,242	41	13	65
	Total	**5,184**	**84**	**26**	**114**

Source: Fertilizer Statistics 1978-79, Fertilizer Association of India, New Delhi 1-230 (1979).

Manuring of Rice Field

Organic manures and chemical fertilizers are both important for rice cultivation. Organic manures improve the physical condition of the soil and supply limited quantities of plant nutrients through enhanced microbial activity. On decomposition, the organic matter is converted into humus in the soil. Humus promotes efficient use of nutrients supplied through fertilizers.

Fertilizers on the other hand contain one or more plant nutrients in concentrated readily available forms. They can be so applied as to supply the nutrients needed by plants and thereby increase crop growth and yield.

A liberal supply of bulky organic manures is necessary, particularly under upland conditions, to maintain the soil in good physical condition and to increase the water holding capacity of the soil. Under low land conditions, the structure is destroyed due to puddling operations and addition of organic matter supplies plant nutrients, particularly the micronutrients. About 10 to 15 tonnes of well decomposed farm yard manure or compost per hectare should be applied 4 to 6 weeks before sowing or planting the rice crop. The field should be ploughed after the manure is applied so as to mix it well with the soil.

For green manuring of rice fields, leguminous crops like sannhemp or *dhaincha* are grown in the fields or the leaves of leguminous trees such as *glyricidia* grown on the bunds of the rice fields are collected and buried. The usual seed rate for sannhemp is 33 to 40 kg/ha and for *dhaincha* 15 to 20 kg/ha. This would provide about 15,000 kg/ha of green matter. The green manuring crops are sown in the rice fields about 6 to 8 weeks before planting of rice. When the crop is about 4 to 6 weeks old, it is turned into the soil. Care should be taken to ensure that about 2 to 3 weeks time is allowed for the decomposition of green matter and that there is sufficient moisture in the soil for its decomposition. If green manure crop is not sown in the main field, about 6,000 to 8,000 kg/ha of green loppings from leguminous trees or forest growth, wherever available, are added 2 to 3 weeks before rice planting and buried in the field. With green manuring, the application of farm yard manure is not essential. Growing of leguminous green manure crop enhances its manorial value. Sometimes the entire dose of phosphate recommended for rice is applied to the green manure crop.

Green manuring is a very common practice in the southern and eastern parts of the country, particularly before planting the main crop (June/July to October/November).

The sharp increase in energy costs has added a new dimension to the use of the fertilizers by increasing its price sharply. Considerable emphasis is, therefore, presently placed on (1) efficient use of fertilizers so as to maximize the output from every unit of fertilizer nutrient applied and (2) application of alternative sources or plant foods like organic manures, bio-fertilizers, etc., to supplement chemical fertilizers. In other words, the current emphasis is on the adoption of an integrated nutrient supply system that judiciously exploits a combination of inorganic and organic sources of plant nutrients including bio-fertilizers, wherever possible, for increased production while simultaneously maintaining or increasing the fertility status of the soil.

Responses to Fertilizer Application

In order to get the maximum benefit from fertilizers used for rice crop, the following three rules should be followed:

1. Apply the correct doses of nutrients with due regard to the needs of the soil and the crop,
2. At the right time, and
3. In the correct way.

The optimum quantity of plant nutrients to be applied to rice depends in the fertility level of the soil and the requirements of the variety planted. One of the best known methods to find out fertilizer requirements is to carry out field tests. The amount of fertilizer to be applied, however, depends on the extent to which nutrients are already available in the soil, which very from soil to soil.

Fertilizer Recommendations

Most Agricultural Departments in the States in India have formulated fertilizer recommendations for rice appropriate to local conditions. These recommendations are based on fertilizer experiments conducted on Government Research Farms and on farmer's fields. Some States have regional and district wise or even variety wise fertilizer recommendations

The recommended fertilizer doses aim at giving the farmer a reasonable return on money spent on fertilizers. These are general recommendations for a wide area.

Soil fertility is a very variable factor. It varies from region to region and even field to field. A generalized fertilizer recommendation does not take into account the variability in soil fertility and can not assure maximum returns from the crop. Therefore, while these general recommendations can give an economic net profit to the farmers for securing higher profits, more precise recommendations on a farm wise basis should be prepared wherever possible based on soil tests.

Fertilizer recommendations based on soil tests take into account such factors as residual effect of phosphate and potash applied previously, and the requirement of lime or gypsum to correct unfavourable soil conditions, and in some cases indicate the kind of fertilizer to be used.

General Recommendation: For Upland **Kharif** *Rice*

Well decomposed farm yard manure (FYM) or compost at the rate of 5 t/ha with chemical fertilizers. The chemical fertilizers should be applied on the basis of soil test recommendation. Otherwise, the general schedule may be followed.

N:P:K @ 40:20:20 kg/ha and 30:20:20 and 60:30:30 for improved, local and HYV respectively should be applied depending on the initial fertility of the soil and the yield potential of the variety. All P and K are applied as basal by broadcasting and mixing at the final ploughing. Wherever possible, these should be preferably placed below the seed with a seed-cum-fertilizer drill. In case of placement, 25% N could also be supplied as basal to have early vigour of the seedlings. When no basal application possible 75% N be applied as first top dressing at the time of hoeing and weeding in the third week from germination and the rest 25% at the panicle development stage 18-20 days before panicle emergence. (*e.g.* 55 DAE in 110-115 days crops). In well drained sandy soil apply all P as basal and N and K in split. In other soils, full P and K should be applied as basal and N in splits.

Medium and Lowland **Kharif** *Rice*

It is necessary to apply FYM @ 5 t/ha at the time of final ploughing for sowing. Besides, adequate amount of N: P: K in the form of chemical fertilizers be applied in splits as per requirement of the different groups of varieties and their yield potential.

In lowland situation, where 30-60 cm of water remains standing for more than 6 weeks or overflows it is not possible to top dress nitrogen. Therefore, all the NPK should preferably be applied in lines below the seeds or to a side with a seed-cum-fertilizer drill or dibbling behind a plough. Slow release nitrogenous fertilizer like USG, LGU or coated urea would prove still better in the situation. However, in extra long duration varieties if the crop shows sign of nitrogen deficiency at the P.I. stage, one or two sprays may be given at 25 days and 10 days before panicle initiation (P.I.)stage (3-4% urea spray to supply 15-20 kg N/ha).

Fertilization for Beusan Rice

Apply all P at the time of seeding. Apply ½ N all K at the time of beusaning. Apply rest 50% N 30 days after. If it is not given at beusaning it can be applied at khelua.

Fertilizer Application in Lowland Rice

In lowland rice where top dressing of nitrogen is not possible, apply ½ N, full P and K of recommended dose at the time of sowing.

Fertilization for Transplanted Rice

Apply all P and K and 25% N at planting, 50% N at 15-20 days after planting and rest 25% N 20-25 days after 2nd dose of application. Forms of fertilizer and methods of application are same as direct seeded crop.

Fertilizer Application for Summer Rice

Apply fertilizer on soil test recommendations. If soil testing has not been done, apply fertilizers as recommended below (kg/ha)

	Low Fertility			Average Fertility		
	N	P_2O_5	K_2O	N	P_2O_5	K_2O
Medium duration	80	40	40	60	30	30
Short duration	60	30	30	40	20	20

Apply 25% N and full P and K fertilizer at final puddling to incorporate the fertilizers thoroughly in the soil. Apply 50% nitrogen 15-20days after transplanting, the rest 25% at 18 to 20 days before heading. Late top dressing will encourage non-bearing tillers.

Time and Method of Fertilizer Application

Time and method of fertilizer application greatly influence fertilizer responses on account of the water-logged condition of the rice fields as well as the nature of the plant itself.

In view of water-logged condition in rice fields, if the basal dressing of N is applied in the nitrate form, the nitrogen will leach into the lower layers, where, due to reducing conditions, it may be lost by denitrification as gaseous nitrogen. Ammoniacal nitrogen, on the other hand, is held by the clay particles and is not subject to denitrification. Thus, for minimizing the risk of nitrogen losses from basal nitrogen application, ammoniacal nitrogen fertilizers is preferable.

Application of nitrogenous fertilizers in standing water leads to considerable losses of added nitrogen. The field should have no more than a thin film of water when nitrogenous fertilizer is applied either as basal or top dressing. Immediately after application of basal dose of nitrogen, the field should be worked in to secure a thorough incorporation. For nitrogen top-dressing the field should be predrained to have a thin film of water, fertilizer applied and reflooded within 24-48 hours.

Water Management

The bulk of the acreage under rice in India depends on rainfall for its water supply and only about 38 per cent of the total rice area in India is irrigated to supplement the rainfall. Since the water requirement of rice is higher than that of any other crop of a similar duration, assured and timely supply of irrigation water has a great influence on the yield of the crop. On an average, about 1.25 cm (½ inch) of water per day is required.

The water requirements are high during the initial seedling period covering about 10 days, the pre-flowering and flowering period covering about 25 days and the grain formation period covering about 5 to 7 days. The crop is adversely affected in case of deficiency of water during the critical periods. The total requirement varies from 38 to 75 acre inches depending on the duration of the variety, type of the soil, its fertility status and the climatic conditions. Application of small quantities of water at short intervals is more effective than giving copious quantities at long intervals. Excess water impedes tillering and maximizes the losses of N when top dressed.

Until the transplanted seedlings are well established, water should be allowed to stand in the field at a depth of 2 to 3 cm. Thereafter, about 5 cm of water may be maintained up to the dough stage of the crop with periodical drainage and re-irrigation. At least a thin film of standing water should be constantly ensured for 15 days before and after flowering. Water should be drained out from the field 7 to 10 days before harvest to encourage quick and uniform maturity of the grain. Heavy soils should be drained 2 to 3 weeks before harvest because of their high water holding capacity.

For high yielding varieties, the planting should be done with a thin film of water and water level maintained at about 2 to 5 cm after seedlings are well established and till the crop flowers. The depth of water should be maintained at 5 cm from the flowering time till grains become hard. After this, it should be drained out. Planting of these varieties in compact blocks facilitates effective water management. Prior to top dressing of nitrogen, the field may be drained for 1 or 2 days, fertilizer applied and the field reflooded within 24 to 48 hours. Where highly reduced conditions occur, the field may be drained and kept dry until the surface cracks appear. The nitrogen should be applied and the field reflooded within 24 hours.

Critical Stages of Water

1. Booting stage: most critical stage of water critical stages,
2. Tillering stage (0-20 days) and
3. Primordia growth to flowering (40-60 days) in dwarf rice. At these stages, submergence (5 cm) of water is must.

Plant Protection

Prophylactic measures like clean cultivation, seed and nursery treatment are cheap and should be widely advocated. Field control of insect pests and diseases should be done.

To get high yields it is essential to prevent pests and diseases from attacking the crop. Earlier, it was recommended that protective sprays of insecticides should be given at various stages of crop growth. However, the increased production cost due to repeated insecticidal sprayings, the possibilities of environmental pollution and the reports of varietal breakdown resistance have all led to a rethinking in the approach to plant protection. The present recommendation involves the limited use of insecticides based on the economic threshold of the insect. In other words, unless the harmful insects are present in sufficient numbers as to cause economic damage to the crop, spraying of insecticides need not restored to. This calls for a constant vigilance on the part of the farmer and extension worker and undertaking expeditiously control measures against specific pests.

Pesticides are poisonous and necessary precautions must be observed in handling these. Contact with skin, eating and smoking during application should be strictly avoided. Instruction regarding their use given by the State Department of Agriculture or by the dealers should be strictly followed.

A number of diseases and pests attack the rice crop and reduce its yield. The common diseases are blast, blight, stack burn disease, foot rot and bacterial blight while the common pests are stem-borer, swarming caterpillar, rice bug, rice grasshopper rice hispa, rice gall midge and rice leaf hopper.

The nursery and the field crop should be protected from the loss due to diseases and pests.

Crop Rotation

In areas of very heavy rainfall, in lowlands subject to flooding and submergence and in the deltaic tracts, rice is grown year after year with a fallow in between. No systematic rotation in such areas is possible.

In tracts possessing irrigation facilities from wells and tanks, rice is roted with sugarcane, banana, betel-vine and other irrigated garden crops such as chillies, tobacco and turmeric. In Northern India, wherever irrigation is possible *berseem* is grown for fodder purposes after rice, or wheat is sown after *kharif* rice, using short duration high yielding rice varieties. In well drained upland rice fields, where the soil is of light texture, potatoes or vegetable crops are grown after rice. In eastern India, in low lying areas with irrigation facilities in the pre-monsoon period, jute is grown during March to June, followed by rice from July to December. In West Bengal particularly, rice-wheat rotation is becoming popular. In the south, where winter temperatures are not low, cash crops like groundnut or cotton are being grown in recent years in sequence with rice under irrigated conditions. Double cropping is extensively practiced in South India and to some extent in Eastern and Northern India. A short duration variety precedes or follows the main long duration variety. In some parts of Tamil Nadu, early and late maturing varieties are sown mixed in a specialized system called **Udu**.

In dry upland rice areas with well drained soils having good moisture retention capacity, a second crop of quick growing legume is grown after the harvest of rice. Such crops are gram, *mung* (green gram), *urad* (black gram), *khesari* (chickling vetch), *kulthi* (horse gram), etc. *Ragi* (finger millet), groundnut and sorghum are also grown in rotation with rice in such areas.

During the last few years, multiple cropping has been adopted in areas with irrigation facilities or assured rainfall with a view to obtaining maximum biological productivity per unit area per day. The multiple cropping sequences, with rice as main crop, which have provided successful, are: 1. rice-rice, 2. rice-wheat, 3. rice-rice-rice, 4. rice-wheat-mung (green gram), 5. rice-wheat-pulse (other than mung), 6. rice-pulse-rice, 7. jute-rice-wheat, 8. jute-rice-potato, and 9. rice-maize-rice.

Harvesting and Threshing

The crop is harvested close to the ground when about 90 per cent of the grains in the panicle are grey in colour (straw colour). Delayed harvesting causes considerable loss by shattering of grains resulting in considerable loss in yield and increased percentage of broken rice in hulling and due to damage by rats and birds. However, early varieties should be harvested 25 days after 50% flowering and medium and low land varieties after 30-35 days after 50 per cent flowering.

The right stage for harvesting is when the ears are nearly ripe and the straw slightly green. If delayed till the crop is dead ripe, grain may be lost due to shedding, lodging and by bird damage. Delayed harvesting may also lead to sun cracks which in turn would affect the milling recovery. The crop is generally cut with a sickle by manual labour, dried in the field for 3-4 days and then threshed by sticks or bullocks. The pedal thresher is economical, time saving and may be used.

After removing husk through threshing, rice is milled and polished to remove the bran layer in order to improve the keeping and cooking quality of rice and to make it attractive for selling.

The early crop is to be threshed within a day or two after harvest otherwise there would be fermentation and discoloration of grain. The moisture content is to be reduced to 14 per cent by alternate drying and heaping for four consecutive days.

Storage of Rice

Rice is stored in the forms of paddy or rough rice, milled raw rice, milled boiled rice, bran, broken and germs either on the farm or in the mills and in commercial warehouses.

Storage of rough rice: Paddy is stored either for seed or for hand pounding or for milling as needed. On the farms paddy is stored indoors in jute bags (gunny bags) or in containers made up of rope, straw, bamboo or mud (Kuthla or Kothi). When the quality exceeds a tonne the paddy is stored out doors and the stores are covered with a sloppy straw roof. The maximum capacity of out-doors is from 1.5 to 10 tonnes. Sometimes these stores lead to losses caused by insects, rodants, sprouting etc. With a view to avoid these losses, use of either cement or steel bins is recommended. These bins are made straight and rice quality of these stores is found to be appreciably better.

In the rice mills, paddy is placed in chambers or in the jute bags for a period of 2-8 months. The paddy stored in chambers without much protection suffers from loss, both in quantity and quality, caused by pests and moisture, while in open weather conditions the grains are damaged with sprouting or discoloration. Birds, rats, monkeys, squirrels and insects attack grains, reduce the quantity, and contaminate the grains with unwanted materials or substances.

In the warehouses which are damp and rodent proof, it suffers a loss only on account of reduction in moisture content and the grains may be stored safely for 2-3 years. The paddy is stored in jute bags is different lots with a free space of 50 cm all around.

Silos are also used in case of exceptionally higher quantities to be stored (500 to 850 tonnes). If these silos are not supervised properly a spoilage and discoloration may take place.

Rice (husked rice), either semi-polished (4-5% removal of bran) or policed (10-12% removal of bran), is stored in jute bags. According to the consumers' liking the raw rice has to be stored for at least over a year before it is consumed while with parboiled rice no such condition is needed and is consumed as fresh. In stores the rice develops rancidity and heating due to either insect attack or excess moisture. The rice can safely be stored in warehouses without reduction in quality and quantity of rice. If the warehouses are not aerated or damp proof, the free fatty acid rises from initial 35 to 320. The rice becomes unacceptable when acid reaches 350 unless it is remilled.

Broken: Brokens are of two types – big brokens are generally over ¼ the size of the kernel and small bokens are less than $1/4^{th}$. These are graded and filled in separate jute bags. The normal storage period for broken is 2 to 6 months because they develop rancidity very soon.

Bran: Bran can be stored for short period because it develops rancidity very fast. It is always better to dispose it off within a week's time after it is obtained from mill and cleaned.

Control of storage pests: The pest control measure should be taken very carefully because antidotes should be applied within their safe limits of residual toxicity to avoid the consequent hazards. Therefore, the pesticides should be applied by trained personals after choosing proper pesticide material.

Important Practices

Karif *Rice*

1. Line sowing of the direct seeded crop and early weeding of the same,
2. Raising of early individual/community nurseries,
3. Growing of short duration high yielding varieties on high and medium land,
4. Shallow, erect and close planting,
5. Use of herbicides to reduce weeding cost,
6. Adequate plant population to ensure 400-500 earheads per square metre,
7. Moderate dose of fertilizer,
8. Fertilizer to direct seeded beusaned rice,
9. Water management,
10. Proper drying and storage.

Summer Rice

1. Cultivation of short duration high yielding varieties.
2. Raising early nurseries in December and early transplanting of seedling in January.
3. Shallow, erect and close planting.
4. Timely weeding and water management in the field to check weed growth.
5. Use of fertilizers and correct method and time of application
   a) Complex or mixed fertilizers may be drilled or incorporated into the soil at seeding or transplanting.
   b) Ammoniacal form of nitrogen may be drilled or incorporated into the soil to a depth of 5-10 cm.
   c) P and K fertilizers should be drilled into the soil at seeding or transplanting.
   d) Avoid application of CAN at transplanting/.
   e) Avoid broadcasting of fertilizers on the surface of soil.

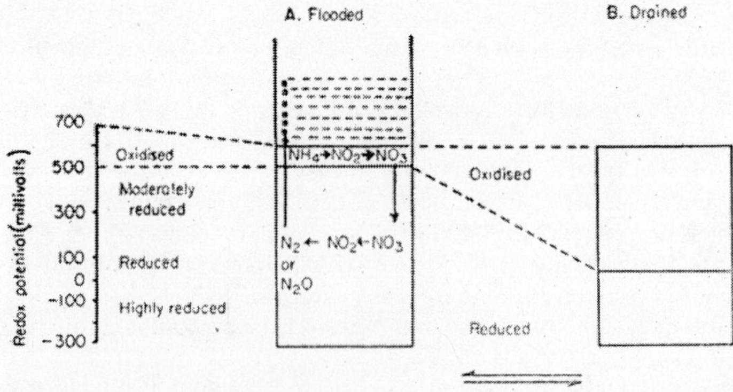

FIG. 6.5:

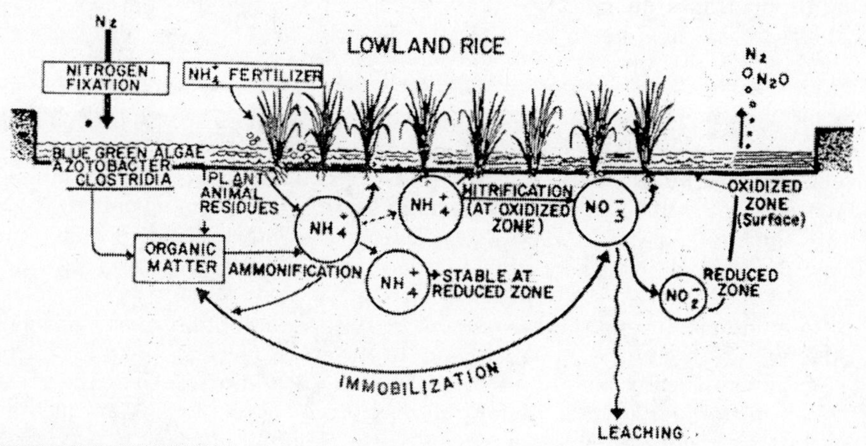

FIG. 6.6: Low Land Rice.

WHEAT (*TRTICUM* SPP.)

Among the cereals, wheat ranks first in area and production in the world. During 1995-96, wheat crop occupied an area of 230.2 m ha with the production of 584.9 m t. Wheat production in the world has steadily increased from 311.6 m t in 1970 to 541.1 m t in 1995. The increase was 73.7% compared to 1920. Productivity also increased from 1.5 t/ha in 1970 to 2.5 t/ha in 1995.

Distribution

South and East Asia greatly contributed towards increase in global wheat production. South Asia's share was only 22% in 1951 which was increased to 30% in 1994. Contribution of developing countries to world production has doubled between 1961 and 1994 while that of the developed regions has remained unchanged. On the contrary, the East European countries and the former Soviet Union which accounted for 35% of the global production in 1961 today account for only 20 per cent. Area and production of wheat in most important wheat growing countries are given in Table 6.6.

TABLE 6.6: Wheat cultivating area and production in different countries, 1996

Country	Area (m ha)	Production (m t)
China	29.5	107.0
United States	25.5	61.2
India	25.0	66.0
Russia	25.0	37.0
European union	17.2	93.8
Canada	12.9	29.0
Australia	10.0	17.5
World	230.2	584.9

The countries such as China, United States, India, European Union, Canada and Australia account for 63.2% of the total area and 70.4% of global production.

Wheat is an important winter cereal contributing about 32% of the total food grain production in India. Since, the introduction of high yielding dwarf varieties in 1995, there has been continuous increase in both the production and productivity. India now occupies the second position in wheat production. About one tenth of the global wheat production is from India. The area under wheat was only 9.7 m ha in 1950-51 which has increased to 25 m ha in 1995-96. This increase works out to 2.6%. However, there was about 10 times increase in grain production. The production was only 6.5 m t in 1950-51 which has tremendously increased to 62.6 m t in 1995-96. The increase in production was about 5.6 times during the first 30 years (1950-1981). During the later period, production was almost doubled. With respect to yield, there was 3.8% increase from 1950 to 1996 (from 663 kg/ha to 2493 kg/ha.).

Continuous varietal improvements and their adoption have pushed the productivity to high levels, almost equal to the world average yield (2541 kg/ha). The projected wheat grain production is around 109 million tones by the year 2020 AD. Wheat productivity can be increased by providing copious water at the right time. Improper nutrient application, poor crop stand and wrong sowing time are some of the major hurdles in increasing the yield.

Wheat is mainly grown in the northern part of India. Major wheat growing states are U.P., Madhya Pradesh, Punjab, Bihar and Haryana. These states contribute 80% of are and 84% of production. Other states like Gujarat, West Bengal, Himachal Pradesh and Karnataka also have significant area under wheat cultivation (Table 6.7).

TABLE 6.7: State-wise area, production and yield of wheat, 1995-96

States	Area ('000 ha)	Production ('000 t)	Yield (kg/ha)
Bihar	2072	4181	2018
Gujarat	506	1124	2220
Haryana	1991	7350	3692
Himachal Pradesh	378	544	1435
Karnataka	218	150	685
Madhya Pradesh	4008	6468	1614
Maharashtra	702	898	1279
Punjab	3282	12724	3877
Rajasthan	2196	5493	2501
Uttar Pradesh	9052	22203	2453
West Bengal	326	850	2611
All India	**25121.8**	**62620**	**2493**

Economic Importance

Bulk of the wheat is produced and consumed in northern and western India. Wheat is consumed in the form of roti, puri, paratha, chapatti or suji halwa prepared from atta, maida or suji. In North India, 76% of the households eat more than 40 kg wheat products per month. Per capita consumption of wheat is 5.35 kg/month for India. It is 12.36 kg/month in the North, 7.55 kg/month in West, 4.82 in the East and 0.40 in the South India. Whet is eaten as atta in the North and West, in the South and East as maida and suji. Rava is consumed mainly in the South. Pasta is the mixture of flour and salt. Pasta products comprise vermicelli, noodles, macaroni and spaghetti. Europe is the major consumer of pasta products. The main bakery products of wheat are biscuits, bread and cake/pastry.

Origin and History

Literatures show that wheat is being grown since pre-historic times. Recods from ancient China show that it was cultivated there by 2700 B.C. and it was known also to the Egyptians and to the inhabitants of Switzerland as early as the Stone Age.

De Condelle believed that wheat originated in the valley of Euphrates and Tigris and spread from there to China, Egypt and other parts of the world. Some American investigators believed that wheat probably originated in Syria and Palestine. N.I. Vavilov, after extensive studies of the origin of the cultivated plants, came to the conclusion that the origin of the durum wheat was probably in the region of Abyssinia, where as the whose group of soft wheat, which include the bread wheat, probably originated in the region of Pakistan, South Western Afghanistan and the Southern parts of mountainous Bokhara.

Classification of Indian Wheats

- *Triticum* spp. and chromosome No.:
- *Triticum aegiloploids* = 2n = 14 = *T. monococcum.*
- *T. durum* = 2n = 28 = *T. dicoccum*
- *T. spelta* = 2n = 42 = *T. macha/T. vulgare/T. aestivum*
- Classification based on the Chromosome number.
- Einkorn series (diploid) : Deploid : having 7 chromosomes : *Triticum aegiloploids, T. monococcum*
- Tetra ploid: having 14 chromosomes : *Triticum dicoccoids, T. timopheevi, T. persicum, T. polonicum, T. turgidum, T. orientale, T. durum, T. dicoccum.*
- Dinkle series: are hexaploid having 21 chromosomes : *Triticum spelta, T. macha, T. aestivum or vulgare,*

In India only three species are economically important:

1. ***Triticum aestivum*: is common bread wheat.**
2. ***Triticum durum*: is macaroni or durum wheat.**
3. ***Triticum dicoccum*: is emmer wheat.**

In India, following Triticum sp are grown

1. **Emmer** (*Triticum dicoccum* schub L): This type was reported to be grown in south *i.e.* Maharashtra, Iamul Naduand Karnataka. This type is believed to be developed from T. *diccoides* koru., a wild form. It is also grown in Spain, Italy, Germany and Russia. It contains 28 chromosomes.
2. **Macaroni wheat** (*T. durum* ***Desf.)***: The durum or macaroni wheat, cultivation in India, is considered to be very old. It is the best wheat for drought conditions or under restricted irrigated conditions of Punjab, M.P., Karnataka, Tamil Nadu, Gujarat, West Bengal and H.P. It is used for semolina (suji) preparation. The important varieties of this group are Raj 911, Malavraj, Mahavika, HD 4530, Jaira, Meghdoot (HI 7483, NP 404, Bansi. It is also grown in Italy, USA, Canada, and Russia. Durum wheats are believed to have descended from Emmer. It contains 28 chromosomes.
3. **Common Bread Wheat** (*T. vulgare* Host.): It is a typical wheat of alluvial soils of Indo-Gangetic plains *i.e.*, Punjab, Uttar Pradesh, Bihar and parts of Rajasthan. The bulk of the Indian crop, therefore, consists of this type. The prominent varieties are K-65, K-68, C-13, Pb-591, C-306 etc. It contains 42

chromosomes.

4. **Indian Dwarf Wheat** (*T. spherococcum* Mihi.): This belongs to the club wheat of western counties. This is found in limited areas of M.P., U.P. of India and in Pakistan. These are characterized by very short and compact heads having shorter grains. They contain 42 chromosomes.
5. **Maxican Dwarf Wheat** (*Triticum aestivum*): This is the type presently grown in India in almost all the wheat growing zones. The wheat is introduced in India by Dr. N. E. Borlaug of Maxico and the wheat is frequently said to be the Maxican dwarf wheat.

It is the bread wheat with 42 chromosomes (*i.e.*, carriers of hereditary determinants) that has their parents – *T. monococcum, Angilops speltoides and Angilops squarrosa, each* having 14 chromosomes. Individually these parents have very poor agricultural plants but in combination they give us the king of cereals.

Chapati making quality depends on **gluten** strength. Strong gluten varieties are most suitable for bread and for expression of high loaf volume, attractive brownish crust, silky and small granular structure of crumbs. Such varieties are HI 977, NI 5439, HW 657 and DWR-39.

Botanical Description

Wheat plant can be divided into two distinct parts, viz., root system and shoot system.

Root System

1. **Primary root system:** It forms at the time of seed germination and arises at the depth where seed is planted.
2. **Secondary root system:** It arises at a point above the primary root system as the principal organ of absorption as the young wheat seedlings progresses to maturity.

Shoot System

1. It comprises of all plant parts visible above the ground. It is composed of stem, leaves and inflorescence.
2. Stem: The stem of wheat plant is round or cylindrical. Stem and inflorescence may be called a culum.
3. Leaves: Leaf consists of four parts.
4. Leaf sheath: It is the basal part of leave. It encircles the culm (stem) and protects the growing point and axillary buds from the weather and provides some support to stem.
5. Leaf blade: The flattened, parallel veined portion of the leaf.
6. Ligule: A membranous or cartilaginous fringe at the junction of the sheath and blade on the side of the leaf next to culm. The continuation of the sheath through the collar is known as ligule.
7. Auricle: Lobes of leaf blade which extend downward on each side at the junction of the blade and sheath. These are horn-like appendages projecting from the collar of the leaf.
8. Inflorescence: The flowering portion of wheat plant which is called ear or head or spike.
9. Rachis: The central zigzag axis is the rachis. Spikelets are borne on alternate sides of rachis, which gives it zigzag appearance.

10. Spikelet: Spikelet is composed of flowers called florets. The number of florets in a spikelet may vary from one to five.
11. Florets: The outer covering of a floret is made up of a lemma and a palea. The lemma enfolds the palea near their attachment point. If an awn is present, it is attached to lemma.
12. Kernel: Wheat has a caryopsis type of fruit. The typical wheat kernel is from 3-10mm in length and from 3-5 mm in diameter. The composition of wheat kernel is:
    a) The germ (the embryo, consisting of plumule, scuteium, radicle and hypocotyls) comptises about 2.5 per cent. It is high in proteins and fats.
    b) The bran (pericarp, testa, nucellus and aleurone layer) accounts for 14 per cent of the grain. It is a by-product of milling and is used in dairy and poultry feeds. Small amounts are used in breakfast cereals.
    c) The starchy endosperm (the storage part of the caryopsis that develops from the union of polar nuclei with the endosperm nucleus) constitutes about 83 to 84 per cent of the kernel. Wheat protein provides a characteristic substance **glutin** which is essential for bread making. Wheat has a relatively high content of protein and thiamine.

Genotype

Cultivation of wheat is older than the history of man. The common wheat would have originated in south western Asia and the durum wheat probably in the region of Abyssinia. Wheat comprises several species of *Triticum* having diploids, tetraploids and hexaploids with 7, 14 and 21 pairs of chromosomes respectively. Out of 12 cultivated species of *Triticum*, four species such as 1. T. *aestivum*, 2. T. *durum* (Desf), 3. T. *dicoccum* (schub L.) and 4. T. *sphaerococcum* are important.

Among them, T. aestivum and T. durum are known as spring wheat adapted to subtropical and temperate regions. The sping types are characterized by a relatively much shorter duration, normally well below 175 days. Under most situations, the duration varies between 100 to 150 days depending on temperature.

The major parts of India, the spring type of wheat is grown. The other type of wheat, T. *dicoccum* and T. *sphaerococcum* are grown in temperate countries and are called winter wheat. They have long duration of around 250 days. These species have an obvious advantage of higher productivity due to the longer duration. However, these types can not be grown in all parts of the world. Because of the longer duration only one crop can be cultivated in a year.

Bread Wheat (*T. aestivum*)

This is the most important species cultivated in India occupying about 90 per cent of the total area under wheat. This is said to be Mexican dwarf type introduced in India by Dr. N. E. Borlaug. This has 42 chromosomes. Varieties 'Sharbati Sonora' was introduced in 1965 and susequately 'Kalyansona' in 1970. Important varieties of wheat developed by All India Co-ordinated Wheat Improvement Project (AICWIP) are: Sonalika, Lerma Roja, Chhoti Lerma, Arjun, E 306, WL 711, UP 2003, UP 262, LoK 1, HUW 206, HUW 234, HD 2189, HD 2285, HD 2329, Raj 3077, PBW 34, Raj 1555, WH 147, HP 1209, Sujatha, VL 421, VL 616, HS 240, HS 295, etc.

Wheat varieties recommended for different situations are listed below:

Irrigated condition	Normal sown	Sonalika, Kalyansona, Arjun, Girija, UP 301, UP 310, UP 319, Moti, HD 2122, Pratap, Choti Lerma, Safed Lerma
	Late sown	Sonalika, Janak, Sharbati Sonora and Shera.
Rainfed Condition	Normal sown	Kalyansona, Sonalika, Girija, Pratap and Janak
	Late sown	Kalyansona, Ridley, Janak, Mukta and Girija

Durum Wheat (*T. durum* Desf.)

It is the second major species of Triticum having 28 chromosomes. About 10 per cent of the total wheat production in the world is from durum type. In India, durum wheat production is estimated to about 3-4% of the total production. The kernels are typically larger, harder and heavier than the kernels of bread wheat. Varying environmental conditions and cultural practices influence the grain quality considerably. The crop cultivated on sandy soils and with less N fertilizers has much higher mottled grains. The crop is cultivated in dry and hot environment in Madhya Pradesh, Gujarat, Maharashtra, Karnataka and Southern Rajasthan. Though its productivity is less, the grain quality is good. Central and Peninsular India are favourable for the production of good quality grain with good hectoliter value, high protein and less yellow berry. Durum wheat cultivated in Punjab has high productivity; however, its quality is not very good. The ambient temperatures during the crop season are comparatively cooler but the humidity is higher. Such a weather condition reduces the lustrousness of the grain and grain mottling increases. The grain is often rated until for exports. Durum wheat is extensively used for making instant foods such as semolina, macaroni, noodles and snack foods.

Varieties: PDW 215, PDW 233, WH 896, Raj 1555, HI 8381, PBW 34.

Wheat Improvement

Real wheat revolution started after introduction of Mexican varieties and breeding materials in the country. The breeders are yet to be satisfied with the evolution of new plant types because the varieties do not express their full yield potentials in the farmers' field and thus the projected target production is not achieved. This may be corrected with several factors viz. selection of non-suitable variety, inadequate inputs, untimely sowing etc. Thus, the main emphasis in breeding was given to produce such morphological architecture of a plant that is capable of converting electro magnetic light energy received from sun into useful yield of chemical energy for maximum production through photosynthetic mechanism. A part from this the introduction of drought, frost, disease and lodging resistant along with high responsiveness to applied inputs, thermo and photo intensity etc. are some of the required characteristics of the newly evolved plant types. IARI and Genetics Department of various Research Institutions and Agricultural Universities in the country are playing a leading role in production of such varieties. Some of the important line of work is mentioned below:

1. **Breeding for dwarf wheats of long coleoptiles:** It is observed that the seeds of Mexican hybrid wheat varieties often sprout coleoptiles which sometimes fail to emerge out of the seed –bed when the seed is sown very dep. Thus the breeders have to produce dwarf wheat varieties with long coleoptile just to ensure better stand of the crop in the field where deep sowing is unavoidable due to low moisture content in the soil.
2. **Breeding for drought resistance:** The Mexican wheat varieties produced

earlier were good for growing under irrigated conditions and occurrence of drought any time during growing period was proved to be hazardous for their production or in other words their cultivation under dry areas was not possible. A number of drought resistant varieties have been produced which are in cultivation viz. Hy 65, Hy 633, Narmad 4, C 306, K 65, K 68, NP 839, NP 862, NP 890, etc. Now the breeders are taking durum as one of the parents for evolving new genotypes for their successful cultivation in dry farming or rainfed areas.

3. **Breeding for evolving triple gene dwarf:** During late sixties and early seventees various three gene dwarf wheat varieties viz. Heera, Moti, Jawahar, Arjun, etc. were evolved but all were found to be susceptible to brown rust. Besides, they had dark-red grain colour which was not liked by the farmers and the plants were so small that their transportation from field to the threshing floor was difficult. Thus the breeders are now keeping these points into account and new varieties are being evolved for general cultivation which is free from the said drawbacks.
4. **Breeding to evolve branched wheat:** Mutational breeding or treatment of NP 797 with gama-rays in gama garden of IARI resulted in evolution of branched wheat in which the ear-head is branched and one ear-head produces 3-4 times more grains as compared to other varieties. This practice is being adopted in irrigated as well as rainfed varieties of wheat to boost its production. However, sterility in some outer spikelets has been noticed that needs due attention in evolving new types.
5. **Breeding Rye and Wheat to evolve Triticale:** Rye, a member of the graminae family, is popularly grown as fodder in foreign countries and the same is found on the North India hills. It has characteristic feature of growing very profuse with exceptionally more number of tillers per plant under poor fertility status of soil having marginal or no irrigation facilities. Breeders took Rye for crossing with wheat and the resultant offspring was named as manmade cereal or rye wheat scientifically known as Triticale. This was done with a view to reduce the required input in wheat production and to increase the yield per unit area. Besides these, triticales have about 20 per cent protein and a very high biological value but the greatest drawback was that the grain colour was dark-red, seeds were much wrinkled with low viability and the plants had a very high degree of sterility. The grains are also susceptible to store grain pests. In the present day breeding these points are being taken into active consideration and probably in near future the farmers would be having a good number of triticale varieties for the cultivation.
6. **Wide Crosses:** Besides Rye and Wheat crosses, scientists are working on barley and wheat crosses. They are also crossing wheat with its wild relatives like Elymus, Angilops, Agropyron, etc. to bring in more viability, disease resistance, drought resistance and improvement in protein and biological values.

Characteristics of New Plant Types as compared with Tall Indigenous Wheats.

New Plant Type (Dwarf)	Tall indigenous
Plants, being dwarf and stiff, do not lodge even under very high doses of inputs viz, fertilizers and irrigation.	The plants are very tall and weak and hence they usually lodge under higher doses of fertilizer and water.
Their sink capacity is greater and hence they are	They are non-responsive to higher doses of

New Plant Type (Dwarf)	Tall indigenous
responsive to higher doses of fertilizers.	fertilizers. They start lodging at N 75 kg/ha or more.
They bear more productive plants per unit area due to synchronous tillering habits.	No. of abortive tillers are more because of a non-synchronous tillering habits.
The leaves are broader, upright and dark green which make an effective and better use of solar energy in manufacture of photosynthesis.	The leaves are long, narrow, drooping and cause mutual shading, Thus they do not make full use of solar energy in terms of phoyosynthesis.
They have a shorter vegetative and a longer reproductive growth period which results in a more number of filled and bolder grains/earheads.	They have a longer vegetative and a shorter reproductive growth period resulting in a more number of unfilled and shriveled grains/earhead.
Grain quality in terms of protein and amino acids is better, but the chapatti making quality is poor and colour is also red which do not appeal to the consumers.	Grain quality in terms of protein and amino acids is poor. But the chapatti making quality and colour being amber the grains have a greater appeal to the consumers.
The varieties being thermo and photo insensitive can easily be included in a high intensity cropping which make farming more profitable.	The varieties being thermo and photo and of longer duration can not included in a high intensity cropping.
They have very high yield potential.	They have a poor yielding ability.

Growth Stages of Wheat Plant

The wheat plant passes through the following four stages to complete its life cycle:

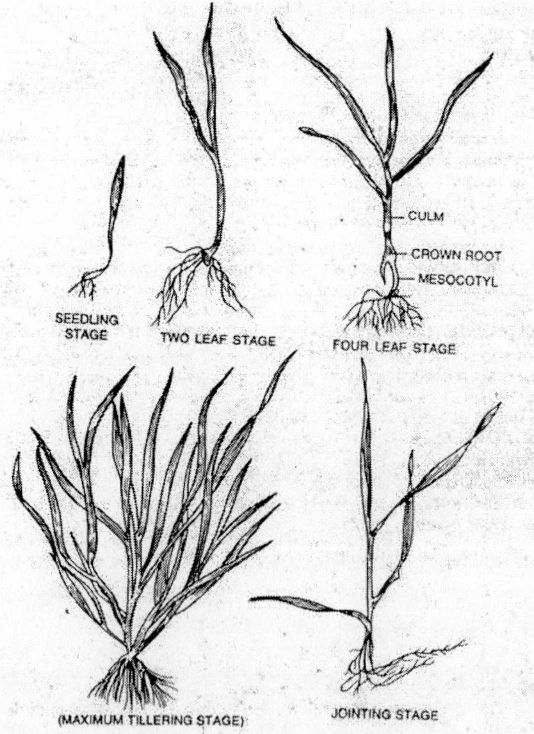

FIG. 6.7: Wheat.

Pre-establishment Stage

a) **Pre-emergence:** Sprouting of seeds by giving rise to seminal roots and coleoptiles.

b) **Emergence:** Appearance of coleoptiles from germinating seeds above the soil surface.

Vegetative Stages

a) **Seedling:** The young plants establish larger root systems in this stage. The stage may be further differentiated as one-leaf, two-leaf, and three-leaf and four-leaf stage.

b) **Crown Root Initiation Stage (CRI):** This coincides with three or four-leaf stage in which the crown roots appear.

c) **Tollering:** Plants develop crown and branch out into tillers from their base of the soil surface.

d) **Jointing:** This is the stage at which the plants start elongating when the nodes start developing above the crown node.

Reproductive Stage

a) **Booting:** In this stage the upper most leaf swells out into flag holding the spike into it.

b) **Heading:** The spikes start emerging out from the leaf sheath at this stage.

c) **Flowering:** Anthesis of florets and fertilization of ovaries take place at this stage.

Post-anthesis Stages

a) **Filling:** The ovaries, after fertilization, start elongating into seeds or ovules pass through milk, soft-dough and hard-dough stages.

b) **Maturity:** Colour of glumes changes and kernels become fairly hard at this stage.

Zones Based on Agro-Climatic Conditions

For convenience of testing and finding out the correct adaptability, the country is divided into five regions or zones based on their agro-climatic conditions. These zones are mentioned below:

1. **The North-Western Plain's Zone:** It consists of plains of Punjab, Haryana, Rajasthan (except south-eastern portion) and western U.P.
   Triticum aestivum is exclusively grown in this zone and is the most important of all the zones. The crop is grown irrigated during November to April and rainfed during October to April.
2. **The North-Eastern Plain's Zone:** It consists of eastern U.P., Bihar, West Bengal, Assam, Orissa, Manipur, Tripura, Nagaland, Meghalaya, Mizorum, Arunachal and Sikkim. This zone is exclusively of *T. aestivum*. The crop is sown in November-December and harvested in March-April.
3. **The Central Zone:** It consists of M.P., Gujarat, South-eastern rajasthan (Kota and Udaipur Divisions) and Bundelkhand area of U.P. About 75% of this zone is rainfed. Both *T. aestivum* and *T. durum* are popularly grown in this zone. In some areas of Gurarat *T. dicoccum* is also grown. The crop is sown from middle to end of October and harvested by end of February or March. The crop suffers from occasional frost in the Narbada Valley, particularly the Tikagarh area of M.P.

4. **The Peninsular Zone:** This zone consists of Maharashtra, A.P., Karnataka and Tamil Nadu, *T. aestivun*, *T. durum* and *T. dicoccum* are normally grown in this zone which *T. durum* is common in rainfed areas. The rainfed crop is sown in October while irrigated in the first fortnight of Novemebr and both of these are harvested in last week of February.
5. **The Northern Hill Zone:** It consists of the hilly areas of Kashmir, U.P., H.P., West Bengal, Assam and Sikkim. The crop is sown in October and harvested in May-June. On high mountains and in valleys it is grown during summer rather than winter *i.e.*, it is sown in April-May and harvested in September-October. *T. aestivum* is the common type which is grown in this tract. In southern hills of Tamil Nadu, two crops of wheat are grown in a year – the first during October to April and the second during May to September.

Climate

Wheat can be classified into two: 1. Irrigated wheat and 2. Barani wheat. It can be grown in tropics, sub-tropics and temperate zones and at times grown from sea level to as high as 3,300 metres.

Soil

Wheat can be grown in alluvial soil, red loam, black cotton soil, clay loam soil. Irrigated wheat comes up well in heavy soil and the non-irrigated wheat is mostly cultivated in Black cotton soil

Production Technology

Cultural Practices

Wheat has been under cultivation since time immemorial but due to introduction of dwarf wheat in India, it is realized that the agronomic practices evolved for the tall indigenous varieties do not suit the growth and development of dwarf wheat. It is, therefore, essential that we always be in touch with the day-to-day changes in the technical at know-how of the improved crop cultivation.

Field Preparation

In villages still we believe in preparing a much pulverized seedbed but recent technology has proved that if the field does not have much of perennial weeds, more than two ploughings are wasteful. Usually, one ploughing with a soil turning plough followed by harrowing and leveling, if necessary, brings about a nice tilth for sowing.

Time of Sowing

Temperature during the growing season and at grain filling is the one amongst several important factors in deciding sowing time. However, the ideal temperature requirement varies from plant type and stages of growth. The dwarf varieties require the following temperature for their growth and development:

Growth stages	Temperature requirement
Germination	20 to 25°C mean daily temperature
Growth and Development	20 to 25°C mean daily temperature.
Tillering	16 to 20°C mean daily temperature.
Accelerated growth	20 to 23°C mean daily temperature.
Proper grain filling	23 to 25°C mean daily temperature.

Minimum temperature which spring wheat can tolerate: - 8 to – 10°C.

Wheat plants are sensitive to very cold or frost injury at any stage of growth particularly at reproductive growth if temperature is below 15°C. Based on above temperature requirement it has been found that for indigenous wheat last week of October, for long duration dwarf varieties like kalyansona, Arjun etc. first fortnight of November and for short duration dwarf wheat varieties like Sonalika, Raj 821 etc. second fortnight is the best sowing time. Under exceptionally late sown condition it may be delayed to latested by first week of December beyond which if area is very small transplanting may be practiced.

It is a *rabi* cereal. Sowing generally is done in October-November and harvested in the month of February-March. On high mountains and in valleys it is grown during summer rather than winter *i.e.*, it is sown in April-May and harvested in September-October.

At the time of sowing care should be taken as there will be sufficient moisture in the field.

Fertilizer Application

a) **Rate of Application:** From the experimental results it is veryinteresting to note that the response of dwarf wheat varieties to applied nitrogen ranges between 35 to 45 kg of grain per kg nitrogen in comparison to 5 kg of grain/kg of nitrogen in case of improved tall varieties. It is observed that lodging susceptible varieties and rainded dwarf varieties respond up to 40 kg N/ha but the economic dose of N for dwarf wheat under irrigated condition is 100 to 120 kg/ha. Under normal soil conditions, both phosphorus and potash have given best response upto 40 kh/ha but beyond this it is a waste to apply any quantity of the said nutrients. In most of the light textured soils Zn has been a key element for getting higher yields and it is proved that zinc sulphate at the rate of 20-25 kg/ha should be applied in soil before last cultivation. The rate of application also depends upon the preceding crop viz. after an exhaustive crop like bajra and sorghum a dose of 120-150 kg N/ha is required while after fallow, any legume crop or heavily fertilized crops like hybrid maize, potato, etc. an application of 80-100 kg/ha nitrogen is quite sufficient.

b) **Method and Time of Application:** It may be concluded from the recent experimental findings that for best results the nitrogen may be applied as broadcasting contrary to this the phosphorus and potash must be basal placed either 2.5 cm below the seed or in either of sides of the seeds. Basal application of Zn has been found to be always better than its spraying after the crop has already suffered a setback from its deficiency in the field.

c) The time of fertilizer application depends upon nature of fertilizer and type of soil. The phosphatic and potassic fertilizers must always be applied as basal while nitrogenous should be applied as basal and top-dressing in split doses. In medium to heavy textured soils application of nitrogen in two splits *i.e.*, half at the time of sowing and half after first irrigation should be done while in the light textured soils, it is better to apply into three splits *i.e.*, one-third at sowing time, one-third after first irrigation and remaining one-third after second irrigation.

Whenever wheat is cultivated after fallow, no manuring is required. Wherever, the green manure crops are practiced no other manure except some amount of super is added. Wherever, nothing is added, FYM or compost or oilcake can be added.

Fertilizers like ammonium sulphate, super phosphate and muriate of potash/potassium sulphate are also added at the time of sowing especially for the irrigated wheat. There is also a practice that all the manures are applied to the *kharif* crops like maize and green manuring crops and the residues are utilized by the wheat.

Method of Sowing

Generally, three methods are applied for sowing wheat *i.e.*, 1. Broadcasting, 2. Behind the plough or line sowing, and 3. Drilling (help of seed drill).

According to resources seed drill or Malabasa (pora or Kera) should be used so as to drop the seed at a proper depth because of shorter coleotile in dwarf wheats.

Seed Rate

Seed rate of wheat depends upon a number of factors:

1. The purity and percentage of germination.
2. If seeds are sown early in season the seed requirement is less. The late sowing, the requirement of seeds is more.
3. Good seedbeds are required less seeds,
4. Drilling requires less seeds and broadcast requires more seeds,
5. If a variety has got poor tillering capacity, the quanties of seeds requirement is more.
6. The seed rate of wheat varies from 100 to 125 kg/ha.

Generally, a seed rate of 100 kg/ha has been found to be sufficient for most of the varieties like Kalyan Sona, Arjun, Janak, etc. which have moderate tillering and medium sized grains. But a higher seed rate of 125 kg/ha is desirable for late sown wheat and normal sown for varieties like Sonalika, Raj 821, etc. which has bold grains and shy tillering habits.

Depth of Sowing

Depth of seeding is governed by soil moisture regimes and soil type, however, a depth 5 to 7 cm is found to be optimum for dwarf wheats. It is preferably better to have 2.5 to 5 cm seeding depth because of shorter coleoptile in new varieties while tall ones may be sown at 7 cm depth.

Rotation

Generally, wheat comes after fallow as it is *rabi* crop. The following rotations are adopted:

Kharif	Rabi
Fallow	Wheat
Rice	Wheat
Maize	Wheat
Green manuring crops	Wheat
Legumes crop-Pulses	Wheat

Irrigation

The high yielding wheat varieties should be given five to six irrigations at their critical growth stages: viz. crown root initiation, tillering, jointing, flowering, milk and dough which come at 21-25 days after sowing(DAS), 45-60 DAS, 60-70 DAS, 90-95 DAS, 100-105 DAS and 120-125 DAS respectively. Of these, irrigation at CRI stage is the most important and delay of every day results in reduction of 1.4 per cent grain yield/day. But, if the facilities are not available to irrigate at the said stages one has to irrigate at the following stages which are the most important. It is also been noticed that if any of the following irrigations is delayed or missed the yield is reduced to the extent of 5-10 quintals/ha.

1. Crown root initiation stage : Between 25-30 days after sowing.
2. Flowering stage : Between 60-75 days after sowing.
3. Milk stage : Between 90-110 days after sowing.

According to availability of irrigation water it may be scheduled as under:

Number of Irrigations				
One	**Two**	**Three**	**Four**	**Five**
CRI (21 DAS)	CRI (21 DAS) Boot (75 DAS)	CRI (21 DAS) Tillering (45 DAS) Boot (75 DAS)	CRI (21 DAS) Tillering (45 DAS) Boot (75 DAS) Milk stage (100)	CRI (21 DAS) Tillering (45 DAS) & Jointing (65) Flowering (85) & Milk stage (100)

Thus, if a farmer afford s to give one irrigation he should give it at CRI, if two then at CRI and Boot, and if three then at CRI, Tillering and Boot and so on according to the above schedule.

Interculture

The crown roots appear in the third week after germination. Weeding, manuring and irrigation at this stage helps root growth and development and also better tillering. Nitrogen top dressing can be done between the lines before hoeing and weeding so that the fertilizer gets incorporated into the soil there by reducing losses and increasing fertilizer efficiency.

Weed Control

For an effective control of weeds, following chemicals should be sprayed by making a solution in 400 to 600 litres of water/ha:

Chemical control of weeds in wheat crop.

Name of chemicals	Rate kg a.i/ha	Time of spraying	Name of weeds to be controlled
2, 4-D	0.40	5 weeksaftersowing (3-5 leaf stage)	*Chenopodium* sp., *Anagallis* sp., *Asphodelus* sp., *Mplilotus* sp., *Lathyrus* sp., etc.
Avadex or	1.00	Pre-plant mixing to top 2-3 cm soil	*Phalaris* sp., *Avena* sp., etc.
Nitrofen (TOK E-25)	1.5	Pre-emergence or post-emergence 10 days After 1st irrigation	- do –
Asulam	1.5	Post-emergence 25 days	- do –
Isoproturon	1.0	after sowing	
Metoxuron	1.5		

Phalaris minor (weed) is controlled by Tribunil/Dosanex/Isoproturon @ 2 kg a.i./ha 32-35 days after sowing (DAS). Dosanex and Isoproturon also control wild Oat (*Avena fauta*). At present isoproturon is ineffective to control phalaris.

Leader @ 33.3 g/ha or sencor @ 250 g/ha 30 days after sowing is used to control phalaris these days.

Plant Protection Measures

Disease Control: Most of the varieties released are resistant to nearly all the diseases but even after selecting such varieties if the diseases are noticed, we should apply the following chemicals for their control:

Preventive and Control Measures of the Commonly Occurring Diseases in Wheat

Name of disease	Preventive measure	Name of chemical	Rate
Smut	Use of healthy seed Rouging and burning of disease affected plants Use of resistant varieties Solar heat treatment	Seed dressing with Vitavax/Benlate	2-2.5 g/kg of seed
Rust	Use of resistant varieties Timely sowing	Zinc/Manganese Dithiocarbamates	-
Bunts	Use of resistant varieties	Seed dressing with Agrosan GN/Ceresan	2 g/kg of seed
Powder mildew	-	Dusting with yellow sulphur, Sodium thiosulphate, Karathane/Miltox or spraying with Benlate.	-

Bavistin, Thiram. Brassical, RH 124 and Diathane Z-78 or M-45 may also be used for control of fungal diseases in wheat.

Harvesting and Yield

The crop is harvested when the grains become hard and the straw becomes dry and brittle. The maturity time of the crop differs from zone to zone and also depends upon whether it is grown under rainfed or irrigated conditions because the rainfed crop matures much earlier than in irrigated one. General time for harvesting may be mentioned below under various zones of the country:

Wheat growing zones	Time of harvesting the crop
N.E.P. Zone	It starts from latter part of March and continues until mid-April.
N.W.P. Zone	Second fortnight of April.
Central Zone	End of February to March.
Peninsular Zone	Second fortnight of February to beginning of March.
Hilly Zone	May-June.

The harvesting is mostly done by sickle but since the recent past the combines are also being used in parts of Punjab, Western U.P., Haryana etc. The crop is threshed by treading with cattle on the threshing floor or by power driven thresher.

The national average yield of wheat grain is about 12 to 13.8 quintals/ha. The average yield under rainfed condition is much below the national average, however, under irrigated conditions it has gone as high as 100 to 115 q/ha. Under National Demonstration Project the same is found 80 q/ha. Usually the straw yield is one and half times to two times the grain yield except in case of triple gene dwarf in which the straw yield is less than that of old varieties.

Storage of Wheat

Wheat grains, soon after their harvest, contain about 10 per cent moisture and become ideal for air tight storage. In case they are not fully dried they may be exposed to sun so as to bring a moisture per cent around 8-10 and then should be stored in especially designed air tight bins like Pusa bin, Pantnagar kothla, PKU bin, Kisan kothi, etc. To keep away the insects it is necessary to fill the conainers to their capacity but in case the container is not full it must be fumigated from time to time for the safety of grains from insect damage. The fumigation may be done by placing either ethylene bromide ampules or almunium phosphide tabelets in the grain on the top and closing the container air-tight for 8 days or more. These chemicals evaporate their vapours and by penetration in grain, kill all the insects that are present in the grain. Thus, the grain becomes insect free that also remain free of fresh attacks in air-tight containers.

Quality of Wheat

Quality of wheat grains is an important factor in fixing the price of wheat or subjective evaluation of wheat varieties. The agronomic practices like time of sowing, irrigation, weed control etc. affect the yield and development of grains which ultimately affect the yield and quality of such grains. The recovery of atta (whole meal) from the grains, thus produced, is also found to be very poor because of larger proportion of bran in them. Therefore, the timely sowing, irrigation at critical growth stages and control of weeds etc. are essential for maintaining the quality of wheat grains.

Important Practices

1. Division of field into small basins,
2. Line sowing and timely weeding at 18-20 days stage.
3. Top dressing, hoeing, weeding and irrigation at crown root initiation stage.
4. Use of the best quality seeds: Single gene dwarf varieties: Sonalika, UP 262, WL 711, Girija, Double Gene dwarf varieties: Kalyansona, Arjun, Janask, UP 215, HD 2204, Pratap Triple gene dwarf varieties: Hira, Moti, Jawahar, Jyoti, Sangam, HD 1941, HD 1941, HD 1977, UP 301, UP 319.
5. Critical stages of irrigation: 1. Crown Root Initiation (CRI) stage, 21 days after sowing, 2. Flowering stage (90 DAS), 3. Late tillering stage, 4. Milk stage.
6. Late sown wheat varieties: Sonora 64, Sarbati Sonora, Sonalika(RR 21), Safed Lerma, UP 301, NP 830.
7. Grain quality should be maintained by applying the agronomic practices.

Important Observations

- **Ligule:** Membranous structure, continuation of the sheath through the collar.
- **Auricle:** Horn or claw like appendages projecting from the collar of the leaf.
- Flower is closed by lemma (outer bract) and palea (inner bract) and awn is the extended part of lemma.
- Test weight: 40 g, Shelling % = 60%.
- Harvest Index = Economic yield/Biological yield x 100 = 40-45%.
- **Norin** dwarfing gene was isolated from Norin series wheat varieties and Nor (Rht) in wheat variety in Japan. Rht --- Reduced height.

- **The first such variety: Norin 10 (dwarfing gene wheat).** Norin 10 was brought to USA by S. C. Salamon in 1948. Using Norin genes Dr. O. A. Vogel developed a dwarf winter wheat variety Grains in USA and many other varieties were developed by Norman E. Borloug, Nobel laureate in 1961-62 at CIMMYT, Maxico.
- **In 1963, Government of India imported 100 kg of Mexican wheat varieties,** Sonora 64 and Lerma Rajo with the help of Rockfeller Foundation and Five varieties by IARI were LermaRojo 64 A, S 63, Sonora 64, Mayo-64 and S-227.
- After extensive tests, huge quantities of Lerma Rojo 64 A (Single gene dwarf variety) and Sonora 64 (Double gene dwarf variety) were imported in 1965-66 from CIMMYT.
- In 1965, both varieties were released for general cultivationin India and brought **Green Revolution** (Wheat Revolution).
- Later on these were replaced by Kalyansona and Sonalika. Later on Kalyansona becomes susceptible to all rust and Sonalika to kernel bunt.
- Source of dwarfing genes (1) Norin 10 (Japan), (2) Tom Thumb (Tibet), (3) Olsen dwarf (South Rhodesia.
- In 1970, triple gene dwarf varieties were released.
- National average yield: 22 q/ha.

MAIZE (*ZEA MAYS*)

Maize or Indian corn is one of the most important cereals of the world. A part from direct human consumption, maize grains form an important ingredient of poultry and cattle feed. Physiologically, maize is grown one of the efficient species domesticated by man. Globally maize is cultivated in an area of 140.1 m ha with the production of 576.8 m t. (1995-96).

Distribution

Among the cereals, the productivity of maize is the highest (4.1 t/ha) as compared to rice (3.7 t/ha) and wheat (2.5 t/ha) Corn is produced largely in the western hemisphere and Europe. It is the most important grain crop in the United States. Other corn producing countries are China, Brazil, Mexico, Russia, India, Indonesia and Philippines (Table 6.8).

In India, maize occupies the third place among the cereals after wheat and rice. In 1995-96, it occupied 6.11 m ha with the production of 9.12 m t. The average productivity of maize in India is much lower (1.5 t/ha) and the world average (4.1 t/ha). The important states growing maize are Uttar Pradesh, Rajasthan, Madhya Pradesh, Bihar, Himachal Pradesh, Karnataka, Andhra Pradesh and Jammu and Kashmir. These states account for more than 79% of the area and 81% of production. (Table 6.9).

TABLE 6.8: Maize production in the world, 1996

Country	Area (m ha)	Production (m t)	Yield (kg/ha)
Brazil	13.4	32.0	2393
China	23.1	119.4	5173
India	6.2	8.7	1408
Indonesia	3.8	8.9	2362

Country	Area (m ha)	Production (m t)	Yield (kg/ha)
Mexico	7.9	17.3	2192
Philippines	2.7	4.2	1518
United States	29.6	236.1	7975
World	140.1	576.8	4117

Source: FAO year book- Production, 1996.

TABLE 6.9: Production of maize in important states of India, 1995-96

States	Area ('000 ha)	Production ('000 t)	Yield (kg/ha)
Andhra Pradesh	308	785	2548
Bihar	717	1248	1739
Himachal Pradesh	308	661	2147
Jammu and Kashmir	295	468	1586
Karnataka	331	1201	3634
Madhya Pradesh	857	1151	1343
Rajasthan	907	808	891
Uttar Pradesh	1068	1470	1376
All India	6106	9118	1493

Eeconomic Importance

Maize is largely consumed as food in various forms. It is also used as animal food, particularly for poultry. Green maize plants are used as succulent fodder. Popping the corn is a method of starch cookery. Maize is a raw material for a number of products viz., starch, glucose, dextrose, sorbitol, dextrine, high fructose syrup, maltodextrine, germ oil, germ meal, fibre and gluten products which have application in industries such as alcohol, textile, paper, pharmaceuticals, organic chemicals, cosmetics and edible oil. Maize-based industry is the driving force of the U.S. economy.

Origin and History

The place of origin of maize or corn (*Zea mays* L) is still uncertain though some workers mention it to be highlands of Peru, Bolivia and Equador while others say it to be Southern Mexico and Central America. The first mention of the crop on the records was November 5, 1492, when Columbus reached the American mainland and his scouts brought to him the maize samples from the Island of Cuba. The sculptures and painting in temple and the maize excavations in the New World show that maize was grown as an important food crop about 2500 years before Columbus landed there.

The maize or the corn was introduced into Spain after the return of Columbus from America and from Spain it went to France, Italy and Turkey. In India, the maize was introduced by Portuguese during the seventeenth century, however the recent evidences show the possibilities of its pre-Columbus introduction through the Atlantic Arabian trade route. From India it went to China and later it was introduced in Philippines and the East Indies. New it is being grown in USA, China, Brazil, Argentina, Mexico, France, South Africa, Russia, Rumania, Yugoslavia and India.

In India, the major area of the crop is confined to Gujarat, Rajasthan, Punjab, Haryana, M.P., A.P., H.P., Jammu and Kashmir and Bihar. The grains apart from food as bread, pops and gruel are used for many industrial products like manufacture of starch, alcohol, acetic and lactic acids, glucose, paper, rayon, plastic, textile, adhesives, dyes, synthetic rubber, resins, artificial leather and boot-polish.

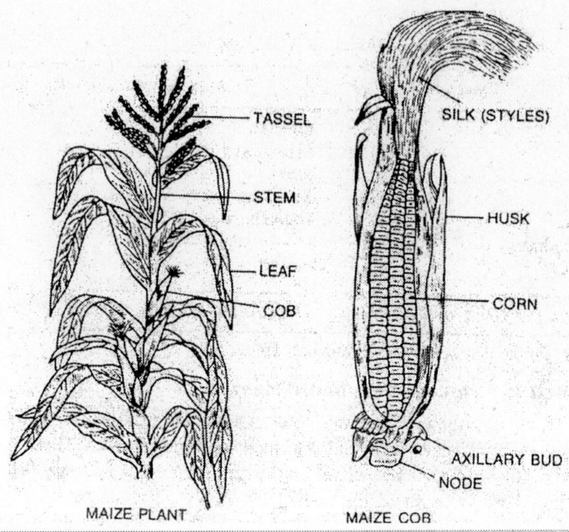

FIG. 6.8: Maize plant and maize cob.

Bobtanical Description

- **Root System:** The root system of maize is deep and fibrous. It consists of:
- ***Seminal or temporary roots:*** Consists of radicle and number lateral roots which arise at the base of the first node of the stem under soil surface just above the scutellar node.
- ***Crown or coronal roots:*** They arise from the basal portion of the stem.
- ***Brace, prop or aerial roots:*** They arise from second, third and sometimes forth nodes above the soil surface. All may or may not enter the soil.
- **Shoot System:** comprises stem, leaves(including leaf sheath and leaf blade) and inflorescence.
- ***Inflorescence***: Maize is a monoecious plant having both male and female inflorescences on the same plant. Male flowers are borne in a tassel at the top of the stem and female flowers are borne inside the young cobs.
- ***Kernel or caryopsis***: Maize kernel is one seeded fruit or caryopsis. Seed enclosed within the pericarp consists of the embryo, endosperm and remnants of seed coat and nucellus.

Composition of Maize Kernel

Protein	=	10.0%	K	=	0.40%
Oil	=	4.0%	P	=	0.43%
Water	=	13.5%	Mg	=	0.16%
Carbohydrate	=	70.0%	S	=	0.14%
Crude fibre	=	2.3%	Other minerals	=	0.27%
Albuminoides	=	10.4%	Ash	=	1.40%

Maize protein **zein** is deficient in tryptophane and lysine, the two essential amino acids. Maize is low in calcium and fairly high in phosphorus.

Maize Type

The genus *Zea* belonging to the tribe Maydeae of family gramineae with 10 pairs of chromosomes which has only one species, *Zea mays*. The maize was classified by Sturtevant in 1899 into seven groups or types based on the endosperm of kernels. They are pod corn, pop corn, flit corn, dent corn, flour corn, sweet corn and waxy corn.

1. **Pod corn (*Zea mays turnicate*):** The kernel is enclosed in a pod or husk, the ear formed is also enclosed in husk. Pod corn is also known as cow corn, forage corn and husk corn. It is not grown commercially.
2. **Pop corn (*Zea mays everta*):** The kernels are small and posses a higher percentage of hard endosperm starch. The ability to pop is due to rapid expansin of moisture in each individual starch grain upon the application of heat.
3. **Flint corn (*Zea mays indurate*):** The kernels consist of endosperm with soft starch in the centre and completely enclosed by a very hard outer layer. The kernels shrink uniformly as they mature.
4. **Dent corn (*Zea mays indentata*):** The kernels are characterized by a depression or dent in the crown due to shrinkage during ripening in a deposit of soft starch at the crown. This group is the most widely cultivated corn.
5. **Flour corn (*Zea mays amylacea*):** The kernels consist almost entirely of soft starch with a very thin layer of hard starch on the sides. They are also known as soft corn.
6. **Sweet corn (*Zea mays saccharata*):** The kernels are translucent, horney and more or less wrinkled in appearance when dry. They have a large proportion of sugar to starch and hence sweeter than other corns.
7. **Waxy corn (*Zea mays ceratina*):** The endosperm looks waxy adhesives and for textile and proper sizing.

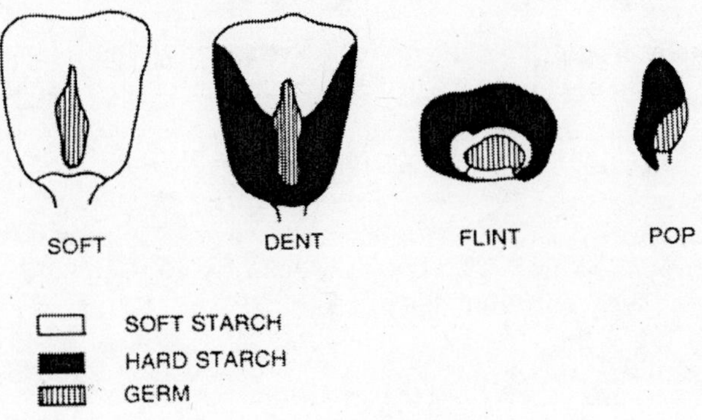

FIG. 6.9: Types of Maize.

Important Varieties of Maize

Very early maturing types	Sathi, Kathri, Teen Pakhia
Medium early maturing	Basi, KT 41, Jaunpur, Rudrapur Local
Late maturing	Malan

Hybrids and Composites

Hybrids	Ganga 1, Ganga 101, Ranjit, Deccan, Ganga 5, Ganga safed-2, Hi-Starch, Ganga 4, Himalayan 123, Ganga 3, VL 54.
Composites	Vijay, Amber, Sona, Kisan, Jawahar, Vikram
Nutritionally superior maize composites	Shakti, Rattan, Protina

The composite seeds can be used for subsequent sowing.

Growth and Development

A growing plant of maize passes through following growth stages:

1. **Seedling stage:** This is the sprouting stage which comes about one week after sowing and the plants have about 2-4 leaves at this stage.
2. **Grand growth stage:** This is knee height stage of the plants which arrives about 35 to 45 days after sowing. Plants need first top-dressing of nitrogenous fertilizer and final mechanical or manual inter cultivation because if this operation is delayed by any day the leaves will be damaged.
3. **Tasseling stage:** This is more technically called as flower initiation stage at which the tassels or male flowers come out. Final to[-dressing should be done at this stage. Fertilizer application after this stage does not give any positive response.
4. **Silking stage:** This is alsoknown as cob initiation stage at which the female flowers or cobs are formed. These are auxillary flowers unlike tasseles that are terminal ones. Usually they are formed in the axis of 11th to 13th leaf whereas male flowers or tasseles are formed at the apex after 14 or 15 leaves have come out,
5. **Soft-**dough stage: This may also be called as milky stage which commences after pollination and fertilization are over. Grains start developing but they do not become hard. This stage may be guessed by seeing the silks on the top of the cob which remain partially green at this stage and the covering of the cobs also remain green. This is the beat stage for using green cobs for table purposes.
6. **Hard-**dough stage: This is maturity stage at which the leaves get dried completely and become very brittle. Harvesting should be done at this stage.

Production Technology

Climate

Maize crop is warm weather loving crop, however it is grown in a wide range of climatic conditions as numerous plant types of varied climatic requirements are now available. Even so about 85 per cent of the total acreage under maize is grown during monsoon because of the fact that the crop stops growing if the night temperature falls below 15.6°C or 60°F. It is observed that the crop can not withstand frost at any stage of its growth.

The crop can be successfully grown in the areas receiving an annual rainfall of 60 cm which should be well distributed through out its growth period. The crop may tolerate an annual rainfall of 250 or 400 cm (as in Assam and West Bengal hills) provided the field does not get flooded as water logging is hazardous for the crop at any growth stage. Under scanty rains the crop must be irrigated for a better growth, development and yield of the crop, especially after tasseling stage.

It is observed that the crop needs more than 50 per cent of its total water requirement in about 30 to 35 days after tasseling and inadequate soil moisture at the grain filling stage results in a poor yield of shriveled grains.

Since it belongs to C-4 group of plant kingdom it needs bright sunny days for an accelerated photosynthetic activities and rapid growth of the plant but this has to be associated with an abundant water supply failing which the crop will start wilting or withering. Prolonged cloudy period is harmful for the crop but an intermittent sunlight and cloud or rain is the most ideal for its growth.

Now the farmers use to grow either hybrids or composite varieties which have larger and wider leaves, hence it may tolerate a very mild wind velocity as the leaves get torn or damaged at high wind velocity. At the grain filling stage severe lodging may take place because of torrential wind velocity and this would result in a heavy reduction in crop yields. The lodging may, to certain extent, be averted by planting the crop rows parallel to the wind direction or by planting the crop little earlier *i.e.*, fifteen to twenty days before onset of monsoon.

Climatic Requirements

Altitude: It grows from sea level to 3000 metre attitudes. It is grown in many parts of the country throughout the year. Growing season exceeds 140 days.

Temperature

Minimum temperature for germination .... 7°C
Most suitable temperature for germination... 21°C
For growth... 32°C
For best crop growth, mean day temperature.. 24°C
Mean night temperature... 15°C
Minimum lethal temperature.. 1°C
Maximum lethal temperature.. above 35°C.

Rainfall: 50-75 cm of well distributed rain is conducive to proper growth.

Soil

Deep, fertile, rich in organic matter and well drained soils are the most preferred ones for the crop, however, maize may be grown on a variety of soil types. The soil should be medium textured with good water holding capacity. The crop is very sensitive to water logging and since it is mainly grown during rainy season, care should be taken to assure that water does not stagnate on the soil surface for more than 4 to 5 hours. From the results of the various field trials it is observed that a loam or silt loam surface soil and a brown silt clay loam having fairly permeable sub-soil are the ideal soil types. Thus the ideal soil is the one which is neither clayey nor sandy and has a pH between 6.5 and 7.5 along with an exchange capacity of around 20 meq/100 g, base saturation of 70 to 90 per cent, bulk density of about 1.3 g/cc water holding capacity of about 16 cm per metre depth.

Field Preparation

The main objective of field preparation is to control weeds, facilitate sowing, provide aeration, and better seed soil contact improves water-holding capacity, etc. These things may be achieved by giving one deep summer ploughing immediately after harvest of *rabi* crop. It helps in an early sowing of the crop after giving one light ploughing by country plough followed by planking. In case of heavy soils two or three deep ploughings are needed whereas in light soils too much tillage is a wasteful. The crop generally needs two to four harrowings, therefore, seed-bed preparation hardly creates any problem.

During *kharif* season the crop has to be protected from water-logging, therefore, it to provide shallow furrows after certain distance and these furrows are connected with main drains. The drains must be laid out against the slope so that it may control soil erosion caused by water. Sometimes ridges and furrows are made in the whole field, especially in case of areas having high rainfall, and maize seeds are sown on the sides of the ridges. In this case the fertilizers are applied only on the ridges; in contrary to this the crop is sown in furrows during *rabi* and spring seasons for a better water use efficiency and higher yields.

Sowing of Seeds

Time of Sowing: With the availability of thermo and photo-nonsensitive maize varieties it has become now possible to grow this crop throughout the year. Usually, three crops of maize viz., *kharif*, *rabi* and *zaid* are grown in the country and sowing time differs accordingly.

***Kharif* Crop:** It is observed that after outburst of monsoon the rains impose a serious problem in sowing of the seed, therefore, it is always better to sow the crop at least a fortnight before rain starts so that the seedling are well established before they get torrential rains. Early sowing also reduces the weed competition, the crop matures earlier and the field becomes free for an early *rabi* season crop. *Kharif* sowing is done at different times in different regions:

Crop region	Sowing time
North-Eastern Hills	First week of March.
North-Western Hills	April end to early May
Peninsular India	May-June.
Indo-Gangetic Plains and A.P.	Mid June to Mid July.

***Rabi* Crop:** This crop is popular in Bihar, parts of U.P., Andhra Pradesh, Tamil Nadu, Karnataka and Maharashtra. The optimum date of sowing is end of October to mid of November. The crop is raised under assured irrigation.

Spring Sowing: This crop is grown in Bihar, Maharashtra, and tarai belt of U.P. and the most optimum time for crop planting is late January to end of February. The crop is grown irrigated.

Seed Rate and Spacing

Maize, being a non-tillering crop, needs to be very properly placed. Seed rate and spacing are inter-related factors and both have their telling effect on each other. However, seed rate and spacing depend upon growing season, variety, size and weight of the seeds. Generally, the crop is line sown except under rainfed conditions and fodder crop where the farmers sow seeds broadcast. Keeping everything in mind it has been observed that a spacing of 60-75 cm x 25-30 cm is the most optimum for higher grain yield. The crop responds to a planting density of 55,000 to 70,000 plants per hectare, though it depends on the variety, growing season and its duration. Early maturing hybrids respond to higher plant population and *vice-varsa*.

The crop is sown with the help of a seed-drill, or by dropping the seeds in furrows behind the country plough. Under low lying areas the seeds are dibbled by the help of a *khurpi* but under no case the depth of seeds should exceed more than 5 to 7 cm depending upon soil texture and moisture status of the field.

With corresponding spacing mentioned above a seed rate of 15 to 20 kg for grian crop and 25 to 30 kg for fodder crop have been found to be an ideal. Usually, a relatively lower seed rate of the crop is used during *kharif* season because of a very luxuriant growth of the plant.

Seed Treatment

Maize seeds should be treated with fungicides like Captan or Thiram or Emisan @ 2 to 3 g/kg of seeds. Seed treatment protects the crop from seed borne diseases viz., leaf blight in maize.

Method of Sowing

The crop is sown in for different methods viz.,

1. **Broadcasting:** This method is still common by some backward farmers but it is unscientific, requires more seed rate, imposes problem in weeding etc. and hence this should be discouraged.
2. **Drilling Method:** This method is adopted by big farmers in which seed drills are used for sowing the seeds. Some times under high moisture condition of the soil it becomes difficult to run the seed drill because the sproughts are chocked with the wet soil and black rows are observed. The method is well suited for sandy soils and is very quick.
3. **Dibbling Method:** This method is very time taking as the seeds are placed by the help of a *khurpi* manually at required distance in the rows that are made 75 cm a part. It needs less seed rate and is the best method for costly seeds.
4. **Hill Sowing:** In this method the seeds are dropped with the help of corn planters. The planters are placed at an appropriate place pressed for dropping three or four seeds at one place which is technically known as a hill. It also needs relatively less seed rate.
5. **Sowing the seeds behind the plough:** This method is very common in almost all the maize growing areas of the country. In this case shallow furrows are opened by the help of a country plough and seeds are dropped in them and planking is done thereafter to cover the seeds. After the seedling has emerged out, they are thinned out for maintaining a required plant population.

Sowing the crop in sun-rise and sun-set direction resulted in a higher grain yield and production of healthier plants because of least no mutual shading of the plants. The response was more pronounced with hybrid and composite varieties as they are most sensitive to shading effects.

Interculture

Thinning should be done keeping one plant per hill within 15 days after germination. Early interculture of maize within 2 to 3 weeks of germination destroys weeds. A second hoeing and earthing at 6-7 weeks stage, is a very useful practice. The hoeing or interculture should be shallow to avoid root injury. Earthing helps to provide better anchorage to the crop and prevents lodging from strong winds in *rabi* and summer season.

Fertilizer Management

The maize is an exhaustive crop and prefers light soil types for its cultivation which further aggravates the nutrient deficiency as they are lost quickly after fertilizer application in the field. The fertilizer management, therefore, is of paramount importance in the crop cultivation. Nutrient requirement and its mode of application is, however, governed by a number of factors viz., soil type, planting season, variety, preceding crop, method and time of fertilizer application etc. It is observed that maize grown after potato needs relatively less doses of fertilizers and so is the case when the crop is grown after any pulse crop. Irrigated crop also needs higher doses of inputs than rainfed ones.

Maize requires a regulated and assured supply of nutrients particularly nitrogen throughout its growing period right from seeding stage to grain filling stage. A judicious application of organic manure such as well rotten compost or F.Y.M. at the rate of 15 to 20 tonnes per hectare about 20 days before sowing the crop, green manuring or application of oil cakes at the rate of 6 to 10 quintals per hectare has been found to be most ideal for an increased crop yields, especially for grain production.

Demand of plants for nitrogen is more than any other nutrient and it is noticed that its deficiency at any stage of growth, especially at tasseling and silking stage, may lead to virtual crop failure. The nitrogen deficiency is characterized by stunted and spindly plant growth with yellowing of green foliage particularly the lower leaves. The nitrogen utilization pattern is found to be increased from seedling to knee height and reaches to the peak at tasseling stage when the plants remove nearly 4-5 kg N/ha/day (the data was recorded in case of hybrid and composite varieties). It is interesting to note that the response of applied nitrogen is the highest in poor fertility conditions than the normal ones. Under an ideal fertility management it is noticed that the new plant types yield 15 to 25 kg grain per kg of applied nitrogen. Plants respond to a dose of 120 to 150 kg N. ha and so is the recommendation for hybrids and composite varieties, however, in certain cases the economical yield of maize of maize grains has been obtained even up to the application of 200 kg N per hectare. The crop is generally grown under rainfed or partial irrigated conditions during rainy season and nitrogen loss through leaching become very obvious from the field. Therefore, split application of the same is needed. From the results of various trials it is clear that the best results from applied nitrogen may be achieved when nitrogen is applied in three splits viz., 1/3 rd at sowing, 1/3 rd at knee-height (about 35-40 days after sowing) and remaining 1/3 rd at tasseling stage. Among the methods fertilization, placement below the seed and side dressing of nitrogen at second and third split applications gave the best results. The crop needs relatively lesser nitrogen when grown during winter season which may be accounted for higher uptake and lower loss of the nutrient from the field. Similarly, during spring season when the crop is taken after potato, peas or gram it needs low nitrogen content. On an average it is found that 100 to 125 kg N/ha becomes sufficient but the mode of application should be be same as in *kharif* season.

Phosphorus, as usual, helps in development of all phases of maize plants and its deficiency leads to purpling of leaves and tends to delay maturity with an imperfect ear formation. It shows its deficiency mainly at the seedling stage, though it is needed most after flowering stage. Indian soils are found to be rich in phosphorus content and, therefore, the crop does not respond well to the applied doses of phosphate. Thus, the response of applied phosphate is directly proportional to fertility status of the soil and the previous history of the fertilization/ Therefore, for an economical application of phosphate and best response of the crop it becomes essential to get the soil tested for the availability of phosphate in the soil and the crop should be fertilized accordingly. Under low or medium availability of soil phosphorus of 50 to 80 kg P_2O_5/ha may result in an excellent crop. Experimental results of phosphate nutritional trials show that the grain yield was increased to the extent of 50 to 200 per cent depending upon phosphate availability of the soil. Basal placement of phosphate in rows about 5 to 8 cm deep in the soil gives the best results as it is less mobile and it does not get lost through leaching. Thus, it remains available to the plants until they need it.

Potassium is essential for translocation of water and photo-synthate within the plant body. Higher doses of nitrogen and phosphorus in hybrids and composite varieties make it essential to apply potassium to the crop. It is observed that the plants need more than half of their potash requirement up to or before flowering stage which makes it a must to apply potassium along with phosphate as basal row placement. Its deficiency is characterized by shortening of internodes, yellowing of leaves and firing at the margins of the lower leaves. Soil is generally rich in potash contents and hence the response is not very linear, however, based on rest test data the crop may be fertilized with 40 to 60 kg K_2O/ha for a remunerative return It is noticed that the potassic fertilizers containing Cl ions into them result in a poor grain filling and lower yield of the crop. Cultivation of new plant types, leveling of fields and light texture of soil lead to zinc deficiency that results in poor grain yield of the crop. Deficient plants show a broad band of bleached tissues on each side of the midrib, beginning from the base of the leaves. Zinc deficiency is very pronounced in the soils of north Indian states viz. Gujarat, Punjab, Jammu and Kashmir, Delhi. U.P. and A.P. Therefore, it is advisable to apply 15 -20 kg zinc sulphate along with basal application of fertilizers. The deficiency of Zn in plants at latter stages of growth, however, may be corrected by foliar application of $ZnSO_4$ dissolved in water with half the quantity of lime (5% $ZnSO_4$ + 2.5% hydrated lime).

Weed Management

The crop is heavily fertilized and sparsely grown; both of these attract asevere weed infestation that result in a drastic reduction in grain yield. Crop responses very well, to mechanical weeding done by bullock or tractor drawn implements between the rows and hand weeding within the rows. The weeding operation may be repeated two or three times but not after knee-height stage of the crop. Losses caused by weeds are more during the early stage than in the later stages. Mechanical weeding is good for root aeration but continuous rains during *kharif* season pose a serious problem in which the soil becomes wet and any agricultural operation can not be done. Under such conditions the weeds grow very luxuriantly and they do permanent harm to the crop before they are controlled. Therefore, an integrated weed control measure is essential for taking care of weeds from initial stage of growth. Mechanical weeding in any case should not be done after knee height stage because it will damage plant leaves.

The crop is found to be infested with grassy and broad-leaved weeds. The important grassy weeds are *Echinicola colanum* (Sawan), *Dactyloctenum aegyptium* (Marka), *Elusine indica* (Jhingri), *Saiaria gluca*, *Cyperous species* (Motha), *Cynodon dactylon* (Doob) and *Pharagmites karka* (Narkul) whereas *Amaranthus viridis* (Chaulai), *Celosia argentina* (Safed Murga ka Phool), *Phyllanthus niruri* (Hazaardana), *Cleome viscose*(Hurhur) etc. are the most common broad-leaved weeds associated with maize crop in the field. Variety of weeds and its infestation varies from season to season and place to place *e.g.* the weeds differ during *kharif*, *rabi* and *zaid* or spring season.

The following weed control measures should be adopted for an efficient and effective control of weeds.

1. Pre-emergence application of Simazine or Atrazine at the rate of 1 to 1.5 kg/ha of 50% W.P. should be done but the field must be free from all established weeds.
2. If the broad-leaved weeds are posing problem a post-emergence application of 2, 4-D or Banvel-D (Dicomba) should be done at the rate of 1.5 to 2 kg a.e./ha.
3. Two or three weedings followed by earthing up for proper standability of crop takes complete care of the weeds. Operations must be finished before flowering starts as it damages the leaves and lateral roots of the crop that ultimately reduces the grain yield. The most appropriate time for first weeding is when the maize seedlings become about two weeks old. Two hoeings at a week's interval should be given after wards to keep the soil friable, clean and free from crust formation. If needed, the crop may be given one additional weeding after about 30 days of sowing. It is essential to thin out unwanted plants from the crop rows just to avoid competition among them. Thinning operation must be done at the time of first weeding or even earlier.

Water Management

Maize is grown in three seasons viz., Kharif, rabi and zaid or spring season of which rabi and zaid crops totally depend on irrigation while kharif crop is mostly grown rainfed. Maize is a vigorous, quick and tall growing plant having broad leaves, therefore, its water requirement is exceptionally high. A vigorously maize plant needs about 2 to 3 litres of water per day during peak-growing period or on an average it's consumptive use of water varies from 2.5 to 4.3 mm per day. However, frequency and depth of irrigation depend upon climatic condition, soil type, type of the variety of the crop grown and organic matter content of the soil. The growth stages of the crop also affect the water management requirement to a greater extent. It is observed that a good crop of maize needs a rainfall or irrigation about 1.0 to 1.2 metres per hectare during its life cycle and most of which is needed during the growth period. There are certain critical growth stages of maize for irrigation in new plant types at which moisture stress results in a considerable reduction of grin yield but wilting of the plants for more than two days at any stage of its life cycle is harmful. The most pronounced critical growth stages for irrigation are seedling stage, knee-height stage, flowering (tasseling and silking stage) and grain filling stage. Of these moisture stress at early growth stages results in a delayed tasseling and silking by 3 to 5 and 5 to 8 days respectively, while stress at latter stage of growth causes in a drastic reduction in yield. Excess water or water-logging at the same time is equally or even more harmful as the crop can not tolerate stagnant water and is very sensitive to water logging. Crop grown under water-logged conditions become pale, grow poorly and the uniformity of the crop is lost. It is observed that water logging in maize field for one day damages the crop, and for three days it almost spoils the whole crop. Therefore, an efficient drainage is of greater significance than the irrigation

Irrigation scheduling during any season of cropping in maize at o.30 mm or on depletion of 25 to 30 per cent available moisture from field capacity has proved to be beneficial. The frequency or interval of irrigation is mostly based on the growth stages of the crop and availability of irrigation water *e.g.* the crop should be irrigated at least four times viz., seedling, knee-height, tasseling. Silking and grain filling stages respectively but under limited water supply it may be irrigated thrice at seedling, tasseling and grain filling stages or twice at knee-height and grain filling stages based on the availability of water. The crop is grown rainfed during *kharif* season and water stress at early growth stages results in an elongation of roots which ultimately induces drought resistance in plants, though it delays flowering of the plants.

Water logging is highly detrimental for the crop as these two stages are most sensitive for excess water and water logging for more than 3 or 4 days may result in a yield reduction to the extent of 50 per cent or even in total crop failure depending upon severity of water logging. The losses of water logging may be averted by growing crop on ridges in poorly drained fields and by giving light and frequent irrigations to the crop. The hardiness in plants during *kharif* season may also be induced by growing the crop at least 15 to 20 days earlier than the onset of the monsoon.

Plant Protection Measures

Diseases of Maize and their Control

Maize plants suffer from the attack of a number of bacterial and fungal diseases which are described below:

Stem Diseases: Maize plants from the seedling stage suffer from attack of soil and seed-borne fungi and bacteria. These diseases are:

1. **Seedling blight:** the young seedlings get infected by various fungi and they start dying. The disease may be controlled by seed treatment with Captan/Thiram @ 3g/kg of seed before sowing.
2. **Soft rot:** The disease is caused by the bacterium (*Erwinia carotovora* var Zeae) in which the plants get infected at the collar spreading up to a height of 75 cm from the ground level at any stage of the crop growth period. The diseased plants show typical wilt symptoms and they collapse to the ground because the organism dissolves the middle/lamella of the cell walls that results into disintegration of tissues. The plants show white oozing at the infected parts that stinks and may be seen by the naked eyes. Poorly drained heavy soils along with hot and humid weather favour the attack of the disease. The disease may be controlled by providing an efficient drainage and by selecting a light textured soil for growing the crop.
3. **Stalk rot:** It is caused by *Cephalosporium sp.* and the plants get infected as secondary infection due to mechanical injury during field operation or by insect borers. The disease spreads from one internode to another of the plants especially in case of susceptible varieties. Disease may be controlled by selecting resistant varieties, control of insect borers and avoiding mechanical injury to the plant.
4. **Charcol rot:** It is caused by *Microphomina phaseoli*. The plants get affected at the final stage in which the effected parts become dry and brittle having innumerable pycnidia. The disease may be controlled by adopting a long term rotation of 4-5 years duration, rouging out diseased plants and by growing crop in properly drained light textured soils.

Foliage Diseases: These diseases damage the foliage of the affected plants:

1. ***Helminthosporium spp*:** *Helminthosporium turcicum* is common in local varieties of Southern India and kills about 75 per cent plants whereas, *Helminthosphorium maydis* is found prevalent in North India that causes devastating blight. Pre-sowing seed treatment with Captan, Thiram or Agrosan G. N. @ 3 kg/kg seedcontrols the disease.

2. **Leaf rust:** The disease is caused by *Puccinia sorghi*. It is very common in Himalayan region during *kharif* season and in southern India during *rabi* season. Spraying of 0.2 per cent Zineb solution 2-3 times at an interval of 10 days soon after appearance of rust sori on the leaves.
3. **Downy mildew:** The mildew is caused by *Sclerophthora philippinensis* which is frequent to the extent of 5 to 6 per cent. A new mildew caused by S. rayssiae was reported in 1967 from West Bengal, western M.P., southern Gujarat, northern Bihar, A.P., parts of U.P. The diseases plants show long parallel stripes, 3 to 7 mm wide, yellowish–tan to purple-gerugemons in colour. Defoliation, malformation or shredding is not common in affected plants.
4. **Smurt:** *Ustilago zeae* causes smut disease in ears and tassels by which they are partially transformed into galls. The disease may be controlled by applying lower doses of nitrogen, uprooting of smut affected plants and adopting long duration crop rotation.

Pathogens found to attack cobs: There are some pathogens that affect the cobs at all stages including storage. Of these *Cephalosporium sp.* infect the grains by showing a characteristic feature of fan-like streaks on the stigmatic side of the endosperm and 5 to 10 per cent grains in each cob are found to be affected by this pathogen, *Rhizoctonia sp.* and *Curvularia sp.*, *Fusarium sp.* are also seen on the cobs during storage condition if the cobs are moist *Aspergillus sp.* are also common on cobs in the storage.

Important Insect-pests

There are many pests attack the crop and cause a considerable damage to the crop, if they not controlled at an appropriate time. Following are the important ones that need due care:

1. **Stem borer (*Chilo zonellus Swinh*):** Stem borers are the most damaging and serious pest of maize and they are common in all the states where maize is grown. Borers attack the crop right from seedling stage to the older plants. In case of seedlings the insect makes dead hearts while in older plants the larvae bores into the stalk through apex and makes tunnels that results into a considerable reduction in the grain yield. The cobs are also seen attacked by these borers and grains are damaged. Following spraying schedule be adopted for the effective control of insect:

Spraying schedule of insecticides on Maize crop.

Spray	Crop age	Insecticides
First	10 DAG	Cararyl (200 g Sevin 50 WP) or trichlorophon(400 g Dipterex 80 SP), or endosulfan(250 ml Thiodan 35 EC) in 200 litres water by manually operated sprayers.
Second	20 DAG	Sevin 300 g or Dipterex 550 g or Thiodan 350 ml in 300 litres water by sprayer or soil application of 5 kg Endrin 2 per cent granules.
Third	30 DAG	Sevin 400 g or Dipterex 750 g or Thoidan 500 ml in 400 litres of water by sprayer
Fouth	40 DAG	Repeat as in third or use granules as mentioned in second spraying at this stage.

DAG: Days after germination.

Other pests like Pink borer, Leaf borer, grass hoppers and Army worms may be nicely controlled by spraying of Thoidan 500ml or 350 ml Methyl Parathion 50 EC or 900 ml of Endrin 20 EC/ ha in 350 litres water (the plants should not be fed to the cattle upto three weeks after application of these insecticides as they cause a serious stomach disorder). Dusting the crop with Aldrin 5 per cent at the rate of 25 kg/ha or malathion 5 per cent 30 kg/ha controls hairy cater pillars.

2. **Shoot fly and aphids:** Following measures should be taken to control them
   a) There should be no water logging in the field.
   b) Application of 10 per cent Phorate (Thimet) granules in the furrows before sowing at the rate of 15 kg/ha should be done and the insecticides should be covered with a thin layer of soil after which the seeds should be sown.
   c) Spraying was 0.01 per cent Metasystox (1 ml chemical in a litre of water) solutionat the rate of 560 litres per hectare after one week of germination. The spraying should be repeated at every 15 days interval for more effective control.

Harvesting and Grain Shelling

The grain crop of maize is harvested when cob sheath turns brownish, grains become hard and they do not contain more than 20 per cent moisture in them. In case of indigenous varieties the plants start yellowing and drying along with maturity of the cobs but in case of new plant types the plants may remain green at the time of full maturity of the cobs, therefore, colour of the plants become misleading in such case. There some symptoms of maturity in hybrids and composites such as the sheath becomes brown, sheath becomes loose, silks dry up completely etc. There are two methods of maize harvesting viz., plucking of cobs and stalk cutting. In case of former method the ears are removed from the standing plants and they are piled up for twenty four hours then they are spread for drying in the sun. In this method the stalks may be used as green fodder. While in stalk-cut method the plants are cut and piled up in shade and the cobs are removed after two or three days of harvesting. The dried plants are used for hay making. The cobs thus removed are dried as in previous method and grains are separated out from them. Common practice in villages is hand shelling but efficiency remains very poor. Now corn-shellers of greater efficiency which are manually driven, tractor driven, electrically operated are available for use. The left over plants are used for hay, the sheath is used for fuel, and cobs after grains are separated out are used as fuel.

Maize grown for fodder are harvested at the milk to early dough stage, the earlier harvested crop usually yields less and is poor in protein content. For silage making late dough is preferred.

Grain yields varies according to variety, place of cultivation and inputs used. Normally in case of local varieties of low fertility a yield of one or two tones per hectare is observed while hybrid/open pollinated /composite varieties grown in Indo-Gangetic plain have yielded about four tones/ha and the same varieties in Peninsular India around 5-7 tonnes/ha.

Important Practices

1. Sowing in lines and early weeding.
2. Maintenance of adequate plant population.
3. Moderate fertilizer use.
4. Critical stages of growth for Nitrogen application: (i) Germination, (ii) Knee high, (iii) Tasseling stage.

5. Critical stages of growth for irrigation: Tasseling to silking stage.

Important Observations

- Three types of roots: (a) Seminal roots, (b) crown/coronal root, (c) Brace/aerial/Prop roots.
- Maize is monoecious plant/cross pollinated crop. Male inflorescence is called **tassel** and removal tassel is called **detasseling.** The style is a very long silky filament. The cluster of which is known as **Silk**.
- Maize protein s called **Zein** and deficient in tryptophane and Lysine.
- Idea of **Hybrid** maize was first conceived by E. M. East and G. H. Shull in 1910 by **Single Cross technique** (mostly used in USA and China)
- **Double cross technique** for hybrid seed production by D. F. Jonese (1920). The double cross technique is mostly used in **India**.
- **All India Coordinated Maize Improvement Project** was started in 1957. First time in 1961, four **double crossed hybrids** were released: Ganga 1, Ganga 101, Deccan, Ranjeet. Later on, other double cross hybrids were released: VL 54, Himalayan 123, Hi-starch, Ganga 2, Ganga 3 and Ganga 5.
- **Composites are Jawahar, Vikram, Kisan, Amber, Sona and Vijay.** Among composites, Amber has the highest yield potential (50-55 q grain/ha**). In 1967, these are released for the first time in the world.**
- **Lysine rich** composites (around 10-11% proiein and 3.5 to 4% lysine) are **Protina, Shakti** and **Rattan,** released in 1971; such varieties are called **Opaque 2 composite.**

BARLEY (*HORDEUM VULGARE* L.)

Barley is believed to be originated from Asia and Ethiopia, Its Sanskrit name **'Yaw'** is mentioned in **'Veda'** and use of barley in religious ceremonies has been described which shows that barley was grown in India since ancient time. It is an important cereal crop in the world cultivated in the world in temperate and sub tropical areas. It ranks fourth with respect to area and production among cereals after wheat, rice and maize. In 1995-96, production was 155.3 m t from an area of 66.6 m ha. The average productivity is 2.3 t/ha. Barley is cultivated in Russia, Canada, France, Spain, U.K., Germany, Turkey, Australia, Morocco and Ethiopia. In India, it is an important winter cereal cultivated in an area of 0.84 m ha with the annual production of 2.2 m t. The average productivity of barley is 1.79 t/ha. Barley is predominantly grown in the states of Eastern Uttar Pradesh, Rajasthan, Madhya Pradesh, Haryana, Punjab, Bihar and Himachal Pradesh.

Barley, a number of the gramineae family, fins its use in the feed industry and brewing. Barley grains are mainly used for the purpose of malting. They are also used as feed for cattle and horses and the straw is also fed to cattle.

Botanical Description

Like other cereals, barley also belongs to the grass family and has all the vegetative characters like wheat except that the auricles of leaf are conspicuous and very much pronounced which clasp the culm.

1. **Root system:** It consists of shallow and deep roots. Shallow roots spread laterally about 15-30 cm at right angles to the tillers. Deep roots go up to 0.75 to 150 cm.
2. **Stem:** Stem is cylindrical and possess five to seven hollow internodes

separated by solid nodes.

3. **Leaves:** Leaves arise alternately and consists of a sheath, blade, ligules and auricle. Auricles are very conspicuous.
4. **Inflorescence:** Inflorescence is called spike or head. Spike consists of spikelets attached at the nodes of a zig-zag rachis and each spikelet has two glumes and a floret. In two rowed barley only central spikelet is fertile whereas in six rowed barleys all the three spikelets are fertile.
5. **Kernel (grain):** Grain of barley is a caryopsis consisting of lemma, palea and a rachilla. Caryopsis is composed of the pericarp, endosperm and embryo. Barley grain is a caryopsis measuring about 8-12 mm long, 3-4 mm wide and 2-3 mm thick. The grains normally consist of hull except that of naked variety of barley grains which become free after threshing. Spikelets may be awned or bearded, awnless or hooded as the case may be because all the three types are under cultivation.

Composition of Grain

Moisture	12.5%	Albuminoids	11.5%	Carbohydrates	74%
Fat	1.3%	Crude fibre	3.9%	Ash	1.5%

Classification of Barley

Fertility of the lateral spikelets forms the basis of barley classification and the cultivated barley may be classified into three main groups viz., (i) Six-rowed barley (*H. vulgare*, L. emend. Lam.), (ii) Two-rowed barley (*H. distichum*, L. emend. Lam.), (iii) Irregular barley (*H. irregulare*, E. Aberg and Wiebe), (iv) Four-rowed barley (a six rowed barley with two rows of lateral florets overlapping to form single row). Of these, the six rowed barley is the most commonly cultivated type.

Agro-Technique for Barley Production

Ecological Requirements

Climate: The tropical climate favours barley crop, though it may be grown under subtropical climatic conditions. The crop requires around 12-15°C during growing period and around 30°C at maturity. It can not tolerate frost at any stage of growth and incidence of frost at flowering is highly detrimental for yield. The crop possesses very high degree of tolerance to drought and sodic conditions. It is commonly grown under rainfed or drought –prone areas but rains during growing period result in good growth of the crop, however, there should be no rain at maturity because it causes discoloration of the grains and also makes them unfit for malting or seeding.

Soils: Sandy to moderately heavy loam soils of Indo-Gangetic plains having neutral to mid saline reaction and medium fertility are the most suited types for barley cultivation, however, it may be grown on a variety of soil types. Being salt tolerant it is the best substitute for sodic soils and with same reason its cultivation has become possible on saline coastal areas of Sundarban in West Bengal and saline black soils of canal irrigated areas of northern Karnataka. A higher spot with an efficient drainage would be the best location for barley growing. The soil should not be very fertile as the crop lodges very severely and a drastic reduction in yield is observed. Acidic soils are not fit for barley cultivation but if it is unavoidable to grow, liming with appropriate quantities of CaO should be done at least one month earlier to sowing.

Land Preparation

Barley, being a shallow rooted crop, responds well to light textured fine seed bed. The desired tilth may be obtained by one ploughing with soil turning plough followed by 2-3 ploughings with desi plough or 2-3 harrowings by tractor or bullock power. The bed should be planked after every ploughing or harrowing as it pulverizes soil and also preserves moisture. In dry land areas the rain-water may be nicely converted by deep summer ploughing, discing after every rainfall, leveling and bunding during monsoon and the moisture thus preserved may be fully utilized by an early sowing of barley seed. The roots and stubbles of previous crop and established weeds should be picked up and thrown out of the field as they attract termites. Before last ploughing 20 kg of 10% BHC dust or 10-15 kg of 5% Aldrin dust should be applied and mixed well with soil to protect the crop from termites. In case of saline soils the land should be prepared and planked nicely and then left undisturbed for nearly a week's time before sowing. This results in an upward movement of soluble salts from root zone and ensures quick germination and better growth of plants.

Seed and Sowing

Time of sowing: Timely sowing of barley ensures good returns from the crop. The sowing of rainfed barley should be adjusted with soil moisture status and prevailing atmospheric temperature. Results of All India Barley Improvement Project indicate that sowing time of barlet differs in accordance with agro-climatic zones of the country as mentioned below:

Zones	Time of sowing
Lower hills (upto 1000 m above sea level)	Last week of Oct to mid. November
Mid-hills and valley areas (1000-1500 m above sea level)	2nd fortnight of Oct. to first week of November
High hills (1500-2200 m above sea level)	First fortnight of October
Dry temperature zone (above 2200 m above sea level)	2nd fortnight of April to first fortnight of May

Seed rate and spacing: The seed rate depends upon its test weight, spacing, sowing time and method as well as fertility status of the soil. However, it has been observed that normally a seed rate of 90-100 kg/ha under timely sown in medium fertility lands and about 120 kg /ha under late sown in poor fertility lands should be used. For an ideal crop a row spacing of 23 cm in irrigated and 23-25 cm in rainfed conditions should be given. The depth of sowing may be 5 cm under irrigated conditions and 6-8 cm under rainfed conditions, however, it depends upon soil moisture status of the field at the time of sowing.

Method of sowing: Line sowing in shallow furrows is superior over broadcasting. The best method of sowing is dropping seeds with a chonga (a tube connected with a funnel like structure behind country plough) attached to a desi plough or with a seed drill. Dropping seeds in open furrows behind desi plough is found to be inferior to drilling. Soil should be made compact by planking or running a roller for better seed-soil contact

Seed treatment: Use of treated seeds is always better than inviting the disease and controlling it later. Therefore, seed treatment is the most important operation in crop production. Following chemicals are recommended for seed treatment to protect the crop from fungal diseases which are mentioned against them:

Diseases and Chemicals used to Control them

Diseases	Name of the chemicals	Rate of chemicals
Helmintho sporium, leaf stripe, covered smut, other seed borne diseases	Ceresan, Agrosan GN	2.5 g/kg of seed
Loose smut and other seed borne diseases	Vitavax, Denlate or Bavistin	-do-

In case of saline and rainfed areas sowing of overnight water soaked seeds results in a quick germination and it also ensures better stand of the crop.

Fertilizer Application

It is believed that high fertility results in lodging of barley crop and hence it is the preceding crop in the rotation which is often fertilized and barley is grown on residual soil fertility. However, evolution of new high yielding dwarf varieties have shown a positive experiments conducted at various centres of All India Barley Improvement Project conclude that there is an increase in grain yield with increasing doses of nitrogen up to 100 kg N/ha however, it depends upon sol test values, climate and variety of the crop. Contrary to this results of experiments conducted at other centres show that barley responds strikingly up to a dose of 20-40 kg N/ha under rainfed and 40-60 kg N/ha under irrigated conditions. The response pattern shows an additional yield of 2-6 quintals/ha by applying 20 kg N/ha under rainfed and 7-9 quintals/ha under irrigated conditions. The source of Nitrogen had no effect on either grain yield or yield attributes.

Indian soils are rich in phosphate and potash contents, therefore the economic doses should be determined only after having the soil test values. It is observed that phosphate helps in increasing response of applied nitrogen and in developing an extensive root system in dry land agriculture. An economical dose of 20 kg P_2O_5/ha has been found to be an ideal, however, the response of potash is not yet recorded but a dose of 20 kg K_2O/ha has been found to be an appropriate dose.

Mode of application: Based on research findings it has been found that in case of irrigated heavy soils half of the nitrogen, full dose of phosphate and potash should be applied as basal and remaining half of nitrogen should be top dressed after first irrigation or 30 days after sowing, while in case of light soils $1/3^{rd}$ N with full dose of phosphate and potash should be applied as basal, $1/3^{rd}$ N after first irrigation and the rest $1/3^{rd}$ N after second irrigation. Rainfed and the barley grown under saline soils should be fertilized by applying entire quantities of fertilizers as basal placed at a depth of 8-10 cm into the soil.

Irrigation

Barley requires lesser quantity of water than many crops because of its higher drought tolerance but under heavy fertilization it responds well to application of irrigation water. However, number of frequency of irrigation depends upon winter rains, soil types and variety. In dwarf mutant varieties following critical growth stages have been recognized for irrigation.

1. Seedling or sprouting stage which requires sufficient moisture at the time of sowing.
2. Active tillering stage *i.e.* 30-35 days after spwing.
3. Flag leaf stage.
4. Milky stage or soft drought stage or grain filling stage.

Of these stages active tillering stage and grain filling stage are most critical ones that come around 30-35 and 60-65 days after sowing respectively. Water logging or heavy irrigation should be avoided especially during February/March because this results in crop lodging. If there are good winter rains coinciding to these stages, the number of irrigations should be reduced accordingly.

Weed Control

The weeds must be effectively controlled in the same way as in case of wheat but if wild oats and *Phalaris minor* are not present in the field, only 2,4-D 500-750 g a.i./ha dissolved in 400-700 litres of water should be sprayed and it takes care of most of the common weeds. It must be sprayed within 30-40 dates after sowing. The volume of water is kept about 300-400 litres in case of irrigated and 600-700 litres in case of rainfed areas.

Plant Protection Measures

Pests and their control: Barley is often attacked by soil insects like white ants, Gujhia weevil and cutworms which may be controlled by soil application of 5% Heptachlor or 10% BHC at the rate of 20 kg/ha.

The second type of pests is foliage pests like aphids that may be controlled by spraying of 0.5% Methylparathion (metacid), or 0.05% Penitrothion (Sumithion), or 0.05% Endosulfan (Thiodan), or 0.05% Lindane over the foliage. The spraying should be repeated 2-3 times at 10-15 days intervals if aphid infestation continues.

Diseases of Barley

Stripe disease: It is caused by *Helminthosporium graminium* and becomes very serious problem for barley. It may be controlled by over night immersion of seeds in 2% Thiram or Captan solution at 25°C. The seeds should be dried, bagged and stored properly. Dithane M-45 can also be used but it has only partial control of the disease.

Stripe rust: This is also very important disease of barley which is common in almost all the released varieties. It is observed that seed treatment with 0.2% Plantvax prevents the attack upto boot stage. If the seed treatment is not done, spraying the crop with 2% systemic fungicide like Plantvax (Oxicarboxin) controls the disease.

Molya disease: It is caused by nematods (*Heterodera avanae*) and the disease has been noticed in Rajasthan and adjoining areas of Punjab and Haryana. In sandy soils the losses can be as high as 70 per cent. The infected plants become pale and stunted. Soil treatment with nematicides like Aldicarb S, DBCP; Aldicarb 10 G, Counter 5G, Cytrolane 5G and Carbofuran 3G control the disease but the treatment is very expensive. Seed treatment with Carbofuran SP and Flowable and oftanol is quite effective in controlling the disease and is less expensive also. Apart from use of chemicals some preventive measures such as rotating barley with non-host crops like sunflower, mustard, potato, gram, onion, etc., 2-3 deep summer ploughings at 10 days intervals and selection of resistant varieties like RD 347, etc. are also found to be very effective. If the chemicals are to be applied, a dose of 1 kg a.i./ha should be drilled. Aldicarb and Carbfuran also control aphids completely.

Smuts: Both covered as well as loose smuts are common in barley crop which are seed-borne diseases. Seed dressing with Organomercurials and non-mercurials like Zineb, Dithane M-45 are very effective. Use of wide spectrum systemic fungicide like Benlate controls loose smut of barley. Vitavax (Carboxin, a systemic fungicide) may control internally seed borne infection of *Rhizoctonia* *spp.* and *Selerotium spp.*, the seed borne pathogens which cause root rot of barley. Mixtures of formulation like Vitavax + Thiram or Benlate + Thiram both at the rate of 0.25% concentration increase the fungicidal activity.

A part from chemicals, some agronomic practices like depth of sowing and time of sowing are also found to be effective in controlling the spread of diseases. It is observed that the incidence of disease infection is directly proportional to increased depth of sowing and also late sowing.

Huskless Barley Varieties

1. **Karan-3:** It is semi-dwarf variety good for Gangetic plains and rainfed U.P. which yields about 40 to 50 quintals of grains per hectare.
2. **Karan-16:** It is dwarf variety good for limited irrigated areas of Eastern Plains Zone and Diara lands which yield about 45-50 quintals grains per hectare.
3. **Kiran-18:** It has all the characters of Kiran-16 except that it is recommended for Gengetic Plains, saline-alkali and Diara lands of Eastern plains Zone of India.
4. **Kiran-19:** It is similar to Kiran-18 for adaptation, habits and growth characters except that it yields about 50-55 quintals of grins per hectare.
5. **Kiran-750:** Identified for Central Zone irrigated areas.
6. **DL 487:** Identified for NEPZ irrigated areas.
7. **K 1155:** Identified for NEPZ irrigated areas.
8. **Kiran 757:** Identified for late sowing in NWPZ irrigated areas.
9. Other huskless barley varieties are Karan-163, Karan-231, Karan-265.

Hulled Barley Varieties

1. **Karan-15:** It is suitable for every part of country. It is dwarf growing variety and yields about 50-60 q/ha.
2. **Kiran-280:** It is suitable for rainfed NWPZ areas. It has semi-dwarf plants and yields 40 q/ha.
3. **NH-87:** It has semi-dwarf plants. It is good for rainfed NWPZ and may yield about 35-40 q/ha.
4. **K-392:** Tall variety identified for NEPZ irrigated areas.
5. **DL-472:** Tall variety identified for NWPZ irrigated areas.
6. **Barley varieties for malting:** C 138, RD 57, RD 137 and RS-6.
7. **Barley varieties for better chapatti making qualities:** Karan 16 (Neelam) and Kiran 19 possess better chapatti quality and nutritive values and digestibility may be improved if barley flour is supplemented by 25-50% wheat flour.
8. **Kanpur-318:** It is similar to NH-87 except that it has tall plants.
9. **Salt tolerant varieties of barley:** This group includes Karan-3, Karan-19, Jyoti(Hulles), C251, Amber, Azad, K 289 as important varieties.
10. **Nematode resistant variety:** Raj Kiran (RD 387) – it grows 85-90 cm tall having erect dark green leaves, straight ears and short awns. It matures within 135-140 days and yields about 40-50 q/ha. The plants do not lodge due to good straw strength.
11. **Six-rowed barley varieties**: This group includes following varieties.
12. **Jyoti:** It was evolved at CSAU, Kanpur by crossing between K-12 x C 251 and subequantly released in 1969.
13. **Amber:** It was evolved at Kanpur by crossing K-12 and CN 294 and released in 1969.

14. **Vijay:** It was evolved at Kanpur by crossing K-12 and C-251 and released in 1972.
15. **Azad:** It was developed by K-12 x K-19 parents at Kanpur and released in 1975.
16. **Drought resistant rainfed barley varieties:** The following varieties have been identifield and recommended for pre-release multiplication.
17. **K-141:** Developed by K-18 x IB 254 in 1977 for rainfed culture in NEPZ and re;eased by State Variety Release Committee 1981.
18. **K-169:** Developed by K-18 x IB-2 in 1977 for NEPZ.
19. **K-252:** Developed from K 12 x K 19 in 1978 for rainfed and Diara lands under NEPZ.
20. **K-273:** Developed from K 752/10 x EB 410 in 1978 for rainfed and Diara lands NEPZ.
21. **K-225:** Developed in 1978 from K-71 x IB 226 and in good for rainfed culture under NWPZ areas.
22. **K-404:** A tall, hulled variety identified for NEPZ.
23. **K-1166:** A tall, hull-less variety identified for NEPZ.
24. **Karan-765:** A hull-less variety identified for Indo-Gangetic Plains and Central Zone.

Barley Improvement in India

In early days barley was believed to be grown on saline-alkali lands or lands with no irrigation facilities having marginal inherent soil fertility. The grain yield used to be very poor due to researns like:

1. The old varieties had very poor response to management practices and when grown under slightly enriched conditions a severe lodging owing to serious losses in yield were observed,
2. The varieties had high susceptibility to diseases like smut, yellow rust, foliar diseases, foot and root rot, Molya and pests like nematode, termites and aphids. These problems cause a high instability in production,
3. It was regarded as second thought crop and was not given due care,
4. Lack of dwarf and input responsive varieties, and
5. Lack of scientific production technology for the crop.

Of these factors lodging alone accounted for a reduction in yield to the extent of 15-60 per cent and attempts were made to recognize the lodging indices (factors responsible for lodging). It is observed that lodging is essentially a loss of balance by the plants and it occurs due to weak root system that does not sustain the load of foliage which is in plenty. The same is expressed as shoot x height/root ratio. The lowest internode is another point of loss of balance which is expressed as strength of the lowest internode/height x mother shoot ratio. Thus, an ideal plant type should have sufficiently developed root system, dwarf habit and strong straw. Present emphasis is given on evolution of strong strawed dwarf or semi-dwarf plant type having good yield potentials under late sown and saline soils.

Efforts were made to utilize the 'UZU' dwarf genes from Japan and other exotic materials from different sources to evolve required plant types but very promising and fruitful results were not obtained as in case of wheat and rice. But even then it did not led to any distress or discouragement to the breeders and now they have a number of promising dwarf varieties that are bred conventionally or through mutational breeding and they have now opened new vistas in barley production. Many new varieties are now at final stage of testing which have an edge over the existing semi-dwarf varieties. Some of the Karan varieties, notably 'Karan 92' have outstanding performance at national level and are destined to bring further spurt in barley production in very near future.

So far barley was cultivated under rainfed and saline-alkali soil conditions but evolution of new varieties have made it possible to grow under late sown conditions also. Results of All India Barley Co-ordinated trials show that the barley CV. DL 36, DL 88, Jyoti and P 103 gave significantly higher grain yield than wheat (Sonalika) and the minimum increase over wheat was about 10 q/ha.

Hulled barley was the general reason for its non-acceptance by the consumers as its flour was very inferior compared to that of wheat, however, from the nutrition point of view barley is not at all inferior to wheat (Table 6.10).

TABLE 6.10: Amino acid composition in Barley grains as compared to other food grains

Cereals	Protein (%)	Lycine (%)	Leucine (%)	Isoleucine (%)	Ratio L/Isoleucine)
Barley	11.0	3.69	7.13	3.83	1.86
Maize	9.5	2.67	12.52	3.68	3.40
Millets	9.7	3.42	9.55	4.09	2.34
Rice	7.5	3.99	8.64	4.00	2.16
Sorghum	10.1	1.25	8.24	2.43	3.39
Wheat	12.2	3.07	7.14	3.50	2.04

Source: Report of the study team of National Commission on Agriculture on wheat, Barley and Triticales, 1973.

Thus it is essential to change the food habit as barley can be grown with relatively lower inputs than that of wheat. The breeders have evolved new barley dwarf strains which are capable of giving higher grain yield/ unit area but they require much less inputs, especially water which is still in short supply in major areas of the country. Thus the bottleneck in its acceptance was the husk and now breeders have successfully come out with huskless, barley varieties and during 1980-81 two dwarf high yielding varieties namely Karan-18 and Karan-19 had been released. Thus evolution and release of two hulls-less barley varieties having greater demands by farmers, makes the landmark in the barley improvement in the country.

Barley mutants developed at IARI: The barley varieties developed through mutation at IARI, New Delhi are BM 4, BM 10, BM 11, BM 14 and BM 21.

Even beyond these the breeders are now doing intergeneric crosses with wheat and barley which would probably solve many other problems.

Breeding Barley to evolve Varieties for Intercropping with autumn Sugarcane

North Indian canes are usually planted in October so as to get early canes for milling. The cane seedlings remain dormant of stunted until February because of lower temperature. To utilize the space between sugarcane rows various crops are being tried with varied success. What has also been tried but intercropping results reveal that wheat had an adverse effect on the growth of cane seedlings as it competes for nutrients from the soil and also causes shading on case seedlings. Trials with interrow planting of double dwarf barley varieties like 'Karan-12' and 'Karan-25' which are only 30-35 cm tall and mature within 90 days have shown much promising results as these varieties are less exhaustive in nutrient uptake and also cast least or no shadow-over come seedlings.

Breeding for Fodder Barley

Barley is the best suited for fodder purposes because of its very vigorous plant growth. Barley may be grown extensively for fodder purposes where two cuttings can be taken or it may be grown for fodder –cum-grain purpose in that one cutting may be taken as green fodder after 60 days of sowing and rations are left for seed production but the crop must be sown in first week of November. The later group is called as dual purpose varieties viz., Ratna, K-125 and BG-25 besides these new strains like Karan 2, Karan 5 and Karan 10 are being tested for dual purpose.

Besides above purposes, breeding in progress to evolve huskless, dual purpose salt resistant, tall growing varieties and also strains that may be grown in diara lands and those having high nutritive values etc.

Uses of Barley

Apart from industrial uses such as malting, brewing, barley is consumed by human being in many forms like *sattu* (flour from roasted barley grains). Dalia, halwa, biscuits and barley pops. Barley is highly digestible compared to wheat because it does not have gluten. Barley chapatti is very palatable and can be easily digested even by the persons suffering from stomach ailments.

Food Value

Barley is very nutritious and a rich source of protein and B-complex vitamins. Its protein quality is superior to the protein of corn and beans. Most of the world's barley crop is converted to malt. It is most important grain used in the brewing of beer. Barley can not be used for leavened bread because of low gluten content. It is much used in soups. Barley is also widely used as feed for live stock.

Food value, Minerals and Vitamins in Barley

	Food Value	Minerals and Vitamins	
Moisture	12.5%	Calcium	26 mg
Protein	11.5%	Phosphorus	215 mg
Fat	1.3%	Iron	3 mg
Fibre	1.2%	Vitamin C	3 mg
Minerals	3.9%	Small amount of Vitamin B complex	3 mg
Carbohydrates	69.6%		
	100.0%		

Barley* value per 100 g edible portion Calorific Value – 336

CHAPTER 7

Production Technology of Crop: *Millets*
(Ecological Requirements, Cultural Practices And Crop Production Management)

MILLETS

A group of cereal crops is called millets. They are warm weather cereals with small grains. They are grouped into major and minor millets. Millets serve as the staple food and pearl millet are known as major millets. Millets serve as the staple food for the people in the arid and semi-arid tropics.

Millets are grown for food fodder. Small millets are the staple food crops of the working class. These crops are risk alleviating for the small farmers. Cultivation of small millets stretches from sea level to almost 1828 metres above mean sea level. Finger millet, Kodo and barnyard millets have the capacity to produce more than 2000 kg/ha. The small millets have a wide adaptability and can be grown in extreme soil and climatic conditions. In all small millet growing states (Tamil Nadu, Madhya Pradesh, Orissa and Bihar), the crops are extremely important to tribal farmers because of their earliness and providing grain as well as fodder. Small millets are practically devoid of grain storage pests and have long storage life. Bread could be prepared from finger millet for diabetic patients. Small millets could be popped or flaked. Finger millet is used for malting and brewing. Malted finger millet is used in infant food and in milk thickness formulations. Finger millet malt brewed with an equal proportion produces beer of commercial standards.

NUTRITIONAL VALUE

Millet grains are highly nutritious and even superior to major cereals (rice and wheat) in many respects. They provide cheap protein, minerals and vitamins to the poorest of the poor where the need for such ingredients is the maximum. The dehuhusked prosomillet has a protein content of 12.5% whereas. Wheat has only about 11.8%. Finger millet contains high amount of Ca (344 mg/100g) compared to wheat grain (41 mg/100g).

SORGHUM (*SORGHUM BICOLOR* L. MOENCH)

Sorghum is called as **'camel of crops'** since it can withstand drought greatly. It is also known as great millet. It is believed that cultivated sorghum originated in east central Africa. Sorghum is mainly cultivated in the Semi Arid Tropics and is extensively grown in Africa, Asia and North America.

Distribution

It is cultivated in an area of 47.2 m ha with an annual production of 69.0 m t. The average yield is 1.5 t/ha. Africa and Asia alone contribute 81 per cent area and 54 per cent of world production. India contributes about 83 per cent area in South East Asia and 30 per cent of world acreage (Table 7.1).

TABLE 7.1: Area and production of sorghum in the world, 1996

Continent/Country	Area (m ha)	Production (mt.)	Yield (kh/ha)
Africa	24.2	20.9	361
North America	6.8	25.8	3786
Asia	14.1	16.7	1180
India	11.7	10.5	897
World	**47.2**	**69.0**	**1463**

Source: FAO Year book-Production, 1996.

In India, the share of sorghum out of total cropped area is about 7 per cent and it contributes about 6.2 per cent of total food grain production. Among the cereals, it ranks third in area and production after rice and wheat in India. Though cultivated area of sorghum has remained static, production and productivity have almost doubled in 1980-81 compared to 1950's. Currently it is cultivated in an area of 11.4 m ha with an annual production of 9.5 m t. Maharashtra, Karnataka, Madhya Pradesh, Andhra Pradesh, Rajasthan, Tamil Nadu, Uttar Pradesh and Gujarat are the important states growing sorghum (Table 7.2).

TABLE 7.2: Important states growing sorghum in India, 1995-96

States	Area ('000ha)	Production ('000 t)	Yield (kg/ha)
Andhra Pradesh	851	626	735
Gujarat	325	233	719
Karnataka	1984	1739	876
Madhya Pradesh	1018	863	848
Maharashtra	5557	4982	896
Rajasthan	594	139	235
Tamil Nadu	505	489	968
Uttar Pradesh	437	420	962
All India	**11444**	**9547**	**834**

Economic Importance

Sorghum belongs to family *gramineae* and ranks third in the major food grain crops of our country. Like maize, sorghum grain is also fed to cattle, poultry and swine. Besides, it is also used as a fodder crop. Sorghum is broadly classified into four agronomic groups: (i) Grain sorghum, (ii) Sorgos, (iii) Broomcorn and (iv) Grass sorghum. Grain sorghum is mainly used for grain purpose and also for forage and silage. Sorgos are otherwise called as sweet sorghum. Broomcorn is used for the manufacture of brooms. Grass sorghum may be annual or perennial and is used for hay or pasture.

Grain sorghum: Sorghum is used as food in India but used as cheap source of feed for poultry and cattle industry. They also serve as row material for industrial starch and alcohol production.

Sweet sorghum (*Sorghum bicolor*): It has a juicy pitch and produces small amount of grain as compared to grain sorghum. Grains are used for making flour. Stalks are rich in sugar and used for alcohol preparation. The straw residue after extracting sugary juice is used as livestock feed. Jaggery and crystalline sugar can also be made from sweet sorghum.

A sweet-stalked variety **'SSV 84'** was released in 1992 with 51.4 t/ha green cane yield, 17.8 per cent brix content and 1.88 t/ha grain yield. This variety could be utilized only for raw sugar (jaggery) production due to crystallization problem. Such sorghum is a best alternative for sugarcane in areas where sugarcane can not be economically grown. It is also possible to produce ethanol from grains and juice of sweet sorghum.

Forage sorghum: The average yield in the demonstration plots in farmers' fields is 40-50 t/ha green fodder with the best yield up to 60 t/ha. Multicut varieties: Pusa chari (PC 23), Haryana Chari (JS 73/53), SL 44, Meeti Sudan (SSG 59-3), Hybrids: Proagro Chari, MFSH-3, Hara Sona, PC 106.

Origin and History

There are different views about place of origin of jowar. Warth (1937) was the opinion that it was originated in India and Africa. De Candolle (1884) and Hooker (1897) said that jowar was originated in Africa. According to Vavilov jowar originated in Abicinia and from there it was brought to U.S.A. and European countries by by slaves. It spread to other countries by the help of visitors, passengers and merchants.

Botanical Description

Sorghum is an annual plant belonging to family gramineae and genus sorghum. The sorghum plant may be botanically described as under:

Root System: Roots are finer and more fibrous than maize. Coronal roots develop from lowest nodes of the stem and are heavily branched. Most of the roots are confined to the upper 15 cm of soil but in later stages of development, they may reach a depth of one or one and half metre. Prop roots may develop from the axillary buds on the lowest nodes of stem above the ground level.

Stem: Sorghum stems are solid, though the centre may become spongy with spaces in the pitch.

Culms: have 7-18 nodes and internodes.

Leaves: The number of leaveson the main stem varies from 7-24. Mature leaves may reach a length of 30 to 135 cm, and a width of between 1.5 and 13 cm at the widest point. The length of the leaf sheath may vary between 15 and 35 cm.

Inflorescence: The inflorescence of sorghum is a panicle also known as heads. The panicle is composed of spikelets which occur in pairs, one of them being sessile and other pedicellate.

Sessile spikelet has two glumes. The glumes enclose two florets, the upper being perfect, the lower sterile and consisting of lemma only. There are two lodicules lying adjacent to base of the spikelet.

Seed: The seeds or the grains are technically called caryopsis which are rounded, pointed at the base with a depression near the end. The grains may be white, pink, yellow or brownish-yellow depending upon variety. The caryopsis consists of endosperm, pericarp, testa and germ.

Composition of Grain

Protein	:	10-12%	Lysine	:	1.4 – 2.8%
Fat	:	3%	Leucine	:	7.4 – 7%
Carbohydrate	:	70%			

Classification of Sorghum

Earlier it was known that cultivated sorghum has a single species, *Sorghum vulgare* Pers. later it was termed as Sorghum bicolor (Linn) Moench (Dogget, 1965). Recently Harlan and de Wet (1971) gave a modified and simple classification based on spikelet type as:

a) **Basic Races:** 1. bicolor, 2. guinea, 3. candatum, 4. kafri and 5. durra.
b) **Hybrid Races**: 1. guinea-bicolor, 2. candatum-bicolor, 3. kafri-bico;or, 4. durra-bicolor, 5. guinea-candatum, 6. guinea-kafri, 7. guinea-durra, 8. kafri-candatum, 9. durra-candatum, and 10. hafri-durra.

Fundamental types of spikelet (Rao and House, 1971): Bicolor, guinea, caudatum, kafri and durra.

Important Varieties

Sorghums in India are almost wholly cultivated as rainfed crops and their yields depend on monsoon fluctuations. The prevalence of tall and long duration varieties which tend to expend most of their energy in vegetative production compelled with lack of fertilizer application and often sown under low population render they low yielder even in normal years and uncertain yielder in abnormal years. It has been realized that 30 per cent of total variation in yield appears to be vulnerable to rainfall fluctuations while remaining 70 per cent is controllable through use of suitable hybrids and new plant type varieties, adequate fertilizer application maintenance of plant population, timely sowing and plant protection measures.

The physiological efficiency of the new hybrids and plant type varieties when grown in different seasons are reflected by the distribution of dry matter between ear and stalks. They have better sink and source ratio than traditional varieties in which 70 per cent of dry matter gets accumulated in the stalk. Most of the hybrids and new plant type varieties have more than 50 per cent of dry matter accumutated in the ear. Genotype alternations involving reduction of duration, reduction of total dry matter and its more efficient distribution between stalk and ear have proved much advantageous in sorghum cultivation.

Varieties: A part from CSH 1 and CSH 2, the two early hybrids, some more promising hybrids have been developed. Two hybrids, CSH 1 and CSH 5 are presently cultivated on a fairly large scale during *kharif*. The new hybrids CSH 5 and CSH 6 (2219 A x CS 3541) offer considerable promise for *kharif* season and for *rabi* season, besides CSH 1 and CSH 4, hybrids like 36 A x 146 and 36 A x PD 3-1-11 are promising and have recommended for prerelease multiplication. Apart from hybrids, some non-hybrid promising strains with new plant type features have also been identified. Seven such varieties with yield levels comparable to hybrids have been developed and six of them CSV 2 (302/303), CSV 3(370), CSV 4 (CS 3541), CSV 5 (148/168), CSV 6(604) and CSV-7R (R-76) were released in the year 1974-75. Suitable hybrids and varieties for different states are presented in the following Table 7.3.

TABLE 7.3: Varietal Recommendation for different states

States	Hybrids	New plant type varieties
Kharif		
Maharashtra	CSH-1, CSH-4, CSH-5 CSH-6	CSV-2, CSV-3, CSV-4, CSV-5, CSV-6, CSV-1 (Swarna)
Karnataka	CSH-1, CSH-2, CSH-5 CSH-6.	CSV-3, CSV-5, CSV-4, CSV-1,

States	Hybrids	New plant type varieties
Andhra Pradesh	CSH-1, CSH-5, CSH-6	CSV-3, CSV-4, CSV-1, CSV-2, CSV-6
Madhya Pradesh	CSH-1, CSH-5, CSH-6	CSV-3, CSV-2, CSV-4, CSV-6.
Rajasthan and Gujarat	CSH-1, CSH-3, CSH-6,	CSV-3, CSV-2, CSV-4, CSV-6
U.P.	CSH-1, CHS-5, CSH-6	CSV-2, CSV-3, CSV-4, CSV-6
Tamil Nadu	CSH-1, CSH-5, CSH-6	CSV-4, CSV-5
Rabi		
Maharashtra	36 A x 148/168, CSH-1 36 A x PD 3-1-11, CSH-4	M-35-1, CSV-7R
Karnataka	CSH-1, 36 A x 148/168, 36 A x PD 3-1-11	M 45-1, CSV-7R
A.P.	CSH-1, 36 A x 148/168 36 A x PD 3-1-11	CSV-2, CSV-4, Csv-5, CSv-7R
Tamil Nadu	CSH-1, 36 A x 148/168 CSH-5, CSH-6 36 A x PD 3-1-11	CSV-4, CSV-5
Gujarat	CSH-1, 36A x 148/168 36 A x PD 3-1-11	CSV-7R
Summer Irrigated		
Tamil Nadu, Karnataka and A.P.	CSH-1, CSH-5, CSH-6	CSV-1, CSV-5

Climate

Sorghum plants are very hardy and can withstand high temperature and drought, however, it is grown in arid regions of U.P., Rajasthan and humid regions of Bengal and Bihar. It may be successfully grown under atmospheric temperature ranging between 15°C to 40°C and annual rainfall ranging from 400 to 1000 mm. The crop does not grow well under low temperature during winter in North India but in Central and Peninsular India, it is grown very successfully during winter because of relatively high temperature than Northern parts. Sorghum can tolerate drought conditions very well because it remains dormant during moisture stress conditions but resumes growth when favourable conditions reappear. It has a high resistance to desiccation, low transpiration ratio and large number of fibrous roots. It can tolerate water logging conditions better than any other cereal except rice. Sorghum is a short day plant (photoperiod).

Minimum temperature for germination .... 7-10°C
Temperature for optimum growth.. 26 – 30°C

Soil

Sorghum is grown on a variety of soil types but the clayey loam soil rich in humus is found to be the most ideal soil. It may tolerate mild acidity to mild salinity under pH 5.5 to 8.0. A good sorghum soil must have an efficient drainage facilities, though, it may withstand water logging more than maize.

Production Technology

Field Preparation

Deep summer ploughing with mould board plough soon after rabi crop is harvested and leaving the soil exposed to sun uptil onset of monsoon is a common practice. With the onset of monsoon the land should be ploughed twice by country plough. But the soil for grain crop should be pulverized for a fine tilth by harrowing and cross planking. If desired, field should be leveled before sowing so that rain water may be well distributed throughout the field.

Sowing

Time of sowing: Sorghum crop is grown in almost all the seasons of the year. In northern Indian conditions it is grown in *kharif* season but in southern India the crop is grown during *rabi* and summer seasons. *Kharif* crop should be sown soon after first break of monsoon rains, *i.e.*, nearly in last week of June. Sowing too early during April/May results in (a) synthesis of dhurrin by plant roots which is translocated to stem and leaves in the form of hydrocyanic acid (HCN) and is deadly poisonous to cattle, (b) poor grain yield because the flowering coincides with rains and pollens are washed off due to which the grains are not filled, and (c) harvesting, threshing etc. becomes difficult due to frequent rains. Therefore, the best sowing time is is in last week of June to first week of July depending upon onset of monsoon. However, in some districts of A/P/ and Gujarat sowing of hybrids like CSH 1, CSH 2, CSH 3 and CSH 4 is done in the end of August to beginning of September.

In Deccan plateau areas of Maharashtra, Karnataka and A.P. the sorghum is grown as winter rainfed crop constituting about 30 its total acreage in the country. Sowing of this winter or *rabi* crop is done in 2nd fortnight of September to middle of October. Since the crop depends upon residual moisture of rains and is normally rainfed, therefore, short duration varieties like CSH 1 and CSH 3 should be selected for cultivation.

In Tamil Nadu, A.P. and Karnataka the summer crop of sorghum is very popular where the sowing is done during January-February by selecting early maturing varieties like CSH 1, CSH 4, PSH 2 and Swarna.

Seed Rate

Required plant population may be obtained by using a seed rate of 8-12 kg/ha. The seeds are sown about 3-4 cm deep in the furrows. The fodder crop needs about 30-35 kg/ha seed. Gap filling or thinning, as required, should be done soon after germination so as to maintain ideal plant population.

Spacing

A spacing of 40-45 cm between the rows and 15-20 cm between the plants is quite satisfactory in giving 1,50,000 to 2,00,000 plant population/ha for an ideal grain crop. Some farmers still adopt broadcasting method of sowing but line sowing either by seed drill or hand sowing in furrows behind the country plough is an ideal method for raising the crop, however, the fodder crop may be sown broadcast

Seed Treatment

Sowing of fungicide treated seeds takes care of all the fungicidal diseases like grainsmut in the crop. The seed treatment may be done by Agrosan G.N., Ceresan, Thiram at the rate of 3 g/kg of seed or sulphur at the rate of 3 g/kg of seed. The seed is treated, dried and sown in the field.

Methods of Sowing

Sorghum seeds are small and therefore, should not be planted deep into the soil.

Deep sowing results into poor germination percentage. The depth of planting should not exceed 3 to 4 cm. Therefore, line sowing either by the help of seed drill or hand sowing in furrows behind the country plough is the most preferable for raising the crop than the broadcast method of sowing.

Planting Row Direction

Sorghum is a tall growing crop and the chance of mutual shading can not be over-ruled. Therefore, sowing in sunrise and sunset direction is the best row direction as it does not cast the shadows over the plants of the same row or on the side rows.

Transplanting in Sorghum

Transplanting method of sowing in sorghum has been found quite remunerative particularly with the hybrid varieties. This method provides a proper stand of the crop and has given more grain yield as compared to the direct seeded crops. Particularly when sorghum hybrids are sown in summer or winter, transplanting method may be adopted. Because it has been observed that summer sown crop is very much susceptible to the attack of shoot fly and direct correlation exists in planting dates and occurrence of shoot fly. The attack of this insect may be controlled or minimized with the adoption of transplanting method during summer and winter.

For transplanting method, nursery beds (4 m x 1 m) are prepared first. While preparing nursery bed application of three baskets of compost, along with 500 g of ammonium sulphate, super phosphate and potassium sulphate each should be applied in each bed. Sorghum seed should be sown at the rate of 100 g per bed in lines 9 to 12 cm apart. Regular care for weeding and watering should be taken. The seedlings will be ready for transplanting after 15 days. The seedlings from the nursery should be drawn from 15 days to 24 days age.

Experimental results have indicated that transplanted crop established within a week. Heading indirect summer sown crop usually occurs in 40 to 45 days, while winter sown crop it takes about 60-65 days. Whereas in winter transplanted crop heading is delayed for 5 to 7 days. The extra few days taken by transplanted crop to head is in consequence of the time taken to establish when transplanted. In younger seedlings heading is earlier as compared to older seedlings. Better yield was obtained when 15 to 20 days old seedlings were used in transplanting.

Date of planting: Early planting of new plant types has been found beneficial. These varieties should be sown as early as possible after the rains begin. The last week of June to the first fortnight of July is the best period for sowing hybrid sorghum. Late *kharif* plantings generally suffer heavy shoot fly attack. The planting dates with local varieties in different zones generally coincide with the shoot fly free periods. As a rule it is preferable to be well within these planting dates. Late September to early October for *rabi* and late January to early February for summer season have been found suitable for sowing the sorghum.

It was also recommended that under rain-fed conditions during *kharif* season, planting of sorghum should be completed within a week of the onset of first monsoon shower. Delay in planting beyond one week of onset of monsoon may cause drastic decline in yield particularly in heavy soils areas like Indore and Prabhani. Under irrigated conditions, the optimum date of planting was the last week of June for areas round about Delhi and Indore. For *rabi* areas Deccan, September sowing result in about 100 per cent increase of grain yield over October sowings.

Interculture

Thinning of sorghum crop at 10 days stage, early interculture of sorghum within 2 to 3 weeks after germination can destroy weeds, a second hoeing and earthing at 6 to 7 weeks stage are desirable. The hoeing or inter-culture should be shallow to avoid root injury. Earthing helps to provide better anchorage to the crop in the rainy season and prevents lodging.

Weed Control

The observations are indicative that the period of most severe crop-weed competition is upto 30 days after sowing and if the weeds are left uncontrolled they cause a drastic reduction in yield. Manual weedings and hoeings help in solving the weed problem but it is possible only during *rabi* and *zaid* cropping seasons while rains do not permit the manual weeding or hoeing during *kharif* season. Application of herbicides like Atrazine @ 0.5 kg a.i./ha or Propazine @ 1.0 kg a.i./ha dissolved in 900-1000 litres of water, therefore, becomes obviously essential to control weeds. These herbicides should be applied before emergence of sorghum seedlings.

For successful cultivation of new plant types of Jowar, effective control of weeds is necessary. The common weeds which are found in jowar field are *Cyperus rotundus*, *Digera arvensis*, *Pluchea lanceolata*, *Euphorbia sp.*, *Striga lutea*, *Chorchorus sp* in non grasses and *Cynodon sp*, *Echinochloa colunum*, *Eleusina indica*, *Sorghum halepense* and *Dactyloctenium sp* in grasses.

The following weed control measures have been found effective.

1. Application of 2, 4-D at the rate of 1.5 to 2 kg a.e./ha. Amine one ester salt should be applied four weeks after the sowing.
2. Semazine or Atrazin application at the rate of 1.5 kg a.e./ha has also been found equally effective.

Fertilizer Application

Sorghum is is generally considered as an exhaustive crop because it improvishes the soil fertility. This may be attributed due to the fact that its root system is concentrated on the upper soil layer and also it renders available nitrogen into unavailable by the action of cellulose decomposing microorganism during the decay of the roots and stubbles. As such crop is a heavy feeder and it is truer particularly with new plant types.

Experimental results in recent years have clearly indicated that sorghum hybrids are very responsive to fertilizer application particularly to nitrogen and phosphorus.

Nitrogen: The economic optimum level of nitrogen for the improved sorghum varieties like Swarna, and hybrids CSH-1, CSH-2 and CSH-3 is about 100 to 120 kg /haunder irrigated conditions and 80-100 kg N/ha under rainfed conditions.

Even under *Kharif* season as rainfed crop, in the areas receiving 400 to 500 mm precipitation during season nitrogen should be applied at the rate of 100-120 kg/ha.

In heavy soils the basal application of total nitrogen has given better response than split application. Split application has been found superior to basal application only in light and medium soils. In case of split application half of the total nitrogen should be applied at the time of sowing and the rest half at about

'knee-high stage' ***i.e.***, about 30 days after sowing under irrigated condition. Under low rainfall areas with insufficient precipitation during early stage of crop, the top dressing of nitrogen is not necessary.

Phosphorus: New plant types of sorghum also responds to phosphorus fertilizers in many soils, but it is always better o apply phosphorus based on soil test analysis. Howver, application of 60 kg P_2O_5/ha should be done in high rainfall areas and 40 to 45 kg in low rainfall areas.

Studies on phosphorus nutrition of hybrid sorghum have clearly indicated that with increase in dry matter production the uptake of phosphorus also increases simultaneously from seedling stage up to harvesting. However, the peak period for phosphorus uptake has been found during 30 to 90 days after sowing, indicating that P in sorghum is required at each stage of growth. In view of this result, theoretically split application of phosphorus may prove more advantageous than basal application in hybrid sorghum. Since, phosphorus has very low mortality, its top dressing will result availability of P only in top soil *i.e.*, 1 to 2 cm deep. On the other hand at later stages of crop growth the requirement of phosphorus is met with the natural soil phosphorus. Notwithstanding this, field experiments at I.A.R.I., New Delhi have indicated the usefulness of split application of phosphorus in soils having low to medium in available phosphorus content. In case the soil analysis data are not available, the total quantity of phosphorus should be applied at the time of seeding.

Potash: Research information regarding application of potash indicates that response to potassium is very infrequent. Therefore, no potassium should be applied except in the areas deficient in potash. However, under usual condition an application of potash at the rate of 40 kg K_2O/ha has given better result. The total quantity of potash should be applied as basal application during land preparation.

Placement of Fertilizers

Direct contact of fertilizer with seed is known to cause injury to the germinating seedlings in sorghum. It is, therefore, recommended that fertilizers for sorghum be applied in a band along the row, 4 to 5 cm deep and 4 to 5 cm to the side row. This is done the best with a seed cum fertilizer seed drill

Irrigation and Water Management

The new plant type varieties of sorghum are drought tolerant but not to the extent of the tall varieties. However, it gives favourable response to optimum moisture conditions. These varieties are known for their stability of performance over a range of environment with unpredictable rainfall and moisture levels. Another feature of the high brides is not only to stand a long period of soil moisture stress but to recover from such adverse effects when the environment becomes favourable.

Generally, the *kharif* sown crop is not irrigated. However, a very heavy pre-planting irrigation to wet the sub-soil is most useful. Supplemental irrigation becomes necessary with these varieties when heavy application of fertilizers has been made and the rain is not sufficient or when there is prolonged drought condition during growing season. If monsoon fails, light and frequent irrigation (5 to 6 cm) should be given. At no stage the crop should be allowed to show wilt condition. Since these varieties are very much susceptible to water logging condition, care should be taken to avoid this situation.

The plants required water most at boot leaf and heading stages and watering before these stages is essential. After dough stage, application of water is not necessary. The yield of sorghum is very much affected when there is rainfall at the time of flowering. If possible, sowing time of these varieties should be adjusted to avoid rainfall during flowering. Higher yields from these varieties have been obtained when irrigations were applied before the available soil moisture fall below 50 per cent in rooting zone. The optimum moisture range for irrigation has been found to be from field capacity to 0.55 atmospheric tension measured at 15 cm depth for hybrids. These observations are useful guide for scheduling irrigation programme when these varieties are sown under *Rabi* season.

Plant Protection

The high yielding varieties of sorghum are very often attacked by insects like shoot fly, stem borer and midge. Effective measures, therefore, are essential to over come them, for obtaining higher yields. The following control measures have been found useful.

Shoot fly: For shoot fly, application of 10 per cent Phorate granules at the time of 1.5 g/metre row or 5 per cent Disulforton (Thiodemeton) at the rate of 3.0 g/meter row in seed furrow at the time of sowing have been found useful. If these insecticide are not available the crop should be sprayed by Carbaryl (50% W.P.) are sufficient for one hectare. The crop should be sprayed twice, the first at 3 to 5 days of emergence of seedlings and the second at 5 to 7 days after first spray.

Stem borer: Uprooting of stubbles and burning them and application of 1 per cent Lindane granules, or 16 per cent Malathion or 4 per cent Carbaryl two/three times at 10 days interval starting from 20 days after emergence, have been found effective in controlling stem boerers.

Midge: Midge can be controlled by spraying Carbaryl (50% W.P.) 3 kg or Lindane (20 E.C.) 1.25 litres or Endosulfan (30 E.C.) 1 litre in 500 to 600 litres of water per heactre. The spraying should be done before flowering *i.e.*, about three days after panicle emergence.

Controlling Diseases

For controlling the diseases pre-sowing treatment of seeds with Thiram at the rate of 2 g/kg of seed has given encouraging response. In case of appearance of diseases in field, the following treatments should be given.

Spraying of wettable Sulphur at the rate of 1 to 1.5 kg/ha from 45 days of sowing upto flowering with an interval of 10 to 15 days. This has been found to check rust, leaf bight and zonate leaf spot.

Dusting the ears just after the emergence with Sulphur 10 to 15 kg/ha.

Sorghum is attacked by Smut, Ergot or Surgary disease and downy mildew; however the extent of damage to the crop is much less as compared to that from insects. Other diseases like leaf smut, head smut, loose smut and long smut are also found to attack the crop but their harmful effects are almost negligible and hence they are not considered as serious problems in sorghum production.

Mixed Cropping

Growing grain sorghum with some legumes like green gram, black gram, red gram etc. had been most prevalent practice in the past, however, with the evolution of high yielding dwarf plant types mixed cropping is no more a common practice. But cultivation of fodder sorghum with cowpea, guar, moth, sunflower, etc. is a customary practice because legume mixture enriches nutritive value of fodder, increases fodder yield per unit area and also minimizes ill effects of sorghum in the soil.

Crop Rotation

Red Soils:	Sorghum – Horse gram
	Sorghum – Niger.
Black cotton soil:	Sorghum - Gram
	Sorghum – Coriander.
	Sorghum – Sugarcane – Groundnut (3 years)

It is also sown as mixed crop with arhar, cowpea, cotton, groundnut, ragi, etc.

Ratooning of Sorghum

Ratooning means allowing the stubbles, left after harvesting to grow into a crop with a view to save cost of field preparation, seed etc. It is observed that rations give equal or even more grain yield than the first crop. However, chance of insect/pest and disease infestation is much higher than in the first crop. Following practices should be adopted for an ideal ratoon crop of sorghum.

1. Harvesting of entire crop at one time from a height of 10-15 cm above the ground level and plants should be harvested when they are green.
2. The left over leaves and broken parts of stem must be collected and burnt to avoid multiplication of insect/pest and diseases from them
3. The field should be irrigated soon after harvesting of the crop.
4. An application of 60-80 kg N/ha in two splits: first soon after irrigation and before earthing, second within 30 days of first application should be done.
5. Only two healthy sprouts/stubbles should be allowed to grow and remaining weaker ones should be removed.
6. Short duration and rationing variety like CSH-1 should be selected for second or ratoon crop.
7. Ratoon crop needs irrigation as it is a post-monsoon crop; therefore, it must not suffer from moisture stress especially from boot leaf to grain filling stage.
8. One need to be vigilant about attack of insect/pest and diseases and all preventive as well as curative measures should be taken against them.
9. The ratoon crop matures earlier *i.e.*, after 80-85 days of sowing after which the crop should be harvested.
10. Second rationing may also be practiced, however, it gives very poor grain yield and second ratoon is possible only when the crop is absolutely free from insect pest and diseases.

Production of Sorghum Fodder

Use of sorghum as fodder is very popular in all the zones of the country where it is grown for grain purposes especially in humid and sub-humid regions. Its luxurious vegetative growth favours its production as fodder. Fodder sorghum is used in various forms like soiling crop, silage crop and hay making depending upon its availability and surplus. For economic and successful fodder production from sorghum following points should be taken into account:

1. Sowing should be done soon after onset of monsoon or even earlier with irrigation facilities to avoid attack of shoot fly and the crop gives higher fodder yield.
2. It is always better to have mung (green gram) as preceding crop so that there is no loss in soil fertility. This also eliminates persistent weed problem in the

crop.
3. A high seed rate of 20-25 kg/ha is used with a view to get about 3-5 lakhs plants.
4. Application of only nitrogen at the rate of 80-100 kg/ha is sufficient. The nitrogen should be applied in two splits *i.e.*, half at the time of sowing and remaining half as top dressing about 30 days after sowing.
5. The fodder crop becomes ready for harvesting right after 60 days of sowing depending upon the purpose of harvesting as for example soiling crop becomes ready after 60 days, silage crop when sorghum starts flowering while hay after the crop has been harvested for grains.
6. A yield of about 300-500 quintals of green fodder. 200-250 quintals of soiling crop and 150-180 quintals of hay is obtained from one hectare area.

Harvesting and Treshing

The high yielding varieties mature in about 100-120 days duration after which they are harvested. Generally, two methods of harvesting *i.e.*, either stalk-cut or cutting of earheads by sickle are employed. However, in foreign countries sorghum harvesters are used. In case of stalk-cut method the plants are cut from near the ground level, the stalks are tied into bundles of convenient sizes and stacked on the threshing floor. After 2-3 days the earheads are removed from the plants, while in later case the earheads, after their removal from the standing crop, are piled up on the threshing floor and after few days they are threshed.

Threshing of earheads is done either by beating them with sticks or trampling them under bullocks' feet. Later method is quicker and is practiced by majority of farmers who use to grow the crop on large scale.

Harvesting of fodder crop is done at different stages depending upon its purpose of use *e.g.* for karvi the plants are harvested when they reach to grain maturity, for soiling the crop is cut after harvested when they are at milky-grain stage.

Yield

The grain yield of improved varieties under assured water supply ranges between 25-35 quintals/ha and that of hay or karvi between 150-170 quintals/ha. The green fodder from soiling crop of hybrid sorghum ranges between 500-600 quintals/ha while that of local varieties it ranges between 300-450 quintals per hectare.

Storage

Grain: It is stored in under ground pits, metal seed bins, gunny bags, etc.

Fodder: The dried stalks are heaped as used as fodder but the green fodder, the green

Fodder: The dried stalks are heaped as used as fodder but the green fodder, the green plants are chaffed and fed to the cattle, if excess, they are stored in sylo pits or sylo stores and used in the off seasons.

Important Practices

1. Popularising high yielding varieties and adequate plant population.
2. Early sowing in lines and early weeding.
3. Maintenance of plant population.
4. Moderate fertilizer use.

Important Observations

- Roots are finer and more fibrous than maize.
- Usually no tillering, prop roots may develop.
- Jowar is poor in lysine but rich in leucine.
- Inflorescence is called Panicle (Commonly called Head).
- Thinning is important operation at 12-15 cm (within row).
- First sorghum hybrid in 1964: CSH-1 (90-100 days, coordinate sorghum Hybrid-1).

BAJRA (PEARL MILLET) (*PENNISETUM TYPHOIDEUM*, S & H)

It is also known as cattail millet or bulrush millet. This is an important crop in arid regions of Africa and Asia. It is probably originated in tropical Africa. In USA, it is raised as a pasture crop. Pearl millet is the fourth important food crop in India next to rice, wheat and sorghum and the second major coarse cereal grain after sorghum. Pearl millet is extensively cultivated in poor fertile and water-deficit soils, mostly under low and erratic rainfall. Globally it is cultivated in an area of 36.5 m ha with the total production of 29.6 m t. The average productivity is 745 kg/ha (1995-96).

Distribution

India is the largest producer of pearl millet in the world. Area under pearl millet was 9.0 m ha during 1950-51 and increased to 12.9 m ha during 1970-71. The area, however, decreased to 9.4 m ha during 1995-96. However, the production increased from 2.6 m t in 1950-51 to 8 m t in 1970-71. There was significant increase in yield from 288 kg/ha in 1950-51 to 658 kg/ha in 1990-91.

TABLE 7.4: Major states growing pearl millet in India, 1995-96

States	Area ('000 ha)	Production ('000t)	Yield (kg/ha)
Andhra Pradesh	133.6	120.3	900
Gujarat	1086.9	1059.2	975
Haryana	574.0	408.0	711
Karnataka	399.6	264.3	661
Madhya Pradesh	140.9	127.1	902
Maharashtra	1664.9	965.4	580
Rajasthan	4257.3	1152.4	271
Tamil Nadu	215.5	251.5	1167
Uttar Pradsh	873.0	1019.1	1167
Total	**9381.9**	**5392.2**	**575**

Source: Fertilizer statistics, 1995-96.

Principal pearl millet growing states are Rajasthan, Maharashtra, Gujarat, Uttar Pradesh, Haryana and Karnataka which account for 90 per cent of the total area and 86 per cent production (Table 7.4). Pearl millet is cultivated in areas with the rainfall of 150 to 700 mm. It is largely grown for home and local consumption and the marketable surplus is limited. Pearl millet can substitute wheat and rice in processed foods. The development of alternative foods adds value to commercial exploitation of this millet.

Origin and History

The exact place of its origin is yet to known, however, by seeing the cultivation pattern it is believed that bajra plant originated either in India or Africa. According to some scientists it is understood that bajra was used in festivals in China in ancient days which shows that it might have originated from China and from there it must have spread to India and later to the Western countries. Contrary to this it is thought that bajra is evolved from elephant grass which originated from Africa, therefore, its place of origin is also probably Africa. Presently it is grown in Arabia, Southern Europe and later on it has been introduced in USA as fodder crop.

Economic Importance

Pearl millet, popularly known as bajra, belongs to the family *gramineae*. It provides staple food for the poor in a short period in the relatively dry tracts of the country. It is the most drought tolerant crop among the cereals and millets. The phenomenon of apical dominance is unique mechanism for survival under water-stress.

Classification

The genus Pennisetum is divided into five sections:

1. Pennicillaria,
2. Gymnothria,
3. Eupennisetum,
4. Heterostarchy, and
5. Brvivalvula

Most of the species (about 32) belong to section pennicillaria.

Botanical Description

Root system: Root system of bajra is fibrous consisting of primary, secondary and prop roots.

Stem: Solid and single, made up of nodes and internodes. Nodes are swollen while internodes are cylindrical and glabrous.

Leaves: Leaves are long and fairly broad and held erect by thick midribs. Leaves consist of blades and sheath. The base of blade is slightly auricled, the legule is short.

Inflorescence: It is cylindrical spike packed with spikelets. The spikelets may vary from 800-3000 per spike. The length of spike varies from 15-60 cm. Spikelets occur in pairs.

Caryopsis (Grain): The seed is 3-4 mm long and 2-2.5 mm wide. The colour of grain varies from whitish yellow to grey or dull light blue.

Composition of Grain

Moisture contents 12.4%
Protein 11.6%
Fat 5%
Carbohydrates 67%
Minerals 2.7%

Genotypes

Bajra is important millet of India and is second to jowar in area and production. It is grown practically all over the country and more particularly in dry and semi-dry zones of northern and peninsular India in soils of poorer fertility. It ranks fourth in order of cereal crops and has been considered as the staple food of the common man in country. Since, it is grown, mostly under marginal condition of moisture and fertility of soil which coupled with low yielding potentiality of varieties often results into very poor per hectare yield. Varieties of bajra available for cultivation, before the introduction of hybrids, were tall in nature and poorly responsive towards improved managements and as such their yielding ability was very low. The introduction of dwarf high yielding new plant type of hybrids has opened a new era in Bajra cultivation and increased the yield 10 to 15 times over local improved varieties.

The introduced hybrids of Bajra have been found to be widely adapted and more stable and higher yielding than the traditional varieties. The change in the plant type and utilization of hybrid vigour are the only factors responsible for making these varieties high yielder. In general, the new plant types of Bajra are early maturing, rapid in early growth with high proportion of grain to straw and have larger photosynthetic area. The most important aspects of these hybrids are their stability of performance over a range of environments and their competitive capacity to stand high population pressure. In addition to this, the hybrids are very responsive to fertilizer application and more particularly to heavy nitrogen dressings. The response of these varieties to nitrogen even under limited moisture conditions have been three to five times more than local varieties. The output of grain per kg of nitrogen is around 20 kg in these varieties as compared to 5 kg in local ones.

Production of new plant type hybrids in Bajra has become possible due to the availability of cytoplasmic male sterile lines. The cytoplasm of such lines is designated as 'B' Cytoplasm. During breeding the dominant gene in hymozygous condition interacts with B. cytoplasm and brings about complete male sterility. It has been observed in meiosis studies that sterility was due to failure of synapsis in majority of homologues. For production of hybrids it is thus necessary to have a series of lines in which genomic constitutions are such that they have dominant gene combination for restoring fertility and at the same time the best combining ability. Genome of such cytoplasmic sterile line, should be substituted with the genome of good combiner, high yielding line and then they are crossed with restorer lines for production of hybrids. The whole procedure may be summarized into three steps:

1. At first stage cytopasmic male sterility is introduced in selected inbreds lines by repeated back crossing. The male sterile line thus obtained is designated as A strains.
2. Cytoplasmic male sterile: A strains qre then crossed with male fertile inbred line which is designated as B strain. These B strains are necessarily male fertile but non-pollen restoring types. They are maintained by self or sib pollination. Crossing between A and B lines are done only to maintain the lines of A strain for futher breeding.
3. While maintaining A and B strains, another line with male sterility but pollen restoring type is also simultaneously maintained. This line is designated as strain R. For producing the hybrids the strain A is crossed with strain R.

Systemetic varietal improvement programme was initiated on the pattern mentioned above in in 1963 under Coordinated Millet Improvement Programme by ICAR in the country. As a result the first promising hybrid HB-1 was released for general cultivation duly approved by Central Variety Release Committee in 1965. Subsequently HB-2, HB-3 and HB-4 were released. These varieties are very much true in their plant type with high yield potential.

Varieties

The adoption of high yielding varieties has stabilized production even during the drought years. Hybrids have a higher yield potential than improved varieties. The hybrids and he improved varieties have given more than 2000 kg/ha grain in Gujarat, Haryana and Andhra Pradesh.

Improved Varieties: CO 1, CO 2, CO 3, CO 4, CO 5, K 1, X 3, for Tamil Nadu, Bajra 207, Bajra 28-15, L 17, Baroda and Babapuri for Gujarat and Maharashtra, RSK and RSJ for Rajasthan, T 55, S350, S 530 and S 28 for Punjab and Uttar Pradesh.

Hybrids: India has the privilege of being the first country in releasing hybrids in pearl millet. HB 3 is the most popular hybrid because of its shorter maturity, bold grains and adaptability to moisture stress conditions. Adoption of these hybrids led to 75 -100% increase in yield over local varieties. Other hybrids are NHB 3, NHB 4, NHB 5, BK 560, MH 179, ICMH 451 and Pusa 23.

Composites: These are recurring problems of downy mildew in hybrids which has resulted in release of several composite varieties which yield as much as hybrids. They are adaptable to diverse environments and have more durable resistance to downy mildew than the generally uniform hybrids *e.g.* WCC 75, ICTP 8203, Raj 171, ICMV 221.

Downy mildew resistance types: Downy mildew (*Sclerospora graminicola*) is the most important biotic factor limiting pearl millet production. Some of the composites (WCC 75, MP 171) and hybrids (ICMS 451, Pusa 23) show resistance to downy mildew.

Climate

The crop has a very wide adaptability as it may grow under different day lengths, temperatures and moisture stress. Most of the varieties developed in India are photo-insensitive which helps in growing the crop during monsoon, *rabi* and *zaid* or spring seasons. The crop requires low annual rainfall ranging between 40 to 50 cm and dry weather. At its grand growth stage the crop needs wet weather, light showers and bright sunshine, at the flowering stage there should be no rain because the pollens are washed off and fertilization is found to be very poor, while at maturity it needs fairly dry weather associated with bright sunny days. The crop may tolerate drought but can not withstand a high rainfall of 90 cm or above. It can not tolerate frost also. Pearl millet is a warm weather crop. The best temperature for growth is between 20 and 28°C.

Soil

Light textured soils of low inherent fertility, good drainage, and mild salinity are the best types for the crop. The soil should be deep and free from stones and concretes. Crop does not tolerate soil acidity. However, it grown in a variety of soils, such as sandy loams of U.P. and Punjab and light soils of Rajasthan and north Gujarat, heavy clays of Andhra Pradesh, Tamil Nadu and very light soils of Maharashtra and shallow black, red and light soils of Deccan and southern India. Since water logging is very harmful to these varieties, soil selected for cultivation must have good drainage facility.

Seeds

The seeds are 3 to 6 mm long weighing about 3 to 10 mg. The seeds are flattened on the sides but they have an oval appearance with one end tapering. The colour varies from light pale yellow to dull light blue.

Production Technology

Field Preparation

The crop needs very fine tilth because the seeds are too small. For raising a good crop of bajra it is essential to do the summer ploughing by mould board plough and after onset of monsoon the field should be harrowed twice or thrice or ploughed by country plough. Care should be taken to remove all the weeds and stubbles from the field and if possible, the field should be leveled and efficient drains should be provided in the field. Clods must be broken by planking the field after every ploughing so that a fine tilth may be obtained to facilitate the sowing and proper distribution of seed at an appropriate depth.

Sowing

The most appropriate time of sowing is middle of July. Sometimes it is sown earlier by giving one pre-monsoon light irrigation so that the seeds have emerged out before onset of monsoon but under such cases the rain water washes off the pollens from the ears and grain filling is found to be poor or adversely affected. Therefore, sowing should be done either by middle or last week of July.

Seed Rate abd Spacing

Four to five kg seed per hectare is found to be enough it it is sown by drilling method but the dibbling method requires still lesser quantity of 2.5 to 3.0 kg/ha. A spacing of 40 to 45 cm between rows and 10 to 15 cm within rows is found to give satisfactory yields of hybrid bajra. The seeds, being very small, should not be sown deeper than 1.5 to 2.5 cm in well prepared seed bed.

Seed Treatment

The organic-mercurial compounds like ceresin, Agrosan G. N. etc. should be used at the rate of the 2.5 to 3.0 g/kg of seed to control the seed borne diseases otherwise these fungal diseases cause a high mortality of young seedlings soon after their germination.

Methods of Sowing

Various methods of sowing like broadcasting, drilling, dibbling and transplanting are common in bajra cultivation. Of these the drilling method is most popular. Transplanting is practiced by only very improved farmers and the method is explained in a separate heading.

Direction of Sowing

Line sowing of bajra crop in east-west direction has proved to be better than north-south direction of crop rows. Experiments conducted at Allahabad Agricultural Institute in planting direction has proved that sowing of bajra in sun-rise and sunset direction results in about 10to 15 per cent increase in grain yield.

Application of Manures and Fertilizers

It was of common belief that bajra needs very low quality of nutrients but experiments conducted under All India Coordinated Millet Improvement Project have proved that new plant types of bajra especially hybrids respond to very high doses of fertilizers. The nutrient requirement of the crop differs according to the variety, irrigation facility and growing season.

Nitrogen: It has been observed that the hybrid varieties show four to five times higher response to nitrogen as compared to local ones and this response is still higher under poor fertility conditions of the soil. High yielding varieties need about 50 to 60 kg N/ha under unirrigated conditions, 80 kg N/ha under limited irrigation and 100 to 120 kg N/ha under assured irrigation conditions. Under poor fertility conditions increase in yield has been noticed up to a dose of 160 kg N/ha but on an average 80 to 100 kg N/ha for hybrids 60 to 80 kg N/ha for composites and 40 kg N/ha for local varieties have been recommended. In rainfed areas application of organic manures such as FYM or compost helps in increasing the crop yield. These organic manures may be applied at the rate of 15 to 20 t/ha.

Phosphorus: Experimental findings reveal that phosphorus has no positive effect on yield of the crop even up to the application of 75 to 80 kg P_2O_5/ha. In case of high initial phosphorus content in the soil the hybrids and composites have failed to give any response over 40 kg P_2O_5/ha. Thus a dose of 40 to 50 kg P_2O_5/ha has been recommended for an ideal crop of hybrid/composite varieties of bajra.

Potash: Generally potash is present in abundance in the Indian soil, therefore, its application in case of potash rich soils is not recommended, however, to make a balanced nutriention to the plants it is essential to apply about 40 to 50 kg K_2O/ha especially when high doses of nitrogen have been applied.

Methods of fertilizer application: The crop is grown mainly in light soils where chances of nutrient loss are very high, therefore, it is always better to apply fertilizers in split doses. The organic manures must be applied and mixed well into the soil about 20 days before sowing of the seeds so that they may get fully decomposed before the crop get started. It is observed that for beneficial effects of fertilizers half of nitrogen, full phosphorus and potash should be basal placed at a depth of 8 to 10 cm before or at the time of sowing. Care should be taken that the fertilizers should not come in direct contact of seeds, therefore, they must be placed either 5-6 cm below the seed layer or by the side of the crop rows. One fourth quantity of nitrogen should be top dressed or band placed at 5-7 cm apart from crop rows after about 30 days of sowing when weeding and thinning are completed. Remaining one-fourth of nitrogen should be band or side placed about 60 days after sowing. Foliar application of urea to the tune of 50% of the total nitrogen requirement has proved to be beneficial in increasing crop yield especially in case of severe drought conditions. It is also observed that the foliar application of zinc sulphate at the rate of 2-5 kg/ha has resulted in an increased grain yield. Zinc sulphate may be dissolved in water along with ureaand both may be sprayed together in order to reduce the additional expenditure of spraying.

Intercultural Operation

The crop needs two inter cultural operations – one about three to five weeks after sowing and another about a fortnight later. At the time of first intercultural operation thinning or gap filling, as needed, should be done and all the weeds must be removed from the field so as to avoid any competition with the crop plants.

Weed Control

An integrated weed control method like hand weeding, inter row cultivation by cultivator followed by pre-emergence application of Atrazine at the rate of 0.5 a.i./ha would take care of most of the weeds.

Irrigation

Bajra is grown raifed and the crop being drought resistant hardly needs any irrigation, however, it is observed that the yield may be significantly increased by irrigating the crop at critical growth stages like maximum tillering, flowering and grain filling stage. Water stress or wilting and water logging at any stage of crop growth is highly deterrent to the yield. Therefore, light irrigation and efficient drainage is very essential for bajra production.

Plant Protection Measures

Insect-pests: Stem borers and grasshoppers are serious pests of bajra crop. The stem borer (*Chlozonellus swin*) and pinc borer (*Sesamia inferens* Wlk) initially feed on leaves and later bore into stems where they make tunnel and fill with frass and excreta. The affected plants remain stunted and often develop dead heart. The borers may be controlled by two spraying with 2 litres of Endrin 20 E.C. or 500 ml of parathion 50 E.C./ha at an interval of 10-15 days.

The grasshopper (*Hieroglyphus nigrorepletus* Bol or H. *banian* F) eats the leaves and the adults feed on soft grains from ears. The pest may be controlled by dusting the crop with 22.5 kg of BHC 5% or Aldrin 5% dust or 12.5 kg carbaryl (Sevin) 10% dust.

Diseases: Following are the important diseases of bajra crop:

Downy mildew or green ear disease (*Selerospora graminicola*): The disease is prevalent in humid areas of all the bajra growing tracts of the country. The affected plants remain stunted, leaves turn yellow and may dry up if the disease at seedling stage and the plants are die. The infected ears are transformed wholly or partially into green, small twisted leaf like structures.

Control Measures

a) Seed treatment with fungicides.
b) Roguing of disease infected plants at the seedling stage around 40 days of sowing and uprooted plants should be destroyed and buried into soil.
c) Spraying of Ziram, miltex or Diathene Z-78 or M-45 at the rate of 2.0 kh/ha in 800-1000 litres of water. Spraying should be done soon after rouging of infect plants
d) Second spraying at the above mentioned concentration should be done at boot-leaf stage.

Smut: Smut is caused by *Tolypesporium penicillariae*. It is common under humid climatic conditions in all the the states where bajra is grown. Seed treatment with fungicides like Agrosan GN or Cerosan controls the disease.

Ergot (*Claviceps microcephala*): Seed treatment with 20% common salt solution followed by washing with fresh water and then treating with Agrosan GN, Ceresan or Thiram at the rate of 1-2 g/kg seed controls the disease. Chrop rotation and deep summer ploughing are found to be effective measures in controlling the disease. The disease affected ears must be destroyed completely.

Harvesting and Threshing

The grain crop should be harvested when the grains become hard enough and contain about 20 per cent moisture in them. Generally, two methods for harvesting the crop are adopted viz. (a) cutting of earhead first from the standing crop followed by cutting remaining plants later, and (b) cutting of entire plant by sickles and stalking the harvested plants for five days in sun for drying.

For easy separation of grains the earheads should be dried well before threshing. Threshing may be done either by beating the earheads with sticks or by trampling the earheads under bullock's feet. The later method of threshing is common when the crop is grown extensively but the case of smaller areas the grains are separated out by beating the earheads with sticks.

The separated grains must be cleaned and dried in sun to bring about 12-14 per cent moisture after which the grains may be bagged and stored in a moisture proof store.

Yield

An ideal crop of one hectare irrigated area gives about 30-35 quintals of grains and 100 quintals of stover while unirrigated one yields nearly 12-15 quintals of grains and 70-75 quintals of stover.

Important Practices

1. Popularising high yielding varieties, hybrids and composites,
2. Adequate plant population to be maintained,
3. Early sowing in lines and early weeding,
4. Moderate fertilizer use, and
5. Proper drying and storage

Important Observations

- The most drought tolerate crop among cereals and millets.
- Sensitive to water logging and acidic soils.
- First hybrid: **HB 1** (Hybrid Bajra 1) released in 1965. HB 1 = Tift 23-A(male sterile) x Bil 3 B.
- **HB-4:** first hybrid developed by using local male parent, duration 80-85 days.
- **Weed control:** Atrazine/propazine @ 0.5 kg/ha as pre-emergence.

Minor Millets

Small millets may be defined as millets cultivated for their small grains which are borne on short, slender, grassy plants. They are also known as minor millets. The list of minor millets is given in Table 7.5. Small millets are mainly grown in arid, semi-arid or mountain zones as rainfed crops in marginal and sub marginal soils. They are distributed in most of the Asian and African countries and parts of Europe. Finger millet is the principal small millet grown in South Asia. Foxtail millet and prosomillet are important in China and proso millet is grown extensively in the South Western Russia. In Africa, finger millet is of local importance. Some of the small millets are grown in the United States and Europe on a very limited scale. The small millets contribute about 1% of the world food production.

TABLE 7.5: List of minor millets

	Name	Botanical name
1.	Finger millet (Ragi)	*Eleusine coracana*
2.	Proso millet (common millet)	*Panicum miliaceum*
3.	Foxtail millet (Italian millet)	*Setaria italica*
4.	Banyard millet	*Echinochloa colona*
5.	Kodo millet	*Pasplum scrobiculatum*
6.	Little millet	*Panicum sumatrense*

Out of the total area of 123.5 m ha under food grains, the small millets occupy just 3.7 m ha or 2.99 per cent. Production is 3.3 m t or only 1.73 per cent to the total food production. The small millets contribute about 1.86 per cent of cereal grain production. The productivity of finger millet is the highest (1327 kg/ha) among the millets. The lowest productivity of small millets (466 kg/ha) is largely due to poor fertility of soils and non-adoption of improved package of practices. Minor millets are predominantly grown in Karnataka, Tamil Nadu, Maharashtra, Andhra Pradesh, Uttar Pradesh, Bihar, Orissa and Madhya Pradesh. These states contribute about 90 per cent of the area and production (Table 7.6).

TABLE 7.6: Important states growing minor millets in India, 1995-96

States		Finger millet	Small millets
Andhra Pradesh	A	119.2	96.9
	P	140.9	66.0
	Y	1182	681
Bihar	A	83.8	44.4
	P	93.8	20.9
	Y	1119	471
Karnataka	A	964.1	90.2
	P	1630	40.9
	Y	1691	453
Madhya Pradesh	A	16.3	1007
	P	4.8	251.1
	Y	294	249
Maharashtra	A	152.9	141.9
	P	160.6	99
	Y	1050	698
Orissa	A	288/7	56.4
	P	184.6	27.5
	Y	80.7	488
Tamil Nadu	A	162.1	138/6
	P	135.7	98.9
	Y	1948	714
Uttar Pradesh	A	137.8	143.0
	P	186.2	137.9
	Y	1351	964
All India	A	1831.5	1859.5

States		Finger millet	Small millets
	P	2430.4	866.1
	Y	1327	466

A = Area in '000 ha, P = Production in '000tonnes and
Y = Yield in kg/ha. Source: Fertilizer statistics, 1996-97.

Varieties of Minor Millets

Finger millet	:	Sharada, Godavari, Kalyani, Simhadri, BR 407, Gujrat Nagi, Indaf-1, 3,8, Dibyasingha, CO 12, Paiyur-1, K7, Nirmal
Kodomillet	:	Jawahar, Kodo 1, Gujrat, Kodra 1, CO 2, CO 3.
Foxtail millet	:	Chitra, SIA 326, CO 2, CO 4.
Little millet	:	Gujarat, Var 1, Indori 2, CO 1, CO 2, K 1.
Proso millet	:	Ramcheena, Shyamcheena, Varada, CO 2, CO 3, CO4
Barnyard millet	:	Anurang, RAU 3, Banti 1, K 2, CO 1.

FINGER MILLET (*RAGI*) (*ELEUSINE CORACANA*)

Finger millet (Ragi) is the principal small millet grown in South Asia. Its origin may be India as it is not cultivated other part of the world. It serves as the staple food for the people in the arid and semi-arid tropics. It is used for malting and brewing. Malted finger millet is used in infant food and in milk thickness formulations. Finger millet malt brewed with an equal proportion produces beer of commercial standards.

Finger millet (Ragi) is an important crop in India. In Orissa, it is mostly grown in Ganjam and Karaput districts and patches here and there in other districts. It is one of the hardiest crops suited for dry farming (farm without irrigation facilities). It can be grown under condition of low rainfall and can withstand in drought, reviving again with a good shower of rain with remarkable vigour. Its grain has nutritive value and is considered fit for the people who do hard physical work. It can be grown both as the dry land crop and also under irrigation. The crop is free from pest and diseases. Its straw is used as fodder.

Distribution

It is grown mostly in Tamil Nadu, Orissa, Karnatake, Andhra Pradesh, Bihar, Uttar Pradesh, Madhya Pradesh and other parts of India.

Rainfall

As already mentioned, it is cultivated as dry land crop where rainfall ranges 50 to 75 cm. The yield of ragi is directly related to amount of rainfall received from June to October.

Soil

Ragi can be grown well in red soil, alluvial soil and black cotton soil. But, black cotton soil is not properly suitable for ragi. It can be grown in light soils, loamy soils, sandy loam soils. Heavy soils are not mostly preferred.

Seasons

Ragi can be grown throughout the year as kharif crops, as a pre-rabi crop, as a *rabi* crop with light irrigation and as a summer crop with moderate irrigation. Flowering will be delayed if it coincides with cold weather. It is moderately tolerant to salinity. When grown as a pre-rabi crop in August-September, it can be raised without irrigation. Amongest the *rabi* millets, ragi is the most important crop. Because of its low irrigation needs, it has great scope for cultivation in the *rabi* and summer seasons in the tail ends of canals and in the command areas of minor irrigation projects in multiple cropping sequences.

Ragi, though predominantly a kharif crop can be cultivated throughout the year if irrigation is available. In Karaput district of Orissa, ragi is grown from January to April and is followed by rice. In Ganjam district pre-rice ragi cultivation is common.

Crop Rotation

Dry land Ragi: For this, there is no definite rotation principle but it is sown as mixed crop with red gram (arhar) in the south. But in south Orissa, the following rotations are followed: 1. Ragi-Horse gram, and 2. Ragi-Back gram.

Irrigated Ragi: It is grown in garden lands or areas having tank irrigation/any other source of irrigation.: 1. Rice-Ragi-Green gram (300% intensity), 2. Ragi-Tobacco, 3. Ragi-Chilly and 4. Ragi-Potato/Onion.

Field Preparation

The land is prepared thoroughly and ploughing starts from the first shower of rain and repeated with succeeding rain. If worked with mould board plough there is considerable saving in number of ploughings. Sometimes cultivators are also used. The soils should be free from weeds and roots and all the clods should be broken. Manure should be applied at this stage and well mixed into the soil. Before the seeds are sown the field should have fine tilth.

Sowing

Direct Seeded Crop: For direct seeding of ragi, a well pulverized seed-bed should be prepared. Application of 5 tonnes/ha FYM or compost should be incorporated well into the soil along with fertilizers before sowing. Sowing the direct seeded crop in lines 20 cm apart is better than the broadcasted one. Seed rate of about 10 kg/ha will be adequate for the line sown crop. The seeds selected for sowing should be free from mixture and under developed seeds. The under developed seeds can be separated out. For the selection of healthy seeds, salt water treatment can be practiced. Generally, broadcasting and drilling methods are used for sowing of direct seeded crop of ragi.

Transplanted Crop

Nursery: Seedlings should be raised in well prepared nurseries for the transplanted crop. Application of small doses of (10 g each of N and P_2O_5 per square metre of the nursery area) helps rapid growth of the seedlings. Sowing of seeds should be done evenly in the nursery beds early in the month of June. The nursery beds may be irrigated, if necessary. The seedlings will be ready within 25 to 30 days. Generally, 8-10 kg of seeds will be sufficient for transplanting one hectare.

Transplanting: After land preparation, manures and fertilizers should be applied. Transplanting should be close. Early varieties should be transplanted 15 x 10 cm apart and medium and late varieties 20 x 10 cm a part. It is however, economical to transplant seedlings in plough furrows in which case both early and late varieties may be planted in plough furrows drawn 20 cm apart. Two seedlings may be placed 10 cm apart in the furrows at each hill. The base of the seedlings will be covered by soil when the next furrow is drawn. Shallow planting within 5 cm depth encourages quicker and better tillering.

Varieties

A number of short and medium duration varieties of ragi are available for cultivation in kharif, rabi and summer seasons. Growing short duration varieties in rainfed uplang will permit double cropping pf ragi with pulses or sesame. The characteristics of the varieties are given below:

Variety	Origin	Maturity in days		Expected yield in rabi (q/ha)	Special features
		Kharif	Rabi		
Dibyasingha (T-20)	Mutant of AKP-7	85-90	95-100	24	Earheads incurved, grain colour brown, tolerate to blast and stem borer.
AKP-2	Introduction from A.P.	90-95	100-105	20	Earheads incurved, grain colour brown.
AKP-3	-do-	105-110	115-120	33	Earheads incurved, one of the best varieties in medium group.
AKP-7	-do-	105-110	115-120	32	Earheads incurved.
Godavari (PR-202)	Introduction from Karnataka	105-110	115-12	34	Top incurved earheads.
Neelachal (B-4-10-56)	Mutant of (IE-642)	105-15	115-125	36	Semi compact top incurved earheads, grain colour brown.

Fertilizer Management

It is better to apply fertilizer as per soil test recommendation. If the soil test is not done the fertilizer may be applied at the rate of 40 kg Nitrogen, 20 kg P_2O_5 and 20 kg K_2O for short duration varieties of 90-95 days and 60 kg N, 30 kg P_2O_5 and 30 kg K_2Oper hectare for medium and long duration varieties of ragi. For rainfed crop, only 20 kg of N and 12 kg each of P_2O_5 and K_2O per hectare may be applied. Otherwise, fertilizer dose can be applied as per the soil test result. Nitrogen may be applied in two splits, *i.e.*, 50% as basal and remaining 50% as top dressing at the first hoeing and weeding so as to incorporate the fertilizer into the soil.

Intercultural Operations

Early weeding of the direct seeded crop is essential for smothering weeds and getting good yields. The first hoeing and weeding should be done within 2 to 3 wees of sowing and the second a fortnight after. One weeding of the transplanted crop, between 2 to 3 weeks after transplanting is adequate. Where necessary, however, a second weeding mat be given.

Water Management

Ragi can not stand water logging but it can tolerate better than maize and jowar. The field should therefore, be drained out during the monsoon if necessary.

Plant Protection

The recommendation of the plant protection schedules of the state may be followed to control the pest and diseases, if the attack of pest and diseases will be noticed.

Harvesting

Ragi is harvested with ordinary sickles. But ib Andhra Pradesh and Orissa only the earheads are cut and brought to the threshing floor for drying and threshing. The wet land ragi, it is ready for harvest in August. After that the land is puddle and transplanted the late rice. Dry land rice is ready for harvest in October and the horse gram or black gram is sown. After the earheads are harvested, they are heaped and covered for sometimes till favourable weather provides for threshing.

Threshing and Storage

Threshing is done by three methods:

1. Treading out the grains under the feet of the the bullocks.
2. Bitting out the grains with the sticks, and
3. Walking a stone roller over the ear heads.

The first method is the most common. After treading, there is always a small proportion of ear head left un-threshed which has to be bitten by sticks. The threshing of ragi is more difficult than the rice or the other grain crops. As the grains are held much more firmly in the glumes and required considerable pressure.

It is stored in the earthen mare receptacle unger ground pits, bamboo containers, gunny bags and metallic seed bins.

Yield

1. Dry land ragi: grain yield 15 q/ha, but the average yield only 10 q/ha,
2. Wet land ragi: grain yield 30 q/ha, but the average yield 20 q/ha.

Important Practices to be Advocated

1. Line sowing of the direct seeded crop and early weeding, use herbicides to reducing cost
2. Thinning,
3. Moderate fertilizer use.

CHAPTER 8

Production Technology of Crop: *Pulses*

(Ecological Requirements, Cultural Practices and Crop Production Management)

PULSES IN INDIAN AGRICULTURE

Importance of Pulses

Indian population is predominantly vegetarian and protein requirement for the growth and development of the human being is mostly met with pulses, the cheapest source as the milk is becoming expensive day by day. A part from the human diet, pulses from an important fraction of cattle feed and fodder as hay, green fodder and concentrates, etc. Being leguminous, the pulse plants take so little and give so much to our soils that their significance in restoring and maintaining the soil fertility can not be ignored. If given good credit in the cropping pattern, they add more nitrogen to our soil than what we add through chemical fertilizers. Thus, every pulse plant is a mini-fertilizer plant in itself. Being deep rooted crop, these pulses are highly adaptive to dry land areas of the country, which constitute a major cropped area and contribute to about 80 per cent or more of total pulse production. The importance of pulse crops in India may be described as under:

1. **Place of pulse in human diet:** Pulses are the most important part of our diet as pulses are the richest source of protein amongst the food grains mentioned below:

Protein content in major food grains

S.N.	Pulses		S.N.	Cereals	
	Name	Protein (%)		Name	Protein (%)
1.	Gram	21.0	1.	Wheat	12.8
2.	Arhar	22.3	2.	Rice	8.5
3.	Urd	24.0	3.	Rice milled	6.9
4.	Mung	24.0	4.	Barley	11.5
5.	Lentil	25.2	5.	Maize	11.1

Pulses are not only the rich source of protein but they consist of many important amino acids than the cereal proteins and hence both the proteins (from cereal and pulses) taken together in food have greater biological values and better beneficial effects on human health.

Pulses as best source of vitamins: Pulses are rich in Vitamin A which is three or four times greater when compared with wheat. In the international unit, 100 g wheat floor contains 108 unit of Vitamin A whereas gram contains 316 units and lentil contains 450 units.

The Vitamin B-1 content in pulses is much higher than in cereals viz., in international units, milled rice has 20 units, wheat 284 units, gram 100 units, urd 130 units, mung 155 units, and soybean has 300 units of Vitamin B-1.

Vitamin C is also found more in pulses at sprouting because it is absent in ordinary state. Thus, the sprouted grains of pulses contain about 10-15 miligram of Vitamin C/100 gram of weight.

Pulses are also better source of minerals, especially Ca, which is the most required mineral by the human system. The Ca content in gram is 0.19%, in urd it is 0.21% and in mung it is 0.15%.

2. **Place of Pulse in Soil Fertility:** Pulses add 0.5 to 1.5 tonnes of organic matter to the soil in the form of their roots left after harvesting of the crop. A part from this, pulses being leguminous in nature bear nodules on their roots which use to fix atmospheric elemental nitrogen into their colonies and leave the foxed nitrogen in available form for the succeeding crop as residual form of nitrogen. Thus, these crops fix atmospheric nitrogen in the soil and make our soil more fertile. These crops add nitrogen to the soil in two ways:

   The symbiotic bacteria are found on the root surface which use to fix atmospheric nitrogen in the soil, however, their efficiency on nitrogen fixation is governed by several factors like inoculation of pulse seeds through bacterial culture, extent and method of phosphate fertilization, presence of calcium and sulphur contents in the soil, weeds' intensity and their growth, plant protection measures, crop and its variety etc. On an average one hectare crop adds about 15 to 35 kg nitrogen in readily available form.

   Pulses add nitrogen to the soil in the form of green manuring also which is done in two ways, viz., green manuring in situ and leaf green manuring by transporting leaves from the place where they were grown but green manuring in situ is more profitable as it adds nodular nitrogen also to the soil. Following Table indicates the extent of nitrogen fixation by some important pulses.

   Extent of organic matter and nitrogen added to the soil by some pulses

S.No.	Crop	Green matter in (t/ha)	N %	Total added/nitrogen (kg/ha)
1.	Urd	12.0	0.41	38.3
2.	Mung	8.0	0.53	34.5
3.	Lentil	5.6	0.70	32.8
4.	Peas	20.1	0.36	59.4
5.	Cowpea	15.2	0.51	52.4

3. **Pulses help in Soil and Water Conservation:** Pulses have deep and very extensive root system and thus they have better soil binding effect and the soil under pulses use to resist more against the soil erosion caused by both water and wind. A part from this profuse vegetative growth and dense foliage provide a protective soil cover which does not allow the rain drops to hit the soil and thus, the soil particles are not detached by beating of rain drops. This way the most fertile portion of the soil is saved.

   Soil under pulses does not allow water to run very fast as they provide a check and maximum absorption of water takes place which increases the soil productivity, especially in case of dry farming zones, and controls soil erosion. Because of their better ground coverage they reduce the water loss through evaporation from the soil surface and crops do not suffer from drought.

4. **Pulses provide a superior quality fodder and feed to the cattle:** The by

product of dal-mills like bran, seed and broken part of the cotyledons make a superior quality of balanced cattle feed. The stem and leaves, whether fed green or dried or as bhusa, are found to be a rich source of protein and minerals for the cattle.

5. **Control Weeds:** These plants grow very vigorously and cover the ground surface, hence, they suppress or smother the weeds and they do not get favourable conditions for producing seeds. This way their multiplication is also checked and the soil remains fertile and trouble free.

Green revolution, multiple cropping, intensive area programme, high yielding variety programme and multi-objective crop improvement, etc. have helped in a real breakthrough in our cereal production. Thus, the problem of human starvation is, to greater extent, solved but that of imbalanced diet due to short supply of pulses and poor human nutrition still remains unsolved. The average yield of pulses is centred around 400 to 500 kg/ha. Since long time and population has tremendously increased which resulted into a progressive decline in per capita availability of pulses from 68.5 g in early fifties to 34.9 g in the fifth plan period. The reasons for such a poor availability had been the higher population and the short supply due to very poor crop yields. The lower yields of pulses are because of following reasons:

1. The acreage under pulses fluctuated from time to time due to more climatic hazards like drought, frost, floods, etc. Thus, the total production of pulses is not stabilized.
2. The pulse crops are mostly grown on marginal and sub-marginal lands without much care and inputs. Such lands are mostly poor in soil fertility.
3. These crops are generally. Grown under rainfed conditions which very often suffer from aberrant weather conditions and they pull on without any protective or life saving irrigation.
4. Pulses are rarely grown as pure and principal crop but they are grown as bonus crops in the form of mixed/intercrop without considering the compatibility of different crops grown together. Thus crops suffer from shading effects.
5. These crops are considered as soil builders and therefore, they are grown without application of manures and fertilizers. Thus, their growth and development is normally very poor.
6. The crops are found to be severely attacked by diseases, insects and pests as plant protection measures are rarely adopted.
7. Because of their poor initial growth, they are often badly infested by weeds which suppress their growth and result in drastic reduction in their yield.
8. Most of the local varieties, which are in general cultivation, possess undesirable characters like:
   (a) They are spreading type and need wider spacing, thus less plant population/unit area can be grown.
   (b) They are of long duration and per day productivity is very poor.
   (c) The local varieties are thermo and photo-sensitive and their adaptability is very limited,
   (d) They are less responsive to inputs and grow profusely if supplied with fertilizers and irrigations.
   (e) They are susceptible to many fungal diseases like wilt, etc.

The problems of short supply of pulses was brought to notice of the I.C.A.R.

and an integrated All India Coordinated Programme was started in 1972-73 in collaboration with U.S. Department of Agriculture for the pulse improvement work in the country. The main emphasis for increasing pulse production was to be laid on following points:

1. Bringing an additional area under short duration high yielding varieties to fit in multiple cropping programmes to be grown as catch crop.
2. Developing new cropping systems like companion cropping, mixed cropping or intercropping for growing pulses between widely spaced crops such as sugarcane, maize, potato, cotton, arhar, groundnut, bajra and jowar, etc. both under irrigated and rainfed conditions.
3. Evolution, multiplication and use of improved seeds of various pulses.
4. Adoption of efficient plant protection measures.
5. Basal placement of phosphatic fertilizers and treating the seeds with *Rhizobium culture.*
6. Growing pulses on relatively fertile lands rather than growing them on marginal and sub-marginal lands.
7. Adoption of improved package of practices like line sowing, control of weeds, harvesting at right time or at physiological maturity to avoid splitting of pods and thereby shattering losses.
8. Granting subsidy on Government loans to the growers and providing improved seeds, fertilizers and plant protection materials on concessional price to the farmers.
9. Evolution of better plant types of pulses for boosting pulse production.

   The varieties thus developed are characterized as under:

a) Non-spreading and erect types.
b) Thermo and photo-sensitive.
c) Early maturing.
d) Responsive to apply inputs with high yield potentials.
e) Fairly resistant to water logging, insect pest and diseases.
f) Resistant to drought and frost.
g) High nutritional qualities.

DISTRIBUTION

Pulses belong to the family *leguminaceae.* The grains are used for staple purpose and hence called as gain legume. In the world, pulses are grown in an area of 71.0 m ha with the production of 56.8 m t (1995-96). China, India and Russia are the important countries growing pulses. India accounts for about 36.0% of the area and 27.1% world production in 1995-96. Yield level in India is much lower (552 kg/ha) than the world average (799 kg/ha). The commonly grown pulse crops in India are given in Table 8.1.

In India, pulse area and production slowly increased from 19.1 m ha in 1950 to 23.9 m ha in 1995-96. The area and production have not increased; rather there is a reduction in the period 1960-1980. There was also not appreciable increase in the yield level over the years. The yield level slightly increased from 441 to 552 kg/ha during 1950-1996. However, there were fluctuations over the years. Production of pulses fluctuates depending upon the monsoon behaviour. Pulses are largely grown on marginal lands, mostly under rainfed conditions. There was virtually no break-though in developing improved and high yielding varieties.

Among the pulses, chick pea and pigeon pea are the important crops accounting for 47.0% of pulse area and 60% of total production. Uttar Pradesh, Madhya Pradesh, Rajasthan, Maharashtra, Orissa, Bihar, Andhra Pradesh, Karnataka, Gujarat and Tamil Nadu account for about 95.3% of production (Table 8.2). Pulses are grown in *kharif* and *rabi* seasons. Out of total acreage, 45% is *kharif* pulse which account for about 34.5% of pulse production. *Rabi* pulses contribute 55% of the total area and about 65% of production.

TABLE 8.1: Pulses grown in India

Crop	Botanical name
Chickpea or Bengal gram	*Cicer arietinum* (L.)
Pigeonpea or red gram or Arhar	*Cajanus cajan* (L.)
Black gram or urd bean	*Phaseolus mungo* (L.)
Green gram or mung bean	*Phaseolus aureus* (Roxb.)
Cowpea	*Vigna unguiculata*
Lentil	*Lens culinaris*
Moth bean	*Vigna aconitifolia* (Jaeq.)
Horse gram	*Dolichos biflorus*
Lab lab	*Dolichos lab lab*
Rajmash	*Phaseolus vulgaris* (L.)
Lathyrus	*Lathyrus sativus*
Pea	*Pisum sativum*

TABLE 8.2: Area, production and yield of pulses for importantpulse growing states, 1995-96

States	Area ('000 ha)	Production ('000 t)	Yield (kg/ha)
Andhra Pradesh	1537.7	639.7	416
Assam	106.9	57.1	534
Bihar	922.7	572.3	620
Gujarat	840.3	456.5	543
Haryana	428.5	416.4	972
Karnataka	1528.4	724.1	474
Madhya Pradesh	5133.4	3102.2	604
Maharshtra	3305.0	1660.9	503
Orissa	2191.9	1176.0	537
Punjab	100.1	82.5	824
Rajasthan	3573.8	1462.5	409
Tamil Nadu	961.9	359.7	374
Uttar Pradesh	2923.5	2251.7	770
West Bengal	196.3	125.9	641
All India	23916.6	13192.3	552

Source: Fertilizer statistics, 1996-97

Pulses represent a group of crops and each crop has its own requirement for inputs and environment. Hence, individual consideration is required for enhancing the production.

Production of French beans and mung bean with a growing export market should be stepped up by intensification of their cultivation. Cultivation of French beans during *rabi* season in non-traditional areas of northern plains and other suitable pockets should receive greater attention. There is ample scope of bringing more area under spring or summer cultivation of mung bean in northern India. High yielding and stress-resistant varieties should get up priority in the pulse improvement programme. There is a good scope of expansion of area under pulses in various inter, relay or sequential cropping systems to increase the productivity per unit area. Rice fallows should get maximum attention to grow pulse crops.

CHICK PEA OR BENGAL GRAM (*CICER ARIETINUM* L.)

Gram is believed to have originated somewhere in south west Asia, probably in Afghanistan and/or Persia. According to Aykroid and Doughty (1964), it originated somewhere in eastern Mediterranean, however, according to De Candolle the gram is mentioned in Sanskrit which indicates that it is cultivated in India from a longer period than in any other country in the world. It is now grown in many countries like Iran, Turkey, Central and Southern Africa, Greece, Rumania, Egypt and some parts of Australia.

In India it is mainly grown in the states of Bihar, Haryana, Madhya Pradesh, Maharastra, Punjab, Rajasthan and Uttar Pradesh etc. It is used in many forms as Dal, Chhole sweets and many attractive dishes and snacks are prepared from its flour. Its leaves consist of mallic and citric acids which are very useful for stomach ailments and it is best blood purifier. Nutritionally it is very reach as it contains about 17-21 per cent protein, 62 per cent carbohydrate, good amount of fat, besides it is a rich source of Ca, Fe and vitamin C (in green stage) and vitamin B_1. Its feed and straw are highly rich in nutrients and are mostly used as productive ration.

India is the largest producer of chick pea in the world, sharing 65 and 70% of the total global area and production respectively. During 1990-95, chick pea occupied 7.26 m ha and produced 6.21 m t accounting for 31.3% and 43.9% of the national pulse hectarage and production respectively. Introduction of high yielding varieties has helped increasing productivity up to 855 kg/ha against a mean global productivity of 768 kg/ha.

Botanical Description

Indian grams have been classified into two broad groups:

- Desi or Brown grams (*Cicer arietienum*): The colour of the seed ranges from yellow to dark brown. Plants are small with good braching ability, chromosome number is 2n= 14 or 16.
- Kabuli or White gram (*Cicer kabulium*): The seed colour is usually white. Grams are bold and attractive. Yield potential of this group is poor as compared to desi or Brown gram.
- Root system: Root system of gram includes central strong tap root and many lateral branches that spread out in all directions in the upper layers of soil. There are many nodules on roots.
- Stem: Stem is branched with granular hairs on it. Main branch in gram usually produces not more than one secondary shoot.
- Leaves: Leaves are pinnately compound with one terminal leaflet. There are 9 to 15 pairs of leaflets.

- Flowers: The flowers are typical papilionaceous consisting of five sepals, five petals comprised of one standard, two wings and two keels, ten stamens, nine fused to form one staminal column and one free and a carpel with the style borne laterally on the ovary. Self pollination is the rule but cross pollination may occur to the extent of 5-10 per cent.
- The pod is about 2 cm long and usually contains two seeds. A single plant produces about 50 to 150 pods.
- Seed: The seeds are spherical in shape, wrinkled and with a pointed beak. They vary a great deal in size as well as colour.
- Gram grows into a small herbaceous plant with a maximum height of 45-60 cm but in case of dense population or faulity management of fertility and water it may grow slightly taller which are branched with granular hairs on them.

Distribution

The crop probably originated from west Asia. This is an important pulse crop in India. About 70% of the world production of chickpea comes from Asia. Among the pulses, chickpea occupies 32.5% of are and 44.0% of production in India. The crop is chiefly grown in cool, dry periods on receding soil moisture. About 55% of the crop is cultivated as rainfed. Chickpea is predominantly grown in North India. Madhya Pradesh, Rajasthan, Uttar Pradesh, Maharashtra, Haryana, Karnataka and Bihar contribute about 91,4% of area and 95.5% of the production. The average yield of the crop is about 700 kg/ha Table 8.3).

TABLE 8.3: Important states growing chickpea in India, 1995-96

States	Area ('000ha)	Production ('000t)	Yield (kg/ha)
Bihar	121.0	85.0	702
Haryana	382.0	385.0	1008
Karnataka	290.4	143.0	492
Madhya Pradesh	2664.5	1988.9	746
Maharashtra	717.1	376.0	524
Rajasthan	1620.2	1090.2	673
Uttar Pradesh	1115.1	784.2	703
All India	**7198.4**	**5019.6**	**697**

Source: Fertilizer statistics, 1996-97.

There are two types of chickpea, **'deshi'** and 'kabuli' **types**. The former has small sized seeds with coloured seed coat. They are mainly cultivated in Easter Africa. The 'kabuli' types have large sized seeds with white seed coat. The seeds are consumed whole after cooking or as dhal. Soaked grain and husk are fed to cattle.

Crop Improvement

In early fifties and sixties, the efforts were made to select high yielding genotypes rom germplasms and landraces without hybridization and selection. Systemic research work on chick pea was started under All India Coordinated Pulse Improvement 1967 having 12 centres in State Agricultural Universities with headquarters at IARI, New Delhi. During 1977, the project was upgraded to project Directorate was and its headquarters was shifted to Regional Research Station of the IARI at Kanpur, which was later termed as Directorate Pulse Research (DRP). In 1993 an Independent All India Coordinated Project was launched with 9 mandatory and 12 verifying centres due to an increased importance of chick pea.

During 1981-90 emphasis was made on developing disease resistant varieties of chick pea. As a result Pusa 212, Pusa 244, Pusa 256, Pusa 261 were developed through hybridization; Pusa 408, Pusa 413 and Pusa 417 were developed through mutation at IARI. The Kabuli variety BG 267 was also developed during that period GL 769 and PBGI were developed at Ludhiana for wilt and Ascochyta blight affected areas. Several other promising varieties viz. Gaurav, Haryana Chana-1, GNG 146, RSG 2, Vikas Vishwas, Phule G 12, JG 62, JG 315, JG 74, Bheema, Kranti, Sweta, CO 3 and PDG 84-10 have been released for various parts of the country. The mutation breeding has developed BGM 408 (for NW Plains), BGM 413 (for NE Plains), BGM 417 (for Central Zone) and RS 11 and RSG 2 (for Raj) which are multiresistant and suitable for rainfed culture.

Climatic Requirement of the Crop

Being the leguminous crop gram needs less rainfall and having a deep root system it does very well under dry tracts which receive an annual rainfall of 60-100 cm. Water logging at any stage of growth is very harmful and the crop may vanish, however, it responds to light irrigations at flowering and grain filling stages. It prefers cold weather but frost is deadly harmful, especially at flowering and grain formation stages. Hailstorm at maturity causes great damage to the crop.

Soil

Gram is grown practically on a variety of soil types ranging from very light to heavy ones. In north Indian states it is cultivated on sandy loam to clay loam soils whereas in south on Deccan Plateau and Central India or Maharashtra gram is grown on black cotton soils. But too light and too heavy soils, both the extremes are not good for the crop. The best type of soil is clay is clay loam with an efficient drainage and free from soluble salts preferably having neutral pH.

Land Preparation

Gram needs clodded and rough seed bed for good aeration in root zones, thus a required seed bed may be obtained by one deep ploughing followed by two harrowings. In diara lands one ploughing by deshi plough brings about cloddy condition and the crop can be sown.

Seeds and Sowing

Certified, treated seeds with 0.25% thiram of high quality should be used for sowing the crop. Second fortnight of October is found to be the best sowing time but it may be delayed by first week of November for higher yield. Broadcasting should be avoided and the seeds may be sown in furrows behind the country plough but the best way is to drill the seeds either by seed drill or by malabasa at a depth of 8-10 cm. A row spacing of 30 cm in desi types (*Cicer arietinum*) and 40-45 cm in Kabuli types (*Cicer kabulium*) should be used. The seed rate is decided by seed size and weight but a seed rate of 75 to 100 kg/ha depending on these factors is used.

Varieties for Different Situations

Important varieties	:	Radhey, G-235, C-104, BR-78, R-11, Gwalior-2, Vikas, Vishwas, Ujjain-21.
Drought tolerant	:	RS 10, G 24, T 3, T 87.
Late sown situation	:	Jg 74, Strain 76, G 235, Pant G 114.
Ascochyta blight, Gaurav, Resistant	:	GNG 146, Pusa 261, PBG 1.
Wilt Resistant	:	Pusa 212, Jg 315, BG 244, Phule G 81-1-1, ICCV 10

Manures and Fertilizers

Being a leguminous crop gram does not respond to nitrogen fertilization whereas the response of potash is very inconsistent but the phosphate plays a very vital role in root development, nodulation, and growth and yield of the crop. However, Kabuli types need about 15-20 kg nitrogen/ha for a good initial boosting of the growth. Therefore, it is recommended to apply about 15 kg N/ha, 50-60 kg phosphorus and the nutrient carriers must be basal placed at a depth of about 10-15 cm in the soil.

Application of sulphur at the rate of 20-40 kg S/ha showed encouraging results in all the zones giving an average nutrient use efficiency of 10-15 kg grains/kg S. The response has been still of high order in light textured soils of the NWPZ, Gypsum, Pyrite and single super phosphate as source of S are equally effective.

Among micro nutrients Zn is the most critical, especially in tarai soil of U.P. and Bihar and light textured soils of Punjab, Rajasthan and Haryana. Its application at the rate of 25 kg/ha as basal or 0.5% $ZnSO_4$ + 0.25% burnt lime mixture gives very promising results. Boron and Mo fertilizer gives better results in acid soils of Chhota Nagar as well as in North Bihar. Foliar spray of 2% DAP and 2% KCL at pre-flowering stage enhanced yield by 8-15%. Inoculation with Rhizobium culture also gives about 10-15% increase in grain yield.

Water Management

The crop is grown mostly under rainfed conditions where irrigations are not possible but researches have indicated that for higher yields it is essential that the crop be given light irrigation. There should be sufficient moisture in the soil while sowing the crop. One light irrigation at flowering and one light irrigation at grain development stage are required for good crop.

Nipping

It is the process of plucking the apical buds of the crop at about 30-40 days after sowing. Nipping stops the apical growth and promotes the lateral branching, thus the plants become more vigorous and produce more flowers and pods and yield per plant is increased.

Weed Control

Gram responds well to hand weeding and hoeing as it improves aeration. The best time for weeding and hoeing is between 25-30 days after sowing and if the weeds are still present in the field, the second weeding should be done about 60 days after sowing. Weeds may be controlled by herbicides also for which Basalin at the rate of 1 kg a.i. per hectare should be used as pre-plant well incorporated in the soil or Tribunil at the rate of 2.5 kg/ha may be sprayed as pre-emergence.

Plant Protection Measures

Control of insects/pests: Cutworm and pod borers are the major pests of gram crop. The cutworms cut the tender stems, branches or damage the leaves by cutting the top, edge or midribs, etc. Cutworms may be controlled by soil treatment 5% BHC dust or 3% Heptachlor at the rate of 25 kg and 35 kg/ha respectively. Aldrin dust could be substitute for these chemicals, if they are not available.

The pod borers use to bore the pods and eat the grains inside. They may be controlled by spraying of 35 EC Endosulfan in the concentration of 2 ml/litre at 15 days interval and about two or three applications are required. Endosulfan also controls aphids and cutworms if present in the field.

Control of Diseases: The crop is prone to wilt, blight and rust diseases which have practically no cure except using resistant varieties like BG-212, Pusa 109 and C-235 for wilt prone areas and G-543 and C-235 for plight. A part from using resistant varieties, seed treatment with Benlate and Thiram (1:1) at 2 g/kg seed, for wilt, soil treatment with Brassicol and Captan at 10 kg/ha rate, and adoption of proper crop rotation should be followed.

Yield

A good pure crop of gram gives about 15-20 quintals yield/ha in case of desi varieties but the Kabul varieties give about 25-35 quintals/ha grain yield. The yield in the case of mixed or intercrop depends upon the plant population of gram.

PIGEONPEA, REDGRAM OR ARHAR (*CAJANUS CAJAN* L.)

The origin of pigeon pea is now considered to be the peninsular India. About 90% of the world's pigeon pea is produced in India. It is the perennial shrub and a short day plant. The crop is grown as an annual crop in India and as a perennial in many other countries where the pods are harvested at regular interval. The crop has a deep root system and hence highly drought tolerant. In India, pigeon pea is the second important pulse crop occupying 14.5% of the area and contributes 15.5% of the total pulse production. It is predominantly cultivated in Maharashtra, Uttar Pradesh, Madhya Pradesh, Karnataka, Gujarat and Andhra Pradesh. These states contribute about 86.8% of the area and 84.4% of the production (Table 8.4).

TABLE 8.4: Important states growing pigeonpea in India, 1995-96

States	Area ('000 ha)	Production ('000 t)	Yield (kg/ha)
Andhra Pradesh	327.4	103.0	315
Gujarat	383.0	279.0	728
Karnataka	426.9	197.8	463
Madhya Pradesh	398.2	325.3	817
Maharashtra	1051.9	627.3	596
Orissa	167.2	127.1	760
Tamil Nadu	165.7	101.6	613
Uttar Pradesh	508.5	492.6	969
All India	3567.0	2361.9	662

Source: Fertilizer statistics, 1996-97.

Classification

The *Cajanus cajan* differs in plant character, pod character and maturity duration, etc. but most of the cultivated types belong to two categories.

1. ***Cajanus cajan* var. bicolor:** They are late maturing, plants grow very tall or probably they are the tallest of both the types which are freely branched and bear flowrs at the end of the branches. The pods are relatively longer and use to contain 4 to 5 seeds in them
2. ***Cajanus cajan* var. flvus:** They have shorter duration and accordingly they fall in early maturing category of plants. Plants are shorter, bushy having flowers at several points along the branches. The pods are also shorter which bear two to three seeds in them.

Ecological Requirements

Climatic Requirement: Arhar needs moist and warm weather during germination (30-35°C), slightly lower temperature during active vegetative growth (20-25°C) but about 15-18°C during flowering and pod setting, however, at maturity it needs higher temperature of around 35-40°C. Water logging, heavy rains, frost are very harmful for the crop. Hailstorm or rain at maturity damages the entire crop.

Soil Requirement: The crop may be grown on any type of soil but sandy loam to clayey loam soils are supposed to be the best. Soil must be very deep, well drained and free from soluble salts in them.

Land Preparation

The land may be prepared as for most of the *kharif* legumes keeping in view that there are no bigger clods, established weeds and left over stubbles of the previous crop. The field should be thoroughly leveled so that there is no accumulation of water in the field.

Seeds and Sowing

Treated seeds of suitable variety having high germination and high real value should be selected for sowing the crop. In irrigated condition the crop should be sown by giving one pre-monsoon irrigation at least a fortnight earlier than the first shower so that plants are well established during rainy season, however, under rainfed conditions the sowing may be done immediately after rains have started. Thus, in no case the sowing should be delayed beyond last week of June. In case of *diara* lands which are prone to flood, the sowing should be done by mid of September.

Line sowing by seed drill or *malabasa* at a depth of about 7-10 cm using a row to row spacing of 60-75 cm and plant to plant 15-20 cm is always better than broadcasting. About 10-15 kg high quality seed would be sufficient for one hectare area but it depends on variety and sowing time.

Varieties of Pigeonpea (Arhar)

Extra-early varieties (120-130 days)	:	Prabhat, Pant A 1, Pant A 3.
Early varieties (140-160 days)	:	T 21, BS 1, Khargona 2, UPAS 120, Pusa Ageti, Pusa 74.
Medium maturing varieties(160-200days)	:	Mukta, Sharda, C 11, BDN 1, SA 1.
Late maturing type (230-260 days)	:	Gwalior 3, Sweta, NP(WR) 15.
Mosaic resistant	:	Bihar, DA 11.

Hybrids	:	ICPH 8, COPH 1, COPH 2, PPH 4.
Wilt resistant	:	NP(WR) 15, BDN 1, BDN 2, Mukta Sharda.

Fertilizer Management

As a practice the arhar crop is grown on marginal and sub-marginal lands of poor soil fertility status due to which the yield is found to be so poor that it does not give much profit. The crop is heavy feeder but due to symbiotic bacteria present in the root nodules, most of the required nitrogen is fixed from atmosphere, however, it needs heavy doses of phosphate. The crop grown from inoculated seeds needs lesser quantity of nitrogen, as the plants emerged from treated seeds are well nodulated hence they have high nitrogen fixing capacity than the plants grown from untreated seeds.

For raising an ideal crop it is required to apply about 25-30 kg N, 50-75 kg P_2O_5, 30 kg K_2O (if the soil is poor in K) and 10-15 kg $ZnSO_4$ in one hectare crop. The entire dose of fertilizer should be basal placed at a depth of 12 to 15 cm or 7 to 10 cm below the seed layer in the same row.

Water Management

The crop is mostly grown rainfed but one light irrigation between flowering and pod filling stage increases the seed yield. In the event of drought or under aberrant weather condition the crop needs life saving or protective irrigation which is in addition to those mentioned earlier.

Weed Management

The crop plants grow very slowly during their early growth period of 45-50 days and the crop suffers from a severe weed infestation which causes a drastic reduction in grain yield. Therefore, it is advisable to keep the field free from weeds and a weed free condition may be achieved by giving two hand weedings once about 25-30 days and another 45-50 days after sowing of the crop. Where hand weeding is not possible use of Lasso (Alachlor) at the rate of 3 kg/ha as pre-emergence or Basaline at the rate of 1.0 litre/ha as pre-sown incorporation in soil may be done. By using herbicides the field will be weed free upto 50 days of sowing and least damage to the crop will take place.

Plant Protection Measures

Diseases: The crop mostly suffers from wilt, stem rot and sterility mosaic diseases. The wilt and stem rot may be controlled by using resistant varieties, providing well drained soil conditions and adopting a nice crop rotation. The sterility mosaic may be controlled by using resistant variety and three to four spraying of Metasystox 0.1% solution when the mites (*Eriophyid mite*) start attacking the crop and first attacked plant is visible in the field.

Insects and Pests: There are several insects which are found to attack the crop like pod borers, tur pod fly, plume moth, hairy caterpillar, leaf hopper, bean fly, etc., but the pod borers are found to cause a considerable loss to the crop. Most of the insects or almost all of them can be killed by spraying of 0.2 to 0.3 per cent solution of Endosulfan 35 EC or Nuvacrone 40 EC soon after they are visible in the field. The spraying should be repeated for their complete control and if needed a higher concentration of 0.4 or 0.5 per cent may be used for spraying.

Yield

A good crop of one hectare gives abut 25-30 quintals of grains and about 50-60 quintals of sticks and 10 quintals of dried leaves in the form of bhusa.

Arhar in Mixed or Companion Cropping

Arhar can be very well intercropped or sown mixed with a number of other crops like rice, groundnut, sawan, til, maize, urd, mung, cowpea and soybean and an additional yield may be obtained. These crops do not harm the arhar crop because by the time arhar starts growing (end of September) the other crops are ready for harvesting.

GREEN GRAM OR MUNG (*PHASEOLUS AUREUS* ROXB.)

Mung originated in India. The crop is widely cultivated throughout Southern Asia and to a smaller extent in parts of Africa. In India, it is the third important pulse crop and mostly cultivated in southern and eastern states in the rice fallow condition (January to April). In north India, it is grown in summer season because of its shorter duration. The whole or split grains are used as a pulse. The straw and husks are fed to cattle. Some of the promising early maturing (60-65 days) high yielding varieties are PDM 11, PDM 54, Pusa 105. Varieties identified as resistant to yellow mosaic virus are Pusa 101, Pusa 105, Hyb 12-4, Hyb 4-3, ML 131, Pant Mung 1.

In India, mung is mostly grown in the states like Andhra Pradesh, Maharashtra, Orissa, Rajasthan, Gujarat, M.P., Punjab and U.P. etc. Mung contains about 25% protein of high digestibility. When sprouted, the seeds synthesise a remarkable quantity of ascorbic acid (vitamin C) in them. They also bear an appreciable amount of riboflavin. The seed coat and broken parts of embryo or cotyledon, while processing make a high quantity cattle feed. It is a soil building crop which fixes atmospheric nitrogen through symbiotic action and as such the green plants, after removing the matured pods, can be used as green manure crop. The green plants may also be used as fodder after removing the matured pods.

Climatic Requirement

The crop needs high temperature, less humidity and moderate rainfall of about 60-80 cm per annum. It is hardiest of all the pulses and hence, it can be very successfully grown in drought prone areas during *kharif* season but during summer, in north India, it requies irrigation. Water logging is fatal for the crop.

Soil and Land Preparation

It does well on loam soils but can be grown on a variety of soil types with proper water management. The soil should be free from soluble salts and a neutral soil reaction is preferred for its better growth and development.

The land is prepared like other *kharif* season pulse crop but during summer it requires a thorough preparation to a pulverized bed. The bed must be free from stubbles and weeds completely.

Seeds and Sowing

Under north Indian conditions the crop is grown during *kharif* and summer seasons where sowing is done in last fortnight of June to first week of July during *kharif* and summer crop is sown during middle of March to middle of April. In southern parts of the country, where mung is grown during *rabi* season, it is sown in October-November.

Kharif season crop requires lesser seed rate and wider spacing because of very profuse plant growth but summer and *rabi* season crops need closer spacing and thereby a higher seed rate is required. About 12-16 kg treated, best quality seeds of higher germination should be used during *kharif* season and the seed should be sown in lines at 40-45 cm apart. Summer crop needs about 20-25 kg seed/ha. The crop is sown in lines at 30 cm distance. The seeding depth should not more than 5-7 cm.

Manures and Fertilizers

In case of light soils of poor fertility it needs addition of organic manures like FYM or compost at the rate of 15-20 tonnes/ha. Therefore, it may be supplied with 15-20 kg N/ha, 50 to 80 kg P_2O_5/ha which should be applied as basal placed at 10-15 cm depth just before sowing. In case of saline soils, where Zn is usually deficient in soil, a basal application of 15-20 kg $ZnSO_4$ after every three years is found to be beneficial for the crop.

Water Management

Kharif season crop is generally given one pre-monsoon irrigation at the time of seed bed preparation to ensure good germination but thereafter it does not need any irrigation until an aberrant weather or drought occurs. In drought condition the crop is irrigated as and when it needs water. Rabi crop of south India requires irrigation at a fortnight interval but the summer crop of north India needs very frequent irrigation and an interval of 7 to 10 days is found to be good for irrigating the crop. There should be no water logging at any stage of the crop growth because anaerobic condition caused by water logging is not tolerable by the crop.

Varieties

Many short duration improved varieties of mung are available for growing in different seasons. There are local varieties which have also established in the pattern of farming in different areas. The varieties have to be chosen according to the cropping patter, moisture retentive capacity of the soil local preference. Some of the information on the improved varieties of mung that are recommended to Orissa state is given below.

Variety	Approximate duration in days of the varieties of mung		
	Sept.-Oct. Sowing	Summer sowing (Mid Feb. to Mid. March)	Remarks
Pusa Baisakhi	-	60	Summer sowing in Inland Districts only
T-44	-	60	- do –
K-851	-	65	- do –
Kopergaon	-	60	- do –
Sujata (Hybrid 12-4)	70	60	Summer sowing, free from yellow mosaic
Jyoti (Hybrid 4-3)	-	70	- do –
DhauliTT-9E	65	60	Tolerant to cercospera and Powdery mildew diseases
PDM 1	65-70	-	- do –

Variety	Approximate duration in days of the varieties of mung		
	Sept.-Oct. Sowing	Summer sowing (Mid Feb. to Mid. March)	Remarks
PDM 4	70	65	- do –
Ratila Selection	70	-	
PDM 54	65	-	Resistant to powdery and YMV.
Pusa 105	60	-	- do –

Weed Management

The initial growth being very slow the crop usually suffers from a severe weed competition up to 35 days after sowing which causes a drastic reduction in gain yield. Therefore, it is essential to do two hand weedings: once after about 20-25 days and second time after about 35-40 days of sowing the crop. Use of Basalin at the rate of 1 litre/ha as pre-plant incorporated into the soil is possible due to continuous heavy rains.

Plant Protection Measures

Diseases: The crop is attacked by various vital diseases, like yellow mosaic, mosaic mottle, leaf curl, etc. and other diseases like seedling rot and anthracnose. Of these the vital diseases may be controlled by using resistant varieties and spraying of Metasystox 0.1% solution at 10 days interval and two to three sprayings will be enough to keep the vectors away or to kill them. The seedlings rot may be controlled by treating seed with fungicide like Thiram or Captan at the rate of 2.5 g/kg of seed before sowing while the anthracnose may be taken care of spraying of Diathane Z-78 or Diathan M-45 at the rate of 2 kg/ha when the diseases appears in the crop.

Insects and pests: The crop suffers from hairy caterpillar, leaf hopper, Jassides, etc. and these may be controlled very effectively by basal soil application of Thimet 10 G at the rate of 10 kg/ha. If they are still visible (which has a rare possibility) one spraying of 1.5 of Endosulfan/ha will control the caterpillar and other pests.

Yield

One hectare good crop of mung may yield about 12-16 quintals of grain and nearly quintal if bhusa.

BLACK GRAM OR URD (*PHASEOLUS MUNGO* L.)

Urd or black gram is a native of India and originated from Phaseolus sublobatus, a wide plant. From India it spread to many countries like African, European, American and Asian countries. Urd being rich source of protein (24%), carbohydrate (60%), fat (1.5%), amino acids, vitamins and minerals, etc. it has become popular pulse crop in Pakisthan, India, Bangladesh, Burma, Srilanka, and most of the African countries. In India, black gram is very popularly grown in Andhra Pradesh, Bihar, Madhya Pradesh, Maharashtra, Uttar Pradesh, West Bengal, Punjab, Haryana and Karnataka

Climatic Requirement

It needs warm weather and due to the same reason the crop is grown during rainy season and during summer season in north India. However, in central parts of the country it is grown during rainy as well as winter seasons. It is a warm weather crop and cultivated in rainy, winter, spring and summer seasons in various parts of the country

Soil and Land Preparation

Though it is grown on a variety of soils like sandy to black cotton soils, but the most ideal one is the well drained loam soil. It can also be grown on saline and alkali soils because it has good tolerance to salts.

The field preparation is done as in the case of mung or arhar. All the weeds and stubbles of the previous crop should be picked up and taken out of the field as their presence will invite termites.

Seeds and Sowing

In north, the approximate sowing time of urd is from middle of June to middle of July. The *rabi* crop of central India is sown in the months of October-November and the summer or spring season crop is sown from middle of February to middle of March. The treated superior quality seeds should be sown in lines made about 25-30 cm apart but is case of mixed crop the rows of urd are adjusted according to the space left between two rows of main crop. About 12 to 16 kg seed during rainy season and 15-25 kg seed/ha during summer season is used for sowing the crop.

Fertilizer Management

The crop requires 15-20 kg nitrogen and about 50-60 kg P_2O_5 per hectare which should be basal placed but when it is to be grown as mixed crop with other crops; the fertilizer meeting the requirement of the main crop should be applied.

Water Management

The crop can not tolerate long dry spell and therefore, irrigations are very useful for urd. The crop should be irrigated at flowering and pod filling stages but the water should not accumulate in the field. One pre-sowing irrigation at the time of seed bed preparation is required to ensure good germination, if the land is dry.

Weed Management

The crop requires two hand weedings at 20 and 45 days after sowing but if hand weeding is not possible, pre-emergence application of Lasso at the rate of 1.5 litres a.i./ha keep field weed free up to 50 days and thereafter the crop smothers the weeds

Plant Protection Measures

The crop is attacked by beetles, jassids, white flies, pod borer, etc. and jassids use to transmit the yellow mosaic virus in the crop. These all may be controlled by soil application of 25 kg/ha of disulfoton at the time of sowing. After about 50 or 60 days of sowing if any insect is visible then spraying of Rogor or Metasystox at a concentration of 0.2-0.3% should be done. Proper control of while flies takes care of the yellow mosaic disease of the crop as it spreads by the white flies which use to transmit it.

Varieties

Andhra Pradesh	PDM 71-1, (rice-urd), PDM 71-2 (upland)
Assam	T 9, and T 27
Bihar	T 9, Pusa Sel. 1-1, and Naveen
Gujarat	Anand 46-5
Haryana	Mash 1-1, Mash 48, No. 55 and T 9
Madhya Pradesh	Khargone 3, Mash 48, No. 55, T 9
Maharashtra	T 9, No.55, Pusa Sel.1-1, Sind Kheda 1-1
Karnataka	No. 55, Khargone 3, D, 67
Orissa	T 9
Punjab	Mash 1-1, Mash 48, T 9
Rajasthan	Krishna, T 9
Tamil Nadu	CO 1, ADT 1
Uttar Pradsh	T 9, T 27, T 65, Pusa Sel 1-1, UPU 19, UPU 26, UP 30
West Bengal	T 9, B 76
Kerala	T 9, Pusa Sel 1-1

Some information on the improved varieties of urd that are recommended for the state of Orissa is given below.

Variety	Approximate duration in days			Remarks
	Feb.-March Sowing	June-July Sowing	Sept.-Oct. Sowing	
T9	65	75	70	Suitable for summer, kharif and pre-rabi
Sarala (B12-4)	70	75	85	-do-
Pant U 26	65		70	
Pant U 30		80	90	Suitable for *kharif* and pre-*rabi*

One hectare good crop of urd may yield about 15-20 quintals of grains and nearly equal quantity of bhusa.

HORSE GRAM (*MACROTYLOMA UNIFLORUM* LAM)

Horse gram, *Macrotyloma uniflorum* (Lam) Verdc is an important crop of south India. It is used for human consumption. Its ration is specially good for draft animals. It may also be used as green manure. Horse gram is mainly cultivated in the states of Andhra Pradesh, Karnataka and Tamil Nadu in India. It is also cultivated in other countries mainly Shri Lanka, Malaysia, West Indies etc.

Horse gram belongs to family Leguminasae sub family Pipilionaceae. Plants are 30-45 cm in height. The leaves are trifoliate yellowish green or green in colour. Seeds vary in colour and shape. Seed coat is hard.

Horse gram is grown on a wider range of soils. However, lateritic soils are the most prominent. Farmers grow it as dryland crop in the areas of low rainfall. It is grown as rainy season crop but in the areas of heavy rainfall it is taken as a post-monsoon crop. The sowing of seed is generally done by broadcasting. Sowing in rows may be done at a row distance of 25-30 cm. A seed rate of about 40 kg is sufficient for one hectare of land. Horse gram responds to phosphatic fertilizers and a dose of 25 kg P_2O_5/ha is the optimum.

Very little work has been done on the agronomic requirement of horse gram. Under All India Pulses Improvement Project, some agronomic experiments have been conducted on horse gram. As regards planting dates, early planting (21 June) produced significantly higher than later dates. In central zone, early planting by June end appears to be the most ideal time. There is a drastic reduction in grain yield as the time of seeding is advanced. The productivity of existing varieties of horse gram is very low. Normally an yield of 5-7 q/ha is obtained.

Variety: The sailent features of some promising varieties are as follows:

- **CO 1:** It is a local selection from Mudukulathur and was released in 1953, The plant are trailing, matures in 110 days and seeds are brown, It is suitable for dry land cultivation. Average yield is 8-9 q/ha.
- **DS-1-2 and DS-2-2:** These varieties are most suitable for high and medium land condion and cultivation in dry lands. Generally, the varieties are grown in September-October and mature in 90-95 days. The seeds are brown in colour. Average yield is 10-12 q/ha. These varieties are most suitable for Orissa.
- **Hebbal Hurali-**1: It is a selection from PLKU 32. It matures in 90-100 days. Yield potential is 18-20 q/ha. It is photosensitive.
- **Hebbal Hurali-**2: It is a selection from EC 1460. Main features are similar as Hebbal Hurali.
- **HPK 2:** It is selection from local germplasm of Himachal Pradesh. It matures in 125 days. It matures in 125 days. It is semi-spreading type. Seed colour is light yellow with the seed weight of 30-50 g/1000 seeds, Average yield is 10 q/ha.
- **HPK 4:** All the features are similar to HPK 2 except seed colour being dark grayish, mottled.
- **Madhu:** It is selection from local germplasm, matures in 100 days, bushy habit, dirty cream seed with red spot, seed weight 30 g/1000 seeds. It is specially suitable for Chhotanagpur region of Bihar. Average yield is 10 q/ha.
- **PDM 1:** It is a local selection from Bhimilipatnam. It matures in 100-105 days. Plants are spreading, flowers yellow with light purple wash, seeds buff coloured. It is suitable for *kharif* sowing. Yield potential is 16 q/ha.
- **VZM 1:** It is a local selection from Vizianagaram. It matures in 100-105 days. Plants are tall, semi-erect, flowers light yellow with purple wash and seeds are black. It is suitable for *kharif* planting. Yield potential is 15 q/ha.

KHESARI (*LATHYRUS SATIVUS* L.)

Khesari is chiefly grown on the residual moisture in rice field as ‘utera’ crop. Seeds of khesari are broadcasted in standing rice crop about 15-20 days before the harvesting of rice crop. Khesari is used as ‘dal’ in human consumption. It is also used as feed and green fodder by the animals. Khesari is most resistant crop and can be grown in adverse situation. It is resistant to moisture stress as well as high moisture condition.

Lathyrus is indigenous to Europe and West Asia. It is mainly grown in the Indo-Gangetic plains comprising of Madhya Pradesh, Uttar Pradesh, Orissa and West Bengal.

Khesari plant is a herbaceous annual with slender, glabrous, freely branched, winged procumbent stems, well developed tap root. Leaves alternate, pinnate terminating into 3-5 tendrils, prominent leaf like stipules, ovate, entire attached to petiole at base, leaflets 2-4 entire, linear lanceolate, sessile 7.5 x 1.3 cm. Flowers are blue or purple with white heel. Stamens diadelphous, anthers uniform, stigma spatulate. Pods flattened, oblong up to 4 cm long with about 5 seeds. Seed colour ranges from white to brown, sometimes mottled. Germination is hypogeal.

Seeds of khesari contain a free amino-acid named as E-N. Oxalyl-amino alkaline (BOAA) which is a neurotoxin, causing disease-lathyrism in human beings for consuming large quantities of lathyrus in daily diet for a long period. Somayajulu *et. al.* (1975) reported that seeds of lathyrus contain 0.1 to 2.5 per cent of BOAA. Efforts have been made to develop varieties with low BOAA content. Low BOAA genotypes developed at IARI contain about 0.2 per cent of BOAA which is safe for human consumption.

In southern eastern zone, trial was conducted at Berhampur (Orissa). Six entries were sown on 19th November after rice harvest. Genotype JRL 43 was found to be top ranking (8.89 q/ha). Next in order was JRL 115 which was significantly superior to other genotypes. In northern eastern plain zone, the trail was conducted at Dholi with four entries. Planting was done on November 10 after smaller millets. JRL 16 yielded lowest. In view of above data, it may be summarized that Nirmal and JRL 6 in EZ, JRL 43 in SEZ and Pusa 24 in NEPZ are the highest yielding genotype of lathyrus.

Very little work has been done on the agronomic requirement of lathyrus. Sowing is done during October and November. In *utera* cropping, seeds are broadcasted in standing crop of rice, 15-20 days before harvesting of rice crop. A seed rate of 40-60 kg/ha is used. Under normal conditions, sowing should be done in rows at a distance of about 30 cm. Lathyrus also responds to phosphatic fertilization and a dose of 20 kg P_2O_5/ha is adequate. It is also cultivated as mixed/inter crop with chickpea, lentil, barley and mustard etc. Crop is raised on residual moisture. One hand weeding after 25-30 days of sowing is sufficient. A yield of 15-20 q/ha is obtained.

Seed yield (q/ha) of lathyrus genotypes in different zones

Genotype	Location		
	Berhampore (EZ)	Berhampur(SEZ)	Dholi(NEPZ)
JRL 4	-	6.92	-
JRL 6	10.33	-	-
JRL 16	6.94	6.59	2.75
JRL 41	5.27	6.92	3.82
JRL 43	-	8.89	-
JRL 115	-	7.41	-
JSD 3	6.85	-	3.33
Nirmal	14.32	-	-
Pusa 24	6.11	-	4.51
RLS 1	-	4.45	-
CD	1.33	0.27	0.86

Variety

B 1: It is a selection from West Bengal germplasm. It is low in neurotoxin content and is suitable for West Bengal. Average yield is 2-10 q/ha.

Pusa 24: It is a selection from germplasm of M.P. and was developed in 1976. It matures in 150 days. Seeds are white round small with the test weight of 70 g/1000 seeds. Seeds are low in neurotoxin content (BOAA). It has wide adaptability and is suitable for major khesari growing areas. Average yield is 8-10 q/ha.

OTHER PULES

- **Cowpea** is a dual purpose crop used for fodder and grain. Green tender pods are used as a vegetable.
- **Lentil** is of two types: large seeded with yellow cotyledons and small seeded with pink cotyledons. The latter is mostly grown in India. Lentil is cooked whole or used as *dhal*
- **Moth bean** is mostly grown as a dry crop and highly useful against wind erosion. It is used as a green vegetable or whole grain as pulse.
- **Horse gram** is grown over a wide range of soils, mostly on poor or laterite soils of India. It is grown mainly as a rainfed crop. The grain is used as human food and also as a concentrate feed for cattle.
- **Lab lab** is a perennial twining or creeping legume. There are two types: garden and field varieties and are used as grain, vegetable and fodder.
- **Rajmash** is cultivated in the sub-Himalayan '**tarai**' and partly in Satara district of Maharashtra.
- **Lathyrus** can withstand extreme moisture conditions. It is grown on the residual moisture under '**paira**' or '**utera**' cropping. The crop is mostly used as fodder. The grain is used as pulse by the poor people during scarcity period. When eaten in large quantity over a long period, it initiated a physiological disease in human beings, lathyrism.
- **Peas are** one of the important pulses in the Indo-Gangetic plains. Unripe pods are used as a green vegetable. Broken parts of the pods are fed to cattle.

CHAPTER 9

Production Technology of Crop: *Forage*

(Ecological Requirements, Cultural Practices And Crop Production Management)

LIVESTOCK AND FORAGE PRODUCTION

Livestock production in India has taken a new turn by venturing itself on commercial basis, consequent to the introduction of Integrated Farming System as well as the cross bred cows and high production targets. The stability and sustainability of such high targets mainly depend on the year round feeding of the animals with adequate quantity of nutritious and balanced feed. To develop suitable feeding system, to alter the fodder production system, to maintain milk production and animal health and to increase the farm income, the integrated farming system is the only way to the farmers. Agronomists pave the way to develop viable package of production practices like fertilizer and water management, spacing, time of application of nutrients and harvest. These are the main points for getting the higher and quality fodder.

The common forages and feeds include green fodder (grasses, cereals and legumes), dry fodder, farm by-products (straws, haulms and stovers etc.) and concentrates (by-products like oil cakes, molasses and synthetics such as feed mixtures). Agricultural prosperity of India is closely bound up with the adequate nutrition of the cattle, which provides the milk and principal motive power in the Indian farming. Hence, to improve the yield a suitable agro-technology is essential. There are number of forage crops under cultivation. Forage crops are classified as follows:

1. Annual graminaceous – summer growing – Sorghum, Sudan grass, Maize, Pearl Millet etc.,
2. Annual leguminous – summer growing – Cowpea, rice bean, moth bean, *Glycine sp.*, Velvet bean and guar.
3. Annual leguminous – winter growing – Berseem, senji, Fenugreek, velvet bean.
4. Annual graminaceous – winter growing – Oats.
5. Perennial graminaceous – summer growing – Hybrid Napier Guinea Grass, Para grass.
6. Perennial graminaceous – inter growing – canary grass.
7. Perennial leguminous – summer growing – *Stylosanthus sp.*
8. Other non conventional legume – *Alysicarpus*, *Clitoria*, *Desmodium*.
9. Non legume and non-graminaceous – Sunflower, Rape Mustard, Turnip, Carrot, Amaranthus.
10. Perennial legume – winter growing lucern.
11. Perennial legume – Browse plants – *Leucaena sp.*

Though India supports nearly 1/5th of the world's cattle population (with only 2 per cent of world's geographical area) we are far away from self-sufficiency in milk, meat, wool, and other animal products. Milk production is a factor of genetic potential, nutrition, husbandry management and health. The critical analysis of the complex reasons reveals the fact that there is a big gap between the actual need and the previsions made to the cattle. Looking into the overall scenario, one can easily see that the animals are not provided with sufficient nutrition. As a result they are not able to express their full genetic potentiality. It has been observed under the All India Coordinated Research Project on the economics of milk production that nearly 62-65 per cent of the total cost of milk production is due to the feeding of animal, when it is fed on fodder based ration the cost could be brought down to 40 per cent but with the concentrate oriented feeding, the cost rises to 80 per cent of the production. Thus, it needs no emphasis that feeding of animal is a single factor, which can be identical as the major constraint in economic livestock production. Addition of green fodder on non-legumes and legumes helps to reduce the cost of concentrate feeds as well as maintain the normal health and reproduction of all ruminants. The projected requirement for 2010 AD is 1015, 600 and 90 million tones of green fodder, dry fodder and concentrates respectively. The corresponding deficit is estimated to be 64, 16 and 80 per cent. With the present feed and fodder resource, we will be able to meet only 46.6 per cent of the animal requirement and that is why we are getting only 50 per cent of desired animal products. For many obvious reasons, the area under fodder cultivation remained at the level of 4.4 per cent of the cultivated area for the past 2-3 decades (80 per cent grain crops, 15.6 per cent other crops). In order to improve the quality and quantity AICORP was initiated up in 1970. In this net work programme, the scientists have been able to identify location specific fodder and species and have developed suitable cultivation technology for economic production.

The forage production technology includes the development of different fodder production system to suit wet, garden and dry land farming systems in addition to forage production in marginal and waste lands including problem soils such as saline, alkaline, acidic and ameliorates lands. The essential components of the system for maximizing fodder production include:

1. Improved forage varieties – for increased fodder production per unit area per unit time.
2. Mixed cropping with legumes to improve fodder quality.
3. Silvipasture.
4. Tree leaf fodder.
5. Non conventional fodder.

The technologies on forage processing, fortification of grade roughage and preservation of seasonal excess by making or by ensiling are essential to regulate continued supply of fodder for the animal.

Forage Crops and Varieties Suited

Irrigated

(i)	Perennial grasses Cumbu-Napier hybrid CO-1 CO-2 and CO-3 Guinea grass CO-1 Water grass or buffalo grass	(iv)	Perennial Pulses Lucerne CO-1 desmanthus
(ii)	Annual/biannual grasses Deenanath	(v)	Annual Pulses Cowpea CO-5, Beans/Field beans, Cluster beans, Soybeans
(iii)	Cereal fodder Fodder cholam CO 27, K 10 Fodder maize, African Tall Fodder combu CO 8.		

Rainfed

Perennial grasses	
Blue buffel	Australia grass
Marvel grass	Congo grass
Rhodes grass	Sudan grass/Johnson grass
Spear grass	Bracharia grass
Setaria grass	Arisipillu
Perennial pulses	
Desmanthus	Centro
Stylo	Rice bean
Desmodium	Moth bean
Sirato	Vigna marine bean
Clitoria	Velvet bean
Cereal fodder	
Cumbu	Cholam
Maize	
Fodder type small millets like tenai, kudiraivali.	
(i) Grasses	(ii) Cereals
Kikuyu grass	Oats
Tuber grass	Barley
Rye grass	Maize
Dactylis grass	Rye
Setaria	
Pulses	
Berseem	Clover
Lucerne	Beans

Forage Trees

1.	*Leucaena leucocephala*	2.	*Acacia planifrons*
3.	*Azadirachta indica*	4.	*Erithirina indica*
5.	*Morinda tinchoria*	6.	*Sesbania grandiflora*
7.	*Ziziphus jujube*	8.	*Albizia lebbeck*
9.	*Pithacellobium* sp.	10.	*Ficus benghallensis*
11.	*Tamarindus indica*	12.	*Ficus religiosa*
13.	*Delonix regia*	14.	*Ceiba pentandra*
15.	*Cassia fistula*	16.	*Albizia amara*

Processing and Economic use of Fodder

For real benefits of white revolution, greater emphasis is to be given to economically feed the animals with good quality roughage around the year. Adequate feeding with balanced and nutritious feed during the lean periods necessitates preserving and processing of fodders.

Processing: Processing facilitates intake and improves palatability and digestion.

Chaffing	Reduces waste
Chaffing and mixing	mixing legumes with non-legumes, and mixing dry fodder with green fodder improve palatability and uptake.
Grinding	improves digestibility of grains, facilitates enzymetic digestion.
Soaking/sprinkling with water	facilitates enzymatic digestion.

Chemical/physical/microbial treatments including hay making, ensiling, enriching poor quality roughage (Urea, urea-molasses), etc.

To minimize the feed cost, the following seven considerations have been recommended.

1. Avoid over feeding.
2. Prevent wastage.
3. Include non-conventional feed ingredients.
4. Process and preserve the feed.
5. Enrich/fortify the poor quality fodder.
6. Include more quality of green fodder.
7. Avoid fodder containing toxic principles.

Mixed Cropping

The beneficial effect of mixing cereals and legumes to enhance the feed value is given below.

Effect of Mixed Cropping with Cowpea

	Crop mix	Crude protein yield (kg/ha)	Increase in CP yield (%)
1.	Sorghum (Pure crop)	680	
	Sorghum + Cowpea	1024	150
2.	Maize (Pure crop)	722	
	Maize + Cowpea	1049	140
3.	Bajra (Pure crop)	632	
	Bajra + Cowpea	971	154
4.	Pennisetum pedicellatum(Pure crop)	658	
	Pennisetum pedicellatum + Cowpea	947	145

In another experiment, Deccan maize + cowpea CO -1 at a ratio of 1:1 resulted in 35 t green fodder yield/ha as against 35.5 t/ha in pure crop of maize and 19 t/ha in pure crop of cowpea. Hence in a year with 5-6 crops of maize + cowpea mixture (60 days crop) it is easier to obtained a green fodder yield of 200 t/ha with increased crude protein content. Similarly bajra napier hybrid + Desmanthus gave highly nutritious green fodder rich in crude protein throughout the year. Desmanthus, Stylosanthus or Clitoria are the best combination for dry lands.

Sylvipasture

This is simultaneous growing of trees and under-story herbaceous vegetation of forage value on the same place of land and in the agricultural system. Such silvipastoral farming systems have been practiced in arid and semi-arid parts. The trees as a group offer wood, fuel, fruits, oils, industrial raw materials, medicinal principles, shade, aesthetic pleasure, and pollution suppression, control of soil erosion and above all leaf and pods as forage to domestic animals. The biodynamic of the silvipastoral system involves four major distinct live forms viz. the herebaceous vegetation (for the most part grasses), the tree component, the domesticated animals surviving on the foregoing vegetation and man who controls the other three components. In addition, soil and climate have their own roles to play and complicate the diversity. Establishing suitable components of such systems deserve consideration. Silvipasture is very useful for developing feed resources in waste lands. *Leucanea leucocephala* CO-1 + *Cenchrus glacus* CO-1 combination seem to be more promising. It is comparable to or even better that the established Acacia sp. + *Cenchrus sp.* combinations in certain parts of Tamil Nadu.

Advantage

1. Both are perennial crops.
2. Both the fodder crops are highly drought tolerant.
3. *Leucanea* can be maintained at 1 m height.
4. They provide balanced diet to cattle, rich in protein and calcium.
5. Fodder yield is fairly high.

Tree Leaf Fodder

Forests provide leaves, flowers and pods palatable to the animals besides the grasses and small legumes growing underneath the trees. The tree leave fodders are a potential resource of protein, energy, fat and minerals for cattle and sheep. Such tree leaves are valuable fodder during lean periods and forest grazing has been practiced in Tamil Nadu from time immemorial.

Quality of Forage

Luxuriant growth 0f the forages does not always indicate its capability to supply adequately the nutritional demands of animals. Presence of nutrients in the plants does not necessarily mean that it is readily and completely available to animals. Therefore, any attempt to describe the quality of forages should be taken into account. The feed components in forage species and factors affecting their variability and the ability of animals to utilize the feed components and affecting the variability are in such utilization. The nutritive value of forage refers to its chemical composition, digestibility and the nature of digested products. However, the amount of forage consumed by the animal is very important, as it affects total nutrient uptake and therefore, the animal response. The assessment of forage quality involves an integrated evaluation level of its nutritive value and its level of consumption by the animal.

Chemical Composition

Studying its chemical composition can assess the nutritive value of a forage. To assess the chemical composition, proximate analytical scheme resolves a given feed into five fractions.

- Crude Protein (CP)
- Fat/Ether Extract (EE)
- Crude Fibre (CF)
- Ash
- Nitrogen Free Extract (NFE)

The forages are made up of two basic dietary fractions.

(i) Cell contents	(ii) Cell wall content
Lipids Starch Sugars Non-protein nitrogen Soluble protein Pectin, Organic acids and Water soluble Matter.	Hemicellulose Cellulose Lignin Silica

Factors affecting chemical composition

- Soil
- Climate
- Fertilizer application
- Maturation process
- Stage of growth
- Genotype

Digestibility

It is an important measure of the nutritive value of forages and can be defined the amount nutrient, which has been absorbed from the digestive tract per unit of nutrient ingested by the animals. It can also be defined as the difference in the nutritive value between the feed eaten and materials voided by the animals expressed as percentage of feed eaten.

Factors Affecting Digestibility

- Soil
- Climate
- Stage of growth
- Physical form of the forage
- Maturation process
- Genotype
- Physiological condition of the animal
- Chemical composition.

Digested Products

For an accurate measurement of the nutritive value of given forage, knowledge of quantity of digestible energy and net energy available to the animal from that forage and the efficiency of energy utilization for maintenance and production is essential. Mulberry forms the basic food material for silkworms and the bulk of the silk goods produced in the world are from mulberry silkworms. Sericulture involves agriculture, art and industry, mulberry cultivation involves various farming practices, silkworm rearing is an art in the hands of rural people, and reeling of the silk from the cocoons formed by the worms is an industry of different financial investments. Production of mulberry leaves on scientific lines is essential for organizing Sericulture on sound economic lines.

OATS (*AVENA SATIVA* L.)

Oats is an important grass fodder of winter season. This crop is well suited to limited water availability situations. The green fodder contains 10-12 per cent protein and 30-35 per cent dry matter. Oats grains are used in animal feeds. Oats are supposed to to be Asian origin and from there it spread to most of the countries of the world like U.S.A., Russia, Canada, Poland, Australia, France and Germany.

Classification of Oats

The oats are classified according to their chromosome numbers and there are three main groups of cultivated oats:

Group I. This group of oats contains 7 Haploid chromosomes and the important of them are *Avena brevis* (short oats), *A. weistii* (desert oats), *A. strigosa* (sand oats) and *A. nudibrevis* (small naked seeded oats), etc.

Group II. The oats belonging to this group contains 14 Haploid chromosomes. The prominent types of this group are *Avena barbata* (slender oat), *A. abyssinia* (Abyssinia oats).

Group III. These types have 21 Heploid chromosomes and the important members of this group are *Avena fatua* (common wild oats), *A. sterilis* (wild red oats), *A. sativa* (common oats), *A. byzantina* (cultivated red oats) and *A. nude* (hull less oat). Genus *Avena* includes following species which are economically important:

- ***Avena abyssinica* (2n=28).** The Abyssinian oats grows wild and is also cultivated in Ethiopia. It is grown mainly for fodder in several arts of North Africa.
- ***A. brevis* (2n=14).** The short oats is grown in southern Europe, mostly for green fodder or hay due to its high yields of stem and leaf.
- ***byzantina* (2n=42).** The red oats is also cultivated in India besides, Mediterranean regions and warmer parts of subtropical world.
- ***chinensis* (2n=42).** The Chinese naked oats is grown extensively in hilly parts of China for its grain.
- ***fatua* (2n=42).** The wild oats is cultivated for grain, green fodder and other purpose in India and part of south-west Asia.
- ***ludoviciana* (2n=42).** The wild red oats is grown for grain in Mediterranean and south-west Asian regions.
- ***sativa* (2n=42).** It is the common cultivated oats sown in several parts of world for grain, fodder and other purposes.
- ***sterilis* (2n=42).** The fly oats is grown mainly for fodder in parts of Mediterranean region and south-west Asia.
- ***strigosa* (2n=28).** The sand oats is cultivated for fodder in the Mediterranean region.

Botanical Description

Root system: It consists of both seminal and adventitious roots,. The seminal roots appear first, followed by several adventitious roots.

Culm: Its stem or culm is composed of a series of nodes and internodes. The nodes are solid and thick. The internode bearing inflorescence called the peducle have a tendency of branching.

Leaf: The leaves are solitary, alternate, two ranked and sessile. The developed leaves consist of blade, the sheath, a basal portion and the ligule. The ligule is thin membranous appendage which is well developed.

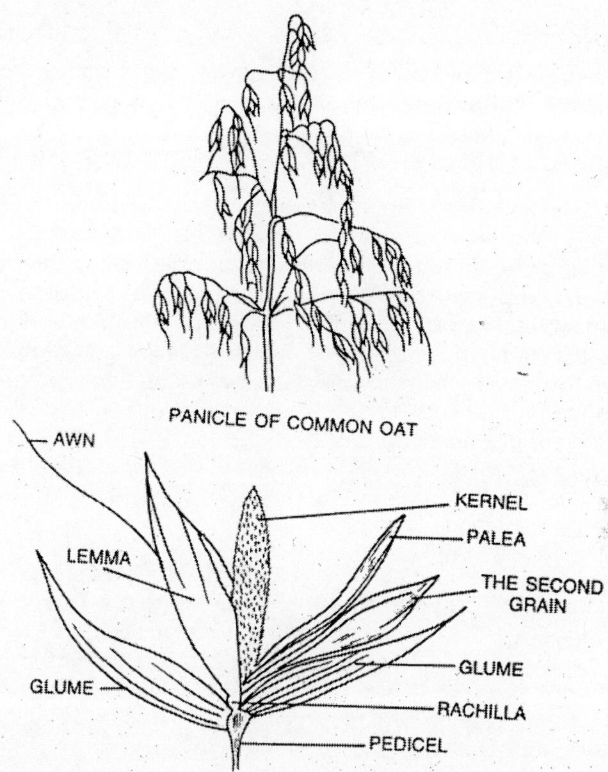

Inflorescence: It is a loose open panicle of spikelets. It consists of a main axis or rachis with lateral branches alternating in a racemose at each node. Each spikelet consists of two empty glumes and the florets. The floret consists of a lemma and a palea, two lodicules, three stamens and one pistil.

Grain: An oat grain is a cryopsis. It contains the endosperm and embryo. The aleurone layer consists of the outer most first and second cell layers of that portion of the endosperm not in contact with the embryo.

Agro-Technique For Oat Production

Ecological Requirements

Oat crop is grown during *rabi* or winter season as an irrigated crop. It requires about 20-25°C temperature for germination and still lower temperature for its better growth and development. Light winter rains followed by bright sunshine and lower humidity and cloudy weather invite many fungal diseases. Rains associated with high wind velocity cause severe lodging of the plants and create obstruction in better growth of the plants. Similarly frost also causes great damage to the crop.

Well drained friable soils of moderate fertility are the most preferred soil type for the crop. Cultivation of oats was limited to only alluvial soil types found in U.P., Panjub, and Haryana but gradually it is gaining popularity on other states like Maharashtra, Gujarat, M.P., West Bengal and Orissa under irrigated conditions.

Land Preparation

The crop needs fine, pulverized bed like that of wheat and barley. An ideal seed bed should not have stubbles of the previous crop and any weed, whether uprooted or established. Therefore, they must be picked up, collected and burnt. The organic manures, if needed should be applied at the rate of 150-200 quintals/ha about 20 days before sowing of the crop.

Sowing

The best sowing time of oats is towards end of October but it may extend upto December; however, earlier sowing helps in getting more cuttings. The fodder crop of oats is mostly sown broadcast but the seed crop needs more care and hence it should be sown always in rows about 25-30 cm apart. About 65-85 kg of oat seed is needed for sowing one hectare area depending upon method of sowing as it is very obvious that line sowing needs less seed. Soils of poor fertility status require higher seed rate because of poor tillering.

Fertilizer Application

It has been a common practice to grow fodder oats on residual fertility of the previous crop; however, the crop does not give good yield. Thus it is always better to grow the crop with additional dose of fertilizers for better growth and development of oats. Crops respond well to organic manures; therefore, a dose of about 150-200 quintals of F.Y.M. or compost should be applied at the time of seed bed preparation. Besides this an additional dose of 40-50 kg of nitrogen/ha should be top dressed in two split doses – once after first cutting and second after about 60 days of sowing. In absence of organic manure a dose of 60-100 kg N/ha and 40 kg phosphorus/ha should be given. Entire quantity of phosphorus and half of the nitrogen should be applied as basal and rest nitrogen should be top dressed about 35-45 days after sowing.

Irrigation

The crop needs three irrigations but there should be sufficient moisture in the field at the sowing time so that better germination may be achieved. Normally the first irrigation is given when the crop becomes about 20-25 days old and second when the crop reaches to the maximum tillering or after first cutting about 45-55 days after sowing and the third irrigation is given when the crop becomes about 75-80 days old. In case of the winter rains any of these irrigations may be skipped which coincides with the scheduled irrigations. In no case there should be water-logging as it is very harmful for the crop.

Harvesting and Yield

The crop needs about 120-150 days to mature. It is a common practice to take two-three cuttings of fodder and then to allow the crop to grow for seed. But normally only two cutting are taken from the seed or grain crop. Of these two cuttings – the first is taken after 60-65 days and second after about 90 days of sowing or at the flowering stage of the plants. The plants are then allowed to grow and set into seeds. One hectare crop gives about 500-600quintals 0f green fodder and about 2-4 quintals of grains. The plants of oats do not become brittle like that of barley or wheat at maturity; therefore, the crop must be harvested soon after the grains have become hard and they split when pressed between teeth because any delay in harvesting will result into a severe shattering loss of grains.

Varieties

- **Kent:** An introduction from U.S.A. which is resistant to rust, blight and lodging.
- **Algerian:** It is an introduction from Algeria and is suitable for irrigated areas.
- **FOS-1/29:** It is recommended for Punjab, U.P., Haryana and Delhi for irrigated and rainfed areas.
- **Brunker 10:** It is suited to limited irrigated conditions and is resistant to loose smut.
- **Weston 11:** An early maturing and tall growing variety.
- **Coachman:** An introduction from U.S.A. which is resistant to rust, blight and lodging.
- **HFO 114:** A dual purpose variety recommended for Haryana.
- **UPO-50:** It is resistant to blight, rust and lodging and recommended for U.P.
- **UPO-94:** It is like UPO-50 variety.

BERSEEM OR EGYPTIAN CLOVER (*TRIFOLIUM ALEXANDRIUM* L.)

Berseem is believed to be originated in Asia Minor and from there it was introduced to Egypt. Berseem was brought to India by Sir Fletcher in 1904 and it was first grown at Manipur Khas farm of Sind Province which is presently in Pakistan. Because of its introduction from Egypt it is famous as Egyptian clover and it is gaining its increasingly importance as *rabi* crop. From Sind it was brought to Punjab and then to other north Indian parts and lastly it is being grown even in Southern parts of the country.

Economic Importance

Berseem, being a multicut and high yielding legume, provides a nutritious and succulent fodder during the winter/early summer season. It is rightly called as king of fodders. It has 20 per cent protein and 70 per cent dry matter digestibility. The crop is grown only in irrigated areas. Berseem fodder contains 20-24 per cent protein, 2.89 per cent Calcium and 0.40 per cent phosphate at the green stage. It is very good soil builder and adds about 45 kg available nitrogen to the soil. During surplus months it can be preserved as silage with all nutritive values intact, to feed to the cattle. But if it is alone fed to the cattle they may suffer from bloating. Therefore, it must be mixed with dry fodder and then fed to the cattle.

Botanical Description

It is a small shrubby annual growing upto a height of 60-80 cm Roots are confined mostly to 40-50 cm soil depth.

Stem: Stems are hollow and succulent, but become fibrous after flowering stage.

Leaves: Leaves are small, tender and trifoliate.

Flower: Flower heads are round and yellowish white in colour.

Classification of Berseem

Berseem belongs to the family leguminosae and genus *Trifolium* which consists of nearly 290 species as the most important forage legumes. The most important species for forage and pasture are - berseem (*Trifolium alexandrium*), Shaftal (*T. resupinatum*), white clover (*T. repens*), red clover (*T. pretense*) crimson (*T. incarnatum*), Alsike clover (*T. hybridum*) and Subterraneum clover (*T.

subterraneum), etc.

Climate

Berseem prefers dry and cool climate for its proper growth and development and therefore, it is grown during *rabi* season with high humidity and cloudy days are not good for berseem crop. It requires 25 to 35°C temperature for germination and 15 to 20°C for the best tillering and vegetative growth however, at flowering slightly higher temperature of about 35 to 37°C is needed. Frosty nights or a minimum temperature of 12°C or lower and a maximum of 40°C are detrimental for the crop and yield is drastically reduced. It may be grown successfully in areas which receive annual rainfall of 150 to 250 cm or even lower but the irrigation must be assured.

Soil

Berseem bears nodule bacteria, for which water-logging is not suitable and hence it is required tht the soil must be well drained and free from harmful salts. Clay loam soils rich in calcium and phosphorus are the best soil types. The crop can tolerate mild acidity.

Land Preparation

The seeds being very small, berseem requires very fine seedbed. After harvest the *kharif*, if the soil does not have sufficient moisture, the field should be given light irrigation and then 3-4 ploughings followed by planking each time helps in breaking the clods and leveling the surface. The field may be laid out into smaller beds of convenient sizes according to topography and source of irrigation water.

Manures and Fertilizers

Berseem being a leguminous crop needs very little quantity of nutrients. It soils is saline and very light textured then about 15 to 20 tonnes of well decomposed organic manure should be applied in the field and well mixed by ploughing and cross ploughing. It must be supplemented with about 80-100 kg P_2O_5 per hectare. The phosphate should be placed for better response or should be applied at the time of final ploughing in the field. If the organic manures are not to be applied then about 30-40 kg nitrogen/ha should be applied at the time of phosphate application. Application of Molybdenum at the rate of 1 kg/ha is found to increase the fodder yield significantly.

Seed and Sowing of Berseem

Sowing time: Berseem needs about 30°C atmospheric temperature at the time of germination, which is found in the first fortnight of October. If the crop is sown earlier then the seeds may be washed off and the seedlings may die due to high rainfall and if sown later the growth of plants becomes poor and yield is reduced.

Seed rate: About 25-35 kg of berseem seed/ha is needed but if the tetraploid berseem is to be sown then about 35-45 kg seed/ha is used for sowing. It is found that the yield in first cutting is very poor; therefore, about 1-2 kg of mustard seed should be mixwd with berseem seed while sowing. By doing so the mustard grows quickly and the yield in the first cutting is increased by about 20-25 per cent. It is always better to mix the seeds of diploid and tetraploid together because the initial growth of tetraploid and later growth of diploid are found to be poor but by mixing both a higher fodder yield from first cutting till last cutting may be obtained. Sometimes berseem seed is mixed with methi seed of fodder quality or even oat seeds are mixed and sown. The seeds of both the types of berseem are mixed in equal quantities.

To eliminate Kasani from berseem crop the seed should be poured in 1 per cent common salt solution and the floating kasani seeds should be taken out and remaining seed of berseem should be sown.

Seed treatment: Berseem is not a native of India and has been recently introduced in the country, therefore, the nodule bacteria of berseem is not present in the soil. It is, therefore, essential to treat the seed with Rhizobium culture (*Rhizobium trifoli*) for better nodulation, good growth and subsequently higher fodder yields.

When the berseem crop is to be grown for the first time in the field then the seeds are inoculated with bacteria culture. For treating the seed it should be first soaked into fresh water for about 8-12 hours, and then treated with culture. For better sticking of culture with seed and also providing initial energy to the bacteria the culture is prepared with jaggery. About 1.5 litre of water is mixed with 150 g of jaggery and boiled. After cooling the jaggery liquid about two cigarette tins of culture mixed with it and then seed is well mixed and dried in a cool shady place. In case the culture is not available, about 60 kg of powdered soil from the field where berseem was grown earlier should be applied in the soil beore sowing of berseem.

Sowing method: The berseem crop may be sown in three ways like:

1. *Sowing in flooded field*: Nicely prepared beds are flooded with 5-7 cm deep water and planking is done in order to level the surface and after 2-3 hours the seed is uniformly spread in standing water. About 24 hours after seeding the water is drained out and re-irrigated after 3-5 days of sowing when germination is complete.
2. *Sowing in dry fields*: In this case the seed in broadcasted or sown in lines 25 cm apart by drilling and then planked. After sowing the field is divided into convenient sized beds and later the beds are flooded with water.
3. *Sowing berseem seed is standing rice crop as relay cropping*: When it is expected that rice will be harvested very late and berseem sowing, therefore, will be too late then in that case berseem seed is broadcasted in standing rice crop. Thus, the berseem germinates in standing rice crop and it takes over after rice is harvesd. This way the cost of land preparation is saved and timely sowing is also done.

Irrigation

The depth and frequency of irrigation is decided by soil type, number of cutting and nature of berseem crop – whether a sole crop or sown mixed with some other crop. Generally, the interval of irrigations during October is about 10 days and during November to January it is 15-20 days and thereafter again 8 to 10 days. Thus it needs about 10-12 irrigations in all.

Mixed cropping of berseem

Berseem is grown mixed with sugarcane, napier, etc. as an intercrop. Both these crops remain dormant during winter when berseem can grow well and give very good yield besides enriching the soil with nitrogen. Thus, the autumn planted sugarcane and napier (Pusa Giant napier) are recommended to be intercropped with berseem.

Harveating and yield

The crop takes slightly more time for first cutting and it is done about 40-45 days after sowing whereas, the subsequent cutting thereafter are done at an interval of 20-25 days. The cutting is stopped after first of February for a better seed crop because the crop starts flowering during March and delayed harvesting results into poor seed development, poor seed yield of inferior viability and low germination. Thus we can take about 6-8 cuttings by which we may get about 600-800 quintals green fodder/ha along with an additional yield of about 3-5 quintals seed/ha.

Varieties

Deploid varieties like Maskavi, Fahali, Saidi, Zaidi, BL-1, BL-2, BL-10, BL-22, BL-30, BL-92, JB-3, JB-4, IGFRI-S-99-1, and UPB-101. UPB-103. UPB-104, UPB-1905 and Khdrabi are very popular but newly evolved high yielding tetraploid varieties like Pusa Giant, T- 526, T-724, T-780, T-529, T-560, T-561, T-674, T-678, T-730 etc. are very promising and give about 50 per cent higher fodder yield.

Seed Production

After taking 3-4 cuttings the plants should be allowed to grow from 3rd week of March. The plants start flowering from last week of April and seeds get matured by end of May when the crop should be harvested and threshed; seeds should be cleaned, dried and stored in damp proof stores.

Sometimes seed crop is attacked by thrips, semi-looper and aphids which may be controlled by spraying 0,02% Malathion. Dustings of at the rate of 25 kg/ha takes case of semi-looper. A healthy crop gives about 6-8 quintals of seed yield per hectare.

LUCERNE/ALFALFA (*MEDICAGO SATIVA* L)

Lucerne, also known as alfalfa, is a valuable leguminous forage and hay crop. The crop can be grown both in the irrigated and rainfed areas. The green herb containing about 15-20 per cent protein, has a digestibility of about 72 per cent.

Alalfa or Lucerne is one of the oldest cultivated fodder crops which is practically grown in almost all the parts of the world because of its adaptability to warm, temperate and cool subtropical regions. Lucerne was first grown in Iran and alalfa, being Arabic word, was given as its name in Iran. It is supposed that Lucerne was brought to Greece from Iran and to other European countries. In USA its cultivation was first started in Georgia in 1736 hereas in India it was introduced around 1900 but now it has become most popuar and wide spread forage crop of both pasture and cultivated lands. Lurcerne has world wide distribution. Presently, Argentina, USA and France are the most important Lucerne growing countries. In India it is being grown in U.P., Punjab, Haryana, Maharashtra, Tamul Nadu etc.

Botanical Description

It is a herbaceous annual as well as perennial plant growing up to a height of 60-150 cm.

1. **Roots:** It has a deep root system consisting of strong main tap root and number of lateral roots.
2. **Stem:** Stem is erect and branches arise from crown, which is a woody base

on stem near ground level.

3. **Leaves:** The leaves are trifoliate; the middle leaflet possesses a short petiole, a characteristic which distinguishes it from berseem.
4. **Flower:** Colour of the flower is usually purple, but it may be blue, yellow or white.

Climate

Lucerne prefers high light intensity and cool temperatures. It can tolerate short spells of drought and remain dormant. Fresh growth starts as soon as moisture becomes available. It does not tolerate water-logging and high humidity. Lucerne is adapted to relatively dry conditions and it may tolerate heat and cold, however, it can not grow under humid conditions with high temperature. The crop needs water but under pasture it may be grown as rainfed.

Soil

Depending upon irrigation facilities it may be grown on a varied range of soil types such as under rainfed conditions it is better to grow the crop on heavy soils but under irrigated conditions it may be nicely grown on light or sandy soils. Thus it needs sandy loam to clayey soil but heavy soils need an efficient drainage system as the crop does not tolerate water-logging. Lucerne prefers a fertile soil which is rich in organic matter contents. Soils of neutral reaction are the best for the crop, however, it can be grown on saline or alkaline soils but acidic reaction is very harmful for the crop.

Land Preparation

Like berseem, Lucerne also needs very fine seed-bed as the seeds are very small. The required tilth may be obtained by giving one deep ploughing followed by two harrowing and cross planking. The established weeds and crop stubbles must be picked up and collected outside the field and burnt.

Seeds and Sowing

Sowing time: The best sowing time for the crop is from middle of October to beginning of November. Any delay in sowing results in a lower germination, poor growth and subsequently fodder yield due to very low temperature in beginning and high temperature in the months of February/March.

Seed rate: The seed rate depends upon method of sowing, type of crop *i.e.*, whether the crop is to be sown mixed, intercropped with sugarcane or napier or it is to be grown as pure crop. The broadcasting obviously needs higher seed rate ranging between 17-22 kg/ha, whereas line sowing at 30 cm distance needs only 11 to 15 kg/ha but in case of inter-cropping or mixed cropping it requires 6 to 12 kg/ha.

Seed Treatment

If Lucerne is sown for the first time in the field then it needs bacterial culture for seed treatment or addition of about 350 kg of well powdered soil from the field where it was grown in the previous years. The soil should be uniformly broadcasted in the field. The seed treatment results in a quick and higher germination of seed and results in a vigorous growth of plants for early cutting and higher yields of fodder.

Method of Sowing

Lucerne is sown in two methods like broadcasting and line sowing. In case of broadcasting the field is prepared, seed is sown broadcast and then the field is flooded or otherwise the field is first flooded and then seed is sown by broadcasting. Both these methods are common but the latter has given very promising results. In case of line sowing the water soaked seed is sown in shallow furrows and then field is flooded from across the furrow direction so that the seed is not collected at one place towards slope.

Irrigation

The crop needs very frequent irrigations during its early growth period at an interval of about one week but once the plants are established and grown little taller they need subsequent irrigations at an interval of 15-20 days during winter and 10-12 days during spring and summer seasons. The most common practice is to irrigate the crop after every cutting or if it is hot then to give two irrigations between two subsequent cuttings, however, during hot months it needs three irrigations between two cuttings.

Harvesting and Yield of Fodder

The October sown crop becomes ready for first cutting in second or third week of December and the subsequent cuttings are taken every after 25-35 days interval uptil May after which it may be left for setting of seeds as the plants start flowering. On an average 6-8 cuttings may be taken which give a total yield of 500-600 quintals of green fodder/ha and a seed yield of about 2-3 quintals/ha. From February the cutting must be stopped for a better seed yield. The seed crop becomes ready for harvesting during end of May or beginning of June.

In case of pasture where Lucerne is to be grown as a pure crop, the crop plants are left in the field and they remain during rainy season and start growing again as the atmospheric temperature falls below 20°C *i.e.*, after October and the crop becomes ready for the first cutting by middle of December. In pasture it is better to sow Lucerne with grasses because feeding Lucerne mixed with grasses does not cause bloat to the cattle which is due to presence of certain glycosides known as *soponins* in Lucerne plant at green stage.

Varieties

1. **Sirsa 8:** It gives 500-700 quintals fodder and 2-3 quintal seed/ha.
2. **Sirsa 9:** It yields 600-900 quintals fodder and 3-4 quintals seed/ha.
3. **NDRI Selection 1:** Yields 1000 quintals fodder/ha continues to grow upto 6 years.
4. **Rambler:** An introduction from Canada and yields 600-900 quintals fodder/ha.
5. **Other varieties:** LL Composite 2, LL Composite 3, LL Composite 5, Anand 2, Co-1, JGFRI 244, SS-627, Atir, KS-243.

NAPIER/ELEPHANT GRASS (*PENNISETUM PURPUREUM*)

Napier is supposed to be native of tropical Africa. The genus Pennisetum consists of about 50 tropical and sub-tropical species which are in cultivation in counties like America, Africa, Asia and Southern Europe under both tropical and sub-tropical parts.

Napier was introduced in India in 1912 and after an observation of its growth behaviour it was named as 'Elephant grass'. It has become now very polpular fodder grass in Assam, West Bengal, Bihar, Orissa, Punjab and U.P. under warm and moist areas receiving an annual rainfall of around 100 cm.

Climate

The crop needs high temperature of over 25°C and high humidity but water-logging is very harmful for the crop as due to water-logging the growth and development of the plants are adversely affected and yield is reduced. The crop does very well under an annual rainfall of 100 to 150 cm but high wind velocity at any stage of growth is not good as the plant lodge down badly because of their profuse growth behaviour. A temperature below 20°C is not congenial for better growth and development of plants and they become dormant till temperature does not rise above 25°C.

Soil

The crop likes heavy soils of high inherent fertility and high organic matter content in them. Napier, being heavy feeder, does very well under sewage irrigated conditions. It requires a neutral soil reaction but it can tolerate mild salinity and acidity however, under acute acidic or saline reactions the crop does not grow well and a poor yield is obtained. The crop requires an assured water supply but at the same time the field must have an efficient drainage system also so that no water-logging takes place.

Land Preparation

The land should be applied with 20-30 tonnes of well decomposed organic manures like FYM, compost or leaf mould about a month earlier of planting. For land preparation it should be ploughed once by mould board plough followed by two harrowings and cross planking. When the required tilth is obtained, the field should be laid out into ridges and furrows each at 50 cm distance so that the distance between subsequent ridges and furrows may come to 90-100 cm. The ridges should be about 20 cm high from ground level. The well prepared beds should not have any clod or stubble and weeds, either dry or green whatsoever in the field. If the field is sloppy the ridges and furrows should be laid out across the slope having checks at various places for proper distribution of irrigation water.

Planting of Napier

The seeds of napier are very small and hence their use is restricted to only for breeding purpose or for initial multiplication of improved varieties, therefore, the crop is commonly planted with vegetative parts like setts as in case of sugarcane or with rooted slips, whichever is fond easy. Planting setts in the field does not always give assured plant population and hence raising rooted slips in the nursery than planting them in the final field is the best solution for the problem, though this method is time consuming and very expensive. The crop can also be multiplied by splitting the rooted stubbles of the previous crop and planting them at the required spacing but these stubbles do not always produce healthy shoots and hence they need slightly more nitrogen for better growth and development. The rhizomes may also be used with encouraging results of higher fodder yields.

If the crop is to be multiplied through setts then healthy shoots are selected and cut into pieces each containing three buds but none of the three buds should be damaged or dry. The average length of each sett should be about 45-60 cm and spacing between two setts should be about 20-30 cm and between two rows it should be about 90 cm. The setts are plant like that of sugarcane in horizontal position but when the crop is to be planted through rooted slips or rooted stubbles of previous crop they are always planted in vertical position. One hectare area requires about 12,500 cutting or about 40,000 rooted slips.

Manures and Fertilizers

Napier, being very exhaustive crop, needs heavy manuring and therefore, it should be given about 250-300 quintals of organic manure like F.Y.M. or compost per hectare. The crop is given about 30 kg nitrogen/ha while planting and thereafter once in July and second time in February *i.e.*, around end of winter with 45 kg nitrogen/ha each time. It does not need additional phosphorus and potash because the required quantity of these nutrients is supplied through organic manures. In case of sewage irrigation the crop does not need even the nitrogen and the entire quantity of all the three nutrients are supplied through sewage alone.

Irrigation

The crop does not need any irrigation during rainy season provided rains are normal and well distributed but under aberrant weather conditions it is given irrigations as and when required. Napier is irrigated at a fortnightly interval during winter and at 10-12 days interval during summer season. Water-logging at any stage should be avoided because it is very harmful for the crop.

Intercultural Operations

Napier is a perennial crop and it is often irrigated with sewage water, thus, it is found to be badly infested by numerous weeds which need to be eradicated so that a better quality fodder may be achieved. Intercultural operations or intercropping with a suitable legume takes care of all the weeds. The legume, when intercropped with napier, enriches the soil besides supplying balanced fodder to the cattle.

Harvesting and Yield

The crop becomes ready for first cutting about three months after its planting and thereafter each subsequent cutting may be taken at about two months interval however, it depends upon the vegetative growth of the crop. On an average 6-8 cuttings can be taken which gives about 400-600 quintals of green fodder per hectare. The yield may go as high as 1200 quintals/ha, through a record yield of as high as 2000 quintals/ha has been obtained in Bangalore with best management and good manuring of the crop. The crop should be harvested from very close to the ground so that better branching and ratoons may come out and a good ratoon is obtained. Cutting of too matured plants is not desirable because too matured plants become very hard and more fibrous and delayed cutting is observed to reduce the fodder yield also. Therefore, it is advisable to take cutting when the plants are in most succulent stage. Napier crop should be allowed to remain in the same field for a maximum of 5-6 years and thereafter it must be replaced by some other crops.

Mixed Cropping

Studies on intercropping was taken up at I.A.R.I., New Delhi during summer when hybrid napier was intercropped with soybean, cowpea and mung. The inter row spacing of napier was 90 cm and that of intercrops was 30 cm. This way three rows if the intercrop were taken between two subsequent rows if napier. Al the operations like fertilizer application etc. were done according to the needs of the napier. In all four cuttings of napier and two cuttings of legumes were taken and it was found that napier and cowpea was the best combination from yield and economics point of view.

Napier remains dormant during winter season when it can be intercropped very profitably with Lucerne or a mixture of diploid and tetraploid berseem. This gives the highest yield of fodder and nurses the napier plants by protecting them from frost and improving soils fertility. The yield of berseem can further be increased if mustard seed is mixed with berseem seed in the ratio of 1:10 *i.e.*, one kg of mustard at every after 10 kg of berseem seed. Such mixtures also improve the fodder quality by increasing protein contain and palatability.

Improvement of Napier in India

During recent past a number of crosses have been made to improve napier for higher yield as well as quality of fodder. It was found that a cross between *Pennicum purpureum* and *Pennicum typhoideum i.e.*, between napier and African bajra and the the cross was found to be best. This hybrid was named as Pusa Giant Napier which requires less water and gives the highest yield. Giant napier is richer in protein and sugar compared to old napier by about 25 and 12 per cent respectively. Other varieties like Hybrid-4 and EB-4 developed at I.A.R.I., New Delhi were found to be very sensitive to low temperature and they remain dormant during winter season. Pusa Hybrid Napier 1 and Pusa Hybrid Napier -2 developed by crossing between Napier and bajra, were found to be resistant to low temperate and give about 25-40 per cent more fodder than Pusa Giant Napier 1 is good for soil conservation point of view and it can be used for grazing unlike others whereas the stems of Pusa Napier-2 are more juicier and sweeter than that of other napier varieties. They may be grown on soils of relatively lower nutrient contents.

Plant Protection

Napier is found to be attacked by stem borers and they cause a considerable damage to the crop, therefore, its control is very essential. Application of 0.05 per cent endosulfan twice through spraying on the foliage helps in controlling the past but it should be applied at least 30 days before cutting of fodder because it persists in the stems and causes stomach disorders in cattle when fed within 30 days of its spraying over the plants.

Giant Napier has got more of oxalate content in its stems and is found to reduce the milk yield of milch cattle but Pusa Napier-1 and Pusa Napier-2 do not have such characters and are good for all the animals.

Varieties

Pusa Giant, N.B. 21, Pusa Napier-1, Pusa Napier-2, E.B.-4, Gajraj, N.B. 5, N.B. 393 etc.

Plantation of Fodder Trees

Social forestry has become a matter for discussion now-a-days and the efforts made during the last two decades in this field have brought spectacular achievements in solving ever increasing demands for feed, fuel and food. Social forestry is plantation of such trees whose leaves can be fed to the cattle and the twigs, after removing the green leaves, can be used as firewood or fuel. Period during which most of the fodder crops become dormant or they are yet to be grown is said to be leaned period. Cattle face a great setback due to imbalanced fodder and their output is adversely affected. Plantation of fodder trees helps in a regular supply of green leaves which are rich in protein, mineral and are very palatable and digestible. Thus, plantation of fodder trees reduce pressure on cropped area as these plants do not need any additional land but they can be planted on the land left around the houses, parallel to fence or farm road. These plants are very useful as they supply good feed to cattle, provide fire wood or fuel, act as air purifiers and help in reducing the atmosphere pollution, provides shade to human and animals and some species of trees can be used for extraction of best fibre after retting of the twigs like Sesbania, etc. The leaves of such plants bring early reproductive phase in mulch cattle. Some species give out gum of very superior quality like Koo-babul and *Sesbania spp.* These trees, in turn, do not need any nutrition or special care, except during seedling stage when they need protection from goats or other cattle grazing.

These fodder trees are also planted in the pastures which remain in association with pasture grasses and the systems of growing fodder–cum-fuel trees in association with grasses in the pasture is popularly known as *Silvipastural system.*

Certain locations like cattle shed or poultry pens need shade. Fodder-cum-shade providing trees are recommended to be planted at a distance of 6-8 metres or even more in and around the cattle shed or poultry pen compounds. A reduced distance of 5-6 metres is kept on the bunds to provide protection by rain beating and erosion or from the soil conservation point of view. While in case of silvipastural system these trees are planted at a distance of 5-7 metres from both the sides and suitable or most compatible pasture grasses/legumes are inter sown where both the trees and grasses grow together and trees come in production after 4-5 years of planting. This combination gives about 45-50 quintals of hay/ha from the compatible grasses besides 400-500 quintals of green leaves from trees every year and good quality of fire wood for fuel. The most compatible fodder trees and grasses are Israeli babool (*Acasia tortilis*) with Anjan (*Cenchrus ciliaris, C. sertigerus*), Babool (*Acasia arabica*) with Dhaula (*Chrysopogon fulvus*), Siris (*Albezia lebback*) with Sain (*Sehima nervosum*) and Unjal (*Acasia amara*) with Dinanath (*Pennisetum pedicillatum*), etc.

Conservation of Fodder During Surplus Months for Lean Period

Berseem/lucrne become surplus during December to February and jowar/bajra/M. P. chari/ maize, etc. become surplus during September to November and get spoiled if not preserved properly. India has about 4.5 per cent of its total cultivated area under fodder crops in addition to grasslands and pasture areas. Even this much area produces so much fodder that it creates a glut and starts spoiling whereas during periods when the next crop is germinating or due to aberrant weather the fodder crops become dormant and supply becomes scarce. This imbalanced supply results into a drastic reduction in milk production by milch cattle and the work output is also reduced by drought cattle. The common practice in village is to store straw of cereal (Bhusa) and pulses but due to absence of green fodder the quality of the feed remains poor. These things make it essential to conserve the surplus fodder for the period when there is no supply at all. The surplus fodder may be preserved and stored in the form of hay or silage according to its suitability.

Hay Making

In this process the green forage is presereved in dry form without affecting the quality of the fodder. The hay quality is largely governed by numerous factors like species of the crop, stage of harvesting, moisture regulation of the store, etc. The best quality hay should have all the leaves intact with green colour and pleasant aroma and it must be free from disease pathogens like molds etc. The best suited species for quality hay are grasses like Anjan, Sain, Dinanath, Marvel, Spear, Blue panic, Dhaulu, Guinea, Thin Guinea, Napier, Rhodes, etc. and legumes like Lucerne, Berseem, Guar, Cowpea, Methi, Senji, etc.

Procedure of Hay Making

The crop raised for hay making should never be grazed and it should be harvested at flowering stage because at this stage maximum protein and carbohydrates and minerals are accumulated in the plants. The harvested plants are left in their respective rows of the field which are technically known as *windrows.* In afternoon of the same day the plants left in the windrows are collected into heaps and left for few days. It must be remembered that there are bright sunny days during harvestin and hay making as high humidity and cloudy weather will result into fermentation and rotting of the hay, thus it will lose its quality. However, during summer the hay making is done in shade to avoid desiccation and bleaching of the plants. The hay must be stored in moisture proof barn having low temperature and low humidity. The hay should be inspected from time to time for taking suitable precaution.

Silage

It is the product obtained by packing fresh fodder in a container where it is allowed to ferment under anaerobic conditions without losing the nutrients. The most suited crops for silage making are those which are succulent and more watery like Lucerne, berseem, cowpea, soybean, jowar, maize, etc. For silage making when they are most succulent. Best quality silage should be free from foul odour and uniform in colour and moisture. Faulty storage conditions result in production of dark brown or black silage because of overheating. Such silage should not be fed to the cattle as they are harmful. For prolonged storage and maintenance of best quality it is required to use some preservatives like sodium metabisulphide, molasses in India but in U.S.A. dilute sulphuric acid, hydrochloric acid, phosphoric acid, sulphur dioxide, etc. are used as preservatives.

Silage Making

The crop should be harvested at its right stage, chopped up in about 2 cm bits and then packed in silos. While filling the chopped forage it should be remembered that the material is trampled well and made compact so that air is not trapped between the layers and best anaerobic fermentation is possible. The pits or silos must be filled in shortest possible time say within 3-4 days so that maximum air may be excluded. Nonlegume plant materials are poor in protein content, therefore, to enrich the silage with protein it is recommended to sprinkle urea at the rate of 3-4 kg per tonne of plant material. The silage pits are filled upto 1-2 metres above the ground level and the top should be given a dome-like shape for draining rain water, when the pits are in the open space, then the pit should be sealed. The silage becomes ready for feeding after 60-90 days but only 2-3 kg of silage/cow/day should be given in the beginning for proper adjustment of cattle to the change of food habit and slowly the quantity may be increased to 15-20 kg/cow/day.

Besides pits the silage is also preserved in the silos. The pukka silo towers are used in U.S.A. and other European countries wherein doors are made at different heights for easy removal of silage. In India silo pits are made in place of towers with cemented brick walls and pukka bottom. Katcha pits having sides carpeted with dry grasses or hay or strw may also be used. Carpeting in bottom and sides is done with the object that the silage or the chopped material may not come in contact with moisture, otherwise, the silage will get spoiled.

Constraints and Strategies of Fodder Production

India has a livestock population of nearly 416 million heads and yet the country suffers from a perpetual shortage of milk, meat, hides and wool however, livestocks contribute about 25 per cent of total income obtained through agriculture.

According to the National Commission on Agriculture (NCA, 1976), India would require about 70.10 m tones of concentrates, 590.1 million tonnes of green fodder and 373.0 million tones of dry fodder to feed it's live-stocks by 2000 A.D. This quantity represents only the maintenance part of the feed which is generally taken as 2 per cent of the body weight of the live-stocks. Taking the projected population of 215 million cattle, 100 million buffaloes, 55 million sheep and 112 million goats by 2000 A.D., the fodder requirements at 2 per cent of the body weight comes to about 796 million tones on a dry weight basis. The projected availability of fodder and concentrates by 2000 A.D. will include about 356.80 million tones of dry fodder, 695 million tons of green fodder and 77.50 million tones of concentrates.

The working Committee on fodder and grasses (1986) reported the availability of 236 million tones of straw (stovers) of cereal crops like wheat, barley, rice, and major millets as crop residues, nearly 137 million tones of grasses and herbs from the grazing areas and 250 million tones of dry grass annually from the forest areas. The availability and consumption of fodder depend on the available areas, climate and edaphic factors, livestock growing pattern and the socio-economic conditions of the farmers. Usually cattle and buffaloes are fed on cultivated fodder and grazing, supplemented with harvested grasses and fodder foliage while donkeys, ponies and mules depend mostly on grazing and harvested grasses. Camels live on browsing and lopping of fodder trees and shrubs.

Green fodder is hardly grown in 4 per cent of the cultivated area in the country and its production is not receiving due attention. The raising demand for grain crops to feed the ever increasing human population is resulting into a further decrease in the land area under fodder crops. The current availability of cultivated green fodder has been worked out to be 208 million tones of sugarcane tops and 14 million tones of edible weeds from cultivated fields and about 24 million tones of fodder foliage from trees mainly from and outside the forest areas especially from arid and hilly regions.

Cause for Lower Production of Fodder in India

1. The alarming rate of increase in human population forced the Government and it's people to give more emphasis on agriculture for fulfillment of

domestic needs due to which the fodder crops could not gain any attention and there could not be any increase in the area and production of fodder in the country.

2. The green revolution/high intensive cropping mostly included grain crops and foder crops were side tracked. During this period a vast area under forest and pasture was brought under grain production. Thus, the fodder production decreased.
3. The country is facing a serious crisis of oilseeds and pulses for which we need to import the edible oils from foreign countries, therefore, in future also the greater emphasis also the greater emphasis will be laid on oilseeds/pulses and the fodder crops will not gain any importance.
4. The existing live-stocks have very low productivity due to which often farmers do not feed them scientifically because they think it to be wastage to spend more on them.
5. During last three decades the fodder crops could not get much attention and place in the researches for boosting their productivity and quality due to higher pressure on grain crops. Thus, these crops remained neglected.
6. It was customary to graze the cattle and the farmers did not think it necessary to grow fodder crops in their fertile soils due to which the fodder production could not increase.
7. Most of the farmers are not aware of fodder preservation techniques like silage making, hay making, etc. due to which there is great wastage of surphus production during peak crop season.

Strategies for Boosting Fodder Production

Keeping higher consumption and nutritional requirements of the cross-bred cattle and live-stocks being evolved in the country, it has been felt necessary to initiate steps for boosting the production and productivity of fodder crops in the country. These steps include following measures:

1. **Increase in the area under fodder crops:** The fodder crops are restricted to only 4 per cent of the cultivated area which needs to be increased to at least 6 to 8 per cent. The wastelands, denuded forest lands and other marginal lands which are prone to erosion problems have to be brought under pasture for regulated grazing so that the stall fed cattle can get better quality and required quantity of fodder. The area is also to be increased by intercropping of forage legumes like cowpea, soybean, etc. with long duration crops like sugarcane, napier and orchard crops.
2. **Evolution of high yielding fodder varieties having superior nutritional qualities:** The existing fodder varieties have very low yield potentials and being very fibrous they possess inferior nutritional characters. Keeping lower yielding ability and poor quality of existing varieties in view several new plant types have been evolved which are capable of giving higher yields of superior quality foder. Some high yielding varieties of various fodder crops have been evolved by India Grassland and Fodder Research Institute (IGFRI), Jhansi like Bundel Anjan 1 of Cenchrus ciliaris; Bundel 2, JHP 4, IGFRI 56-1 of Dinanath grass (*Pennisetum pedicellatum*); IFC 8401 of cowpea and oat varieties like JHO 822 for central India, and JHO 829 for multiple cuts in hills, UPO 94 for areas restricted supply of water.
3. **Growing short duration fodder crops:** Growing short duration and high yielding varieties of fodder crops like cowpea, guar, jowar, maize, etc. during

summer under irrigated conditions.

4. **Popularising fodder crops in rainfed areas:** Considering the vast area under rainfed conditions work on the forage component is being inserted the level of green forage production from such areas ranging between 20.4 and 38.5 t/ha for cowpea, 17.1 and 25.6 t/ha for guar and 24.1 and 38.5 t/ha for mothbean. Thus, in the place of the conventional single crop, a short duration (60 days) fodder crop, namely cowpea (JC 21), Jowar (M. P. Chari/HS 20) and guar (HFC 244) yelded 20-30 t/ha of green fodder, followed by *rabi* crops like barley (Ratna), gram (T-2) and safflower (7-13-3).
5. **Arranging for high quality seeds:** Availability of high quality seeds at the time of sowing forage crops is the greatest bottleneck in boosting fodder productivity. The quantity of breeder seed produced in various crops the year 1990 was in all time high of 35,154.78 quintals. If put into the seed production chain, it will subsequently increase the production of certified seeds.

Breeders seeds of crops	Quantity in quintals
Oilseeds	8,547.76
Pulses	3,703.42
Cereals	7,210.63
Coarse grains	325.89
Fibre crops	272.28
Forages	38.51
Potato	15,000.00

6. **Adoption of Agri-silvi-pasture and Agroforestry system:** At Jhansi, a 3-tier system consisting of fodder trees, shrubs and grasses + range legumes increased the productivity from 6.1 to 11.7 t/ha. The system also reduced rainwater runoff and soil erosion considerably. On undulating terrain, *Albizia lebbak* produced a peak biomass of 7.6 t/ha when planted at 4 m x 4 m distance. On eroded plains, *Acasia amara* gave 1.5 t/ha of fodder. Practicing agroforestry on sodic soils at Karnal it is reported that the Karnal-grass (*Diplachne fusca*, syn. *Leptochlea fusca*) produced 41to 50 t/ha of green fodder without any amendments in 4 years when grown in the interspaces of mosquite (*Prospis chilensis*, sun. *P. juliflora)* planted at a spacing of 5 m x 3 m in sodic soils. Mosquite gave 4-5 t/ha of firewood when its side branches were cut. This land use system ameliorated the sodic soils. The available N in the soil increased from 82 to 139 kg/ha and the organic carbon in the top 15 cm of soil from 0.19 to 0.40 per cent.
7. **Developing pasture in tropics:** Grasslands in the tropics are generally low in quality and introduction of legumes viz. Siratre, Atylosia, Style, etc. have been found to increase the protein content of the herbage from 2 to 7 per cent. Reseeding of grasslands with the improved strain of Bothriochloea using good management practices increased production from natural grasslands Similarly the temperate alpine type cover extending to the high hills of the northern mountain belts comprising Jammu and Kashmir, H.P., Punjab, U.P., West Bengal and Assam offers vast potential for livestock development. At many spots it suffers from denudation and lacks vegetation which requires reseeding with grass and legume forage seeds followed by application of phosphatic fertilizers. Forest department is looking after this systemic approach of developing pasture in such areas.

8. **Adopting silvipastoral system:** The IGFRI, Jhansi is actively involved in identification of suitable species of forage and forage trees for growing together in various types wastelands. Experiments conducted at the Institute show that fodder production, under dry land conditions, can be raised for existing 10-15 t/ha to about 100-120 t/ha as green fodder. Growth of *Acasia tertilis* was least affected when grown with *Cenchrus ciliaris – Stylosanthes* mixture compared to *Albizia amara* and *Hardwiokia binata.* There are mainly such systems which are tried to find out the possibilities of getting fodder from wastelands and thereby reducing pressure on cultivated agricultural lands.

Research Needs in Future for Boosting Fodder Production in the Country

The quality aspects and some of production aspects are still not gaining much attention and they need immediate attention. Some such factors are explained here:

1. **Research needs on quality aspects:** Some fodder crops contain chemicals which are harmful or even fatal to the animals who consume them; therefore, it is needed that these chemicals are reduced to negligible extent by adopting genetic means. Jowar contains phenol, toxin HCN (Hydrocynic Acid)/Dhurin/Prusic acid; bajra contains oxalic acid, etc.), which reduces the milk production in milch animals. Cotton seed contains Gossypol which is harmful for milch animals. Thus, these aspects need immediate attention so that they can be used effectively for feeding to the cattle.
2. **Research needs on nutritional aspects:** The nutritional aspects and foliage characyers like rapid growth behaviour, leaf-density, nutrient content in plants, extent of lignification, digestibility, biological value, etc. need to be improved. The fibre content needs to be reduced and succulence has to be improved.
3. **Evolution of ideo**-types and their quality seeds multiplication: Evolution of high yielding, quick growing, short duration varieties of thermo and photo-nonsensitive varieties should be done so that farmers can choose varieties to suit their soil and climatic conditions and they can get higher fodder yields.

   The availability of quality seeds of high yielding suitable varieties is the greatest hurdle in boosting fodder production, therefore, steps must be taken to multiply the seeds of improved plant types and to make them available to the cultivators.
4. **Studies on agro-forestry,** silvipastoral management and mixed/intercropping systems need to be done more systematically so that right technology could be communicated to the farmers.

Situation of Forage Production in Orissa

Green forage plays a pivotal role in improving the animal health. Deficit is about 53 per cent at National level. The situation is worst in Orissa. There is a great scope to grow cultivated fodder, besides the existing pastures, forest lands and wastelands. But the farmers hardly realize its importance. Hence, the area is too less (11,263 ha) with minimal productivity to meet the requirement of 22.3 million livestock population. The present day climatic aberrations with low productivity in cereal and other grain crops and non availability of feed and fodder resources is one of the considerations to go for at least one cultivated fodder in the cropping system. Moreover, the dairy industry has become a more sustainable area to meet the daily expenses of farmers by fetching higher returns in milk and milk products. Now, it is high time for us to increase the area, production and productivity of rainfed cultivated fodders such as sorghum, maize, hybrid-napier bajra, cowpea, ricebean and sylo. Oat has become a very popular fodder with high palatability in coastal and inland districts of the state. Inclusion of paragrass in high rainfall areas under waterlogged situation is most remunerative.

All India Coordinated Research Project on Forage Crops is operating since 1988. Sustained efforts have been made to develop production technology for cultivated fodders. Series of experiments on various production aspects such as intercropping of cereal and legume fodders, developing appropriate seed rate, spacing and variety for rice been, use of biofertilizer in economization of nitrogen in cereal fodders, weed management strategy for cereal fodders, development of packages for sulphur nutrition in Oat, use of vermicompost, nutrient management of paragrass in wasteland waterlogged ecosystem and identification of crop varieties of cereal and legume are the thrust areas. The technology developed on the above aspects and its utilization by the forge growers of the state is the need of the days. The scientists of this centre although have proper coordination with the state department of Animal Husbandry and Veterinary Sciences, targated achievements have not been made to cover the larger area under forage crops, to involve at least one forage crop in the cropping system by the farmers and provide forage seeds in time. Therefore, time has come to work in a coordinated spirit by involving various departments associated with the cropping programme of the farmers to provide substantial quantity of green fodders to the livestock population. The fodder area (11265 ha) and its productivity being the lowest among the fodder growing states of the country, attempts have been made to improve the productivity through horizontal expansion to meet the requirement of 124.31 and 17,61 m t of green and dry fodders, respectively to feed 22.3 million livestock population of the state. The present day availability of 2.65 and 11.75 m t of dry and green fodder, respectively is too less. Therefore, there is pressure to utilize the pasture land, forest and waste lands including water-logged areas for forage production.

CHAPTER 10

Production Technology of Crop: *Oilseeds*

(Ecological Requirements, Cultural Practices and Crop Production Management)

OILSEED PRODUCTION

Oilseeds are grown widely in countries/regions such as India, China, Argentina, Canada, France, Nigeria, North Central America, USA and Western Europe. Soybean, groundnut, rapeseed and sunflower are the important oilseeds worldwide. Area, production and yield of different oilseeds are given in the Table 10.1.

TABLE 10.1: World production of major oilseeds, 1995-96

Oilseeds	Area (m ha)	Production (m t)	Yield (kg/ha)
Castor	1.6	1.4	896
Groundnut	22.5	29.2	1299
Linseed	2.8	2.3	838
Rapeseed	21.8	30.4	1395
Safflower	1.1	0.8	796
Sesame	6.7	2.5	376
Soybean	62.6	130.3	2082
Sunflower	20.9	25.1	1251

Source: FAO yearbook-Production, 1996.

In India, oilseeds are cultivated in an area of about 26 m ha. Compared to 10.7 m ha in 1950-51, the area increased to 26.4 m ha in 1995-96. Production also showed accordingly. Production was 5.2 m t in 1950-51 and increased to 9.4 m t in 1980-81. Before 1985, the aggregate output of oilseeds did not cross 12-13 m t. To bridge the gap between demand and supply and considering the mounting imports of vegetable oils, the Government of India set up a Technology Mission on Oilseeds in 1986. The objective was to integrate all the facts and sectors of oilseeds all the facts and sectors of oilseeds under a common programme for breaking the stagnation in the oilseeds front. The current oilseed production in India is 22.4 m t. The yield level has also increased substantially from 481 kg/ha in 1950-51 to 851 kg/ha in 1996. Important states growing oilseeds in India are Madhya Pradesh, Maharashtra, Karnataka, Uttar Pradesh, Tamil Nadu, Haryana, Orissa and West Bengal. These states contribute 94.4% of the area and 95.2% of production in India (Table 10.2). The average yield of oilseeds in India is 851 kg/ha. However, compared to world average (1350 kg/ha), the country's average yield is very low (about one-third only.

TABLE 10.2: Oilseed production in important states of India, 1995-96

States	Area ('000 ha)	Production ('000 t)	Yield (kg/ha)
Andhra Pradesh	3093.4	2839.7	918
Assam	306.0	157.7	515
Bihar	223.2	139.4	625
Gujarat	2921.7	2164.4	741
Haryana	636.3	815.6	1282
Karnataka	2612.6	1789.2	685
Madhya Pradesh	5632.4	4901.4	870
Maharashtra	2555.2	1980.8	775
Orissa	463.0	241.7	522
Punjab	228.3	274.2	1201
Rajasthan	3839.0	3068.3	799
Tamil Nadu	1387.5	2042.3	1472
Uttar Pradesh	1718.2	1469.4	855
West Bengal	532.7	415.5	780
All India	26353.6	22428.3	851

Source: Fertilizer statistics, 1996-97.

Nine oilseed crops primarily grown in India are groundnut, rapeseed-mustard, soybean, sunflower, sesame, castor, safflower, niger and linseed. Among them, groundnut and rapeseed-mustard are the two oilseeds that dominate the edible oil economy of the country, contributing about 66% of production. The other oilseeds play a relatively lesser role. Currently soybean and sunflower have emerged as the major oilseeds with greater potential. Oilseeds are mainly cultivated during *kharif* season. However, important *rabi* season grown oilseeds are rapeseed-mustard, safflower and linseed.

Reasons for Low Productivity

About 90 per cent of the oilseed cultivation in India is under uncertain and abnormal weather conditions. Irregular rains and inadequate alternate irrigation sources at maturity largely affect the final yield. Unlike food crops, oilseed crops are grown under low management situations. Full package of practices is not followed since the major category of oilseed growers are small and marginal farmers. Poor plant population, inadequate plant protection, improper fertilizer application and poor adoption of post-harvest technology also contribute to low productivity.

Dietary need of Fats and Oils and Use of Oilseeds

Fats and oils are valuable food ingredients in human diet. They contain saturated and unsaturated fatty acids. Vegetable oils predominantly contain polyunsaturated fatty acids (PUFA). The two PUFA, linolenic acids, can not be synthesized by the body and hence must be supplied by the diet. They are called essential fatty acids. Vegetable oils of groundnut, sesame and sunflower are rich in PUFA (Table 10.3). The country's per capita consumption of edible oils per annum is 7.5 kg which is far below the world's average of 15.3 kg.

Oilseeds are mainly used for extraction of edible oil. They also serve as an important raw material for the manufacture of soap, paint, varnish, hair oil, lubricants, etc. Commodities such as oilcakes, castor oil, high protein selects (HPS) groundnut, sesame seed and extractions of rapeseed, sunflower and groundnut are the main export items. Oilmeal export during 1995-96 was 39.52 lakh tonnes with a value of Rs 2,200 crores.

TABLE 10.3: Fatty acid content and iodine value of oilseeds

Crops	% Composition							
	Oil (%)	Palmitic	Stearic	Oleic	Linoleic	Linolenic	Erucic	Iodine value
Groundnut (Kernel)	44-50	6	5	61	22	-	-	101
Rapeseed	42-46	4	2	19	14	8	40	146
Soybean	17-22	12	2	24	54	8	-	134
Sunflower	36-38	8	5	21	66	11	-	134
Sesame	48-52	10	5	40	43	2	-	114
Safflower	30-32	8	3	13	75	1	-	146

Source: Weiss, 1983

GROUNDNUT (*ARACHIS HYPOGAEA* L.)

The word *Arachis hypogaea* (groundnut) has been derived from two Greek words, Arachis meaning a legume and hypogaea meaning below ground (referring to the formation of pods in the soil). Groundnut has never been found in the wild state anywhere and its origin has, therefore, been a matter of considerable speculations and even controversy. There are two schools of thoughts about its origin – one supporting the view that groundnut had originated in Africa and the other tracing its origin to Brazil in South America. According to some literatures it is clear that from the beginning of the sixteenth century, Portugese ships going to and returning from Brazil always touched the west coast of Africa and naturally they introduced it from Brazil into Africa. De Gandolle(1825) stated its origin to be Brazil(South America). Its introduction to India is considered to be through Jesuit Fathers (Missionaries) who followed Vasco De Gama shortly after his first landing in India *i.e.*, in the first half of the 16th century.

Distribution

Groundnut is cultivated in tropical and sub-tropical countries. Brazil in South America is considered to be the country of its origin. Groundnut is primarily cultivated in USA, Nigeria, Senegal, Sudan, China, West Africa, Indonesia and India. It is grown in an area of 22.5 m ha with an annual production of 29.2 m t in the world. India is the largest producer of groundnut. The crop is introduced to India during 16th century. It ranks first both in respect of area and production contributing about 40% of the area and 36% of world production. Over the years, the area under the groundnut has increased from 4.5 m ha in 1950-51 to 8.3 m ha in 1990-91. Similarly production has also doubled during this period from 3.5 m t to 7.8 m t. The increase in yield is due to adoption of high yielding varieties and improved production techniques. Average productivity of groundnut is about 850 kg/ha while world average is 1100 kg/ha

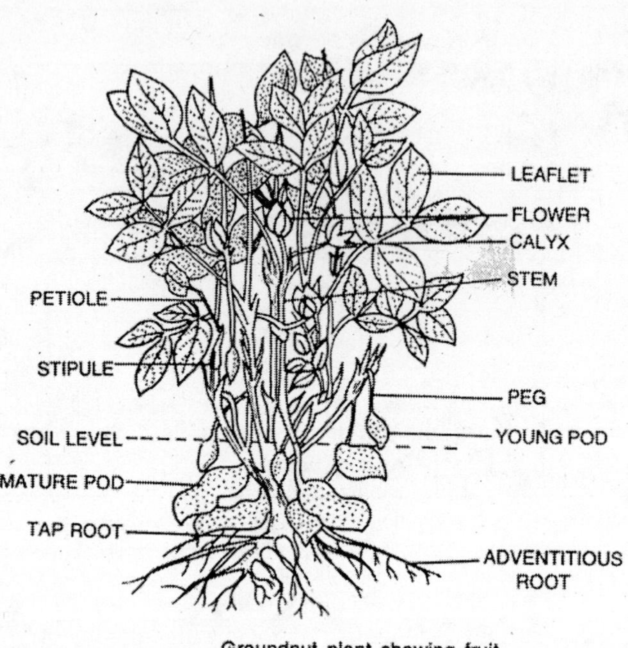

FIG. 10.2: Groundnut plant showing fruit.

Groundnut is predominantly grown in Rayalseema area of Andhra Pradesh, Saurashtra in Gujarat, parts of Tamil Nadu, Karnataka, Orissa, Maharashtra, Madhya Pradesh and Rajasthan. These states contribute 96% of total area and production (Table 10.4). The crop is mainly grown in *kharif* under rainfed conditions. However, in Southern and Western India cultivation of *rabi* groundnut is usually raised with irrigation support. However, the season is not suitable for North Indian conditions due to cold temperature which adversely affects the crop growth. In India, groundnut oil is properly used for cooking. The HPS type of groundnut kernels are exported to foreign countries.

TABLE 10.4: Groundnut production in important states of India, 1995-96

States	Area ('000 ha)	Production ('000 t)	Yield (kg/ha)
Andhra Pradesh	2209.3	2426.2	1098
Gujarat	1902.9	1028.3	540
Karnataka	1178.4	1156.0	981
Madhya Pradesh	280.9	290.6	1035
Maharashtra	511.0	576.4	1128
Rajasthan	214.7	163.6	762
Tamil Nadu	1116.1	1911.6	1713
All India	**7704.6**	**7813.8**	**1014**

Source: Fertilizer statistics, 1996-97

Climate: Groundnut is a tropical crop which requires a long and warm growing season while high rainfall, drought and cold weathers are extremely detrimental for better crop growth. Normally it needs about 70-90°F temperature during its growing period with cold nights at maturity. It can be grown well in tracts which receive an annual well distributed rainfall of 50-125 cm.

Soil: The best type of soil for groundnut cultivation is the one which is well drained, light coloured, loose, friable, sandy loam, well supplied with calcium and moderate amount of organic matter. Soils with poor drainage high acidity or alkalinity must be avoided for groundnut cultivation. A soil pH of over 5.0 or below 8.5 is supposed to be an ideal for groundnut production. In U.P., it is grown in alluvial sandy to loam soil while in M.P., Maharashtra, Gujarat, Andhra Pradesh and Karnataka the crop is taken in black cotton and red soils. Heavy clay is not fit for groundnut production because soil becomes very hard during drought which restricts pod formation and development. Even peg penetration becomes difficult.

Land Preparation: For good germination and higher pod yields, it is essential to get a weed free pulverized, open and aerated seed-bed for sowing the crop. The field must be thoroughly leveled to avoid water logging in any part of the field. A required tilth may be obtained by ploughing twice with mould board plough followed by two harrowing and planking. If the field is infested with white grubs, chemicals like Heptachlor or Chlordane should be drilled at the rate of 25 kg per hectare before the final harrowing. Organic manures, if to be applied, should be spread and mixed well in the soil at least 15-20 days before sowing.

Application of manures and Fertilizers: Groundnut, being a leguminous crop, does not require very high doses of nutrients. In general, it needs about 10-20 kg N, 40 kg P_2O_5 and 40 kg K_2O/ha under rainfed conditions while irrigated conditions an application of about 20 kg N, 40-90 kg P_2O_5 and 20-40 kg K_2O/ha is sufficient. It is observed that Andhra Pradesh soils are deficient in Bo and molybdenum whereas North Indian soils are deficient in sulphur, zinc and calcium. These deficiencies may be corrected by application of 5 kg Borax, 1.0 kg Ammonium molybdate, 15-20 kg zinc sulphate and 200-500 kg of gypsum at per hectare rate. Of these borax and ammonium molybdate should be applied as basal while gypsum should be band placed near roots 30 days after sowing. These nutrients should be applied in the zone of pod development as they are directly absorbed by the developing pods.

The crop responds well to application of FYM or compost at the rate of 6.25 t under rainfed and 12.5 t /ha under irrigated conditions. But under unavailability of organic manures the above mentioned NPK may be met with inorganic fertilizers. So far as sources are concerned it is always better to apply ammonium sulphate/calcium ammonium nitrate, single superphosphate and muriate of potash. In the case of very light soils the nitrogen should be applied in two split doses *i.e.*, half at sowing and the rest half (top dressing as band placed) 30 days after sowing or at the time of last inter-cultivation.

Seeds and Sowing

To ensure good stand, uniform maturity, high yield and better quality produce one has to be careful about the seed quality. The seed must be pure, viable, uniform in size and free from the seed borne diseases. The seed should not be broken or damaged by any means. It is better to use hand shelled kernels but if large area is to be sown, groundnut sheller may be used. The germination of the seed should be checked and the seed lot having over 90 per cent germination should be selected for sowing. It is observed that the germination in case of bunchy type is always higher (90-95%) than the spreading type (85-90%).

To ensure the freedom of seed from seed borne diseases the kernels must be treated with captan 75% dust or Thiram 75% WDP(Slury made by mixing 125 g Thuram/100 kg kernel in 500ml of water) or Difolotan dust at the rate of 2.5 g, 1.25 g and 2.0 g/kg of groundnut kernel respectively.

It is observed that some of the spreading varieties have a dormancy of 60-75 days before they germinate. Therefore, in case if the freshly harvested kernels are to be used for sowing within the dormancy period, the kernels should be treated with some germination promoting hormones like G.A., etc.

It is advised that, when groundnut is to be introduced in some new areas where it was never grown the kernels should be treated with Rhizobium culture for better nodulation, growth and development of the crop. The Rhizobium culture treatment may be done as under:

For 50 kg of kernels, prepare 10 per cent solution of gum arabic in water; add gur (jaggary) to make a solution of 5 per cent. When it is dissolved, add 2-3 packets (200 g each) of peat based culture pour the groundnut kernels and agitate the content to make slurry. When all the kernels have had an uniform coating of the slurry, spread them out on newspaper sheets in the shade. Reject the left over slurry after use and sow the seeds when they have dried completely.

Sowing

1. **Time od sowing:** In north India, groundnut is grown as a *kharif* season crop where sowing is done between April-July under irrigated conditions and between June 20 to July 31, under rainfed conditions. In Tamil Nadu, the crop is sown from May to July where as in Andhra Pradesh and Gujarat, it is sown between January to March.
2. **Seed Rate:** The seed rate depends upon boldness, spacing and germination percentage of seeds. On an average the seed rate for spreading type comes to 60-70 kg/ha while for bunch type it is about 85-90 kg/ha. In the case of very bold seeded variety like T-64 it would be about 100 kg/ha.
3. **Method of Sowing:** Groundnut is sown either by dibbling the seeds behind the plough or by using seed drill. In the case of irrigated crop it is better to prepare ridges and furrows with a country plough and dibble the seeds on ridges as this helps in inter-cultivation and in better pod setting of groundnut. In the case of rainfed light soils the crop is sown on flat beds. For good germination depth of seed sowing is another important aspect which needs due attention. Usually, groundnut seeds should be sown at a depth of 5-8 cm in light soils and 4-6 cm in the case of moderate to heavy soil types. To ensure a good and proper germination a light covering with soil is required over the seed.

   Sometimes rodents and crows are noticed to take away the seeds from the field, therefore, use of some repellants like pinetarand kerosene for seed treatment are recommended to keep the intruders away. But care should be taken to avoid any injury to the kernels.
4. **Spacing:** This varies according to irrigation facilities and type of the seed. In the case of bunchy type and rainfed crop the spacing ranges between 20-30 cm from row to row and 8-10 cm from seed to seed. The spacing in the case of semi-spreading and spreading types with irrigation facilities ranges between 30-40 cm x 10-15 cm and 50 cm x 15-20 cm respectively.

Groundnut Types and Varieties

Groundnut is broadly grouped into two types: Spreading (runner) and Bunch varieties. The bunch type of groundnut has upright growth habit with long and erect main axis and the laterals are oblique to the main axis. The pods are clustered on the base of the plant. The spreading or runner types have very short and erect main stem and the laterals spread horizontally along the ground. In these types, the pods are scattered along the procumbent rooting stems.

Recently released varieties	:	BAU 13, Chitra, CSMG 84-1, DRG 12, GG 3, GG 20, IGGV 86325, IGGV 86590, K 134, M 335, R 8804, TG 22, TAG 24, VRI 2, VRI 3, VRI 4
Rainfed/drought condition	:	AH 32, TMV 1, Spanish Improved, S 230, Jyoti, GAUG 1, DH 3-30, JL 24, Kadiri 71-1
Export purpose (HPS)	:	Polachi 1, BP 2.
Tikka resistant	:	TMV 9, C 501, DH 3-30
Rust resistant	:	TMV 9

Generally, there are three plant types in Groundnut: (a) Erect, (b) Semi erect and (c) Spreading. Erect types are shorter in duration and non-dormant, spreading types are longer in duration and have dormancy. Semi erect types come in between these two types. The following varieties are recommended for cultivation in the state of Orissa. The erect and semi erect types are suitable to be grown in all seasons.

Variety	Habit	Duration	Shelling (%)	Oil content (5)	Dormancy in days	Special characters
AK-12-24	Semi erect	110	73	48.5	Nil	
Kisan (OGI 13-3)	Bunchy type	95	71	50	15	Suitable both for irrigated and non-irrigated conditoin.
JL 24 (Phule Pragati)	Bunchy type	95	75	50.7	-	Non dormant, Large dark green leaves with smooth pods.
J 11 (SB-II)	Erect	110	76	49	Nil	Drought tolerant.
Kadri-3	Semi-spreading	110	75	49	-	Suitable for Ganjam areas where spreading varieties are grown.
ICGS 44	Spanish but semi-erect	120	70	49	-	Response better to irrigation, suitable for all areas of Orissa. Bold kernel 2 seeded pod.
GG-2	Spanich bunch	100	70.5	49	-	Early flowering, dark green leaves, reticulated and seeded pods.
ICGS-II	Spenish bunch	110	70	48.7	-	Dark green leaves, dwarf habit, compact plant and seeded pods

Variety	Habit	Duration	Shelling (%)	Oil content (5)	Dormancy in days	Special characters
						constriction and reticulation less prominent.
Jawan (OG-71-3)	Bunch type	95	70	46.7	-	Resistant to Tikka.

Interculture

First flowering comes 16-18 days after germination. First hoeing and weeding within two-three week stage should be done. Earthing is necessary in erect and semi-erect types to help pegging. Second weeding in spreading types may be given within one month, if necessary. Pod formation starts after 30 days of growth. The soil should not be disturbed after flowering commences. Collar rot and bud necrosis affected plants should be destroyed no sooner then are observed.

Irrigation

Generally, groundnut is grown as a rainfed crop during *kharif* season but if the crop is caught in a long shell of drought, especially at the pod formation stage, supplemental irrigation is given. In the case of irrigated crop, the frequency of irrigation depends upon soil texture and the interval between the irrigations ranges between 8-12 days. The critical stage for irrigation is branching, flowering and pod formation. There should be sufficient moisture at the sowing time in the field, thus if the crop is to be sown before onset of monsoon the field should be given one light pre-sowing irrigation for better germination of the seeds.

Weed Management

Weeds result in a severe competition with the crop and cause a considerable reduction in the yield to the extent of 20-40 per cent based on nature weed infestation in the field. To keep the soil loose and friable and the field free from weeds the crop should be given a hand weeding 20-25 days after sowing and one or two hoeings- the first at the time of weeding and second about a fortnight later. The intercultural operations should be stopped after the pegs have started going into the soil. The bunch and semi-spreading types should be given a light earthing to facilitate the maximum penetration of the pegs into the soils.

For an integrated approach to the weed control problem it is advised to apply TOK-E-25 or Lasso at the rate of 5 litres of commercial material dissolved in 500 litres of water as pre-emergence soil spray within two days of sowing. The chemical weed control should be combined with above mentioned mechanical control measures to take care of the weeds effectively. But on top of everything it is essential to sow the seed in a completely weed-free seed-bed.

Mixed Cropping

The groundnut crop may successfully be grown mixed with several erect growing crops like bajra, arhar, castor, sesame, cotton, maize, sunflower etc. The number of groundnut rows between two rows of any of the said crops depends upon their inter-row spacing and type of groundnut, *i.e.*, bunch, semi-spreading or spreading to be grown.

Harvesting and After Care

Yellowing of leaves is the prominent symptom of maturity. The leaf yellowing is associated with leaf shading (particularly, the older ones), development of proper colour of shell and a dark tint on the inner side of the shells. Usually, the crop takes about 120 to 140 days time to mature depending upon the variety and the harvesting is done in the month of October-November accordingly. At maturity the crop is badly damaged by crows, Jackals, pigs etc

Harvesting and After Care

Yellowing of leaves is the prominent symptom of maturity. The leaf yellowing is associated with leaf shading (particularly, the older ones), development of proper colour of shell and a dark tint on the inner side of the shells. Usually, the crop takes about 120 to 140 days time to mature depending upon the variety and the harvesting is done in the month of October-November accordingly. At maturity the crop is badly damaged by crows, Jackals, pigs etc. which need vigilance. After maturity the bunch and semi-spreading types are generally harvested by hand pulling at an approximate soil moisture, while the spreading types are harvested by digging out the plants with the help of khurpi, spade or by ploughing the field. The left out pods in the soil are collected by hand, later. The pulled out plants are stacked in a safe place for a few days to dry and are stripped afterwards. The stripped pods are cleaned nicely and dried to a safe moisture content of not more than 5 per cent before they are stored because dampness will cause formation of pods and allow developing the poisonous moulds like *Aspergillus flavus* in the kernel. These moulds lead to contamination with aflatoxin which create a health hazard to both human and cattle who consime the kernels.

Plant Protection

There are certain insect pests and diseases which attack the crop and cause a considerable damage to it. The plant protection measures should be taken up as per the recommendation of the sate to control insect pests and diseases of the groundnut crop at proper time.

Yield and Storage

The pod yield is controlled by several factors like climate, soil and varietal potentiality. In general, the irrigated groundnut crop produces about 30-35 quintal pods/ha and rainfed about 15-20 quintal pods/ha. The yield of haulms is usually two to two and a half time to that of pods' yield. The shelling percentage ranges between 70-75% and the kernels have on average 45 to 50% oil in them.

Storage: All the damaged or injured pods must be sorted out before they are stored. Well cleaned and dried pods to 5% moisture level should be stored after filling in gunny bags. These gunny bags are stacked in a store room in tiers comprising not more then ten bags in each tier. The tiers must be stacked on wooden planks in such a way that the air keeps on circulating to avoid damage from dampness, rats etc. The room should be inspected periodically and proper control measures of rats and pests should be taken. The store should be fumigated, if needed and made airtight. The groundnut should be stored in the form of pods rather than kernels but the broken and damaged pods must be taken out and discarded. Apart from this the undersized/underdeveloped and unfilled pods (pop pods) should also be discarded because their presence reduces the market price.

Rapeseed and Mustard

Ayurvedic Samhitas describe the use of 'Sarson' in India. Records show that Europe had oilseed rape and other brassica seed crops since at least fifteen century and for a much longer period in India. It is further recorded that rai (*Brassica juncea* Coss) was introduced in India from China. Later it spread to Afghanistan and other countries from India. According to another report, Rape or Mustard seed was grown from about 3000 B.C. in Indus valley and use of its oil is recorded from the last few centuries of the pre-Christian era. China is the largest producer of rape and mustard and together with India and Pakistan; they grew over 90 per cent of the world production.

Distribution

The crop requires cool temperature (18-25°C) for satisfactory growth. Hence, it is mostly cultivated in the temperate and warm zones of Asia and Europe. The crop is predominantly cultivated in China, India and Pakistan and these countries contribute about 90% of the world production. In India the crop occupies a prestigious position accounting for 25.4% of the total oilseed acreage and 31.1% of total oilseed production. These crops are mainly grown in the northern part of India. About 60% of the area and production is from Uttar Pradesh. Other states growing rapeseed–mustard are Rajasthan, Gujarat, Madhya Pradesh, West Bengal, Punjab, Haryana and Bihar.

Classification of Rapeseed and Mustard

There are two main sub-divisions of the oilseed rape, belonging to the species: rape (*Brassica napus*), turnip rape (*Brassica campestris*). Brown sarson (*Brassica juncea*), is also grown in considerable are in India and Pakistan. The inter-relationship between the principal *Brassica sp* is complex and there are three primary species. The species have been evolved by natural and artificial selection over a long period of time and most of them have developed several types *i.e.*, sub-species which differ morphologically and physiologically, *Brassica oleracea* shows the greatest development and contains knoll-khol, cauliflower, Brussels sprouts etc. *Brassica napus* and *Brassica campestris* both have annual and biennial varieties which differ physiologically in that the later does not flower until exposed to a very long temperature. *Brassica juncea*, a brown Indian mustard, is mainly adapted and grown in considerably large areas for oilseed purposes. Thus, the nomenclature is not fully standardized and understood due to which there are many synonyms and its classification is in a confused state *e.g.* the *Brassica napus* varieties were formally described as *Brassica campestris* and are so described in recent literature.

Characteristic Features of Some Important Types

1. **Toria (*Brassica napur* var, toria):** It is grown as an autumn crop. This variety is susceptible to cold and is sown early in middle or late September and takes about 75-100 days to mature. The variety is obviously low yielding but it responds to irrigation and adequate fertilization.
2. **Yellow and brown sarson (*Brassica napus* var. glauca and var. dichotoma).** It is widely grown in north and central India. There are two main types yellow and brown sarson so named because of its seed colour. It is high yielding than 'toria' and brown sarosn and also from yellow ones. The crop is sown in October and harvested in March/April after about 150-160

days time.

3. **Brown or Indian mustard (*Brassica juncea*):** It is very widely grown in India. The crop is sown in October/November and harvested in March/April after about 110-160 days period. This type gives better yield than *Brassica napur*.
4. **Black mustard (*brassica nigra*):** It is grown in very limited area. It has very low oil content into it but the main purpose of its cultivation is to use it as condiment.
5. **Taramira (*Eruca sativa*):** It is relatively of recent introduction into India. It is believed to be native of south Europeand North Africa. It is relatively a low yielding cruciferous oilseed crop grown in northern India and very often grouped with rape and mustard crops. The variety is particularly adapted to poor soils and low rainfall areas.

Sarson, toria and taramira are commonly called as rapeseed. About 70% area is grown with brown or Indian mustard. Yellow and brown sarson are widely grown in North and Central India. Toria is mostly cultivated as an irrigated crop. Rai, sarson and tarmira are usually cultivated under rainfed condition.

Climate

In India, rape and mustard is grown during winter season and it is observed that the crop needs about 18o to 25o C temperature, low humidity, practically no rain especially at the time of flowering. Rainfall, high humidity and cloudy weather are not good for the crop during winter as it invites aphids and the crop gets spoiled completely. However, under rainfed conditions one to two pre-flowering rains help in boosting the grain yield. Excessive cold and frost are harmful to the crop.

Soil

Generally, the rape and mustards thrive well in medium or heavy loam soils except taramira which is grown on lighter soils but heavy soils subjected to water logging should be avoided as the crop can not tolerate such conditions. Though the crop is grown during winter season and there is very little chance of water logging but still due to heavy winter rains the water may get accumulated and cause a temporary water logging. Very light soils usually cause a serious moisture stress and a poor crop growth is observed. Saline and alkaline soils are often not fit for the crop though it has good tolerance to such conditions.

Land Preparation

The rapeseed and mustards require a fine, firm, moist seedbed so that a reliable supply is assured for germinating seeds and young seedlings. For achieving this type of tilth the field should be given one pre-sowing irrigation if there is less moisture in the field. The field should be given a deep ploughing soon after the *kharif* crop is harvested in middle of September. Therefore, it may be ploughed for 3-4 times with desi plough and planking after each ploughing. The crop and weed stubbles along with established weeds should be picked up and thrown out of the field. In case of dry land areas where pre-sowing irrigation may not be given, the seed may be spread in damp place at night and next morning it should be sown. It increases the germination percentage of the seeds. In case when mustard is to be sown mixed it gets the same tilth as that of the main crop with which it is to be associated.

Sowing

1. **Time of sowing:** Sowing time is very important as the attack of aphids and the extent of damage can be reduced considerably by sowing the crop earlier or before middle of November. Toria must be sown between mid to last week of September as the crop suffers badly from cold if sown late and the crop duration is increased without any additional yield due to slow growth that occurs due to lower temperature.
2. **Seed rate:** It depends upon the nature of the crop, such as pure crop or mixed crop. A pure crop of mustard and rapeseed needs about 6-7.5 kg/ha and toria seeds 4-5 kg/ha while mixed or intercrop of mustard requires 2.5-3 kg/ha seed per hectare, however, seed rate in the case of mixed crop depends on its proportionate area to the main crop.
3. **Method of sowing:** Line sowing has always given better results than broadcasting although broadcasting in case of high moisture conditions is common especially for 'toria' crop. Thus, it may be sown in shallow furrows behind desi plough or through seed drill or 'mala basa' as it is good and economical to adopt any of these devices for sowing the crop.
4. **Spacing:** From the spacing trials it was observed that in the case of pure crop a closer spacing has been better results. On an average a spacing of 40-45 cm between rows and 15 cm between plants is recommended for *rabi* and local sarson but 30 cm between rows and 10-15 cm between plants is recommended for yellow sarson and toria brown sarson. In case of mixed cropping with wheat sowing mustard in parallel rows 150-200 cm apart alternating with the main crop is recommended, however, broadcasting of mixed seed is still in practice but it gives a poor seed yield.

Varieties

Indian mustard	:	Prakash, Varuna, Seetha, Kranti, Krishna, Vaibhav, Vardan, Rohini, 'Patan Mustard 67' Durgamani, Pusa Bold, Bhagirathi, RLM-619, RH 30, RH 81, RH 819, T 59.
Brown sarson	:	BSH 1, Pusa Kalyani, KOS 1.
Toria	:	Agrani, Bhavani, Jowahar Toria 1, PT 30, PT 303, Sangram, TH 63, TL 15, RH 68, TCH 2.
Yellow sarson	:	Rajendra sarson 1, Subinoy, Ys 92-3.
Taramira	:	RTM 314, T 27, TMC 1.

The following varieties of rapeseed and mustard crops are recommended for cultivation in the state of Orissa.

Variety	Duration in days Rabi/Summer	Suitable seasons	Remarks
M-27	75	Rabi	Toria type, suitable for plain areas.
T-9	85	Rabi	- do –
PT-303	80	Rabi	Toria type, resistant to Alteneria blight, white rust and downy mildew
Benoy (YSB-9)	90	Rabi	Yellow sarson, yellow seed.
T.S-29	75	Rabi	Toria type, stem palegreen coloured covered with white waxy deposits.

Variety	Duration in days Rabi/Summer	Suitable seasons	Remarks
Pusa Bold	105	Rabi	Non shattering pods, bold seed.
Pusa Bahar	100	Rabi	Suitable for pure cropping under Irrigated conditions.

The Indian mustard has a very high level of erucic acid in oil and glucosinolates in the meal. Erucic acid is to be removed before it can be used as edible oil for human consumption. Removal of glucosinolates in the meals has made it acceptable as cattle feed in the international markets. Some of the exotic varieties of rapeseed have low levels of erucic acid and glucosinolates ('0' or '00' varieties).

Manures and Fertilizers

Rapeseed mustards are grown rainfed with normal use or no use of fertilizer. But the nutritional investigations and general experience show that use of fertilizers gives substantial increase in yield. However, the response patterns differ very largely according to crop, soil type, climate, and crop and fertilizer practices. Most of the studies have been concluded on primary nutrients and hence, the response of micronutrients is ye to be studied.

Nitrogen Fertilization: Nitrogen is an important constituent of protein for which the plants take inorganic nitrogen in the form of ammonium or nitrate. Higher the nitrogen greater would be the protein and protoplasm which would increase, in turn, greater cell size, leaf area index resulting into greater photosynthetic activity. Thus the nitrogen helps in formation of a larger frame on which more flowers and eventually more pods can develop. This shows a positive link between larger nitrogen supply and higher seed yield

Phosphate Fertilization: Phosphorus fertilization improves the growth of rape-seed crops where it is of little or no effect on growth is observed in the case of cereals. Besides growth, phosphorus reduces the adverse effects of excess nitrogen fertilization. Phosphorus deficiency restricts growth of roots and the aerial parts and sometimes it prevents even flowering. The deficiency symptoms are characterized by a dark bluish-green colouration of leaves, often accompanied by purplish tinges. The stems become bluish-green having purple or reddish coloration. Under severe deficiency marginal necrosis of leaves or premature withering of older leaves may take place.

Potash Fertilization: The role of potash in oilseed rape crops is to activate a wide range of enzyme systems. It also controls other physiological functions like water economy through regulation of stomatal opening and closing, photosynthesis by influencing chloroplast formation, transport of photosynthesis, carbohydrate and nitrogen metabolism.

Quantity and method of fertilizer application in mustards

Variety	Nutrients in kg/ha			Method of application
	N	P_2O_5	K_2O	
Early Toria/Lahi	30-50	25	-	All basal
Mustard/Sarson (Irrigated)	80-150	60	-	Half basal+half top dressing at 30-35 days after sowing.
Mustard/Sarson (Rainfed)	40-60	25	20	All basal placed

Variety	Nutrients in kg/ha			Method of application
	N	P_2O_5	K_2O	
Taramira (Rainfed)	30-40	20	-	All basal

In general, the following doses are found to be optium:

Rainfed: 40 kg N/ha without P and K for all rapeseed and mustard crops.

Irrigated: 40, 60 and 80 kg /ha N are considered optimum for 'toria', 'sarson' and 'raya' respectively, however, for taramira 20 kg N/ha is found to be the best dose. Phosphate and potash have not given any positive response; hence, they are not needed until the soil is deficient in them. The micronutrients are not found to be affecting either yield or quality of the produce; therefore, their use may be done under very special conditions.

Plant Protection Measures

Weeds: In case of 'toria' or early mustard it is observed that late kharif season weeds viz., *Trianthema sp.*, *Phyllanthus sp.* etc. are found while the main crop is infested with *Chenopodium sp.*, Anagallis sp., *Spergula arvensis*, *Asphodelus sp.* etc. Orobanche, a phanerogamic parasite, has been found to be a very serious weed in the crop. The chemical control of these weeds is not very much in practice, hence manual weeding once about 25-30 days after sowing has been found to check the weed competition and yield is not reduced. An effective control of Orobanche may be achieved by their hand pulling and growing mustard after 2-3 years.

Diseases: Amongst several diseases the most destructive ones are Alternaria blight (caused by *Alternaria brassicae* Berk), rust (caused by *Cystopus candidus*), downy mildew (caused by *Peronospera brussicae*), powdery mildew (caused by *Erysiphe poligoni*) and viral diseases. All of these, except viral disease, may be controlled by the application of Diathane M 45 at the rate of 1.5 kg/ha or 2-3 applications of Bordeaux mixture.

Insect and Pests: There are many pests which have been seen attacking the crop viz., pea leaf-minor, mustard sawfly, painted bug, cabbage butterfly, Bihar hairy caterpillar and mustard aphids. Of these mustard aphids are the most serious which alone cause 66-96 per cent reduction in yield in case of B. *campestris* and 27-69 per cent in B. *juncea*. The average loss due to aphids comes to 62 per cent. The pest attacks under high atmospheric humidy and cloudy weather conditions. It may be controlled by spraying of Methyl Demeton 0.02 per cent or Demethoate 0.03 per cent or Ohosphomidon 0.03 per cent two to three times. Aldrin/Heptachlor/Chlordane at the rate of 25 kg/ha controls cutworms and BHC 25 kg/ha controls sawfly.

Harvesting

The crop should be harvested as the pods turn to yellow colour. Amingst the types 'toria' is the earliest maturing variety that takes about 75-90 days time, 'Rai'needs 110-180 days, Yellow sarson 130-160 days and Brown sarson needs 105-145 days for maturity. The crop is harvested with hand sickles and the threshing is done by beating the pods with wooden sticks or by trampling the plants by bullocks. The winnowing is done with the help of natural air current but the wind velocity should not be very high as the seeds, being very small, are blown with the air.

Yield

Rapeseed/mustard is usually grown in neglected conditions and due to to this reseason a very poor average yield of 4-6 q/ha has been reported. The following yields have been obtained from different groups of the crop.

- **Toria:** Average of 5 q/ha and the highest 8-10 q/ha.
- **Yellow mustard:** Average of 10-12 q/ha and the highest of 30 q/ha or more.
- **Rai:** Average of 12-15 q/ha and the highest of 25-35 q/ha.

SOYBEAN (*GLYCINE MAX*)

Soybean is predominantly cultivated in USA, Brazil and Argentina in South America. Soybean occupies an area of 62.6 m ha with the production of 130.3 m t. In the Asia-pacific region, soybean is cultivated in China, Pakistan and India. The crop is cultivated in an area of 4.9 m ha with the annual production of about 4.9 m t in India. Among the major states, about 80% of the area and production is from Madhya Pradesh, other states include Rajasthan, Maharashtra. These states together contribute about 97% area and production (Table 10.5).

Growth of a large number of soybean processing industries in Madhya Pradesh is the primary reason for steep rise of soybean output in that state. Soybean is a short day plant and thrives best in varied climate and particularly in shaded environment. It is generally cultivated in *kharif* season.

Soybean contains about 38-42% protein and 17-22% oil. Soya-protein is a low cost protein and is highly nutritious. Soybean contains all essential amino acids and is rich in lysine. It can be used as a supplement to cereal diet. It is an important constituent in some of the infant foods and milk substitutes. Soya milk and *tofu* (in Japan and Indonesia) are the important preparations from the soybean. It is utilized as a raw material for a number of protein and oil based industries. Soybean oil is rich in poly-unsaturated fatty acids and contains more of linoleic acid (54%). This results in rapid oxidation and deterioration of oil quality. The soil is extracted by solvent extraction and is used after refining for the production of hydrogenated fats, margarine and shortenings. The by-products are utilized in paint industry. The meal is an important source of vegetable protein and can be used for culinary purpose. The meal can also be used for the production of livestock concentrate.

TABLE 10.5: Production of Soybean in important states in India, 1995-96

States	Area ('000ha)	Production ('000 t)	Yield (kg/ha)
Madhya Pradesh	3703.8	3781.6	1021
Maharashtra	635.4	718.8	1131
Rajasthan	427.7	400.8	937
All India	**4887.2**	**4992.7**	**1022**

Source: Fertilizer statistics, 1996-97.

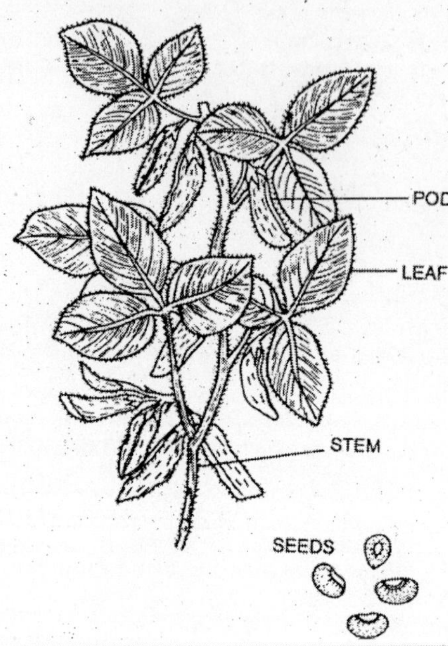

FIG. 10.1: Shoot of soybean plant (Glycine max)

Climate: Soybean needs about 15 to 32°C temperature for germination but for rapid growth the crop needs slightly higher temperature. It is observed that crop does not grow if the temperature falls below 10°C but at the same time if it goes above 40°C the crop gets a greater set back in its growth, flowering, seed formation or even in the seed quality. It is reported that cold temperature lowers the oil content and higher temperature during seed formation increases oil content in seeds. The crop requires about 60-65 cm annual rainfall. It needs irrigation if the rainfall is lower than 65 cm, however, higher rainfall is equally or more harmful than lower one because higher moisture at germination causes rotting of seeds and seedlings, and at later stage of growth excess water causes anaerobic conditions, poor nodulation and poor growth and yield of the crop. Drought, at flowering or just before flowering, results in flower and pod drops. While rain during maturity, impairs the grain quality of soybean.

Soil: Soybean prefers neutral soil having a pH between 6.5 and 7.5. It can tolerate mild salinity but acidity is deadly harmful to the crop. The soil should be light with open structure so that there is better aeration to the plants. It grows well on most of the soil types ranging from light to black cotton soils of M.P. but the best soil type is sandy loam having good organic matter content. The field should have an efficient drainage during growth period of the crop.

Land Preparation: There should be sufficient moisture in the field, therefore, if there is less moisture in the field, one pre-sowing irrigation should be given. The field should be given one or two deep ploughing by mould board plough following which 2-3 light ploughings by desi plough or harrowing should be done so that the clods are broken and soil becomes well pulverized for sowing. The field should be leveled by planking and cross planking. The weeds and stubbles should be picked up, collected and thrown out.

Fertilizer management: Seeing the performance of the crop to different nutrients, it has been found that nitrogen 60-80 kg, phosphorus 60-80 kg and potassium 40-60 kg/ha are **appropriate** doses of fertilizers for soybean crop. Single super phosphate, potassium sulphate and ammonium sulphate are proved to be the best sources of nutrients and they have given the highest grain yield of the best quality seeds. The entire quantity of fertilizers should be applied as basal placed in medium to heavy soils whereas in the case of light soils half of the nitrogen may be top dressed after interculture and thinning about 20-30 days after sowing.

Sowing

1. **Time of sowing:** There are two cropping seasons of soybean viz., *kharif* and spring. In case of *kharif* season, the most common time of sowing is onset of monsoon or last week of June to first week of July. But in the case of irrigated conditions sowing of crop by mid of June results in a better crop growth, less weed problem and poor weed competition because the soybean plants get established by the time of onset of monsoon and weed sprout, while the spring sowing is done between 15th February and 15th March.
2. **Method of sowing:** The conventional method of soybean sowing is still broadcasting but this method had certain drawbacks *e.g.*, plants are not properly spaced, agricultural operations such as weeding, hoeing etc. can not be conveniently done. This method may be the most convenient for fodder crop or for green manure crop of soybean.
   Line sowing by seed drill or malabasa would be most ideal because it needs less seed/ha, weeding and hoeing may be done conveniently and plants are placed at uniform spacing. Therefore, this method should be adopted. Sowing may be done in furrows behind the plough in the case of the seed drill is not available.
3. **Seed Treatment:** Soybean is a leguminous crop which has rhizobium bacteria into their roots. The crop is a recent introduction in the country; therefore, its specific bacteria are not present in the soil which necessitates the use of culture to treat the seed for better nodulation and good growth of the crop. There are two ways to treat the seeds viz., the first way is to bring some soil from the field where it was grown earlier and mix the soil with seed at the time of sowing and the second way is to use the Rhizobium culture and treat the seed with it. Now-a-days bacterial cultures are easily available at reasonable prices; hence it is good to use the culture.
   To treat the seed with culture it is advisable to moisten the seed with little water and pour the culture in the seed container after which it should be stirred vigorously for proper mixing. When the culture is coated over the seeds it should be spread under shade for 30-45 minutes and then sown in nicely prepared field. Jeggary solution may also be used but there is not much need for it.
4. **Spacing:** Usually *kharif* crop is given wider spacing but that of spring is given slightly closer spacing. A spacing of 45-60 cm x 2.5 cm is good for *kharif* crop and 30-45 x 2.5 cm during spring season is recommended. Thinning or gap filling, whichever is required, should be done soon after germination so that there is not much difference in plant growth and entire plant population is uniform.
5. **Seeding depth:** It depends upon factors viz., texture of the soil, moisture

percentage and level of compaction provided after sowing. Under normal conditions, a depth of 2 to 3 cm in heavy soils and 3-4 cm in light soils should be given. Planking should be done after sowing to cover the seed and provide better soil contact.

6. **Seed Rate:** It is governed by numerous factors such as seed size, seed weight, spacing, cropping season, germination and purity percentages of the seed, method of sowing etc. Soybean grown for grain purposes needs about 20-30 kg seed/ha but the fodder crop needs about 70-75 kg/ha during *kharif* season and 100-120 kg/ha during spring or summer season which is because of closer spacing.
   The seed rate is much affected by the nature of the crop viz, sole crop needs more seed rate, intercropping with maize, jowar, bajra and arhar either in alternate rows or two rows of soybean between two rows of main crop and this way 60 per cent of the standard seed rate is needed. An ideal sole crop should give about 5 lakh plants/ha.

Varieties

The crop is an introduction from USA into India and based on growing season they may be divided into following groups:

- ***Kharif* varieties:** Bragg, Lee, Harde, Hood, Sams, Clerk-63, Punjab 1 and Gaurav, etc.
- **Spring varieties:** Hark, Herosay, Merit, Clerk-63, Ankur, Punjab-1, Black Hark, etc.
- **Local improved:** For plains – T 1, T 49; For hills – T 33 and T 1.

Irrigation

In case of *kharif* crop irrigation is not needed and it grown rainfed provided rain are well distributed, otherwise, the crop needs water at critical growth stages. However, during summer the crop can be grown only under assured irrigation and it needs about 5-6 irrigations. The crop should be irrigated at the following critical growth stages to minimize water:

1. Sprouting stage,
2. Flowering, pod initiation and bean filling stages are important stages from yield point of view. It is observed that maximum reduction in yield, due to moisture stress, occurred during the last weed of pod development and filling stage. Studies on this aspect further indicate that water stress at the end of bean filling stage or at beginning of flowering has only small effects on yield but stress during flowering and pod development appeared to be responsible for flower and pod abortion. Stress during bean filling reduced side size

Therefore, it is essential to apply water to the field for better germination and then to irrigate the crop between pod development and filling stages for obtaining higher grain yield.

Weed Control

Presence of weeds in the field causes a drastic reduction in the soybean yields ranging between 6-27 per cent depending upon weed density and their composition/unit area, therefore, it is essential to create a weed free situation upto 40 days of sowing. Adoption of an integrated weed control approach will help in keeping down the weeds. It includes the preventive as well as curative measures.

Preventive measures to control weeds: For an efficient weed management it is essential to start with a weed free crop which is possible by adopting the following preventive measures:

1. Acquiring weed free seeds from some certified seed agency.
2. Cleaning of uncertified or home produce seeds before sowing.
3. Picking up all rhizomes, bulbs, tubers and stubbles of weeds and burning them after they are collected.
4. Cleaning of all agricultural implements and tools before and after their use.
5. Existing weeds, if any, should be uprooted before they have set into seeds.
6. Proper crop rotation should be followed.
7. Line sowing and wider spacing should be adopted to allow inter cultivation.
8. Fertilizers should be placed in crop rows rather than broadcasting.

Control measures: Normally preventive measures are enough to check the weeds infestation in the field but if the weeds are still there then the crop should be given weeding, hoeing or even chemical treatment. One weeding 20-25 days and another 40-45 days after sowing keep the field free from weeds and later the crop itself smother the weeds.

Chemical control of weeds: Sometimes weeding becomes difficult due to unwanted weather, wet soil, unavailability of labour, etc. and the use of herbicides becomes unavoidable. The herbicide application may be divided into following groups:

a) **Pre-planting treatment**: Trifluralin and nitralin are applied as pre-planting incorporated herbicides but they appear to inhibit cell division in roots and occasionally injure the root system and reduce nodule count/unit length of roots. A quantity of 0.6 kg/ha of Trifluralin and 0.8 kg/ha Nitralin are found to be safer doses and are also effective in controlling the weeds. Vernolate is most selective chemical for all types of weeds and a dose of 2.2 kg/ha has been recommended.

b) **Pre-emergence treatment:** Chloramben at the rate of 2.2 to 3.4 kg/ha has been found to be very effective in controlling most of weeds. Linuron is another herbicide which may be used at the rate of 1.0 to 3.4 kg/ha. Other pre-emergence herbicides are Fluorodifen (3.4 to 5.0 kg/ha), Metribuzin (2.0 to 2.5 kg/ha) which may be used for good results.

c) **Post-emergence treatment:** Failure of pre-emergence herbicides or their effectiveness only for a short period makes it necessary to apply some post emergence herbicides like 2, 4-DBat the rate of 0.2 kg/ha, Chloroxuron at the rate of 1.1 to 1.7 kg/ha, Dinoseb at the rate of 1.7 kg/ha, etc.

Plant Protection

Insects: The crop is attacked by several insects such as hairy caterpillar, flea-beetle, jassid and white fly. Of these the flea-beetle and caterpillar can be controlled by spraying of 1.25 kg/ha of Savin/Hexavin or 300 ml of Folithion. But jassids and white fly may be controlled by spraying of 700-800 ml of Malathion or 600-700 ml of Formothion/ha. The pea stem fly or stem borer can be controlled by soil application of Thimet 10 G at the rate of 10 kg/ha Disyston 5G at the rate of 20 kg/ha at sowing time.

Diseases: Yellow mosaic and rust are the most common diseases of soybean. A mixture of 0.1 per cent Metasystox and 0.1 per cent Thiodan at the rate of 400 ml and 100 ml respectively per 100 litres of water/ha sprayed at 20, 30, 40 and 50 days after sowing controls the mosaic completely. Diathane Z-78 at the rate of 2.5 kg/ha when sprayed at 60, 70 and 80 days after sowing, takes care of the problem of rust in the crop.

Harvesting and Threshing

Early varieties usually take about 90-120 days and late ones take about 140 to 150 days time to mature. The crop should be harvested after it has fully matured *i.e.*, when the leaves have become yellow and start drooping. The pods start drying and they break open when pressed. Usually shattering of grains from pods is not common in most of the commercial varieties but the harvesting should not be delayed. The matured crop is harvested manually with the help of sickles. The fodder crop of soybean is harvested even at flowering but the quality of fodder remains inferior.

The harvested crop is kept on threshing floor where it is dried foe a few days and when the crop is fully dried, it is threshed by trampling with tractor or bullocks Grains are separated by winnowing.

Storage

Soybean are not very much damaged as in case of other pulses like arhar, gram, pea, mung, urid, etc. but still all the storage precautions must be taken. The seeds are dried to have about 11 per cent moisture and then they are poured in clean and pre-fumigated bins or stores. In the case, the grain is to be stored in gunny bags, and then the bags are well cleaned, treated with insecticides and dried before filling of the seeds. Soybean, to be stored for seed purposes, should be treated with Thiramin or Captan at the rate of 3 to 4.5 g/kg of seed and the treated seed should be dried and stored but it should never be used for consumption purposes.

SESAME (*SESAMUM INDICUM*)

Sesame or sesamum is cultivated extensively in America, Asia, parts of China and Latin America. It is cultivated in an area of 6.7 m ha with a production of 2.5 m t. Sesame is an important oilseed crop in India occupying about 2.4 m ha with a production of 4.8 lakh tones. India is the largest producer of sesame in the world. Major sesame growing states in India are Uttar Pradesh, Rajasthan, Madhya Pradesh, Maharashtra, Andhra Pradesh, Orissa, Tamil Nadu, Gujarat and Karnataka. Sesame can be cultivated on a wide range of soils. The crop thrives best in medium textured soils with good drainage. Sesame is grown in all the three seasons viz., *kharif*, rabi and summer. It is mainly grown in kharif as a rainfed crop. The seed contains all essential amino acids and fatty acids. It is a good source of vitamins (pentothenic acid and vitamin E). The seeds also contain important minerals such as calcium (1450 mg/100 g) and phosphorus (570 mg/100 g). Sesame oil is mainly used (about 80%) as a cooking oil, particularly in south India. The oil cake is used as a valuable and nutritious feed for milch cattle.

The sesamum (Til) seeds are rich in edible oil (46 to 52 per cent) and they make good sweetmeat when fried and mixed with sugar. After extraction of oil the sesame cake makes very good feed as it is a rich source of protein, carbohydrate, and minerals like calcium and phosphorus. Pure cake is also consumed by human being. The oil is used as coking oil, for anointing the body and for manufacturing perfumed hair oils and medical purposes.

Land Preparation

Sesame is grown as a rainfed crop on high landsand as a pre-rice crop in low lands. Well drained soils are suitable for growing sesame (Til). The land is prepared to fine tilth and applied compost or FYM at the rate of 5 t/ha. Excess water has to be drained from the field as sesame can not stand water logging.

Sowing

Seeds are small and should be sown shallow, within a depth of 2 cm. Deep sowing may affect germination. The seeds with fine soils should be mixed when the crop is sown by broadcasting to ensure uniform sowing. The seeds are sown in lines 30 cm apart and the thinning of plants to 10 cm between plants is done. Line sowing permits easy interculture and weed control. In line sown crop the plants come up uniformly. The seeds at the rate of 7.0 kg/ha is used in case of drilling in lines. The seeds are treated with Thiram at the rate of 3 g/kg of seed before sowing.

Varieties

S-14 (Vinayak), Kalika (BM-3-7), Kanak (BS-6-1) and Krishna (RAUSS 17-4) are recommended for cultivation in kharif and summer season.

Variety	Duration (Days)	Seed colour	Oil %	Grain yield (kg/Ha)	Plant height (cm)	General remarks
Vinayak	85-90	brown	49	600	105	glabrous
Kalika	85-90	brown	49	700	90	glabrous
Kanak	85-90	gray	47	700	95	pods pubescent
Krisna	80-85	Faint black	48	700	105	-

Fertilizer

30 kg N, 20 kg P_2O_5 and 20 kg K_2O per hectare are recommended. Sesame is an exhaustive crop and responds to fertilizers. All the P_2O_5 and K_2O and half Nitrogen should be applied as basal before sowing and the remaining nitrogen at first hoeing and weeding.

Interculture

First weeding, thinning and hoeing should be done at 15 days stage. Thinning should be done to keep plants around 10 cm apart.

Plant Protection

Leaf and pod caterpillar is a serious pest. Phyllody disease is also common. The plant protection measures should be adopted to check the pests and diseases as per the recommendation of the state. Application of excessive nitrogen should be avoided which encourages phyllody disease.

Harvesting

The crop is harvested when the plants start yellowing and drying of capsules commence. This will take place at about 3 months stage. The plants are harvested and stacked them for a week. The plants should be dried keeping the capsules towards the sun and basal portion at the bottom. Threshing may be done after 3 to 4 days of drying.

SUNFLOWER (*HALIANTHUS ANNUUS*)

Sunflower is an important oilseed crop after groundnut, rapeseed-mustard and soybean. The crop is grown in an area of 20.9 m ha with the production of 25.1 m t in the world. Sunflower is predominantly grown in Russia, Argentina and Turkey. It was introduced in India during 1965 as an edible oilseed crop. Although commercial production of sunflower began in 1972, it had emerged as an important oilseed crop only in the 1980s. The crop was cultivated in an area of 0.1 m ha with the annual production of 0.07 m t in 1980-81. However, the area and production has increased significantly. The area increased to 2.1 m ha and the production level to 1.2 m t. The average yield is around 550-600 kg/ha compared to global average of 1250 kg/ha. Sunflower is mainly grown in the states of Karnataka, Maharashtra and Andhra Pradesh. Now this crop has become an important one in Punjab, Haryana and U.P. The yield is the highest in Haryana (1750 kg/ha) and Punjab (1800 kg/ha). The crop comes up well in any type of soil. Since, it is not season-bound; it can be grown through out the year. The crop generally requires 90-100 days to mature but some new varieties come to harvest within 75 days.

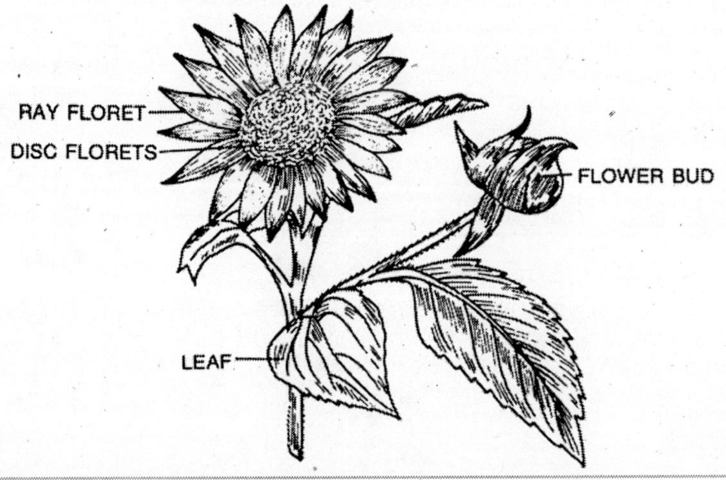

FIG. 10.3: Stalk of Sunflower (*Helianthus annus*).

Sunflower oil has an ideal combination of saturated and polyunsaturated fatty acids. The oil is rich in linoleic fatty acid (66.0%) and hence preferred for the reduction of high serum cholesterol levels. The oil is used for making margarine. The oil cake contains higher amount of protein (40-44%) and balanced amino acids and hence used as cattle and poultry feed.

Ecological Requirements and Land Preparation

Generally, sunflower is grown as a rainfed crop in *kharif*, hence to make use of rain water, it is necessary to plough the land once by a mould board plough followed by harrowing soon after the onset of the rains. The crop prefers deep fertile soil adequately supplied with moisture without producing water logging condition that adversely affects the germination of seed.

Fertilizer

After leveling the land, fertilizer at the rate of 60-40-20 kg/ha N P K for low fertility and 40-30-20 kg/ha N P K for high fertility soils should be applied and mixed properly with the soil. The Indian Agricultural Research Institute however, recommends 40-60-40 kg/ha NPK. To increase the efficiency of nitrogen, 60% of nitrogen may be applied as basal and the remaining N top dressed 40 days after sowing. Increase in the yield and oil content have been reported by application of gypsum and sulphur especially in saline and alkaline soils.

Seed and Sowing

Seeds having over 70% germination and 80 g test weight (wt of 1000 seeds) should be sown at 8-10 kg/ha rate in rows 45-60 cm apart. The spacing between the plants (within the rows) should be around 20-25 cm. In order to have a stand of 80,000 healthy plants/ha thinning or gap filling is required during early stage of crop growth.

It is of utmost importance that all weeds be removed from the field once after 15 days of sowing and again 10-15 days later as during the first few weeks after germination weed competition poses serious problems and reported to cause quantitative and qualitative loss in yield.

Irrigation

The crop being drought resistant, may very well be grown under rainfed conditions. However, under irrigated conditions one or two irrigations are required to increase the yield, although germination, flowering and dough stages are the critical for irrigation.

Varieties

Mordern, EC 6841, EC 68415, CO 1, CO 2, GAU-SUF 15.

Hybrids: BSH 1, KBSH -1, KBFH 8, MSFH 17, Jwala Mukhi.

The following varieties may be taken according to the season and region mentioned against each:

Season	Region	Varieties
Kharif	Central region	
(Sowing-Jan-July)	Maharashtra, Gujarat,	Viniimk or Perrdovik
(Harvest- Oct-Nov.)	Rajasthan, Bundelkhand	Armavirskil
Rabi	South Maharashtra	- do –
(Sowing- Nov.-Feb)		
(Irrigation if needed)	Ahmedabad, UP (Tarai)	Viniik
	West Benal	Perrdovik
	Coimbatore, A.P.	Armavirskij and
	Tamil Nadu	Armaverts
Summer (Irrigated)	Gangetic Plains of Central and West U.P.	Armaveris, BC (101495 Cernianka)

Plant Protection

The crop does not have serious insect-pest and disease problems at present. However, according to a report from Coimbatore Centre, the attack of grass hoppers at seedling stage and green bug at flowering stage are serious. They can be controlled by dusting 10% BHC at the rate of 30 kg/ha. Use of 10% sevin is recommended where leaf eating caterpillar are noticed Sunflower is essentially a cross pollinated crop. In view of this, selection of suitable insecticide that is not harmful to pollinators needs to be made through investigations. Birds are a menance during seed formation stage and constant vigilance to scare them away is necessary.

Harvesting, Yield and Storage

At maturity the back of the floral heads and outer bracts turn yellow and into brown colour respectively. The heads harvested from the plants and seeds are separated. Seeds, when harvested at 8-10% moisture level, have better keeping quality and higher oil content. The stalks may be used as fuel or for making high quality compost. One hectare should yield 20-25 quintals sunflower seed. The seeds should be bagged at 6% moisture level for better keeping quality and germination.

CASTOR (*RESINUS COMMUNIS*)

Important countries where castor is cultivated are Brazil, India, Russia, Latin America and the Far East. Castor is cultivated in an area of 1.6 m ha with an annual production of 1.4 m t in the world. In India, it is grown in Andhra Pradesh, Gujarat, Tamil Nadu, Karnataka and Orissa, according for about 90% of the area and production. The crop is mainly cultivated as a rainfed annual crop. It is also raised in the bunds and irrigation channels.

The oil content in the seed varies from 35-58% with a average of about 47%. The oil is used for varied purposes such as lubricant in high speed engines and aeroplanes, manufacture of soaps, transparent paper, printing inks, varnishes, linoleum and plasticizers, etc. The cake is used as a manure. The leaves are fed to eri silk worms in the areas of eri silk production.

Land Preparation

The ploughing is done to get a fine tilth. Well drained soils are suitable. Compost or FYM at the rate of 5 t/ha may be applied before final land preparation. The drainage channels should be provided in between strips to drain out excess water.

Seed and Sowing

Sowing of seeds should be done in lines 60 cm apart behind the plough with a spacing of 45 cm between plant to plant. The seed has a hard outer coat. Pre-soaking of seeds for 12 hrs is advisable if field moisture level is not sufficient for germination. Two seeds per hill should be sown.

Fertilizer

Phosphates and potash requirements of the crop are high. Application of 50 kg N, 75 kg P_2O_5 and 50 kg K_2O is required per hectare. Half of the nitrogen, all the phosphorus and potash can be applied as basal in the furrows. The balance nitrogen should be applied just before second hoeing so that the fertilizer is worked in to the soil.

Varieties

- **Varieties:** TMV 3, TMV 5, Aruna, Bhagya, Sowbhagya, SA 1, GAUC 1, GAUC-HI, GCH 3, RC-8, SHB 18, HC 6, H 8.

- **Hybrid:** GAUCH 1, GCU 5, DCH 30.

N.P.H. 1 (Aruna) and SA 2 are the recommended varieties in the State of Orissa. NPH 1 and SA 2 take about 110 days from seed to seed. The oil content varies from 50% to 52%. The former can stand clipping which encourages more branching. The latter is suitable for dry land cropping and can withstand drought for 14 to 20 days.

Interculture

Thinning to leave one healthy plant per hill and gap filling at 10 days stage should be taken up. Hoeing and weeding should be done at 3 weeks stage followed by earthing. A second hoeing and earthing should be given at 6 weeks stage which will provide better anchorage to the crop and facilitate easy drainage.

Plant Protection

Plant protection measures should be taken up as per the recommendation of the state when the diseases and pests will occur.

Harvesting

The racemes should be picked up when drying starts. They should be dried in the sun. After threshing the seeds are collected.

Linseed

Linseed is an old world crop which used to be grown both for fibre and oil extraction. The Egyptian records mention its cultivation from about 5000 years back to the Bronze age. Vavilov mentioned linseed and flax to be originated from separate places and according to him the oilseed type of linseed was probably native of south western Asia consisting of India, Afghanistan and Turkey, whereas the fibre type *i.e.* flax was originated somewhere in Mediterranean region, including Asia Minor, Egypt, Algeria, Spain, Italy and Greece.

Linseed is mostly grown in the countries of temperate and tropical zones. It is cultivated in an area of 2.8 m ha with the annual production of 2.3 m t. The average yield of linseed is 838 kg/ha. Argentina, Russia, India, USA, Canada, Pakistan and Australia are the major linseed growing countries in the world. India ranks first in the world acreage, however, the productivity is the lowest (300 kg/ha). The crop is extensively cultivated in Madhya Pradesh, Uttar Pradesh, Maharashtra and Bihar. About 60% of the acreage is from Madhya Pradesh and sub-marginal rainfed lands. The crop is cultivated in *kharif* fallow as well as in a very large areas as 'utera' or 'paira' in rice fields.

The oil content is in the range of 40-45%. The actual yield fluctuates 3-15 q/ha. All parts of the crop are used for commercial purpose either directly or after processing. Oil is used for edible purpose and about 80% of the linseed oil is utilized for industrial purposes such as in manufacturing of paints, varnish, oil cloth, linoleum, pad-ink, printing ink and medicinal purpose. The oil cake is a good cattle feed especially for the milch animals. The stem gives a good quality fibre.

Soil and Land Preparation

The crop may be grown on a variety of soil types; however, it does very well on deep cotton soils of central India and alluvial loam soils of North India. The soil must be well drained and nearly free from soluble salts, though it may tolerate moderate acidity and salinity.

Linseed needs a weed free and fine textured seed bed. Termites and cutworms usually attack the crop, therefore, BHC 10% or Aldrin or Chlordane 5% dust should be mixed in the soil at the rate of 25 to 30 kg/ha at the time of last ploughing.

Seed and Sowing

In rainfed areas the crop is sown earlier in the last week of September or mid of October when the rains have stopped so that best use of residual moisture of rains can be made but in case of irrigated areas the crop may be sown little later in October-November by giving one light pre-sowing irrigation for ensured moisture supply to germinating seeds. In Bihar and eastern parts of U.P. the linseed is sown broadcast in standing rice crop as risky crop during September-October.

Broadcasting of the seeds should be avoided and it should be sown in lines either behind the country ploughs in shallow furrows or by the seed drill. Sometimes it is also sown broadcast in the case of relay cropping in standing rice crop but in the case of mixed cropping of linseed with wheat, gram and barley it is sown in rows after every few lines of aforesaid crops. In case of pure crop of linseed a row to row spacing 20-30 cm is given and a seed rate of 20-30 kg/ha in case of line sowing and 35-40 kg/ha in case of broadcasting is required.

Types of Linseed

There are two main types of linseed namely peninsular varieties and alluvial varieties. The former has deep root system, bold seeds with a few branches. These types are highly suitable for the areas where the surface soil moisture dries up quickly. Rather, the alluvial types have shallow root system having small seeds and many branches. They are best suited for the Gangetic alluvium where the upper layers of the soil are constantly moistened by the rise of underground water. These types give better seed production compared to the peninsular types. If these two types are grown at either condition, they fail to give better performance.

Varieties: K2, Jawahar-7, Jawahar-17, Jawahar-552, Hira, Mukta, Neelam, BS 44, Gaurav (LCK 152), Jeewan (DPL 21), Kiran (RLC 6)Seed and fibre type: Gaurav, Jeevan, LC 8528, Nagarkot. 'Utera' (rice fallow): Jawahar 7, LC 185, Surabhi.

Fertilizer Application

Normally the crop is grown rainfed under unfertilized condition of residual fertility or under very poor fertile doses. The experimental findings indicate that the application of fertilizers in linseed crop can increase the yield significantly. In case of rainfed crop the placement of 20-30 kg each of nitrogen and phosphate but in case of irrigated crop a dose of 30-40 kg of each of nitrogen and phosphate per hectare had been recommended. When linseed is to be sown as relay cropping in standing rice crop, top dressing with only 10-15 kg nitrogen per hectare would be enough. In irrigated condition, top dressing with half of the required nitrogen is more beneficial.

Weed Control

The crop suffers from a very severe weed competition upto 25 days of sowing and crop growth is badly suppressed by the presence of weeds which reduces the crop yield, therefore, it is essential to provide two hand weedings- once after 21 days and second after 35-40 days of sowing.

Plant Protection

Linseed crop gets attacked with gall midge and a severe damage to the crop is observed. The insects may be controlled by spaying of any insecticide like Metasystox, Dimecron or rogor at 0.3% concentration. The spraying should be done two-three times at 10 days interval starting from appearance of the insect in the crop.

Amongst the prevalent diseases wilt and rust are most commonly occurring ones for which resistant varieties should be selected and grown. Early sowing of short duration varieties can reduce the incidence of insect-pests and diseases and good crop yield can be obtained there from.

Harvesting, Yield and Storage

The crop should be harvested at red ripe stage but when fibre and grains both are to be taken from the same crop then harvesting at physiological maturity or pod maturity should be done when the plants are little green. The crop is traditionally harvested by the help of sickles and the threshing is done by beating the plants with wooden mallets or by trampling them under bullock feet. One hectare crop usually produces about 4-5 quintals under mixed cropping but the yield is about 10-12 quintals/ha when grown as pure crop.

The seeds posses about 36-42 per cent oil in them and they may be stored at moisture of 10-12% in a moisture proof store.

NIGER (*GUIZOTIA ABYSSINICA* L)

The crop is grown in many parts of the tropical and temperate zones. Important niger growing countries are India, Germany, East Africa, West Indies and Zimbabwe. In India, niger is cultivated in the states of Madhya Pradesh, Orissa, Maharashtra, Bihar, Karnataka and Andhra Pradesh. The crop occupies an area of about 6.0 lakh ha with an annual production of 1.5 lakh tones. Average yield of the crop is 240 kg/ha. The crop is mainly grown in *kharif* season as a rainfed crop. It is also cultivated in *rabi* season. The oil content varies from 37 to 43 per cent. The oil is used for edible purposes and for the manufacturing of paints, soft soaps, cosmetics, etc. The cake is a valuable feed for cattle.

Varieties

Alasi-1 is the important variety which is suitable for sowing early in July as border crop and also for late sowing in August to early October. Oil content of this variety ranges from 42 to 43 per cent. There are a number of local varieties namely Keonjhar-1, Bada Rasi, Pala Rasi and Kuji Rasi which can also be grown in different parts of the state of Orissa. Better yield varieties like IGP-72, IGP-76 and GA-10 have been identified in the AICRP on Niger.

Land Preparation

The land has to be prepared to a fine tilth for sowing niger. The field should be completely free from weeds. The land where niger is to be sown should be perfectly well drained.

Seed and Sowing

The seeds should be treated with Thiram, Captan or Thiride at the rate of 3 g/kg of seed against seed borne diseases. The seeds should be selected from the fields free of Cascuta weeds. Niger should be sown in July as a border crop around upland kharif crops as cattle do not relish this crop. The main crop is sown from middle of August to early October. Sowing of seeds should be done in lines, 25-30 cm apart with a seed rate of 10 kg per hectare.

Fertilizers

FYM or compost at the rate of 5 tonnes per hectare should be applied. Application of 20 kg of nitrogen and 25 kg of P_2O_5 per hectare is desirable which will give good yields. All the nitrogen and P_2O_5 should be applied at final land preparation before sowing. For improved varieties fertilizer dose 40:20:20 kg/ha of N, P_2O_5 and K_2O respectively, is recommended. If basal application has not been given, top dressing of Nitrogen at the rate of 10 kg/ha should be applied.

Interculture

Inter-culture of line sown crop should be done within 3 weeks of sowing. Early control of weeds is essential for successful harvest. A second weeding may be taken up wherever, necessary.

Plant Protection

Cuscuta parasite is a major problem in Karaput district and partly Keonjhar as well as Mayurbhanj. Application of Furloe-20 G (C.I.P.C.) at 2 kg per hectare five days after inintiation of germination controls the parasite. Alternatively, 2 kg active ingredient of Kerb 50 W in 1000to 1100 litres of water per hectare sprayed 15 days after germination will control the parasite. Use of clean seed and crop rotation will reduce the coscuta parasite. Growing niger in the same plot every year should be avoided.

Cercospora leaf spot is the major disease of niger. The crop should be sprayed with Dithane Z-78 at the rate of 2.5 kg/ha or Dithane M-45 at the rate of 1.9 kg/ha to control the disease.

Harvesting, Yield and Storage

The crop should be harvested in proper time like other oil seed crops and threshed. Then the seeds should be stored properly.

SAFFLOWER (*CARTHAMUS TINCTORIUS*)

Safflower is grown both in tropical and temperate regions. The crop is cultivated to an extent of 1.05 m ha with a production of 0.8 m t in the world. The average yield of safflower is 744 kg/ha. Safflower is grown in Ethiopia, USA, Mexico, Argentina, Australia, Russia and India. Safflower is a minor oilseed crop in India since its cultivation is restricted to Maharashtra, Karnataka, parts of Andhra Pradesh, Orissa and Tamil Nadu. India is the largest producer of safflower which is cultivated in an area of 0.9 m ha with an annual production of 0.4-5.0 m t. The average productivity is 0.5 t/ha. The crop possesses deep and efficient root system and hence utilizes the soil moisture efficiently. The crop is sown during *rabi* season and mostly cultivated as rainfed crop. The crop, being spiny, does not require protection from the stray cattle and birds. The oil content of whole seed ranges 28 to 32 per cent. Safflower oil is highly nutritious rich in essential fatty acid (linoleic acid). The oil is used for edible purpose and highly useful in checking cholesterol level. The oil is also used for various industrial purposes. A number of products could be developed from safflower such as carthamin pigments. This could be used as food additive, for making a type of tea and extracting protein and amino acids.

Varieties: Annigeri-1, K-1, S-144, Nag-7, Tara, Manjira, CO-1, Sharda, Girna, Bhima.

Minor Oilseed Crops

The minor oilseeds of tree and forest origin are available in abundance and offer a good scope to be exploited commercially. It has been estimated that 11.7 m t of minor oilseeds of identified plant species can be collected annually. This will result in the production of 1.7 m t of oil. As many as 100 different oil bearing plants have been identified in the forests of India, out of which only a dozen species have been exploited so far. They not only posses oil but also are well adapted to growing in varied types of waste lands. Important oil bearing tree species and their oil content in the kernel/seed are given below.

Minor oilseeds and their oil content in the seed.

Name	Oil content (5)
Kusum (Schelechera oleosa)	57-60
Mahua (Madhuca indica)	33-43
Neem (Azadiracta indica)	40-50
Persian lilac (*Melia azedarch*)	25-30
Pongam or Karanj (*Pongamia glabra*)	27-30
Sal (Shorea robusta)	12-13

There are a number of other species that they also have good sources of oil. They are Citrulus colosynthis, Jatropha curcas, Pricepia utilis, Basia butrycea, Salvadora spp, Simmondra chinersis, Simarauba glauca, Cuphea spp and Cucurbita foetidissima.

CHAPTER 11

Production Technology of Crop: *Fibre Crops*

(Ecological Requirements, Cultural Practices and Crop Production Management)

COTTON (*GOSSYPIUM* SPP.)

The important fibre crops are cotton, jute, mesta. Other fibre crops include sisal, hemp, flax, roselle. The cotton is the most important vegetable fibre cultivated in about sixty countries in the world. The countries such as Russia, USA, China, India, Brazil, Pakistan, Turkey, Egypt, Mexico and Sudan account for about 85 per cent of the total cotton production. Cotton is cultivated in an area of 34.2 m ha with the total production of 54.1 m t in the world. Cotton is one of the principal commercial crops in India. About 60 million people depend on cotton cultivation and allied enterprises. About 20 per cent of the entire industrial output in the country is from the textile industry. Currently there are around 3.6 million handlooms and 1.1 million power looms which provide direct employment to about 14 million families. The organized textile industry with over 1500 mills is the second biggest in the world. The export materials from cotton include raw cotton yarn, textile finished products including garments and by-products of seed cotton. Foreign exchange worth of about 60 billion rupees is earned annually, accounting for about 25% of the countries' total exchange earnings.

India has the largest cotton area (9.1 m ha) with about 27 per cent of world cotton acreage and 14 per cent of world cotton production. India is the third largest cotton producer, next to USA and China. India consumes 14 per cent of world cotton output and is the second important consumer of cotton next to china. During the last 35 years, cotton are remained static, however, production has increased from 5.6 million bales to 13.1 bales (1 bale = 170 kg) in 1995-96. The yield level also doubled during the last three decades (125 to 246 kg/ha), but still the average productivity is the lowest among the major cotton growing countries (world average is 583 kg/ha).

Out of the world extra long staple cotton (ELS) production of 5.8 lakh tones, India is ranked third important producer next to the USA. But in the world ELS cotton export, the country shared very negligible quantity. Cotton is grown in different states and the major cotton growing states are Maharashtra, Haryana, Madhya Pradesh, Rajasthan and Tamil Nadu. These states contribute to about 99 per cent area and production. Among the states, yield of cotton is the highest in Punjab (442 kg/ha) (Table 11.1).

TABLE 11.1: Important states growing cotton in India, 1995=96

State	Area ('000 ha)	Production ('000 bales)	Yield (kg/ha)
Andhra Pradesh	1056.8	1626.6	262
Gujarat	1410.4	2201.4	265
Haryana	646.0	1283.0	338

State	Area ('000 ha)	Production ('000 bales)	Yield (kg/ha)
Karnataka	673.8	849.1	214
Madhya Pradesh	537.0	422.4	134
Maharashtra	3065.4	2799.2	155
Punjab	750.0	1950.0	442
Rajasthan	606.2	1338.3	375
Tamil Nadu	265.4	546.8	350
All India	**9063.4**	**13093.3**	**246**

Source: Fertilizer statistics, 1996-97.

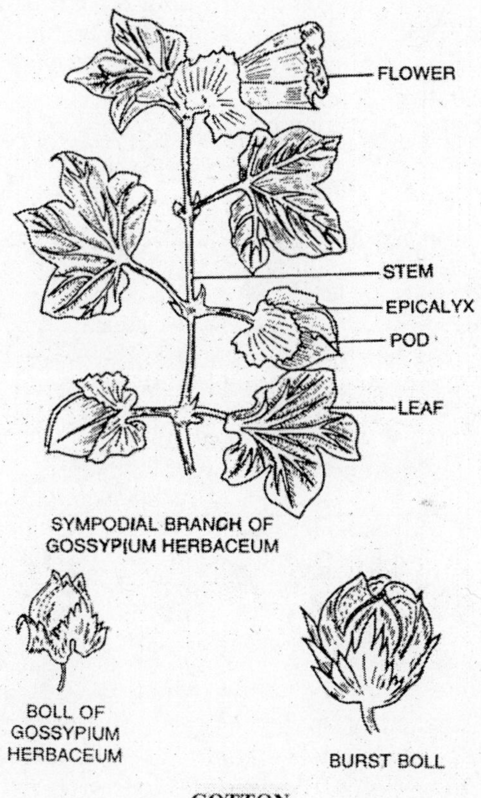

FIG. 11.1: Cotton.

Production Constraints

The major constrains in cotton production are the vagaries of monsoon battering a major parts of the cotton growing area. Cotton is cultivated on marginal lands with low input use. The pest incidence is high, though cotton occupies only about 5% of total cropped area but it consumes nearly 55% of the pesticide used in our country. Adverse factors that affect cotton productivity in the Northern zone are raising water table, salinity, high temperature in the early phases, pest and diseases problems like Helicoverpa. However, pink bollworm, whitefly, root rot and the emerging problem of leaf-curl virus are also causing great concern. In the Southern zone, vagaries of monsoon and extensive damage due to pests are the real problems.

Production Technology

Climatic requirements: Cotton is a crop of sub=tropical climate. It is uncommonly sensitive to vagaries of weather. Continuous rains or a long spell of dry weather may prevent the sowing of the crop at the proper time, hinder seed germination or retard crop growth. Heavy rains associated with high winds or excessive drought at flowering and fruiting times may cause heavy shading of buds and young bolls, thereby ruining the crop. An early frost may kill the plant prematurely. Untimely rains and a heavy humid weather during later stages of cotton season may spoil the produce, lower its ginning properties or promote the attack of insect, pest and diseases.

Cotton needs, on an average, a minimum temperature of 60°F at germination, 70-80°F for vegetative growth, 80-90°F with cool nights during fruiting period.

An annual rainfall of at least 50 cm (well distributed) is the minimum requirement for cotton unless it is grown on irrigated soils. Rain in excess of 125 cm per annum needs proper drainage facilities. It is necessary that the weather should be clear at harvesting or maturity because rain at the time will discolour the lint and thereby reduce its quality.

Soils: Cotton needs a soil with an excellent water holding capacity, aeration and good drainage as it can not withstand excessive moisture and water logging. Under conditions of moderate rainfall cotton is grown on retentive clay loams and with irrigation facilities it may be cultivated even on sandy loam to loam soils. The major groups of soil on which cotton is grown are the alluvial soils, black cotton soils, red sandy loam and red laterite soils. Alkaline soil is not good for cotton cultivation but a soil with 5 to 8.5 pH can be suitably used for the crop cultivation. The soils well supplied with organic matter or humus are the most productive soils. The lighter soils need liberal application of balanced manure and fertilizers for profitable crop cultivation and on heavy soils the crop does not need only fertilizers but also adequate organic matter to maintain optimum drainage and aeration conditions.

Land Preparation

For conserving soil moisture and securing good germination of seed and optimum growth of plants, proper preparation of land is necessary. The preparary tillage varies with the crop, the nature of the soil, previous crop grown, the season of growing, the system of cultivation and weather conditions.

In black cotton soil areas 3-4 summer ploughing are given to keep down the weeds and after onset of monsoon the land is prepared by **Bhakar**.

In alluvial soils, if the crop is grown under irrigated condition, one light irrigation is given and the land is prepared but if it is grown under rainfed condition the preparation starts only after onset of monsoon. Generally, the land is prepared by giving one ploughing with mould board plough followed by two harrowing so that the weeds are controlled, soil moisture is well preserved in the surface layers for first few weeks after sowing and a good surface drainage is provided to prevent water logging in the field.

Fertilizers

It has been estimated that an average rainfed cotton crop of one hectare area takes away from the soil about 23 kg N, 18 kg P_2O_5 and 78 kg K_2O, while in case of irrigated crops, the uptake is two or three times more. It has been observed that the soil rich in organic matter content (being between 1-5%) yields more cotton. Therefore, addition of organic manures such as compost/ farm yard manures/oilseed cakes (sufficient to supply half of the requirement N) and green manuring or heavy does of P applied to Berseem in the field results in a higher yield of cotton. About the mode of application it has been observed that basal fertilization with full dose of P and K followed by N half just after thinning and rest half near flowering phase of the plants has given the best results.

The rate of application depends up on the type of soil, type of the cotton to be grown, crop rotation, irrigation facilities, etc. However, a general rate of fertilizer application has been described in the table given below.

The micronutrient deficiency has not been noticed but it is always better to get the soil analysed and if needed their application should be made. In case of alkali soils the crop must be supplied with Zn and S once after 3 years for better production of cotton.

In case of rainfed crop the organic manures are applied at the above mentioned rates but inorganics are applied at half of the above mentioned doses.

Zone-wise fertilizer recommendation for cotton crop

Region	Type of cotton grown	Rate of application				Time of application	Source of N	Remarks
		Org. m. (t/ha)	N	P_2O_5	K_2O			
			(kg/ha)					
Northern region	Hirsutum and Arboreum	4-6	60	30	30	½ N at thinning ½ N at flowering	A/S or Urea	Addition of organic manures is needed after every 6 years of its application.
Central region	*Herbaceum*, *Arboretum* and *Hirsutum*	4-6	80	25	30	½ N at sowing ½ N at 6-8 was*		
Southern region	*Hirsutum* and *Arboreum*	4-6	56	20	20	½ N at thinning ½ N at flowering		
Western region	*Herbaceoum*	4-6	50	20	20	-do-		
Eastern region		4-6	40	20	20	-do-		

WAS*: weeks after sowing.

Seed and Sowing

Time of sowing: Sowing of cotton after 15th May results in a lower crop yield, but sowing of the crop depends upon water resources. If irrigation facilities are available the crop should be sown in April, especially American cottons, otherwise it must be sown just after the monsoon starts, but not later than 15th

July.

Method of Sowing: The most commonly adopted sowing method by the farmers is to broadcast the seeds and plough and plank the field after sowing. But this method creates problem in controlling weeds and inter-cultivation in the cotton field. The other method is to sow the crop in suitably spaced rows (either skip/ double rows or single rows). Sowing in rows can be done either by dibbling, drilling or placing the seeds in furrows behind the country plough. Sowing in rows facilitates farmers in regulating seed rate, sowing depth, plant population etc.

If the cotton is to be inter cropped or sown mixed with other crops the distance between the cotton rows has to be adjusted accordingly. Generally, cotton crop is intercropped or sown mixed with black gram, green gram, soybean, chillies and groundnut. The mixed cropping with cotton has gained popularity in case of dry land agriculture areas, however, it can be practiced profitably in any of the cotton growing regions of the country.

Seed rate: It depends up on type of cotton, test weight, spacing, and purity of seed and lastly method of sowing. Normally a seed rate of 12-15 kg/ha for Indian cottons and 15-20 kg/ha for American cotton is recommended but in case of intercropping or mixed cropping the seed quantity is reduced. In case of broadcasting, one and a half times of the recommended seed rate is taken.

Spacing and depth of seeding: The spacing between plants and rows depends on the type of cotton to be grown. The Indian cotton is sown at a spacing of 45-50 cm x 50-60 cm and American cotton is sown at 50-75 cm x 60-90 cm. This spacing, however, can be reduced if the sowing is delayed or the soil is poor in fertility.

The depth of seeding depends on soil type, soil-moisture content, organic matter content in the soil. The depth may be lesser in case of high moisture content in the soil, heavy soil type, low soil organic matter content and vice-versa. Generally, a seeding depth of 5-8 cm is used for better germination and plant growth.

Seed treatment: The seed of most of the cotton varieties, and particularly the American types is covered by short fibre called **'fuzz'.** The fuzz makes the seeds cling together, thus creating a problem in free rolling and separation of individual seed for sowing. For facilating sowing, it is necessary to separate the seeds by some seed treatment.

a) **Non-chemical treatment:** In this case the seeds are not subjected to any chemical treatment but they are soaked in water for 10-12 hours and then they are rubbed with fresh cow dung and wood ash or soil dust. After rubbing, the seed is spread in the shade for few hours and when they are dried they can be used for sowing. The rubbing helps in separating out the seeds affected by pink boll worms because they are crushed while rubbing the seed. It also helps in quick germination and a better start of young seedling.

b) **Seed treatment with chemicals:** The seeds are soaked in strong commercial sulphuric acid (which burns the fibre) for two minutes. The seeds are then taken out and washed in running water to remove the chemical. They may be also treated with zinc chloride. The seeds are immersed in the solution of zinc chloride for 10-15 minutes and then washed. These chemicals dissolve the fuzz.

The chemical treatments not only remove fuzz, but they are also said to improve the germinating capacity of the seed, hasten germination, burn the affected seeds with insect, pest or disease and make handling of seeds easier. The chemical treatments are costly and require caution for use to avoid injury to the seed and hence they are not recommended for the common use.

Type of Cotton

There are four cultivable species in cotton. The primary distinction is based on chromosomes number (Hutchinson, 1962). The species are *Gossypium hirsutum, G. barbadense, G. arboreum and G. herbaceum.* Among them, *G. herbaceum* and *G. arboreum* (n=13 chromosomes) are called the old world cotton and cultivated in tropical and sub-tropical regions. The other two species are with n=26 chromosomes and grown in the new world. The species *G. hirsutum* (American upland types) is commercially grown through out the world. The longer staple Egyptian and Sea Island cotton (*G. barbadense*) is cultivated to a much smaller extent. In India, all the four cultivated species are grown with the highest area under G. *hirsutum* (35.9%), followed by *G. arboreum* (17%) and *G. herbarium* (11,0%). The remaining ones are with hybrids (36%).

Varieties

Rainfed	Khandwa-1, Khandwa-2, K-8, Mahalakshmi, GS 23.
Rice fallow	Krishna, MCU 7, LRA 5166.
Extra staple	MCU 5, MCU 8.
Pest and disease resistance:	
Verticillium wilt	MCU 5 VT, Savita.
Whitefly	Kanchana
Boll worm	Abhadita

Hybrid Cotton

There is a phenomenal change in cotton improvement after the introduction of hybrid cotton in the 1970s. The major break-through in cotton production in India has come from the large scale commercial cultivation of high yielding hybrid cultivars of long and extra long staple cotton. Over one fourth of the total cotton area in India is cultivated with hybrid cultivars. The spread of hybrid cotton has been mainly in Maharashtra, Gujarat, Karnataka, Madhya Pradesh, Andhra Pradesh and Tamil Nadu. Maharashtra has the distinction of growing the largest area of hybrid cotton, mostly under rainfed conditions.

Important Hybrids of Cotton

- Superlong staple: Hybrid-4, Varalaxmi, Suvin, DCH-105.
- Medium staple: Dhanalaxmi, TSH 289, CDHB 1, DHB-105

Water Management

India grows hardly 13-15% of its cotton under irrigated conditions. The plants, being very sensitive to water, react strongly to adverse moisture conditions in the soil. The proper amount of soil moisture during the different stages of plant development is the most potent single factor making for a high crop yield. In areas of low rainfall, where irrigation facilities are not available, the following practices should be adopted to safeguard the crop against a very severe drought condition.

1. Bunding, leveling and dividing the field into smaller checks (pieces) to retard

the runoff loss water.
2. Application of compost/Farm yard manure at the rate of 6-8 t/ha (as basal).
3. Early sowing of crop.
4. Use of relatively low seed rate.
5. Hoeing of the crop more frequently.
6. Keeping the field completely weed free.

In case of irrigated cotton of north India one pre-sowing irrigation is given for an assured germination. The second irrigation (usually light) is given 3-4 weeks after the seeds have germinated. Further watering is scheduled according to rainfall. In the southern regions (Tamil Nadu and Kerala) the interval between successive irrigations is usually one week to one fortnight but it varies greatly with the season and the availability of water in the wells. Normally the crop needs 6-8 irrigations. The crop is irrigated till it stops flowering because irrigation after this encourages vegetative growth and the maturity is delayed.

In case of excessive water the field should be provided an efficient drainage so that there is no water logging because the plants can not withstand water logging for any period.

Four critical growth stages have been observed for irrigation in high yielding cotton varieties *i.e.*, commencement of sympodial branching (60-70 days after sowing), flowering (90-100 DAS), boll formation stage (125 DAS) and boll bursting (140 DAS). More frequent irrigations before flowering is detrimental for yield. On an average the quantity of each irrigation should not increase more than 7 cm. Soil moisture stress during flowering and boll formation causes a drastic reduction in cotton yield.

Inter-Cultivation and Weed Management

The cotton crop needs a weed free field up to 30-40 days after sowing. The first inter-cultureis always a manual operation which ensures a complete removal of weeds at a very early stage of crop growth. Subsequently hoeings are given whenever weeds reappear or there is a break in the rains. These hoeings may be done either by hand or by bullock drawn implements. Usually, in case of irrigated crops hoeing after every irrigation, particularly in early growth stages, is deemed necessary for keeping down the weeds. Chemical control of weeds is possible through application of TOK-E-25 or Ansar-521. Application of Diuron at the rate of 400 g as pre-emergence and 600 g as post emergence takes care of most of the weeds. In case of Diuron is not available other herbicides such as flumeturon (cotoran 80 W), Fluchloralin (Basalin 48 EC, pre-plant incorporated), Alachlor (Lasso) and Dibutalin (Amex 820) may be applied as pre-emergence soil spray to cotton crop.

Plant Protection

Like all other crop plants, cotton is also subjected to attack by a number of fungal, bacterial, viral and physiological diseases which take a heavy toll of the crop every year. Like wise the insects and pests also cause a drastic reduction in crop yield and lower down the quantity of the produce. The plant protection measures as per the recommendation of the state should be adopted to control the incidence of diseases in cotton crop.

Cotton pests are greater menace to the crop and they result in a more severe damage to the yield and quality of the produce. Prophylactic measures should be adopted to control cotton pests. Two-three weeks after germination, the crop should be sprayed with metasystox 25 EC or sumithion 50 EC at the rate of 1.0 litre/ha. The chemicals may be substituted by Zolone 35 EC or Thiodan 35 EC or Rogor 30 EC at the rate of 1.25 lit/ha or Savin 50 at 2.0 kg/ha rate.

Spraying with any of the above mentioned insecticides should be repeated at 50-55 DAS, 70 DAS, 85-90 DAS and 110 DAS along with Diathane Z-78 or Diathane M-45 at 1.5-2 kg/ha rate.

Topping

It is desirable to remove the terminal growing point (topping) once from each plant at a height of 1 to 1.2 m or between 80-90 days of crop growth. This practice further terminal growth and encourages sympodial branching and boll development. It also facilitates spraying operations and picking of cotton.

Harvesting

Cotton is harvested in three or more pickings taken at suitable intervals. The number of pickings varies with the maturity habit of the variety. The season of harvesting varies with of sowing and duration of the variety. Generally, the crop sown in April–June is harvested in October-December, while those which are sown in June-September and in September –October are harvested from November to March and from March to June respectively. It can be further listed that the harvesting, in northern and central parts of India, is done from October to December, in Gujarat from January to March/April, in Tamil Nadu, Andhra Pradesh and Karnataka from November to June.

Production of Quality Cotton

There is urgent need to enhance cotton productivity and quality if demands for domestic and export market are to be met. The Central Government of India has sanctioned a special programme on technology mission on cotton with the object of improving production, productivity and quality of cotton. The raw material, seed cotton, is generally affected resulting in sub-standard quality of fibres. Sustainable increase in the production of quality raw cotton and its availability at reasonable prices would be vital for the textile mills. Cotton growers should also realize remunerative prices. Because of the stringent quality standards set by International buyers, the quality of raw cotton has assumed paramount importance.

Now-a-days, too many varieties of cotton are being grown in the country which has led to admixtures at various stages, creating problems of varietal impurity, difficulty in selecting appropriate variety and quality of cotton for the count spun. It had become a problem that cotton farmers were not able to realize stable and sustainable yield even under the best management practices.

Reduction in contamination of raw cotton by extraneous matters during the course of its harvest and transfer from the field to the mill are another priority area in view of the advent of modern automated, high speed spinning mills where manual cleaning of cotton is not possible. The issue of contamination – free cotton has also become crucial to cater to the international buyers in particular, who would not tolerate even a tiny spot on the yam. Hence, there is a need for creating awareness among the farmers to keep cotton free from contamination and improve the systems of handling cotton at market yards, ginning and pressing factories, infrastructural facilities and transport to spinning mills. Grading-cum-price systems based on contamination of cotton should also enable the farmers to get premium price for their produce.

JUTE AND ALLIED FIBRES

Jute and allied fibres consist of the crops like jute, kenaf, roselle, sun hemp, sisal, and ramie and these are called natural bast fibres. Fibre is extracted from the stem of jute, mesta (kenaf and roselle), sun hemp and from the leaf of ramie, sisal. Except sisal and ramie, others are annual crops. Sisal and ramie are plantation crops of 6-10 years duration. Jute and mesta are extensively cultivated in India accounting for more than 53% world production.

JUTE (*CORCHORUS* SPP.)

Jute is a bast fibre obtained from the bark of the two cultivated species of the genus *Corchorus* viz., *C. capsularis* and *C. olitorius* belonging to the family tilliaceae. Jute is known as "Jew's-mallow" in English which is translated from the old name Olus judiacum (Royle, 1855). Royle further explained that jute is called patta in Sanskrit and pat in Bengali. In Maharashtra presents of garments are mentioned as pattanjam (patta-produced) for jute fibre products and Kitajam (insect produced) for silk products. In the Ain i-Akbari (1590) there is mention of sack-cloth (tat) made from jute in Rangpur district. Thus it is worthwhile mentioning that the jute fibre was known in India from comparatively ancient times while jute cloth at least from beginning of sixteenth century. According to Kundu (1951) the place of origin of *Corchorus olitoriusin* Afrrica and *Corchorus capsularis* is Indo-Burma.

TABLE 11.2: Production of jute in India, 1995-96

States	Area ('000 ha)	Production ('000 bales)	Yield (kg/ha)
Assam	88.6	844.2	1715
Bihar	119.8	1112.6	1672
West Bengal	515.8	5671.3	1979
All India	738	7743	1883

Source: Fertilizer statistics, 1996-97 (bale = 180 kg)

Distribution

World production of jute (nearly 95%) is confined to India, Pakistan, Bangladesh, China, Indonesia, Thailand and Nepal. Pakistan and India are the most important producers of jute fibre in the world. Production of jute in India has increased tremendously from 3.3 million bales in 1950-51 to 7.7 million bales in 1995-96. However, the cultivated area has not increased much. The area under jute was 5.7 lakh ha in 1950-51, which increased to 7.4 lakh ha in 1995-96. The yield of jute has significantly increased over the period. It was only 1043 kg/ha in 1950-51 and increased 1245 kg/ha in 1980-81 and the average productivity in 1995-96 is 1883 kg/ha. The increase in yield is due to the development of high yielding varieties. In India, West Bengal has more than 60% of area and production of jute. Other states growing jute are Bihar, Assam (Table 11.2).

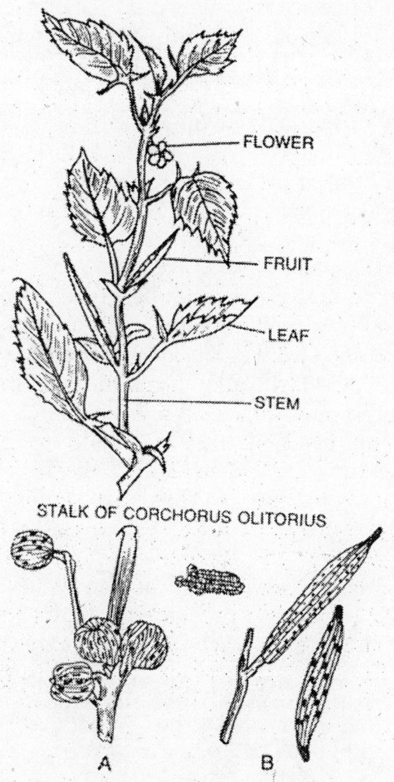

FIG. 11.2: Fruits of *Corchorus capsularis* (A) and *C. olitorius* (B).

Economic Importance

Concerted efforts have been made by the Government to promote the usage of jute fibre. Jute is now hailed as a "golden fibre", which would be the "fabric of the future". The demand for fine fibre has increased recently. The fine varieties having a denier of 10 or less are in great demand. Fine quality fibre is used for the manufacture of the value added products. The different value added products are household textiles, industrial textiles, re-inforced plastics, handicrafts, paper pulp and geotextiles. It has been proved that from jute fibre, fabrics of all designs and patterns could be woven on automatic looms and the products would be akin to those made of cotton. These products are emerging as major foreign exchange earners. Foreign exchange to the tune of 700 crores rupees had been earned during 1996-97 by the export of jute products. About 7 million small and marginal farm families, industrial employees and traders are supported by jute trade.

Ecological Requirements

1. **Climate:** Jute needs a humid, tropical to warm subtropical climate, with high humidity and about 300 mm of precipitation per month during the vegetative growth period of about 3 to 5 months after sowing. It has been noticed that an annual rainfall of 100-150 cm ia an ideal for the crop of which 25 cm rainfall during March, April and May is most suited. Alternate sunshine and rainy days

are most conducive to growth. The young plants can not withstand water logging; therefore, it is necessary to adjust the sowing in such a way that the plants attain a height of about 1 metre before the heavy and constant rain starts. A dry spell after break of monsoon enables the essential weeding and thinning operations to be carried out. The plants thrive best in damp heat where the humidity ranges between 55 to 95 per cent. The most ideal temperature for the crop is between 24°C and 37°C, however, the optimum is around 34°C. A temperature below 20°C is harmful for the crop and it restricts the growth of the plants completely, however, they may flower and set into fruits.

2. **Soil:** A good jute crop can be grown on nearly all types of soil, which are deep enough provided they are neither very sandy nor very heavy. The soils with good physical conditions and well aerated of loams or sandy loams nature are the best. An extreme acidity or alkanity is harmful for the crop and a soil pH of 5.0 to 8.0 is tolerable range of soils of reaction. The crop is grown successfully on old alluvial soils of Bihar, mild acidic soils of Assam, Orissa and light alkaline soils of Tarai districts of U.P. It is observed that clay loam for capsularis and sandy loam for olitorius are the most suitable soil types.

Preparation of Land

Jute seeds being small require very fine tilth. The plants bear tap roots which go deep enough, sometimes 30-45 cm into the soil that requires deep ploughing and removal of weeds and the stubbles of the previous crop. To get a clean, clod-free with fine tilth and well leveled surface it is required to plough, cross-plough and planking is necessary in order to get the required seed bed.

Application of Manures and Fertilizers

A good crop of jute, yielding around 40-50 tonnes of fresh crop weight and 20-25 quintals of dry fibre which is roughly 5 per cent of the fresh weight, requires the following quantity of nutrients from the soil:

Nutrient uptake from the soil by jute crop/ha

Nutrients	Nutrient uptake in kg/ha		Remarks
	C. *capsularis* (JRC 212)	C. *olitorius* (JRO 632)	
Nitrogen	84.0	111.0	Acid soils require 3-7 t of lime as soil amendment after every 3 years
Phosphorus	16.0	28.0	
Potash	137.0	164.0	
Calcium	84.0	124.0	
Magnesium	29.0	25.0	

Source: Handbook of Agric, 1984

The quantity of nutrients may be supplied by an application of 4 to 7 t of compost or FYM under rainfed conditions or in absence of fertilizers. In case the entire quantity of nutrients is to be supplied through fertilizers the following doses are recommended.

- C *capsularis*: 80 to 100 kg N, 40 to 50 kg P_2O_5 and 60 to 80 kg K_2O per hectare.
- C. *olitorius*: 20 to 60 kg N, 15 to 30 kg P_2O_5 and 20 to 50 K_2O per hectare.

In addition to above quantities, the crop needs about 20 to 40 kg of magnesium oxide/ha which should be applied as basal along with P_2O_5 and K_2O. Split application of nitrogen half, as basal and half as top dressing, gives better results. Under short supply of nitrogen a foliar application of 15 kg urea/ha by a low volume power sprayer in two installments between 35 and 60 days of age at a concentration of 10 per cent (as urea) is found to be best.

It is observed that under unirrigated conditions application of nitrogen results in 8-16 times higher yield than control. In potash deficient soils application of potash gives 2-3 times higher yield. Application phosphorus increases resistance of plants against lodging and the fibre quality is also improved.

Seed and Sowing

1. **Time of sowing:** Usually, the sowing of jute starts from end of February for *capsularis* and beginning of April for *olitorius* that continues till the middle of June in some areas. The young jute plants are very sensitive to water logging, therefore, the sowing time must be so adjusted that the crop is able to attain a height of about 1 metre before heavy rain starts. Sowing must be done in February under low lands to avoid water logging at early growth stage. In case of midlands and uplands the sowing is done in March-April and May-June respectively. In Bihar and U.P., the sowing may continue up to mid July because of the monsoon.
2. **Method of sowing:** The most common method of jute sowing is broadcasting. For broadcasting and even distribution of seeds it is mixed with loose soil because the seeds are very small. For still better results it is advisable to broadcast the seeds cross-wise. After broadcasting the field is harrowed and planked for covering the seeds and bringing them in close contact with moist soil that helps in better and quick germination. Weed control becomes very difficult and cumbersome in the broadcast sown crop as the plants are not properly placed.
   Dropping the seeds in shallow furrows 3-5 cm deep behind the plough or using seed drill is always preferred over broadcasting because the line sowing helps in inter cultivation, weeding and sowing. The seeds must be pressed by planking, if sown in pre-monsoon conditions, for a better seed soil contact.
3. **Seed rate:** The seed rate depends up on method of sowing, spacing and variety. Broadcasting method needs about 10 kg seeds/ha in case of *capsularis* and 6 kg/ha in case of *olitorius*. A seed rate of 8 kg and 4 kg/ha is needed in case of line sowing of *capsularis* and *olitorius* respectively which may give a required plant population 2.5 to 5 lakh per hectare.
4. **Spacing:** Two thinning are done to maintain proper spacing. The first thinning is done when the plants attain a height of 7-10 cm in which the plants are thinned to a distance of 4-5 cm. In the second thinning the plant height should be 14-15 cm, and a final spacing of 8 to 9 cm between the plants is allowed. For line sowing a single row seed drill is used to sow the *capsularis* varieties 30 cm apart and the *olitorius* varieties 20 cm apart.

Type of Jute

Jute is a bast or phloem fibre obtained from the stems. There are two cultivated species, *Corchorus capsularis* (white jute) and *Corchorus olitorius* (Tossa jute) (n=7). The former produces small, almost globular and wrinkled capsules. The fruits of C. *olitorius* are long, cylindrical and ridged capsules. About 75% of the jute area is with C. *capsularis*. This species is highly adaptable and cultivated in lowlands and uplands. However, C. *olitorius* is basically an upland crop and can not withstand flooded condition.

Varieties

Capsularis	:	JRC 206, JRC 212, JRC 321, JRC 1108, JRC 4444, JRC 5854, JRC 7447, UPC 94, BZ 2, BZ 22.
Olitorius	:	JRO 514, JRO 632, JRO 878. JRO 7835
Resistant to pests	:	JRO 524, JRC 321, JRO 36 E.

Following Capsularis and Olitorius varieties are recommended for Orissa. The characteristics of the varieties are given below.

Capsularis

JRC 212	:	Entire plant is green, medium duration (140 days, flowering).
(Sbuja Sona)	:	Suitable for medium land.
JRC 7447	:	Selection from JRC 212 after X-Ray radiation. Duration 130 days.
(Shyamali)	:	High yield potential. Responds well to high fertilizer dose, suitable for medium land.
JRC 312	:	Coppery red pigmented variety selected from local material
(Sonali)	:	"Hewti". Early in maturity, 130 to 140 days. Suitable for lowland and early sowing.
JRC 4444	:	The variety was developed by JARI (Barrackpore) from the cross
(Baldev)	:	JRC 212 x D 154, grows to a height of 330 cm., matures 135 days, found suitable for jute growing areas in the state, less suitable to diseases and insect pests compared to JRC 212, moderate susceptible to semilooper and stem weevil.

Olitorius

JRO 632 (Baisakhi Tossa)	:	Entire plant is green. The variety should not be sown before 15th April as it flowers early within 15 days after sowing. When sown after April 15th, it flowers in 130-140 days.
JRO 878 (Chaitall Tossa)	:	Red pigmented stem. Also suitable for early sowing in March as pre-rice crop. It flowers in 130-140 days.
JRO 7835 (Basudev)	:	Plant green. Suitable for early sowing in March. It takes 130-140 days to flower.
JRO 524 (Navin)	:	Plant green, suitable for early sowing in March. Has higher yield potential, Flowers in 130-140 days. Suitable for medium land.

Irrigation

The crop is usually sown rainfed after onset of monsoon but under assured irrigation facilities one pre-sowing irrigation is given to ensure good moisture. The crop normally depends upon rain water but under drought conditions it may be irrigated twice or thrice depending upon the rainfall. The most critical stage for irrigation is germination and knee height and it is given 2-3 pre-monsoon irrigations in addition to post-monsoon irrigations for sustaining good crop growth.

Intercultural Operations

Early and timely interculture is necessary. First weeding, hoeing and partial thinning should be done within 21 days of germination. The operation is repeated fifteen days later and the final thinning is done to maintain a spacing of 5-7 cm between plants in the lines.

In broadcast crop a rake is to be drawn 15-20 days after sowing to thin out the plants and loosen the soil slightly. This is to followed by weeding and thinning to maintain a spacing of 8-10 cm between plants.

Early interculture and partial thinning is very important. Because:

1. Weeds are destroyed early and they do not compete with crop for food and water.
2. Helps early root growth and development.
3. Prevents evaporation losses from the soil.
4. Partial thinning helps in better growth of the crop.
5. Fluchloralin (Basalin) at the rate of 2.2 lit in 1000 lit of water is applied one day before sowing as spray to control weeds and is mixed with soil by laddering.

Plant Protection

Indigo caterpillar, semilooper stem weevil and mites appear on jute. Clean cultivation and trimming of bunds help in reducing pest incidence. Stem rot, root rot and collar rot are common diseases. These diseases are severe in ill drained soils and highly acidic soils. Improving drainage, correcting the acidity through lime application and application of organic matter greatly reduce the incidence of these diseases. The control measures should be adopted as per the recommendation of the state.

Harvesting

Right stage for harvest is the small pod stage (130-135 days age). Thick and thin plants may be bundled separately. This will ensure uniform retting. Leave the bundles in the fieldfor 2to 3 days to allow the leaves to shed in the field and then ret. The colour of the fibre is darkened if the leaves are allowed to stick to the plants during process of retting. Drying of the plants before retting also facilitates the separation of the fibre.

Retting

It is a biological process by which the bast fibre (fibre in the bark) gets loosened for an easy separation from the woody stalk. During this process the gums, pectin and other mucilaginous substances are removed from plants by a combined water and microbiological action. For an ideal retting the Jack should be kept submerged at least 20 cm below the surface of the water because incomplete submergence results in under –retting and produces "croppy' fibre of extremely low value, while over retting 'dazed' weak fibre. It is observed that tying of few plants of 'Dhaincha or Sunhemp" in each bundle of jute causes an early retting. In case of stagnant water addition of little amount of ammonium sulphate or bone meal specially when atmosphere temperature is low, accelerates retting process. Gently flowering, fairly deep, clear and soft water are congenial for an ideal retting. The retting needs around 34°C water temperature, and the retting completes within 10-15 days during July while about 21 to 30 days during Septemeber or afterwards. Most of the defects in fibre quality are due to incomplete or faulty retting. For finding out an exact end point of retting, the Jacks must be examined at least once every day after 10-12 days of steeping.

When retting is over the fibre comes out after pressing the plants between thumb and finger. Soon after the end point, the Jacks should be taken out and extraction of fibre must be completed as soon as possible because any delay results in a poor quality fibre.

Extraction of Fibre

The fibre is extracted when it comes out easily. Over retting leads to weak fibre. Single plant extraction gives good quality fibre. The fibre should be washed thoroughly in clean water to remove dirt and other foreign matters like jut leaves etc. Beating the stalk against wooden pole with wooden mallet is avoided as it spoils the fibre quality. The fibre loses strength and gets tangled. Then it is necessary to squeeze (wring) out water from the bundles and dry on bamboo racks under partial shade for 4 to 5 days. If the fibres are dried in bright sun, they become brittle. Then the fibre should be graded and bundled separately for sale. Grading helps in fetching better price.

Yield

The olitorius being an efficient user of nitrogen gives more fibre yield than capsularis in general. The national average fibre yield is 13 q/ha. However, the maximum yield potential is 40 and 37 quintals/ha in case of olitorius and capsularis respectively. The fibre yield has been found to be directly correlated with plant height and stem base diameter and therefore, a taller and thicker plant gives higher fibre yield than small and thin plant. However, the quality of fibre is better in case of taller and thinner plants and hence, it is preferred. The seed production is about 4-5 quintals/ha from capsularis and 2-3 quintals/ha from olitorius.

Quality of Jute

The quality of jute is influenced by several factors such as variety, environmental conditions and method of processing. Among them, the method of retting is the most important factor that affects the quality. Retting is a microbial process which requires a large volume of water. Over-retting weaken the fibre and incomplete retting makes the fibre coarse. With perfect retting, the fibre gets fineness and strength. Recently advocated method of retting is 'ribbon retting decortications' requires less water, and the green ribbon rets faster than the whole plant.

MESTA (*HIBISCUS SUBDARIFFA/H. CANNABINUS*)

It is a fast growing fibrous bush, native to Africa, this fibre crop has been successfully introduced into many parts of Asia and now is widely cultivated in China, India, Pakistan, Russia and Thailand, A variety of kenaf known as mesta has been grown over large areas in many parts of the country. It is mainly cultivated in Andhra Pradesh, Bihar, Orissa and Maharashtra. These states contribute about 75% of the area under this crop. Mesta is a hardy crop capable of growing under a wide range of climate and soil conditions. It is wholly cultivated as a rainfed crop. High yielding varieties of mesta are HS 4288, HS 7910, HC 583, AMV 1, AMV 108.

In trade, jute and mesta are together categorized as raw jute. The fibre of mesta is called as kenaf, deccan hemp, mesta and Bimlipatam jute. The fibre is used principally for textiles, mainly for sacking and to manufacture of twine and rope. It is an excellent substitute for bamboo and wood pulp. It has been proved at the laboratory level that mesta is well suited for the production of newsprint. It has been estimated that about 10 tonnes of paper pulp can be produced from one hectare of mesta. In India, the fibre is used on a fairly large scale in the jute mills in conjunction with jute. In other countries, the fibre is used for making ropes, twines and fishing nets. The seeds contain about 21% of oil. The oil can be used for culinary purpose and also for soap making.

Field Preparation

Well drained conditions particularly in the early stages are necessary. It comes up on all types of soils except stiff clays. It is also necessary (a) to plough the land to get good tulth. Good tilth helps quick establishment of crop, bettere stand and growth, and (b) to apply organic manures at 5 t/ha before sowing.

Seed Treatment

The seeds should be treated with Thiram or Captan at 3-5 g per kg of seeds. Seed treatment protects the crop from seed borne diseases viz., stem rot and foot rot.

Sowing

1. Pre rice mesta crop is sown in March-April in lowlands with supplemental irrigation,
2. The main crop is sown in May-June.
3. The crop is sown early. Adequate moisture is necessary at sowing time to facilitate germination of seeds. A seed rate of 15 kg/ha for broadcast crop and 12 kg/ha for line sown crop is required to get the recommended plant population in the field. Early sowing helps in quick establishment of crop and facilitates vigorous growth with good pre-monsoon showers or monsoon rain.
4. **Line sowing:** The seeds of the crop are sown in lines 30 cm apart and 8 or 10 cm between plants
   a) Line sowing ensures uniform growth.
   b) Interculture is easy. Weed control is efficient.
   c) Requires less seed.
   d) Yield higher.

Interculture

As in jute, the first weeding, hoeing and thinning should be done when the crop is 15-20 cm (21-30 days age), This operation is repeated a fortnight later (35-45 days age). Early hoeing, weeding and thinning 4 weeks is essential for good growth. The advantages are:

1. Weeds which compete for food and water are destroyed early.
2. Top soil is loosened which reduces evaporation losses in water stress condition.
3. Plant population is not affected.
4. Root growth is hastened.

Fertilizers

Fertilizers should be applied on soil test recommendation. If the soil has not been tested, application of 20 kg N, 25 kg each of P_2O_5 and K_2O per hectare should be done before sowing. Application of chemical fertilizers, in addition to organic manures, greatly increases yields. Top dressing of an additional dose of 20 kg N/ha at 21-30 days stage just before hoeing and weeding will greatly raise the yield.

Foliar Spray with Urea

Foliar spray with urea at the rate of 20-25 kg/ha is given twice between 45-60 days of the crop. The crop responds immediately to urea spray due to quick absorption of nitrogen through leaves and stem. During drought spell, foliar spray of urea revives the crop. During rainy season when soil application of nitrogen is not possible urea spray can be done if 3-4 hours of bright weather is available.

Plant Protection

Foot and stem rot are major diseases of mesta. These diseases are found in years of heavy rainfall and in fields where drainage is poor. The drainage is to be improved to prevent the said diseases. Mealy bug becomes a serious pest particularly under drought condition. The control measures should be adopted as per the recommendation of the state.

Harvesting

Harvesting of the crop should be done at small pod stage. In cannabin us types it will take 130 days to reach this stage. In subdarifa types however, the stage is reached at about 160 days. The plants are harvested by uprooting, when there is adequate moisture in the soil. Leave the plants in the field for 3 days. During this period the leaves shed and then stems are bundled. Thick and thin plants may be bundled separately and put under water as this helps is uniform retting.

Retting

As in jute, keep the bundles in 60-100 cm of water in an upright position for 3 days. Thereafter steep the bundles completely under water arranging them in jacks and weigh down the jacks with stone slabs or dried wooden logs. The jacks should be about 30 cm below water surface and 30 cm above ground level. Contact with soil, affects quality of fibre. Gentle flowing water is good for retting and improves fibre quality. Banana stems or mud as covering material should not be used as the fibre quality will be affected.

Extraction

Retting is complete when the fibre tissues come off the stem as a continuous ribbon. Depending on the water temperature, it may take 15 days to 30 days. After extraction, with least effort wash the fibre in clean water and dry for 4-5 days in mild sun over bamboo racks. It is necessary to bandle the fibre grade wise before sending to market for sale.

Important Practices to be Advocated in the Field

1. Line sowing.
2. Early interculture and weed control.
3. Fertilizer application at moderate doses.
4. Foliar spray with urea.
5. Apply nitrogen in two splits at 3 and 6 weeks.
6. Retting in running water.

OTHER FIBRE CROPS

AGAVE (*AGAVE* SPP.)

These are short stemmed half woody plants bearing a rosette of long, erect, pointed, fleshy leaves. Six species of Agave were introduced in India during the 15th century by the Portuguese. They are *A. anguistifolia, A. Americana, A. lurida, A. cantala, A. sisalana* and *a. vera* – cruz. They are cultivated in the dry zones. Generally, it is grown as a life fence. The wide spread root system serves as a good soil erosion controlled by planting across small gullies and slopes. Mature leaves can be harvested from 3-4 years old plants and from the fifth year onwards leaves can be harvested regularly and used for extracting fibre. The fibre is called to as 'hard' fibre. Retting and mechanical methods are used for fibre extraction. In Maharashtra, cantala fibre is extracted on a commercial scale by retting process. The fibre is used for ropes, mats, cordage, twines and nets. Cantala fibre is pure white in colour, whereas sisal fibre is light yellow. Sisal (*Agave sisalana*) is cultivated in Andhra Pradesh, Assam, West Bengal, Bihar, Orissa, Tamil Nadu, Maharashtra and Karnataka.

Average annual production of sisal fibre in India is estimated at about 12,000 tonnes. The total demand of agave is about 50,000 tonnes. Annually India imports avage worth of Rs 2.5 crores from Tanzania, Kenya and other countries. However, India exports sisal ropes and twines to Nepal and Fiji Islands.

SUNHEMP (*CROTOLARIA JUNCEA*)

Its fibre are used for manufacturing tissue papers for currency besides making ropes, twines, nets, canvas and screens (tat-patti) but it is not used for textile purposes. The crop occupies approximately 1,40,000 ha and produces about 50,000 tonnes of fibre. Sunhemp is grown during *kharif* season in north but during *rabi* season in south. The crop may be sown broadcast by using 25 kg seed/ha or drilled at 30 cm row spacing which requires 15 kg seed/ha. A plant to plant distance of 5-7 cm should be maintained. Generally, the crop is fertilized with 20 kg P_2O_5 and 20 kg K_2O/ha without N.

The crop becomes ready in 120-150 days or at pod formation stage however, few varieties like T 6 is harvested at pod maturity stage in order to get both seed as well as fibre. Its commonly grown in W.B., Bihar, U.P. and M.P. The plants are harvested from close to the ground tied into small bundles and kept in the field for 2-3 days for leaf shedding. The process of retting is similar to that of jute and fibres are stripped off stalk by stalk method from lower end upwards. The average fibre yield comes to about 3 to 5 quintals/ha having 2-4% plant weight on wet on wet weight and 8-14% on plant dry weight basis. Moreover, a highest yield up to 14 q/ha have been obtained from K-12. A seed yield of 1-3 q/ha may be obtained besides fibre when plants are harvested at pod maturity stage.

Varieties

U.P.	-	Kanpur 12 (K-12)
M.P.	-	M-18, M-19, M-35
Bihar	-	BE-1
Tamil Nadu	-	Bellary
Maharashtra	-	DIX
W. Bengal	-	ST 55

HEMP (*CANNABIS SATIVA*)

The hemp fibre is a bast or phloem fibre. It is strong, durable and more rigid than flax. The fibre is used for the manufacture of ropes, sacking, carpets, nets, tarpaulins and webbing.

FLAX OR LINSEED (*LINUM USITATISSIMUM*)

Flax is meant both for fibre and oil. The fibre is soft, lustrous and flexible. The flax fibre is suitable for strong yarns such as are used for sewing thread. This type of linseed produces fibre from stem and oil from seeds, though its fibres are inferior in length and quality. The fibres are soft, shining and possess high water absorbing quality but less elasticity. Flax fibres are spun and woven into fine fabrics like lawn, cambrics and drills, canvas and buckrams besides its use in linen stitching, making twines and nets for fishing, ropes, etc. It acts as best raw material for tissue paper, fire fighting hose pipes, knapsacks and water bags. The flax used for fibre is an introduction from Belgium and Italy which is cultivated during *rabi* season. The seed is drilled in rows at 20 cm distance between plants (within the rows) which requires 40 kg seed/ha. It needs 1-2 irrigations and mtures in 100 days.

The stems are cut from close to the ground; small bundles are prepared and steeped in water for retting which takes three days. After proper retting the fibres are stripped off from individual stem.

Varieties

F 895 and F 896 (exotic collections)

ABACA OR MANILA HEMP (*MUSA TEXTILES*)

The fibre is obtained from the leaves of the plant which is a member of banana family. The Philippines is the chief producer of abaca. It is the strongest of the cordage fibres and is particularly used for marine cordage. The fibre is resistant to salt water and has low degree of swelling when wet.

SABAI GRASS (*EULALIOPSIS* (RETZ.) CE HUBB.)

Sabaigrass is a cultivated perennial grass commercially used for rope making, mats, papochh and paper pulp. It is vegetatively propagated by slips and well suited to acidic degraded marginal and sub-marginal rainfed uplands of Mayurbhanj district of Orissa. It is propagated by vegetative slips. Its initial growth is very slow for 2-3 years during which it suffers from growth and development due to serious weed competition resulting thereby in reduction of yield and quality. It is not generally fertilized. Sabai is a hardy grass to both drought and frost with no pest attack and is unpalatable as a fodder.

The Sabai grass is well suited to hot and dry sub-humid climate with mean rainfall varying from 1200-1400 mm in lateritic soils of the district. Very little information is available on area, production, productivity and cost of cultivation of Sabai grass.

Area, Production and Productivity

The Mayurbhanj district of Orissa has total cultivated area of 4, 47,214 hectares. The study revealed that sabai grass is being cultivated in 11,174 hectares in 17 blocks. This constitutes about 2.5% of the total cultivated area of the district. Again, a high segment of population belonging to Scheduled Tribe resort is cultivated this grass. Three blocks are found promising on the basis of their area, production and productivity of sabai grass. They are Suliapada, Baripada and Morada in orderly sequence representing more than 65% area and production of sabai grass. The block productivity varies between 15 to 33.7 q/ha. Baripada recorded the highest yield (33.7 q/ha). The present level of total production is 33,488.6 tonnes. The value of this raw dry sabai grass at the rate of Rs 2,000 per tonne comes to Rs 6.6 crores. This indicates sabai grass serves as an income to the poor tribal farmers in their risk prone farming system.

Cost of Cultivation

Sabai grass economically survived in the field upto 12 years. The major cost incurred during the establishment year was Rs 3,500 per hectare. This includes preparatory cultivation, cost of planting materials, gap filling, labour wages for opening pits (15 cm deep), planting of slips and watching etc. In the succeeding years till the 12 years, the cost is restricted to harvesting, drying and bundling charges only. The cumulative production is 245.2 q/ha with a total net profit of Rs 38, 968/ha in course of 12 years life cycle of sabai grass cultivation. The return is negative (- Rs 3,160/ha) in the first year due to high cost of cultivation and less receipt. Subsequently, it goes on increasing and the maximum productive year is from 4th year to 10th year period. Thereafter, yield declines drastically. Thus, an average net return of Rs. 3,245/ha/year is assured from sabai grass.

The study brought out the block-wise area, production and productivity of sabai grass with net profit obtained in different years of its establishment. Appropriate technology like scientific cultivation and establishment of marketing facilities for tribal farmers need due consideration for future growth of sabai grass cultivation of Mayurbhanj district of Orissa.

Research Results

The research results so far obtained are enumerated below:

1. Intercropping of green gram, cowpea or black gram with sabai grass is highly remunerative in the initial 3 years of its establishment.
2. Basal application of 40 kg N+ 20 kg P_2O_5+ 20 kg K_2O/ha with 2 cuttings/year is useful to maximize productivity and net return (Barik, 1996).
3. High yielding sabai grass germplasms like 'Chitrada Local' having tall plant height and more tillers/clump should be cultivated for rope making and paper pulp purposes (Barik, 1997)
4. Row x plant spacing of 90 cm x 60 cm, 90 cm x 90 cm or 60 cm x 60 cm is necessary to obtain higher yield and net return of sabai grass under rainfed situation in Mayurbhanj district of Orissa (Barik, 1997).
5. Major weed flora observed in the field were Dactyloctenium *aegypticum, Borreria spp., Heteropogon spp., Cyperus iria, Cynodon dactylon, Sida rhombifolia, Tephrasia purpurea, Digitaria setigera* and *Dichantjium annulatum* accounting for nearly 60 to 70% of total weed population.
6. Crude fibre content of sabai grass was the highest with use of 2,4-D (sodium salt) which might be due to increased translocation of foliar applied 2,4-D (sodium salt) acting as a plant regulator promoting the growth of vegetatively propagated cuttings (CAST 1975)
7. Thus, either 2,4-D (sodium salt) at the rate of 0.5 kg a.i./ha as post-emergence spray at 20 days after planting or basalin at the rate of 1.0 kg a.i./ha as pre-planting soil application with incorporation with one hand weeding should be used for effective control of weeds in sabai grass (Barik, 1998).

CHAPTER 12

Production Technology of Crop: *Commercial Crops*

(Ecological Requirements, Cultural Practices and Crop Production Management)

SUGAR CROP

SUGARCANE (*SACCHARUM OFFICINARUM*)

According to the existing records the sugarcane was grown even before the Vedic period. It is mentioned in Sanskrit (one of the oldest language in India) by the name of Ikshu Ikshura and the sugar is called as Sarkara or Sakkara. From India it spread probably into China, Arabia and Egypt and after the Crusades, it was introduced into Sicily, Portugal, the Canary Islands and later into the New world. Now it is properly grown in various parts of the world such as Burma, Srilanka, Pakistan, Indonesia, Hawaii, Philippines, Indo-China, Thailand Central America, Egypt, South America, Mauritius, Australia, India, etc.

Distribution

Sugarcane is the native of India which is a major producer of sugarcane and sugar in the world. Other important sugarcane producing countries include Brazil, Cuba, Pakistan, Thailand, Philippines, Argentina, Colombia, Indonesia, South Africa and Egypt. Sugarcane is cultivated in an area of 19.5 m ha in the world with a production of 1192.8 m t. In India, sugar industry is one of the largest processing industries, next only to cotton textiles. There are 414 sugar mills in India and the sugar out-turn is at 239 m t. Production of cane has increased from a low level of 57.1 m t in 1950-51 to 241.0 m t in 1990-91. Currently cane production is at 282.9 m t. Between 1950-51 and 1994-95, the area under sugarcane recorded a compound growth rate of 1.81%, yield by 1.19% and production by 3.02% per annum. The contribution of area in increasing the output was more than that of productivity.

Sugarcane area has spread to tropical and sub-tropical belts of India. The area under sub-tropical beltis more than 70 per cent, however, productivity is less due to climatic disadvantage. Hence, the national average productivity is lowered to 68 t/ha which is only a fraction of the actual potential. In India, Uttar Pradesh is the major sugarcance growing state contributing about 48% of area and 40% of total production. Other important states growing cane are Karnataka, Maharashtra, Andhra Pradesh, Tamil Nadu and Bihar (Table 12.1).

TABLE 12.1: Sugarcane production in different states of India, 1995-96

States	Area ('000 ha)	Production ('000 t)	Yield (t/ha)
Andhra Pradesh	213.5	15216.3	71.3
Bihar	123.2	5667.3	46.0
Gujarat	161.6	10511.2	65.0
Haryana	120.0	7010.0	58.4
Karnataka	313.2	24917.9	79.6

States	Area ('000 ha)	Production ('000 t)	Yield (t/ha)
Maharashtra	580.0	46656.1	80.4
Punjab	132.0	8620.0	65.3
Tamil Nadu	314.3	34576.4	110.0
Uttar Pradesh	1974.4	119830.4	60.7
Total	**4138.5**	**282945.4**	**68.4**

Source: Fertilizer statistics, 1996-97.

Economic Importance

In India, sugarcane is mainly used for the production of sugar, gur and khandasari. About 50% of cane production is used for production of white sugar, 40% for gur and 10% used for seed, feed and chewing purpose. Numerous producers of gur and khandasari are the main feature of the milling sector in India. Between white sugar and jaggery, the later contains more calcium and iron (Table 12.2). The by-products of sugarcane include molasses, bagasse, filter mud, sugarcane wax and cane trash. Molasses is used for alcohol production.

TABLE 12.2: Nutritive value of sugar (per 100 g)

Produce	Energy (k. cal)	CHO (g)	Ca (mg)	Fe (mg)
Sugar	398	99.4	12	0.15
Jaggery	383	95.0	80	2.64

Source: Srilakshmi, 1997.

Classification of Sugarcane

According to old classification the cultivated sugarcanes were named as *Saccharum officinerum* and the wild 'Kans' as *Saccharum spontaneum.* But in the recent classification, many plants originally under *Saccharum*, have been changed to separate allied genera viz., Erianthus, Sclerostachya and Narenga. Thus now the *Saccharum* has five spp., namely *S. officinarum, S. barberi, S. sinense, S. sponteneeum* and *S. robustum.*

Varietal Improvement

Flowering is essential for seed set. The natural climatic conditions at Coimbatore were found ideal for flowering and seed set and so Sugarcane Breeding Institute was started at Coimbatore in 1912 with a objective of evaluating suitable, improved varieties for sub-tropical India. The first hybrid variety 'CO 205' was successful in North India. The release of CO 312 for sub-tropical India and CO 419 for tropical India were the major achievement in varietal improvement. Early ripening varieties had an appreciable impact on sugar production. Short duration varieties play an important role to enhance sugarcane productivity per unit time and input used.

Important Varieties

Short duration varieties	:	CO 997, CO 6808, CO 7704, CoS 510, CoC 671, BO 120, CoS 687, CoC 92061, CO 85004.
Mid-late varieties	:	CO 740, CO 1148, CoS 767, CO 86032.
Red-rot resistant varieties	:	CO 1148, CO 6907.

For high sugar and tonnage	:	CO 7201, CoC 671, CO 7634, CO 6304, CO 8153
Low water requirement	:	CO 1158, BO 70, CO 997.
Water logged area	:	CO 1007, CO 1157, CoS 633, CO 975, CO 853.
Ratooning ability	:	CO 1148, CO 419, CO 740, CoJ 46.
For jaggery manufacturing	:	COG 93076
Low temperature and stress conditions	:	CO 1148, CoJ 64.
Saline or alkaline	:	CO 453.

A number of improved varieties of cane under different maturity groups tolerant to drought or water stagnation are recommended for growing in Orissa. In choosing varieties for factory areas, early, medium and late maturing varieties should be planted in staggered manner in order to ensure a continuous supply of cane to the factory. The recommended varieties with their characteristics are given below. Any variety planted early in November or December will, however, take longer time to mature. If planted late also beyond March, their duration (peak maturity period) will be shortened.

CO 997	:	Chrome yellow with light green, Drought tolerant, matures in 10 months, Best suited for well drained high land.
COC 671	:	Yellowish green, Mature in 10 months, good yield high sugar.
CO 527	:	Ashy green with yellow, Quick growing, matures in 10 months, good Ratooner, high juice quality.
CO 975	:	Light green with chrome yellow at joint, Initial vigour slow, maintains high juice quality longer, tolerates water stagnation, mature 10-12 months.
CO 6304	:	Greenish pink, Thick and high yielding variety with high sugar recovery per cent, matures in 12 months, moderately resistant to smut and can tolerate drought condition. Good ratooner non lodging type.
CO 419	:	Green with purple tinge, turning complete purple on exposure, Late maturing, brittle tolerates water logging, matures in 12 to 14 months.
CO 740	:	Green, Profuse tillering, heavy yielder, good ratooner, drought resistant, maintains quality longer, prone to lodging, matures in 12 to 14 months.
CO 62175	:	Yellow green, turns dark with purple tinge on exposure, Non-brittle, thick canes, less susceptible to smut and red rot, heavy yielder, matures in 14 months, suitable for late crushing.

Besides above CO 7508, CO 7704, CO 8401, CoA 7601, CoA 7602, Co 7219, CoT 8201, Co 8402 are also recommended for commercial plantation.

Field Preparation

For flat planting of cane or for planting in furrows, thorough land preparation is necessary. But for trench planting it is not necessary to plough the land more than two times. Planting of cane in trenches is the best method of planting. Trenches should be 40 cm., 20 cm deep and 90 cm apart from centre of one trench to the centre of the other. Drainage should be provided where necessary. For early maturing varieties spacing between trenches may be reduced to 75 cm. Crops in case planted after March, the spacing may be reduced to 60-75 cm.

Trench method of planting has the following advantages,

1. Drainage is facilitated.

2. Weed growth is considerably reduced.
3. Early shoot borer infestation is reduced.
4. Irrigation becomes easier and also less water is required thereby lowering irrigation cost.
5. Better anchorage is provided which prevents the crop from lodging.

In saline soil and low lying areas likely to remain water logged during early stages of the growth of the plant, trench method of planting should not be done. In shallow soils also trench method of planting may not be followed. In shallow soils also trench method of planting may not be followed.

For intercropping furrow method of planting is to be followed.

Climate

In general, January to March is the period of planting and December to March is the period of harvesting. After the harvest, generally a ratoon crop is cultivated. In many counties, 2-6 ratoon crops are allowed. Sugarcane requires large sunny days which should synchronize with sufficient amount of water, with an optimum temperature on a reasonably longer growing period. But there must be at least a check in growth (after maximum growth) for attending maturity.

Soil

Heavy but friable soils with high water retaining capacity, good drainage and high in organic matter content and soil fertility are the best soil types for sugarcane. It can tolerate moderate acidity and alkalinity but soils with slight acidity or having pH of 6.0 are preferred for higher crop yields. Deficiency of N and P retards crop growth. Thus soils rich in P and Ca are good for better juice quality, lodging resistance, soils rich in iron are good for higher sugar recovery and soils rich in P, Ca, N, K and organic matter are the best in all respects.

Seed Rate

a) Early varieties: 12 t/ha (ensure one lakh millable canes/ha of one kg each)
b) Medium varieties: 10 t/ha
c) Late varieties: 8 t/ha (87000 two budded sets or 40 to 50 thousand 3 budded sets: 8-10 t/ha)

Seed Treatment

If possible, the top one third of the cane for planting may be selected. This portion of the cane gives almost 100% sprouting. The whole cane from 6 to 8 months crop can also be used for seed purposes which will give high percentage of sprouting. The cane setts may be dipped for 12 to 24 hours in lime water. Generally, 500 g of quick lime is mixed with 200 litres of water and the setts are soaked in this solution. This treatment with lime water hastens sprouting.

After treatment with lime water, the setts are dipped in 0.5% Emisan or 0.25% Aretan solution for 30 minutes just before planting. By this treatment the setts do not dry quickly and the buds are protected from fungal attack.

Planting

The requirement of setts depends upon the thickness of the cane, distance between the rows and length of the internodes. About 30,000 to 40,000 setts containing 3 buds will be required for planting one hectare (Ensure one lakh millable canes/ha of one kg each).

1. After planting the setts, 5% Aldrin dust is applied over the setts at the rate of 30-35 kg/ha. BHC 10% dust can also be used but it is inferior to Aldrin. The treatment protects the setts from white ant attack and also against early shoot borers. Alternatively, 0,2% Aldrin or Dialdrin or o.1% Lindane solution may be sprayed on the setts and soils around at the time of planting.
2. Setts are planted end to end if 6 to 8 months full canes or only top portion of the cane have been selected for planting. In other cases bud to bud planting is advisable to ensure adequate plant population as all the buds in the middle and bottom portions of mature canes may not sprout.

Different methods of sowing are followed in the county such as (1) Planting in flat beds, (2) Ridge and furrow method, (3) Trench method or Java method, (4) Rayungan method and (5) Distant planting method. Other planting methods are also adopted like (a) Sablag or sprouting method, (b) Tjeblock method, (c) Bud transplanting and (d) Algin method of sugarcane planting.

Plant Population

Adequate plant population should be maintained in the field for getting high yields. There should be about 40,000 sprouts/ha for thick canes and 45,000 sprouts/ha for thin canes in the field which will give a stock population of about 1,00,000 and 1,15,000 per hectare respectively. Planting top portion of the cane will give high germination.

The spacing should be maintained as follows:

(a) Early varieties .... 60 cm row to row
(b) Medium varieties .... 90 cm row to row
(c) Late varieties .... 90 cm row to row.

Fertilizer

Sugarcane is a heavy feeder. The best results are obtained by using fertilizers in conjunction with organic manures. It is advisable to apply fertilizers on the basis of soil test results. Where this is not done, FYM or Compost at the rate of 10 t/ha should be applied along with 200 kg of N, 60 kg of P_2O_5 and 60 kg of K_2O per hectare.

All the P and K are applied in trenches at the time of planting. In short duration varieties, 50% N is applied at 30 days after and the rest 50% N 60 days after planting. In long duration varieties, N is applied in three equal splits *i.e.*, 30, 60 and 90 days after planting (For February planted crop 90 days after split application to be delayed if irrigation is not available). N should not be applied to sugarcane crop 90 days after sowing as late application reduces sucrose content and total yield.

Intercultural Operations

The sprouting of buds is completed within 25 to 30 days. A light hoeing and partial filling of trenches should be done at this stage. This practice prevents attack of early shoot borer. The first hoeing and earthing operation is done within 30th day of planting The first top dressing of Nitrogen may be done before earthing so as to cover the fertilizer under the soil. The second hoeing and earthing operation should be done at 60 days of planting. The second dose of N should be incorporated into the oil before second earthing. Weeds can be controlled by the application of herbicides like Atrazine at the rate of 1 kg a.i./ha as pre-emergence spray.

Wrapping and Propping

Keeping erect the canes, results in better juice quality. For this purpose wrapping and propping are useful practices. When the crop is 4 to 5 months old, remove borer affected tillers and late formed tillers, tie the can shoots in two's or three's with partially dried lower leaves. So also late tillers and water shoots are formed from October onwards should be removed from the main stem. They do not mature in time and spoil the quality of juice it crushed along with the main crop. The practice of wrapping may be repeated two or three times, each time interlocking more cane shoots. The upper portion of the shoots should be tied as the canes grow in height. The recent method of wrapping sugarcane is chain method. The canes are wrapped by chain method each row separately without any support with bamboo to help the canes to stand erect. Stripping of the dried leaves of bottom portions of canes will not active the set roots and suppress the development of buds.

Water Management

For ensuring quick germination, the trenches should be irrigated before planting of the cane sets. This should be followed by light irrigation periodically to keep the soil moist which will ensure germination within three weeks and result in uniform growth and stand. The crop is irrigated till the onset of monsoon. In the hot summer season irrigation of the crop at 7 to 15 days interval will be necessary depending on the soil texture. The critical period for irrigation is between 45 to 60 days of planting and irrigation at this stage should be ensured. Water logging should be avoided. Proper drainage is necessary.

Plant Protection

The cane crop should be protected from pests and diseases. The common pests are early and late shoot borers which cause dead hearts, pyrilla that suck the juice of the leaves in all stages, white ants which damage germinating sets and the mealy bugs. The most common diseases are red rot and smut. The selection of the disease free sets is very important to prevent disease infestation through seed sets. Recommended plant protect measures of the state should be taken up to protect the crop from pests and diseases.

Harvesting

The mature cane should be harvested when the brix reading reaches 18 or above. Hand refractometers have been provided in all the important cane growing centres and this should be used for testing of the juice and for advising the farmers for harvesting the crop at the right stage.

For cane supply to the factory for crushing cane should be transported immediately after harvest. Maximum recovery takes place when the fresh cane is crushed within 24 hours after harvesting.

Inter-cropping

Green gram, black gram and groundnut can be grown profitably as an inter crop are sown between the rows of sugarcane at the time of planting. As the intercrops are harvested earlier, they do not suffer from the shading effect of the luxurious growth of sugarcane.

Extraction of Juice

Use 3-roller bullock drawn crusher for extraction of juice (Extraction is about 50%). Power crushers can be used for this purpose to increase the efficiency to about 60%. Crushing by 6-roller hydraulic crusher with a crushing capacity of 4 t/hr gives 80% extraction and can increase recovery of gur from 10 to 12 per cent.

Purification of Juice

Boil the juice for concentration and removal of dissolved foreign material in a circular or rectangular pan made of iron over a furnace. Normally bagasse is used as fuel. Remove the suspended impurities and gummy colloidal substances in juice during first heating lasting for about 30-45 minutes. Then the juice is clarified by adding chemical or vegetable clarificants. Liming to pH of about 6.4 for immature juice, to about 6.0 for mature juice and to 6.8 to 7.2 for deteriorated juice is optimum for obtaining gur of good colour and hardness. Generally, 50 g of lime is made to solution and added to 40 litres of juice. Crushed Bhindi bark at the rate of 20-25 g or 1 kg of crushed groundnut kernel is added in 40 litres of juice for this purpose.

Concentration

The striking point temperature ranges from 118°C to 123°C. Bad quality juice requires a higher striking point. The correctness of striking point is judged in the following manner. When the boiling progress sufficiently, a small quantity is taken out of the pan and put into a busket of cold water. If the mass forms a ball with a metallic sound when thrown against the side of the pan, the boiling is taken as complete and gur is collected into containers.

Storage

Gur may be collected into galvanized iron drums, tins and alkathene or alkathenelined gunny bags and stored in closed rooms. Desicating agents like calcium chloride or CaO in sufficient quantity, if available may be kept in the room to absorbed excess moisture and keep the room dry (RH. 50-60%). Initial moisture in gur should not exceed 6%.

Quality of Gur Deteriorates Due to

1. Its hygroscopic nature,
2. Absorbs moisture.
3. Inversion.

Ratooning Sugarcane

Ratooning of sugarcane is one of the important methods of reducing cost of production through elimination of seed cost and preparatory cultivation charges. Ratooncrops in general mature earlier than the plant-crop.

The following practices for raising a successful ratoon crop.

1. Harvesting of canes at ground level to avoid upper buds to sprout.
2. Stubble shaving operation with a spade within a week to allow lower buds to sprout effectively. Irrigate the field and dismantling the ridges as to encourage the lower buds to germinate.
3. Necessary gap filling to obtain adequate plant population.
4. Trash mulching to help quicker germination of buds, conservation of soil

moisture, and suppression of weeds and reduction of incidence of early shoot borer.

5. Hoeing of the land for suppression of weeds and better aeration.
6. Use of recommended doses of fertilizers. Ratoon crop requires 25% more nitrogenous fertilizers than plant crop. In general, 250 kg of nitrogen, 60 kg of phosphorus and 60 kg of potash per hectare may be applied. All P and K along with 50% of N may be applied at first hoeing (*i.e.*, 15-20 days of harvest of plant-cane crop). The rest 50% N may be applied as top dressing at 60 days of harvest of plant-crop followed by earthing.
7. Providing irrigation immediately after fertilization and then as and when required at an interval of 15-20 days depending on the type of soil and moisture retentive capacity of the soil.
8. Detrash the leaves as required.
   a) After harvest of the previous plant crop the field may be made clean.
   b) The stubbles are to be sprayed with 0.2% Bavistin or Benlate immediately after harvest of the crop.
   c) 1500 ml of Quinalphos or Monocrotophos per hectare may be sprayed on the standing crop to check the attack of shoot borers.
   d) Dithane M-45 or Dithane Z-78 at the rate of 3.5 kg or Blitox-50 at the rate of 5 kg per hectare may be applied in the standing crop to check red rot and smut disease. Spraying of 0.2% Bavistin or Benlate towards middle of June will also effectively control red rot and smut diseases.
9. Wrapping and propping operations should be done to keep the canes erect.
10. Harvesting of canes on the basis of hand refractometer reading (more than 18).

SUGARBEET (*BETA VULGARIS* L.)

Sugarbeet belongs to family Chenopodiaceae and accounts for about 50 per cent of the total gugar production. Beet roots contain 15-16 per cent sugar. The sugar recovery, however, does not exceed 10-12 per cent. The rich sugarbeet tops, containing 10-12 per cent protein and 60 per cent total digestible nutrients, make nutritive feed for livestock. The beet pulp containing 60 per cent carbohydrates and 5 per cent crude protein is also a good cattle feed.

There is no distinct record regarding the origin and history of sugarbeet but in 1747 it was found for the first time that beet roots are rich source of sucrose and it took nearly 50 years for developing a successful method to extract beet root juice and its crystallization in the form of sugar. F.C. Achard was the first man to set up a sugar factory for processing of beet-roots and making sugar out of their pulp. From Europian countries it was introduced in USA as a sugar crop in early 18th century and first sugar factory was built at Northampton, Massachurits with great problems. Thereafter another factory was established in California (USA) in 1870 after rectifying the defects and it started working successfully since then.

Sugarbeet is essentially a crop of temperate countries like USA, Russia, Sweden, Denmark, German, Belgium, Netherlands, France, etc. Now in countries like Syria, Iraq, Iran, Algeria and Israel sugarbeet is fast assuming its commercial importance. It is now being introduced in Pakistan. Todat sugarbeet is contributing to about 44 per cent of the world sugar production.

Until recent past sugarbeet was famous as a 'Salad' crop for table purposes in India but it is now in the experimental and semi-commercial testing in sub-tropical parts of the country. In Kashmir Valley sugarbeet is grown as a promising rainfed spring season crop. In Sriganganagar area of Rajasthan it is grown on commercial scale after producing first beet-sugar in 1968 by Ganganagar Sugar Mill.

In considerable research work has been done on its varietal performance on yield at G.B. Pant University of Agriculture and Technology, Pantnagar and the National Suar Institute, Kanpur, has been actively involved in processing and manufacture of sugar from beet-pulp, although sugarbeet is yet to be accepted as a commercial crop in India because it is at testing stage in some of the suitable Northern parts of India viz. Jullundhar, Karnal, Muzaffarnagar, Shahjahanpur, Lucknow, Gorakpur, Pusa, etc. The crop is being tested for its suitability as a sugar crop and its performance as compared to sugarcane. In Kashmir Valley and Himachal Pradesh it holds out a promise as a rainfed spring sown crop. The crop, in general, may be very successfully sown in Punjab, Haryana, Rajasthan and U.P.

Botanical Description

It is a biennial plant and does not produce seeds until the second year. But it is usually cultivated as an annual for its sugar.

Roots: The sugarbeet plant has a larger tap root. The top root forms the tuberous root containing sucrose. There are vertical grooves along opposite side of the root from which lateral roots arise. The mature root is divided into three regions.

- **The crown:** This is a convex cone.
- **Neck:** This is broadest part of beet.
- **The root:** the lowermost part which terminates in a long tap root.

Stem: It is very much reduced in size. The leaves arise in basal rosettes on the much condensed stem. The leaves are simple and smooth with a large blade and a petiole.

Inflorescence: It is a terminal open panicle with spicate branches. Flowers are imperfect, regular and without petals.

Fruit: The sugarbeet has a typical aggregate fruit that yields a seed ball with two or more viable seeds, or germs.

Production Technology

Climate

It is successfully grown in any region with temperature ranging from 12-45°C during the crop season. The favourable soil temperature for germination is around 15°C. For proper growth and sugar accumulation an average temperature of about 20-22°C is ideal. Temperature above 30°C retard sugar accumulation. The crop is, however, highly tolerate to frost and cold. The sugarbeet needs high moisture associated with cold climatic conditions throughout its growing period. The commercial cultivation of the roots is possible on the plains during winter season but the seed formation becomes impossible because of unfavourable climatic conditions. Therefore, seed production is done on the Hill of Himachal Pradesh by the Government of Himachal Pradesh with associated efforts of the National Seed Corporation. The seed production programme is given a full impetus by a Danish firm 'De Danske' 'Sukkersabrikker'.

Soil

Sugarbeet prefers loam to clayey loam soils with good inherent fertility and sufficient organic matter having a near neutral reaction though it has a high tolerance to soil alkalinity and salinity. The acidic soils are not al all good for the crop. The crop can not withstand water-logging and hence poorly drained soils are unfit for beet-root production.

Land Preparation

The land is prepared to a good tilth by repeated ploughing and planking operations. After obtaining a fine tilth the field is finally divided into ridges and furrows or simple flat beds based on method of sowing to be adopted. Usually sowing on the ridges is preferred over flat bed planting. Ridges of 10 to 12 cm height are prepared at a distance of 50 cm between them by the help of ridger after mixing manures and fertilizers in the field.

Manures and Fertilizers

Organic manures such as compost or F.Y.M. at the rate of 20 to 30 tonnes per hectare should be applied before the ridging is done. The crop should be supplemented with 100-120 kg of nitrogen, 80 kg P_2O_5 and 100 kg of K_2O per hectare. It is advisable to nitrogen in three equal splits, namely first at sowing, second after thinning and third after earthing up operation but never beyond December as it lowers the quality of beet-roots.

After every top–dressing the crop should be given a light irrigation. From the experimental results it is observed that the crop responds equally to all the sources of nutrients but, if possible, fertilizers containing chloride ions should not be chosen for the application in the crop. Deficiency of micronutrients has not been noticed but the studies are in progress at various sugarcane research centres in North India.

Sowing

In the North Indian plains sugarbeet is grown as a *rabi* season crop and the most ideal time for sowing is October. It is observed that if sowing is delayed beyond October the yield and quality (sugar content) are found to be affected adversely. There are two commonly adopted methods of sugarbeet sowing viz. ridge sowing and flat bed sowing. The depth of sowing should be 4-5 cm with row to row 50 cm and plant to plant 10-15 cm spacings. This requires about 10 kg seed/ha. Generally, sowing is done by dibbling manually or by drilling with cotton seed-drills. Indian Institute of Sugarcane Research, Lucknow, has designed and fabricated a push-type drill for sowing sugarbeet.

Seed Treatment

It is noticed that seed soaking for 4 to 5 hours in plain water helps in getting higher germination of beet seeds. To get still better germination the seeds may be soaked over night in 0.25 per cent solution of a mercurial compound or Aretan or Agallol. This is done by putting the seeds in a cloth bag and dipping the seed bag in the chemical solution until the seeds soak up the chemical. The seeds are then spread open for drying after which they may be used or sowing.

Irrigation

The crop can not tolerate drought and hence it is very sensitive to inadequate water supply but at the same time excessive irrigation is highly detrimental to the root quality of the crop. The frequency and the number of irrigation depends on various factors viz. seasonal variation, type of soil, organic matter content of the soil, etc. in general, a 2-2-4 irrigation schedule at the critical growth stages like formative, leaf growth and root development stage respectively holds a promise in beet-root production. On an average, 15-20 days interval between two subsequent irrigations is supposed to be an ideal. This way the crop needs about 8 to 10 irrigations during its growing period.

Interculture

The field should be kept free from weeds as the crop has a very poor tolerance to weed competition. Therefore, control of weeds during first two months of crop sowing is essential. Use of herbicides like 'betanol' and 'Rowneat' has given a good control of weeds in sugarbeet. The soil should be little loosened and earthing up of the crop should be done by December after finishing the last top-dressing of fertilizer.

Thinning

The sugarbeet seed is a 'multigerm' which gives out four seedlings per 'glomerule' (seed ball). To avoid the competition among themand to remain a single robust plant per glomerule it is necessary to pluck off remaining ones. The thinning should be done when the seedlings are having three to four leaves in them. Thus a uniform spacing of 20 to 25 cm between the plants should be maintained. For keeping a uniform plant population, sometimes gap filling is needed which should be done by sowing the fresh seed instead of transplanting the seedlings because transplanting results in production of malformed or 'fongy' roots having poor weight though transplanting is common in some areas. Gape filling should be done soon after germination.

Crop Rotation

Sugarbeet is susceptible to various soil borne diseases and their chemical control becomes very costly and difficult, therefore, a suitable crop rotation of long duration should be adopted. It is observed that any rotation of three to five years duration is the best for production of a healthy commercial beet-root crop in the country. Some of the best rotations are: Cowpea (fodder)–sugarbeet-cotton-sugarcane(plant)-sugarcane (ratoon); Maize (early) or early rice-sugarbeet-Sanai or dhainch (Green manure)-potato-sugarcane(plant)-sugarcane (ratoon); Guar+sugarcane (companion cropping)-sugarcane (ratoon)-sugarbeet-jowar(fodder)-maize-sugarbeet. Intercropping of sugarbeet with autum planted sugarcane is economically feasible and good for higher yield of both the crops.

Plant Protection

Pest: Young sugarbeet plants are often attacked by cutworms which cause a serious damage to the crop if they are not controlled at right time. Spraying of 5 litres of gamma BHC (20%) or Hectachlor of same strength diluted in 500 litres of water/ha along with rows takes care of the pest very effectively. The spraying may be repeated at four weeks interval in March and April, if needed. Beet leaf hopper (*Circulifor tenellus*) and beet army worm (*Laphygma exigna*) which damage the beet leaves may be effectively controlled by application of parathion. White grubs and beet-root maggot are the other serious pests which must be given due attention.

Diseases: Sugarbeet is susceptible to various diseases and the important of which are leaf-sport (*Cercospora beticola*), *curly* top (a viral disease), seedling rot (*Pythium spp*, *Sclerotium rolfsii* and *Rhizoctonia solani*) and downy mildew etc. Of these, the seedling rots are most important which may be partiallt controlled by the application of 20 kg of PCNB (Penta chloronitrobenzene) per hectare as a soil –drench before the appearance of the diseases. Seed-polishing and seed treatment is also important measures to be adopted as a prophylactic measure for the control of seedling rot.

Harvesting

An October sown crop becomes ready for harvesting in Aprol/May. At the time of harvesting there must be enough moistire to facilitate the harvesting but the field shouls not be wet. The harvesting may be done by running a country plough along the rows on both the sides and lifting of the roots by a slight pull of *Kudali*. The adhering soil should be cleaned of the roots by a thorough shaking but the roots should not be washed in water for cleaning as it causes a rapid deterioration. The roots should be topped at the leaf-crown because allowing the leaves to remain on the roots affects the recovery of sugar adversely. A good crop of a hectare area yields about 300 to 500 quintals of beet-root having a sucrose per cent of 15 in March and 17 in May.

Seed Production in Sugarbeet

The seed production of beet-root is not possible on the plains because of very high atmospheric temperature. On commercial scale the seed production is undertaken in Kashmir and Kalpa Valleys by National Seed Corporation and Government of Himachal Pradesh. For producing the seed healthy roots are selected, they are cut-off about 1/3rd or their tapering length and again planted in nicely prepared and fertilized plots. An efficient plant protection measure and appropriate isolation from other variety are maintained and the seed is produced. The seed production industry has a very wide scope in India.

Sugar Manufacture from Sugarbeet

Sugarbeet gets spoiled very fast and hence it has to be processed in sugar mills within 48 hours of its harvesting, therefore, a well coordinated plan of harvesting and processing has to be followed. Thus, the beet roots must be grown near and around the sugar mills only. The best sugar is extracted through extraction process by using special diffusers based on the counter current technique. The beet juice characteristics necessitate the adoption of carbonation process only for clarification and hence clarification with diffusers and related accessories are essential to process sugarbeet in the traditional sugarcane based sugar mills in the country. Huge quantity of fuel is needed for heating the juice.

Sugarbeet by-Products and their Utilization

One hectare beet crop produces about 5 to 10 metric tones of beet tops which may be used as cattle feed for the milch cattle to increase the yield of milk. Freshly harvested tops contain Oxalic acid and hence they are contraindicated when fed fresh. This bad effect may be counteracted by the addition of 60.0 g of finely ground lime per 100 kg of tops. Under abundant supply of beet tops they may be used as green manure and the tops of one hectare usually add 199 kg of nitrogen to the soil.

Beet pulp, after extraction of sugar may be used as cattle feed, either fresh or dried. The beet pulp may easily replace the grain from the feed concentrates. Mixing of molasses with pulp improves its palatability and taste of cattle. Sugarbeet molasses is extensively used as a raw material for various special fermentations. Besides, being a rich source of lactic acid and vitamins it is also used in many pharmaceutical preparations.

Varieties

Sugarbeet, being a recently introduced crop into India as a commercial sugar crop, has most f the foreign varieties. These varieties are mostly from Germany, Russia, U.S.A., U.K. and Denmark, etc. Following are the important varieties which have been observed doing well under Indianconditions:

1. Erotype E from West Germany.
2. Romoskava from Russia
3. U.S. 35 from U.S.A., U.S. from U.S.A.
4. Maribo Anglo Poly from Denmark, Maribo Magna Poly from Denmark.
5. Maribo Resista Poly from Denmark
6. Triplex from U.K., Bush E from U.K.

Now the diploid multigerm variety 'Romoskava 06' is presently recommended for commercial cultivation. The variety gives higher yield with superior quality besides a better resistance to important beet-root diseases. In addition to this two other Danish polyploids multigerm varieties viz., 'Maribo Magnapo' and 'Maribo Magapoly' have also proved to be the most promising. The seeds of the varieties are under multiplication by a joint venture of N.S.C. and the Government of Agriculture, Himachal Pradesh.

NARCOTICS

The product of these crops are used for stimulation *e.g.*, Tobacco.

TOBACCO (*NICOTIANA* spp.)

Tobacco is originated from America. India is the third largest tobacco producing country in the world, next only to USA and China. Tobacco is grown in an area of 4.8 lakh ha sharing 0.26% of the total cropped area. India's production of leaf tobacco is 544.8 m kg in 1992. Average yield is 1085 kg/ha. Tobacco is grown on a wide range of soils under different climatic conditions. Andhra Pradesh, Gujarat, Karnataka, Bihar, Maharashtra, Tamil Nadu, Uttar Pradesh and West Bengal are the important states producing tobacco in India. There are two species of tobacco: *Nicotiana tabacum* and *Nicotiana rustica.* The former is grown in a large area all over the country. *N. rustica* requires cooler climate and hence, its cultivation is confined mainly to the northern and north-eastern regions of the country. The varieties of *N. tabacum* are called as 'desii types. The plants grow taller, produce broad leaves and the flowers are usually pink in colour. The *N. rustica* types are known as 'vilayati' and Calcutta' tobacco. They are short plant types with round puckerered leaf and yellow flowers. Specific varieties in *N. tabacum* are used for cigarette, cigar, cherrot, bidi, hookah and sniff tobacco. *N. rustica varieties* are used for chewing and snuff tobacco.

Production Technology

Ecological requirements: Tobacco is a tropical crop in origin but it is grown in sub-tropical and temperate regions of the world. It grows from 60oN latitude in Sweden to 40oS latitude in New Zealand, China, India and USA.

The crop needs around 28°C as an optimum temperature for germination but it may germinate at a temperature ranging between 15°C to 35°C. The crop can not withstand frost and starts withering or wilting at a temperature above 35°C.

Tobacco needs 50-100 cm rainfall which should be well distributed throughout growing period. Areas having a rainfall of more than 125 cm are not fit for tobacco production, whereas a rainfall of less than 75 cm results in production of leathery leaves having high nicotin content in them. The leaves remain smaller and possess very poor burning quality.

Water logging at any stage of growth is deadly harmful because the plants can not withstand water logging. The hailstorm causes a serious injury to the leaves any time during growing period.

Relative humidity is one of the deciding factors for yield, quality and curing of leaves and it is observed that a relative humidity of 85-90 per cent is an optimum humidity for the crop because at this level of humidity the leaves are easily used and they do not become very brittle.

Wind storm at any stage of growth after seedling establishment is very harmful because the leaves split and reduce the quality especially when the plants have a large exposed leave surface.

Tobacco, in general, needs an open aerated, well drained light textured, rich in P_2O_5 and K_2O and Fe but poor in organic matter contents. Sodic soils are unfit for tobacco production because the plants absorb a lot of chloride ions which result in a poor buring quality of leaves and a mild acidic soil reaction is always better for the production of superior quality leaves. The suitability of soil, as mentioned under differs according to the types of tobacco to be grown. Tobacco finished products and suitable soil types

Finished products	Soil types
Cigarette	Light sandy soil
Filler (Cigar)	-do-
Binder (Cigar)	Heavy soil
Wrapper (Cigar)	Distinctly heavy soils.
Bidi	Light to medium loams of old alluvium origin
Hookah, Chewing and Snuff tobacco	Sandy to silt loam alluvial soils.

N. rustica is grown on heavy, low lying soils and contrary to this the *N. tabacum* is grown on light and high lands. Thus a good tobacco soil is one which has following characteristics:

1. Low inherent fertility.
2. Good drainage.
3. Light sandy surface soil.
4. Friable sandy loam sub-soil.
5. Low humus or organic matter content.
6. Acidic reaction (pH 5.0-6.0 preferably).

Raising Tobacco Seedlings

The seeds being so small can not be sown directly in the field, therefore, raising of healthy seedlings is an important aspect of tobacco cultivation and it requires a due thought over following points.

Selection of Site for Nursery Bed

The nursery beds should be made always on a higher spot so that they do not get flooded any time but there must be an assured irrigation facility. It is necessary that the beds have sandy texture so that they are open and well aerated for young seedlings. To get healthy seedlings it is advisable to change the place of nursery every year because it reduces the chances of occurrence of seed borne diseases and pests and their subsequent attack on the seedlings. It is always better to prepare raised beds on heavy soils for draining out any excess water from the beds.

Preparing the Nursery Beds

The seeds are sown either on raised beds (10-15 cm high beds having slopes on all the sides) or flat beds of any convenient length and 120 cm width with channels of 45 cm width between the beds. The seeds are very small, therefore, they need very fine tilth. The beds must be free from stubbles, weeds and soil borne diseases.

Application of Manures and Fertilizers

For steady and uniform growth of seedlings in the nursery beds it is essential to apply organic manures in the soil. Addition of organic manure improves physio-chemical properties of the soil besides checking the cracking of the soil surface. For this purpose addition of six to eight tones of well rotten FYM /Compost or 3 to 5 quintals of well powdered neem cake or mustard cake to the soil is recommended which will supply 25 to 50 kg N to the soil. Apart from the organic manures addition of urea, calcium ammonium nitrate or ammonium sulphate to supply 25 kg N is added in cases of very light or poor soils for maintaining the seedling vigour. The dose of N should be split into 4-5 equal amounts and should be topdressed at uniform intervals. The requirement of N, P and K by seedlings differs from type to type of tobacco as mentioned below:

Cigaretter tobacco .... 80 kg N, 60 kg P_2O_5 and 30 kg K_2O
Bidi tobacco .... 160 kg N, 80 kg P_2O_5 and 30 kg K_2O
Hookah tobacco................................ 100-150 kg N, 80 kg P_2O_5 and 40 kg K_2O

Seed and Sowing

This includes the following operations:

a) **Seed selection and seed treatment:** The seed should be of high purity and germination for economic and healthy seedling production. To sort out shriveled seeds or those which are not properly developed, one should pour the seed in a bucket containing water and the floating ones should be taken out. Sometimes to protect the seedlings from seedborne diseases the seed is treated with 2.5 per cent formaline solution or 0.25 per cent solution of Diathene Z-78 or Diathene M-45 before sowing them in nursery. Pre-sowing seed soaking in 50 ppm solution of Gibberellic acid for 48 hours increases and hastens the seed-germination.

b) **Seed Rate:** A seed rate of 3.5 kg /ha is found to be optimum for all types of tobacco. Under hot weather conditions a high seed rate 4 to 5 kg/ha is recommended, otherwise usually 2 to 3 kg seed/ha is sufficient.

c) **Sowing time of seed in nursery:** The sowing time of seed in nursery varies from area, type to type which may be mentioned below:

Sowing time	State	Type of tobacco
August-September, July-October	Andhra Pradesh	Cigarette, Cheroot, Bidi (Nipani area), Chewing and cigar type

Sowing time	State	Type of tobacco
May-June	Gujarat	Bidi
April-May	Tamil Nadu	Snuff, Cheroot, Chewing, Wrapper.
June-July, September	U.P.	Chewing
December	U.P.	Hookah
August-September	Bihar	Hookah and Chewing

d) **Method of sowing the seed in the nursery:** Usually the treated or water soaked seeds are sown broadcast on the surface of the beds and later they are mixed in the soil up to a depth of 1.5 to 2.0 cm with the help of rake. Before sowing, the seeds are mixed with ash or fine sand in 1:15 or 1:20 ratio for their proper or uniform distribution on the beds. Sometimes the seeds are also sown in shallow furrows of 2.0 cm depth prepared at 5.0 cm distance from each other. Later the seeds are covered with sand and leaf mould mixture and are pressed.

e) **Care of the seedlings in the nursery:** Proper care should be taken up to keep the nursery free from weeds, insect pests and diseases with adequate moisture.

Field Preparation

Tobacco seedlings need well ploughed, nicely pulverized and deep soil foe their quick establishment and better growth. For having this type of tilth the field should be ploughed once by a mould board plough followed by 3 to 4 ploughings by country plough and two to three harrowings along with planking after every harrowing. The field should have no weeds and left over stubbles.

Application of Manures and Fertilizers

The organic manures such as FYM or compost (at the rate of 10 to 25 tonnes/ha) and oil cakes (at the rate of 4 to 8 quintals/ha) should be applied at the time of first ploughing so that they are mixed properly in the field before planting the seedlings. The manuring should be supplemented with the fertilizers.

Transplanting of Seedlings

When the seedlings attain their required age, *i.e.*, 7 to 9 weeks in case of tobacum and 5 to 6 weeks in case of rustica, they become fit for transplanting. The nursery beds should be given light irrigation about 7 to 8 hours before uprooting the seedlings from the beds so that the roots are not damaged during their uprooting process. Healthy seedlings with intact roots are selected for transplanting and they should be transplanted soon after they are uprooted. Transplanting is done in rows which are made by help of a rope in the finally prepared field. If possible, the transplanting should be done in the evening and the field should be given a light irrigation for quick establishment of seedlings. The field should be watered after every 3-4 days so that the seedlings do not desiccate in the sun. The gaps created because of death of seedlings or because of their non-establishment should be filled 8-10 days after transplanting freshly uprooted ones. The tome of transplanting and plant spacing varies from variety to variety and place to place. Generally, the spacing from row to row should be 60 to 100 cm and from plant to plant 50 to 80 cm

Method of Transplanting

After seedlings are uprooted from the nursery beds they become ready for transplanting. The methods such as (1) General method of planting, (2) Furrow or salia method and (3) Flat ridge or Ghora method are adopted for planting the seedlings in the field in different places.

Irrigation

Tobacco can not tolerate water logging at any stage of growth but at the same time the plants may not withstand drought also. Therefore, a compromising situation would be to give light and frequent irrigation to the crop. However, the interval and number of irrigation depends upon soil type, weather of cultivation and type of tobacco to be grown. The field should be irrigated when the soil moisture falls about 20 per cent field capacity because higher soil moisture reduces the leaf quality. Usually hard water (saline) from wells is found to give better results but it must not have more than 50 ppm of chloride, otherwise the leaves get burnt and they have poor burning quality.

Intercultural Operations

About 20 days after transplanting, when the plants get established their young rootlets need a very fine and pulverized sub-surface soil for their rapid growth and further expansion. Besides, inter cultivation helps in mulching, removal of weeds, proper aeration, etc. In case of flue-cured, tobacco mulching with 3,600 kg of rice straw after inter-culture helps in preserving moisture and keeping down weeds but care must be taken to protect the crop from termites. The intercultural operations may be done effectively by khurpi or hand hoe in smaller areas and by harrows or cultivators in the large areas. The crop should be intertilled at an interval of 15 to 20 days or after every irrigation till the leaves are not damaged. This way the crop needs 3 to 4 inter cultural operations for proper moisture preservation, better crop growth and yield.

The tobacco crop is very often infested by Orobanche, a total root parasite. This weed is found to infest the field at the topping stage or at the later stage of the crop growth. Orabanche is found to be attached with the tobacco roots and they grow up through the soil round the tobacco plants. It is advisable to collect them or destroy them as soon as they are seen in the field. For chemical control of Orobanche a pre-planting if EPTC at the rate of 6.0 kg/ha (one month before planting) followed by post planting application of 2-4, 5T, 2,4D and 0.1% allul alcohol have been recommended.

Topping

Topping is the process of removal of flower heads either alone or with few upper leaves from the plants. The main objectives of topping are to improve the size, body, texture and quality of leaves. The height of topping and number of leaves at which topping is done vary from type to type of tobacco. The number of leaves in the plants to be topped, are as follows:

Type of tobacco	Number of leaves left on the plant
Cigarette and Cheroot	8-9
Wrapper (cigar)	10-12
Hookah, Chewing and Bidi	10-14

The energy and nutrients absorbed by the plants are diverted to the leaves rather than flower heads. Thus, it helps in the full development of the top leaves or otherwise they will remain relatively shorter when the lower leaves are going to mature. It increases the leaf size, thickness and dry weight of the leaves to the extent of 84, 24 and 48 per cent respectively. The effect of topping is more pronounced in younger leaves than older ones.

De-Suckering

Following topping the dormant buds become active and they put forth branches technologically known as suckers. To get full benefit of topping, the suckers are to be removed periodically. The processes of removal of these suckers commence after 7 days of topping and two-third successive de-suckerings are necessary, otherwise, the topping becomes non-effective. The manual de-suckering is a time consuming, laborious and more expensive process; therefore, use of chemicals may be practiced.

Use of chemicals for de-suckering: Application of coconut oil to the top six axiles soon after topping suppresses emergence of suckers in flue-cured Virginia tobacco. In case of cheroot tobacco of Tamil Nadu application of 2% Naphthalene, Acetic acid in triethenolemin to the topped portion suppressed all the suckers. Use of 2% of Maleic hydrazide or Indol Butyric acid is also recommended for an effective de-suckering.

Piercing of the stem after topping with 20 to 25 cm long needle is found to be beneficial in improving yield and quality of tobacco leaves.

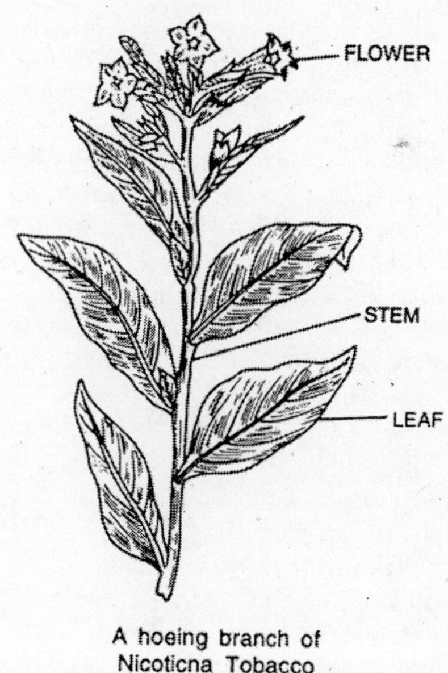

FIG. 12.3: A hoeing branch of Nicoticna Tobacco.

Harvesting of Tobacco

Tobacco is grown for the production of quality leaves which largely depend upon the stage of the leaf at which they are harvested. The sign of maturity and method of harvesting differs from type to type of tobacco. However, the sign of maturity, in general, may be mentioned as: change in colour of leaves from normal green to yellowish, thickening of leaves due to more starch accumulation to them, accumulation of gummy substances on them, flowering of plants, brittleness of the leave, etc.

Thus the leaves possessing all above signs are the matured ones which may be harvested but all of them do not mature at one time because of which the method of harvesting differs from type to type according to their pre-requisite for quality. In general, there are two methods of harvesting tobacco, namely – priming and stalk cut method. In case of priming the leaves are harvested as and when they are fully developed and matured. Priming is supposed to be the best method of harvesting as every leaf gets sufficient time and nutrients for its full development but if the plants are cut the younger leaves get lesser time to grow and as a result they remain under developed and immatured which ultimately affect the quality. Thus for the production of quality leaves priming is more appropriate method of harvesting, though it is more time consuming and expensive.

In case of stalk cut the plants consisting of all the leaves intact are cut from close to the ground. It is essential to see that the maximum number of leaves is matured so that the quality is assured.

Curing of Tobacco Leaves

Curing is a process by which the harvested leaves are made ready for marketing. In more technical terms it may be stated that 'Curing is a process of drying, decomposition of chlorophyll until the green colour disappears from the harvested leaves, hydrolysis of starch into sugars and respiration or fermentation of sugar thereafter'. In curing the mineral salts also crystallize out producing the grain of the leaves. The sole object of curing is to dry the green leaf in such a way that the final product had required colour, texture and aroma. Thus, curing is probably the most important operation in the production of tobacco and the method and efficiency of curing determines, to a large extent, the quality of the product. The methods of curing adopted in India are governed with several factors viz. tradition, convenience, market value of the finished product, economic considerations, etc. However, depending upon type of tobacco, the method of harvesting and maturity of leaves at harvesting, there are four distinguished methods of curing which are prevalent in India namely, (1) Air-curing, (2) Fire curing, (3) Sun curing and (4) Flue curing.

Curing is essentially a **drying process** where by most of the moisture of leaf is removed to impart required colour, texture and aroma to the final product.

1. **Air/shade curing:** practiced for wrapper tobacco in West Bengal and Sri Lanka.
   Tobacco in Andhra Pradesh three main methods –
   a) **Ground curing:** Hookah tobacco in U.P., Chewing tobacco in karnatake.
   b) Rack-curing:
      i) Tamil Nadu: Cigarette, cigar, cheroot, chewing-snuff
      ii) Maharashtra: Bidi and chewing
      iii) Andhra Pradesh: Cheroot and bidi tobacco
      iv) Assam: hookah
   c) **Pit curing:** not very common but mainly in Punjab, T.N., Maharashtra and Andhra Pradesh.
2. **Fire curing/smoke curing:** chewing tobacco in Tamil Nadu.

3. **Sun curing**: cigar and chewing, Bidi tobacco.
4. **Flue curing:** cigarette curing process consists of three stages-
   a) **Yellowing of leaves:** temperature 32-35°C and high humidity 85-90% for 30-40 hours.
   b) **Fixation of colour:** Any delay in drying of leaves, after yellowing is over, results in a development of a brownish tinge on the yellow leaves due to oxidation of tannin like substances called **sponging**.
   A sudden rise in temperature when leaves are still wet, results in a bluish black discoloration is called **sealding**.
   c) **Drying:** at 160°F to dry veins and mid ribs.

About one third of the county's total production of tobacco is used for the manufacture of cigarette. The cigarette types comprise flue-cured Virginia (f.c.v), natu and Burley tobacco. The f.c.v. contributes approximately 95% of the total production of cigarette types. Because of shift in the consumer's preference for cigarette smoking, the area under flue-cured Virginia tobacco has shown considerable increase in recent times. Tobacco as an important commercial crop contributes the highest revenue of Rs 2500-3000 crores to the government and a foreign exchange of Rs 500 crores is earned annually. India exports tobacco mainly to Russia, U.K., Bulgium, Nepal, Germany, France, Czch Republic, Slovakia, Poland, Netherlands and U.S. In India, the Central Tobacco Research Institute (CTRI) is established in 1947 at Rajahmundry in the state of Andhra Pradesh for the improvement of tobacco research and production.

Yield

Yield in tobacco refers to the weight of dry and marketable leaves of various grades pooled together. The yield of leaves differs from type to type as mentioned below:

Type of tobacco		Yield of leaves in kg/ha (cured leaves)
Cigarette		750-900
Bidi		650-900
Chewing type	In Tamil Nadu	1600
	In Bihar, U.P. and West Bengal	800
Hookah	Deshi	& 50-850
	V Jayati	1200-1600

Varieties

The commercially accepted varieties which are grown for different purposes in various parts of the country are mentioned below:

Type of tobacco	Improved varieties	Area for which recommended
Cigarette (Flue cured Virginia)	Harrison special, Chatham, Jayasri, TMVRR 2, Decrest, TMVRR-3, Virginia gold, Kanakaprabha.	Flue-cured Virginia tobacco growing areas of Andhra Pradesh and Karnataka.
	Hicks	Flue-cured tobacco growing areas of Gujarat.
	White gold CTRI special	Light soils of East and West Godavery districts of Andhra Pradesh

Type of tobacco	Improved varieties	Area for which recommended
	402 GSH-3	Small isolated pockets in Nellore and Khamman districts of A.P.
	Dhanadayi	The East and West Gadavari districts of Andhra Pradesh
	Kellu 49, Bhojgund, Annekevi, Keliu 20, Surati 20, Anand 2, Anand 3, Anand 23, Anand 119	Kaira and Baroda districts of Gujarat and Belgaum district of Karnataka.
	Kunkumathiri, Akolgund	The unsure tract of Karnataka
Natu	DG 3 and DG 4, Prabhat	Guntur district of A.P.
Lanka	DR1	Islands of Godavari river, Khamman and War.
Cigar (Filter)	OL 10, Maryland, VV 2, KV 1, Havana, 1-462, Swanbileshman	Madurai and Tiruchirapalli district of Tamil Nadu
Cheroot	OK 1, Bhavani Special, Line 2331	Coimbatore, Salem and Dharmapuri districts of Tamil Nadu
Chewing	Vairam, Sona, Thangam, S 1, Bhagyalakshmi	Madurai, Coimbatore, Tiruchirapalli South Arcot and Thajavur district of Tamil Nadu
Hookah and Chewing (N. tabaccum)	NP 70	Bihar and U.P.
	NP 34, DP 401, DD 413, DD 414, DD 415, DD 417	Darbhanga and Muzaffarpur districts of Bihar, Jalpaiguri and Coach Bihar Districts of West Bengal.
Hookah and Chewing (N. rustica)	NP 18, NP 20	Darbhanga and Muzaffar districts of Bihar
	T 238, C 302	Punjab and U.P.
	DD 405, DD 407	Jalpaiguri and Coach Bihar districts of West Bengal.
Wrapper	Dixie Shade	-do-
Burley	Ky 16, Ky 58	Guntur district of A.P.

Modern crop production technology has considerably raised output but has created problems of land degradation, pesticide residues in farm produce, gene erosion, atmospheric and water pollution. The natural resource base is degraded and diminished and the quality of the environment sustaining human life is adversely affected. Agricultural production has sustained man and great civilizations. The history of the world reveals that great civilizations flourished along irrigation sources and mismanagement of these resources saw the extinction of these civilizations. With exploiting population and rapid depletion and degradation of the natural resources base, Sustainable agriculture has assumed very great significance. The task of meeting the needs of the present generation without cording the ecological assets of the future generation is receiving top priority by environmental planners.

ROOT AND TUBER CROPS (STARCHY)
POTATO (*SOLANUM TUBEROSUM* L.)

Potato is a *solanaceous* tuber crop. It is relished as a vegetable either alone or in mixture with a number of other vegetables. Potato is mainly domestically consumed product. There had been gradual increase in the acreage but the productivity increased to the extent of around 250 per cent. Potato crop, though commercialized but not to the extent it is requireddue to shortage of storage space, lack of processing plants, lesser availability of quality seed-tubers, severe viral, fungal and bacterial diseases at production level and excessive glut during post harvest period.

Origin

Potato is supposed to be originated either in the breeless altiplano of higher Andean region of Peru and Bolivia, South America at elevations between 3500 to 4500 metres above sea level, or/and in Chile and on Island of the coast of Chile, on western coast of South America.

Potato plants were brought into continental Europe very soon after discovery of America. Some people believe that John Hawkins, the introducer of the slave trade, first brought them into Europe, probably in the year 1545 or 1562, while others mention that a Spanish Monk brought them from Equador in 1580. Moving to its new home in Europe in the 16th century it earned new name as poor man's crop in England, Ireland and Scotland. It caught the eyes of Frederick, the Great, in seven years, war during food shortage to meet the demand of the army. During the last two world wars Great Britain, Germany and many other European counties could solve their food problem largely through increased production of potato. It was probably the Portuguese who brought the potato to India, the first mention of it in this country began in the account of a banquel given by Asaf Khan to Sir Thomas Roe in Ajmer in 1615.

Classification

The commercial potato has two important species:

Solanum tuberosum: It is very common. Plants have shorter and thicker stem, larger and wider leaflets.

Solanum andigenum: It is not very common. Plants have thin and long stem, small and narrow leaflets. It has long stolons and mostly coloured deep eyed tubers.

Solanum demissum and *Solanumstenotonum* are two more species which are somewhat important as they are resistant to some type of virus and diseases but they are also not in cultivation commercially.

Botanical Description

The potato is a perennial but as a crop it is treated as an annual. It is mostly vegetatively propagated by means of tubers. Potato cultivation through true potato seed is a recent introduction in some countries, *e.g.*, China.

Roots: The root of potato is adventitious arising from the base of a sprout. The root growth is restricted to top layers.

Stolons: Stolons are the underground stems at the tips of which tuber formation takes place.

Tuber: The tuber is an enlarged underground stem produced on the end of a stolon and not on the roots proper. These tubers are morphologically stems. They process eyes, *i.e.*, buds and lenticels.

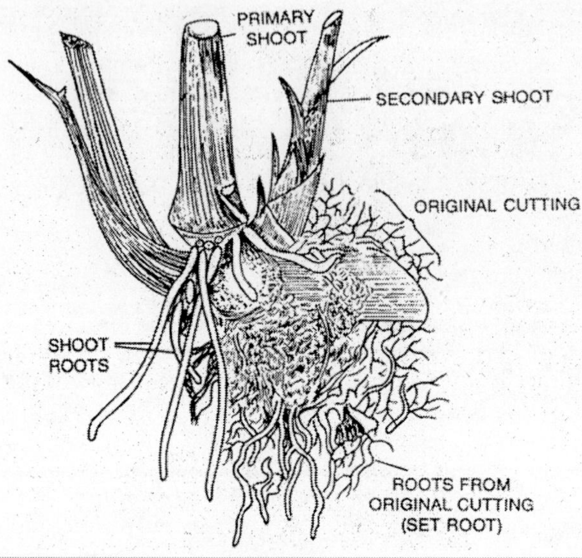

FIG. 12.1: An sprouted sett showing position of roots and shoots.

Tuber Composition (on Fresh Weight Basis)

Carbohydrates..........20.6%
Protein..........2.1%
Fat 0.3%
Crude fibre..........1.1%
Ash..........0.9%

Good amount of leucine, tryptophane and isoeucine also.

Aerial stem: The aerial stems are round or angular, pubescent or glabrous, green or purplish. The stem is of a branching type. It is erect when young and spreads as plant grows.

Leaves: Leaves of potato are alternate and compound. These arise along the stem in a spiral manner.

Flowers: The flowers of potato are in terminal clusters.

Potato Growing Zones of the Country

India has been divided into following agro-climatic zones for raising potato crop:

Western Himalayan Zone

It consists of hilly areas of H.P. J and K. and U.P. and the crop occupies an area of about 26,550 ha under potato cultivation since the zone comprises hills only, the zone has been sub divided into following three zones according to the height of hills from sea level:

a) **Very high hills zone:** Here the crop is grown at the altitude of 2500 to 3000 m above sea level. The soils of this zone are acidic in nature due to excessive leaching and washing of soluble salts having different depths and texture. The area is characterized with occasional snowfall, hail, storm and frost during winter due to which the crop is planted during summer (from May-June to September). The crop suffers critically from moisture stress during early growth stages especially in case of eroded soils of shallow depth and from excess moisture at tuber bulking stage. The highest recorded yield in this zone had been 4508 kg/ha, in Lahaul Spiti area.

b) **High Himalayan Hills Zone:** The zone is situated at 1,800 to 2,500 m height from sea level. The soils here also are acidic and sandy loam to loam in nature. The crop duration is same as that of the first sub zone. The crop here also suffers from moisture stress in early growth stages and excess moisture during tuber bulking stage.

c) **Mid Hill Zone:** This is situated at 1000 to 1800 m height. The soils are acidic and sandy loam to sandy clay loam of varying depths. The cropping period is January-February to May-June. The crop suffers from low temperature associated with occasional frost and hails during early growth stages and high temperature during tuber bulking and maturity.

Plains Zone of India

This zone is again sub divided into three sub-zones:

a) **North-western plain zone:** It is spread over Punjab, Haryana, Rajasthan and parts of neighbouring states accounting for 4.3% of total area under crop. It is situated at about 300 m above sea level consisting of deep soil ranging from neutral to slightly alkaline in reaction. Three crops of potato viz., early (autumn planted), main and spring crops are grown in this region. This is the prime zone for growing seed potato and the seed is supplied to North-eastern zone and peninsular area of the country.

b) **North-central Plains:** Western U.P. and M.P. belong to this zone occupying 25% of total area under the crop. The main crop of potato is mainly grown (October-November to February-March) which gives very high crop yield. Potato seed degeneration is caused due to aphids which carry virus X and S at 15-20°C on plains appearing like yellow mosaic. The rate of degeneration is 15-20% in first year, 25-60% in second year and over 60% in third year in the seed quality resulting into drastic reduction in potato tuber yields.

c) **North-eastern Plains Zone:** Eastern U.P., Bihar, Orissa and West Bengal plains belong to this zone covering about 30% of the total area in country. The soil varies from loamy sand to clay loam ranging from neutral to mild alkaline or acidic in reaction. The crop is raised during November-December to March-April.

North-Eastern Hills Zone

It comprises West Bengal, Assam, Meghalaya, Tripura, Nagaland, Sikkim, Arunachal Pradesh occupying about 10% of total area under crop in the country. The zone has got two potato growing seasons viz., January-February to August in higher hills and August-September to December in lower hills. The temperature during winter falls below freezing point and the crop also suffers from wart disease.

Low Hills and Plateau Zone

The zone situated at about 600 to 1000 m above sea level spreads over a vast area of central and peninsular India of M.P., Maharashtra, Gujarat, Karnataka and Orissa. The zone occupies about 7% of the total area under the crop in India. The soils are red to black in nature. Two cropping seasons viz., July to September as rainfed and November to February as irrigated crop are common. Constantly higher temperature does not suit raising of seed crop and the zone has to depend upon other zones for its seed requirement.

Nilgiri and Kodaikannal Hills Zone

It is situated at about 2200 m above sea level in Tamil Nadu. In this zone the potato is grown in three seasons (round the year). Summer crop (March-April to August-September) and autumn crop (August-September to December-January) are grown as rainfed however, the winter crop is irrigated which is taken in valleys during January to May.

Ecological Requirement

Climate: Potato is a temperate or cold loving crop which needs low temperature, lower humidity, less windy and bright sunny days. The crop needs fertile soil with high moisture content but is must not be wet because wet soils become compact which is not good for tuber growth and development. Potato needs about 25°C at the time of germination, around 20°C for vegetative growth but between 17-20°C for tuberization and tuber development. It is noticed that higher temperature has an adverse effect on the tuber growth whereas a temperature above 30°C stops tuber formation completely. It is probably because at higher temperature the rate of respiration increases and the carbohydrate formed by the process of photosynthesis is consumed rather than stored in tubers. Under higher temperature the plants are elongated, leaves become wrinkled and plants have sickly appearance. Tubers do not develop under such situations. Cloudy days, rains and high humidity are very congenial for spread of fungal and bacterial diseases; hence, such conditions are not good for the crop. In various parts of the country it is grown under conducive climatic conditions however, in the Nilgiri hills, the potato is cultivated throughout the year.

Soil: Potato prefers sandy or medium loam soils rich in organic matter content. The soil of the seed bed should be loose, friable, well aerated with good drainage. Light textured soils having friable soil structure and high humus content tend to promote more uniform soil temperature and better aeration which favour tuber development and harvesting becomes easier. Most favourable soil pH should be between 5-7.0 as the crop does not grow well in saline-alkali soils and under highly acidic conditions it suffers from scab disease.

Cultural Practices

Land Preparation

Tuberisation prefers ti have well pulverized tilth which may be obtained by giving two ploughings with mould board plough followed by two harrowing, plankings and picking up the weeds and plant residues or stubbles. Before planting the tubers it must be ensured that the soil has sufficient moisture required for germination of the crop.

Seeds and Sowing

Selection of seed stock: Potato is vegetatively propagated, hence, the disease pathogens are carried from the mother plants and the crop gets spoiled. Therefore, pure and healthy seed is the basic necessity for successful cultivation of the crop. Following points must be taken into account at the time of seed selection.

1. Seed tuber must be uniform in shape and size.
2. The tubers showing surface diseases like wart, scab, brown rot and nematode, infection must be separated out.
3. The tubers should not be shriveled.
4. The tubers should be in their right stage of sprouting.
5. The sprouts should be about 1 cm long at the time of sowing.
6. The tuber size should be preferably 3.5 to 5 cm in diameter.
7. The seed tubers should be purchased from a trusted agency like State Department of Agriculture, Seed Corporation. etc.

Seed treatment: The seed tubers are treated to protect them from an attack of fungal diseases when sown in the field and to get higher germination there from. The seed tubers should be treated with Agallol, Aretan, Captan, Captan or Dithane M-45 (by taking 1 kg in 45 liters of water) for protecting them from various fungal dieases. The period for dipping in solution may be from 20 to 30 minutes. While Ethylene-Chlorohydrin, Carbon disulphide, Rindite (a mixture of ethylene chlorohydrin, dichloroethane and carbon-tetrachloride in 7:3:1 by volume) thiourea, sodium thiocynate and Gibberillic acid are most effective chemicals for breaking dormancy. In Nilgiri the Great Scot Variety is popularly treated with carbon disulphide @ 1 0z for 32 cu ft or 56.7 g/1.8 cu m space in airtight container for 10 days followed by 7 days storage. At Patna, soaking of seed tubers in 1% thiourea for 1 hour followed by dip treatment in 1.5% solution of ethylene chlorohydrin then storage of treated seed pieces in air tight container for 24 hours is needed. Thiourea and ethylene chlorohydrin supplemented with GA gives best results.

Tubers should not be used for sowing just after taking out from cold store but they must be spread in cool, shady and ventilated place having diffused light for accelerating the sprouting of tubers as sprouted tubers give higher yield than unsprouted ones.

Sowing time: The optimum time of planting potato synchronizes with the period which maximum temperature is about 30-32°C and minimum around 18-20°C. The planting and harvesting time for different regions of India is given below:

Planting and harvesting time for different regions of India

Region	Crop	Planting	Harvesting
(A) Hills			
North-western	Summer	January-February	July-August
North-eastern	Summer	March-Aril	September-October
Southern	(a) Summer	March-April	August-September
	(b) Autumn	August-September	December-January
	(c) Spring	January-February	May-June
(B) Plains			
Northern	(a) Autumn	September-October	December-January
	(b) Winter	October-November	February-March
	(c) Spring	December-January	March-April
(C) Plateau	(a) Kharif	June	September-October
	(b) Rabi	October-November	February-March

Source: Potato in India (Tech. Bul. No.1, CPRI, 1971).

Seed size, spacing and seed rate: The ideal tuber size for planting should be 2.5 to 3 cm in diameter and approximately 50 g in weight. It is an established fact that potato yield increases with increase in seed size and decrease in spacing. However, the seed size and spacing depend upon the variety and the purpose for which the potato is grown. Higher seed rates are used for raising a good crop but lower (preferably cut pieces) for ware potatoes. A correlation of all these factors is given below.

Seed size	Spacing	Seed rate	Crop
2.5 cm x 3.0 cm in dia	45 cm x 15 cm	10 q/ha	Early crop
4.0 cm x 5.0 cm in dia	50 cm x 20 cm	25 q/ha	Seed crop
2.5 cm x 4.0 cm in dia	60 cm x 25 cm	20 q/ha	Ware crop

Method of Planting

There are three popular methods of planting potato tubers:

1. **Flat bed planting:** This method is common in areas having light sandy soils. In this method the tubers are planted on flat beds in very shallow furrows and planking is done. The method needs two earthings – first 25-30 days after sowing and second 40-45 days after sowing. It is not scientific and is not recommended to farmers.
2. **Planting in furrows:** The method is adopted in irrigated and light sandy soils where also two earthings are given on the days mentioned above in flat bed planting.
3. **Planting in ridges:** In this method the ridges of 30 cm height and 60 cm width are made and tubers are planted opposite on these ridges.

Manures and Fertilizers

Potato is a heavy feeder and needs high doses of fertilizers for best yieldof tubers. Role of organic manure is well established and according to organic matter content of the soil a dose of 10-15 tonnes/ha of well decomposed farm yard manure, leaf mould or compost gives better results. An application of 3 to 5 q/ha of well powdered neem cake helps to control the insect/pest and it also supplies plant nutrients. These doses of manures should be supplemented with required quantities of fertilizers as mentioned below.

Fertilizer Scheduling for Potato Crop

Type of the Potato crop	Rate of nutrients to be applied (kg/ha)			Time of application		Remarks
	N	P_2O_5	K_2O	Basal	Top dressing	
Early crop	100	60	100	50N+P+K	50N	Top dressing should be done at the time of earthing.
Main crop	120	80	100-120	60N+P+90K	60N+30K	-do-
Late crop	120	80	100-150	60N+P+120K	60N+30K	-do-
Hill crop	120	60	150	60N+P+120K	60N+30K	-do-

Type of the Potato crop	Rate of nutrients to be applied (kg/ha)			Time of application		Remarks
	N	P_2O_5	K_2O	Basal	Top dressing	
Mid hill crop	80-120	100	150	Whole amount	-	Ammonium sulphate, CAN, Ammonium sulphate nitrate, are better than urea. Super phosphate, DAP, and sulphate of potash are recommended. KCl should not be applied.

A part from major nutrients the crop needs micronutrients also which must be applied as the soil is losing them very fast. The amount of micronutrient salts recommended for potato production is given below.

Rate of micronutrient application in potato

Micronutrient salts	Soil application (kg/ha)	Spray application g/100 lit. of water	Tuber soaking treatment g/lit of water
Zinc sulphate	25	200	50
Manganese sulphate	25	200	50
Ferrous sulphate	10	200	50
Cupper sulphate	10	100	50
Ammonium molybdate	10	50	20
Sodium borate	1	50	20

Water Management

Water requirement of potato is supposed to be very high (500-600 mm) but scheduling irrigation at optimum moisture level during critical stages and sub-optimum level during rest of the period may result into about 30% saving in water requirement of the crop. The critical stages have been regarded to be (i) germination, (ii) tuber initiation, (iii) early tuber enlargement and (iv) late tuber enlargement. Though it is recommenced to irrigate the crop at a soil moisture of about 0.2 to 0.3 atmospheric tension at 15 cm soil depth (which comes to nearly 63% available water has been found to be optimum for better crop yields. The depth of irrigation must be limited to about 25 mm which is equal to CPE (Cumulative Pan Evaporation) for scheduling irrigation but such a shallow irrigation can be done through drip and sprinkler irrigation systems only. The farmers practice is to irrigate the furrows by 5.0 to 7.5 cm deep water which results into hardly 46% water use efficiency or even low. A saving of 38% water could be obtained by irrigating alternate furrows but it caused a reduction of about 10.7% in tuber yield (Bhan and Dharma, 1982).

Potato crop requires frequent and light irrigation at low moisture tension, irrespective of the varieties. This is particularly important with new varieties which produce large tubers. It is therefore, important to keep the ridges moist but not wet which may be attained by giving irrigation at regular intervals. Accordingly the first irrigation should be light and given 7-10 days after planting or even earlier depending upon soil moisture content. The subsequent irrigation should be moderate to heavy which should cover 2/3 rd to 3/4th height of ridge. The intervals in subsequent irrigations could be about 15 days in heavy soils and

10 days in light soils. Over flooding results in soil compaction of the ridges which results into poor aeration and poor tuber development. The crop should be given light irrigation if frost is expected.

Intercultural Practices

Potato tubers are modified under ground stems which use to synthesize anthocyanin and chlorophyll if exposed to sunlight, therefore, it is essential to cover the tubers fully with soil because with chlorophyll formation the tubers stop accumulation of starch and remain smaller and the solanin formation in green tubers results into bitter taste of tubers which is harmful, if consumed. Number of earthing and height of ridges after earthing depends upon method of planting and soil type. In case of flat bed planting two earthings –one at 25-35 days and second 45-50 days after sowing are given. In case of furrow planting method light ridging is done up to the height of 10-15 cm soon after sowing and second earthing is done about 30-35 days after sowing.

While earthing it is essential to loosen the soil around the plants, then fertilizer, needed for top dressing, is mixed with soil after which the ridges are made. Since, the roots are disturbed and partly damaged it is required to provide one light irrigation.

Weed Management

Weeds, like in other crops, are great menace in the field of potato crop and they cause a significant reduction in potato tuber yields, if they are not controlled at an appropriate time. They are many kinds of weeds associated with potato depending upon time of planting as for example in case of early planting *Trianthema spp. Phyllanthus niruri*, etc. are found but in case of timely or late planting *Cyperus dactylon*, *Cyperus rotundus*, *Aspergula arvensis*, *Launia splanifolia*, etc. are very common weeds. Of these weeds *Cyperus spp* and *Cynodan spp.* are very harmful to the crop and they cause a drastic reduction in tuber yield, if they are not controlled at the right time. Generally, in the villages weeds from the potato fields are manually eradicated as it results in loosening of ridge soil and also provides aeration to the crop plants but the method. Being very slow, expensive and cumbersome is being replaced by use of herbicides. However, for the best results it is advisable to adopt an integrated method for control of weeds.

Therefore, preparation of weed free seed bed followed by one weeding about 15-20 days after sowing and one weeding –cum-hoeing at the time of earthing about 30-35 days after sowing helps controlling for the weeds very effectively. The herbicides commonly used to replace mechanical weeding in potato production are given below:

Herbicides, their rates and time of application for controlling weeds in potato

Herbicides	Rate (kg/ha)	Time of application
EPTC	1.2-2.0	Pre-planting
Fluchloralin	0.7-1.0	Pre-planting
Atrazime/Simazine	0.3-0.5	Pre-emergence
Ametryn	1.0	Pre-emergence
Alachlor	1.0-1.5	Pre-emergence
Isoproturon	0.5-1.0	Pre-emergence

Herbicides	Rate (kg/ha)	Time of application
Linuton	0.5	Pre-emergence
Oxyfluorfen	0.15	Pre-emergence
Metribuzin	0.7-1.0	Pre-emergence
Dichlormate	1-2.0	Pre-emergence
2,4-D (80% sodium salt)	0.50	Pre-emergence
Metoxuron	1-1.5	Pre and early post emergence
Prometryn	0.5	Early post emergence(5-10% Germination)
Propavit	1.0	-do-

Source: Agro-techniques for intensive potato cultivation by Grewal, J. S., Sharma, R. C. and Saini, S. S. (1992) an ICAR Publication, New Delhi.

Plant Protection

The crop is found to be attacked by a number of insect/pests and diseases which damage the crop completely. Most of the fungal diseases originating from the seed are controlled by proper seed treatment as explained under the seed treatment heading. Under field condition attack of leaf blight is very common which can be controlled by spraying of Dithane Z-78 or Dithane M-45 by dissolving 1 kg chemical in 450 litres of water. The spraying should be started in late November to early December as prophylactic measure; Application of these chemicals will take care of leaf spot disease also. In absence of these chemicals blitox, blue copper L, Zineb or even Bordeaux mixture (5:5:50) can be used. Mixing of Brestan @ 0.6 kg/ha with Dithane has been found to be more effective.

Among the insect and pests, the sucking insects and cutworms are very harmful. The sucking insects may be controlled by one spraying of Metasystox at 1.2 litres/ha or Rogor at 1.0 litre/ha rate soon after their visibility in the crop. Cutworms may be controlled by soil application of 5% Aldrin dust at planting time (before last ploughing or harrowing so that the chemical is mixed well into the soil). Heptachlor may be used as substitute for aldrin dust. The rate of application should be 30-45 kg aldrin dust/ha or 15-20 kg of Heptachlor/ha. If the soil application is not done, spray application of Aldrin 30 EC at the rate of 6.0 litres/ha should be done over ridges soon after their attack.

Lifting of Potato Tubers

Lifting of tubers is done when haulms die down naturally and tuber skin is hardened, but before the soil temperatures rise above 27-28°C to prevent charcoal rot of tubers which is common in late-harvested crops in plains.

Post Harvest Care and Grading of Potato Tubers

Drying and curing: During harvesting tubers undergo a considerable skinning, bruising and other damages leading to rottage, if not handled properlyand these damages are much greater when tubers are immature at harvest. The potato on the hills is normally harvested in wet weather favouring greater susceptibility to bruising and rot–causing organisms like *Pythium*, *Phytophthora* and *Erwinia spp.* Where as the crop on the plains suffers from heat injury due to direct exposure to sunlight and temperature above 25°C for long periods favouring bacterial soft-rot (*Erwinia* and *Bacillus spp.*) and charcoal-rot (*Macrophomina phaseolina*).

Potato curing can be very properly done by heaping the tubers under shade in a ventilated place at round 15-20°C. The heaps can be made about 1-1.5 m high and 3-3.5 m wide for 10-15 days in Indo-Gangetic Plains. In the hills the potato is harvested under wet weather where the tubers should be spread in thin layers for proper drying of skin before heaping.

Sorting: Sorting leads to removal of diseased, damaged or rough tubers where as grading leads to classification of healthy tubers into sizes of accepted standards. These operations are essential in order to fetch good price in the market and to the uniform emergence of the crop.

Grading of tubers: In order to assure uniform quality of superior grade tubers and to get remunerative returns, it is essential to grade them according to their respective sizes. The potatoes of the plains may be grades as follows:

1. Seed size tubers,
2. Large size tubers, and
3. Chats

The potato tubers obtained from hills are classified or graded as under:

1. Special size with a diameter of 8 cm and above.
2. Phool size with a diameter of 3 to 5 cm, and
3. Ration size with a diameter of 2.5 cm and below.

While grading the tubers it must be kept in mind that damaged, diseased and tubers not true to the types should be separated out and discarded.

Storage

Potatoes are perishable in nature and have to be kept in proper stores to avoid rotting of tubers. In plains the seed tubers may be kept in cold stores between February-March to September-October. At places where cold stores are not available the potatoes may be stored in country stores. A thatched house be built in a cool place, preferably under the trees. The potatoes may be stored init on wooden or bamboo racks. Potatoes may be kept in layers 2—30 cm high and covered with a layer of sand. The ventilators of the store should be closed during the day time and late in the evening. The potatoes may be periodically examined to remove the rotting tubers and to break the large sprouts.

Potato tubers in the hills are subjected to 4.8 to 9.4% loss in tuber weight and poor emergence where as in the plains; seed tubers are stored in bags in cold stores at 4-5°C and 90-95% humidity. The installed storage capacity in India is around 4.5 m tones which is not enough to store even one-third of potato produce. The CPRI has developed store run on passive evaporative cooling which is more efficient and economic at high temperatures and lower humidities.

Quality Factors in Potato

Potato is no more a vegetable but it has become a staple food crop of India. Therefore, potato tubers have to be evaluated for seed quality and cooking quality separately. These factors have been described below:

A. **Seed qualities in potato:** No aspect of potato growing is more important than the selection of best quality of seed as the yield depends more on quality of seed to be planted then any other single factor. The following plants should be taken into account while evaluating the seed quality of tubers:
   a) True to the type.
   b) Adaptability of the variety in the region of which it has to be purchased.
   c) Abnormalities in the tubers.

d) Desirable size.
e) Sprouted conditions of the tubers.
f) Proper grading.
g) Freedom from insect/pest and diseases, etc.

B. **Cooking quality in potato:** This is most desirable quality in potato for which the consumer pays money. The factors responsible for a better cooking quality are mentioned below:

a) **Mealiness:** It depends on the case with which the cells of potato separate from one another on cooking so that they become piled up in a loose flavour, with air between them rather than remaining in a solid mass. Disruption of potatoes in cooking in partly explained by the turgidity of tissues, the more turgidity they are the more easily do they fracture and permit the ready penetration of boiling water. The mealiness may be calculated from the following formula:

$$\text{Mealiness} = \frac{\text{Albumin old matter}}{\text{Starch}} \times 100$$

Or

$$\text{Meanliness} = \frac{\text{Nitrogenous matter}}{\text{Starch}} \times 100$$

The mealiness may be interpreted as follows:

Based on Albuminoid matter		Based on nitrogenous matter	
Values	**Quality**	**Values**	**Quality**
8.5-14.0	Normal	20-25	very good quality
6.8-8.4	Slightly mealy	17-19	Good quality
4.3-6.5	Mealy	16.0	Medium
		Below 16.0	Poor

a) **Consistency:** The potato may be mealy or sometimes consistent also because the starch does not integrate due to high temperature and more starch. Thus the mealiness and consistency go together.

b) **Colour:** There are two main discolouration which appear in cooked potatoes:
   i) The browning of the chipped, fried and crisped potatoes brought about by the process of caramilization of sugar.
   ii) The blackening of potato pieces is developed due to the oxidation of tyesine and O. Dehydric enymes. In immature tubers more free tyrosine is present which gets blackened on wilting or skin drying of tubers. Sometimes blackening is due to ferrous compounds present in the tubers.

c) **Flavour:** The flavour is found to be related with oil content, mineral salts and ratio between contents of potassium and nitrogen. The burnt taste which occurs in chipped, fried and crisped potatoes is due to presence of sugars. The potatoes are graded into three classes viz., mealy, soggy and waxy.

Factors Affecting the Quality of Potato Tubers

There are numerous factors which affect the potato quality and some of them are discussed below:

1. **Hereditary:** This is also known as varietal or genetic characteristic which includes mealiness, consistency, colour, flavour, sugar content and tendency to blackening.
2. **Edaphic factors:**
   (a) **Soil Conditions**: Sandy soils produce better quality tubers than heavy soils. Both flavour and consistency of boiled potato are found to be better in potatoes produced from light sandy or red soils than those from clayey, silty or any other heavy soil types.
   (b) **Topography**: Potatoes produced on hills are of superior quality than those produced on the plains because of least attack of insect, pest and various types of diseases.
3. Climatic conditions
   (a) **Rainfall:** The crop affected by rains is found to be inferior in quality than the tubers produced under absence of rains. This is because of the fact that wet season/weather favours the attack of some insects/pests and diseases and the quality is thus found to be impaired.
   (b) **Temperature:** Air temperature governs the blackening of tubers. It has been found that little or no blackening developed in tubers when the average air temperature was between 70-80°F, it was medium when temperature was between 60-70°F and it was too much when temperature was between 50-60°C.
   (c) **Light:** The bright sunny days or more light intensity results in a high specific gravity, dry weight and mealiness of tubers.
4. Cultural practices
   (a) **Earthing operations:** Earthing helps in covering the tubers which are otherwise exposed and form Silanin, an alkaloid (non edibleand bitter tubers in taste). These exposed tubers start sprouting and remain smaller in size. Thus the quality of tubers produced under proper earthing condition are better than those not earthed properly.
   (b) **Effect of plant protection chemicals:** Use of insecticides had no effect on quality but application of 2,4-D + M.H. (Malleic Hydroxide) at full blume stage was found to increase the starch and nitrogen content of tubers and prevented sprouting loss in weight of tubers during storage conditions.
   (c) **Effect of manures and fertilizers:** Organic manures are found to improve the quality. The potato plants prefer ammoniacal form of nitrogen in their early growth. Similarly chloride ions interfere with starch formation and finally the yield is reduced.
5. **Biotic factors:** Attack of insect, pest and diseases not only reduce the tuber yield but also impair the quality of the produce.
6. **Storage conditions:** Storage of tubers at 6°C temperature increases sugar content of the tubers while at 4°C temperature a considerable loss in ascorbic acid (vitamin C) takes place. The lower temperature in store inhibits the sprouting, results in a loss of water (at 0-5°C) and the tissues become soft.
7. **Cooking process:** The cooking process also controls the quality of tubers such as steaming potatoes show more or losses than boiling in water. Vitamin B and C are soluble in water; hence, if the water used for boiling is thrown the quality is reduced. A part from this, boiling potatoes in presence of salt,

causes further reduction in quality. The losses can be reduced if the potatoes are kept first in cold water and then put for boiling instead of putting them directly in boiling water.

The seed plot technique was evolved in 1959 at CPRI based on aphid surveys (CPRI, 1954 for the seed production on plains the surveys revealed the weakness of traditional seed production pattern and delineation of Low-aphid periods. Most safe period for growing healthy seed crop on plain was found to be from October to February.

Potato in Different Cropping Systems

Potato may be grown in types of cropping systems *e.g.*, as intercrop, companion crop and relay crop. These may be described as under:

Relay cropping systems: In relay cropping system, the potato is very often relay cropped with cucurbits, coriander, barley, maize, etc. 45-50 days after sowing of potato crop. The success of this cropping system would depend upon the following management factors:

1. The crop selected for relay cropping should be relatively slow growing.
2. The relay crop must be sown when potato has almost completed the vegetative growth and only tuber bulking is left.
3. Sowing of the relay crop should not be disturbing the potato crop.
4. Management practices like irrigation, plant protection measures required for relay crop should not be harmful to potato crop.

A companion cropping with mustard, wheat, barley can be done very effectively without much reduction in the tuber yield and an additional profit of rupees six to eight thousand/ha can be made. In this type of planting the seed tubers of fourth rows are planted in three rows by reducing plant to plant spacing and fourth row is used for sowing of mustard. But if barley, wheat or maize is to be sown as companion crop, then in every alternate furrow of potato any of these crops can be sown, after an early earthing up of potato (25 days after potato sowing).

Success of inter-cropping or companion cropping would depend upon following management factors:

1. The crop selected for inter-cropping or companion cropping must not be compatible. This involves considerations of canopy and rooting characteristic of the crops and their relative growth rates. Crops of different growth rates, duration and rooting habits but of similar canopy characteristics would appear to be more compatible than others.
2. The planting or sowing times and spacing of the crop should be suitably adjusted to prevent smotherning effect of the aggressive crop on the weaker crop as far as possible.
3. The irrigation requirement of the companion crops should be more or less the same. If one of the crops has higher water needs, the other crop should be such that it is not adversely affected when irrigation is scheduled to the former.
4. The fertilizer needs must be assured for each crop in as much as the intercropping should be reckoned as a measure for saving the space and effective use of land area, but not as a measure of economy on inputs in terms of fertilizers etc.

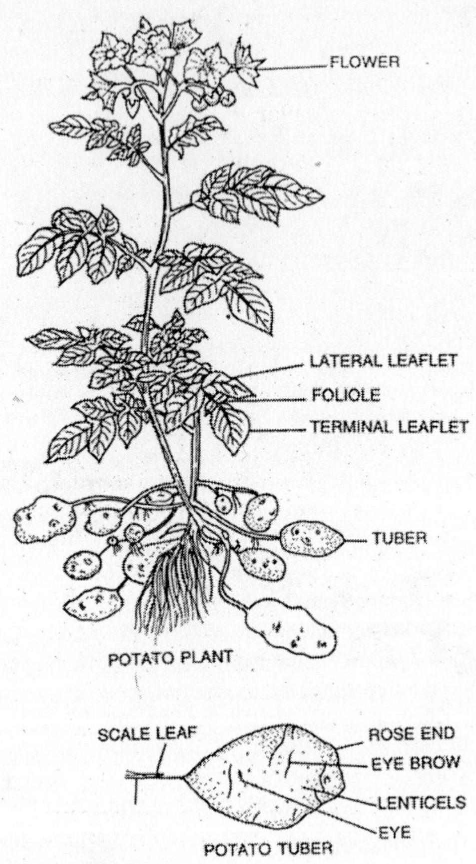

Potato Based Cropping Systems

Cropping system consisting of potato may be divided into four major groups viz.

1. **Potato Dominated Cropping:** This system is prevalent in river beds of Patna City and Deesa (Gujarat) where the area gets submerged in rainy season due to back flow and this is cultivated only during *rabi* season permitting potato monoculture. Some areas of U.P., Punjab and Nilgiri use to have double cropping of potato under intensive cropping programme system but this system is not desirable as it tends to increase the risk of spread with the seed and perpetuation of new potato diseases and pests as cyst-forming nematodes in Nilgiris, brownrot in mid-hills and plateau, tuber moth in plains and plateau, common scab and black scurf in north-western plains are the few examples of such problem. Therefore, adoption of crop rotation consisting of suitable cereals and millets is the only way to eliminate such problems.
2. **Potato Cropping Incorporating Legumes and Green Manures:** At some places potato is taken in very intensive cropping system viz., maize-potato-wheat-mung/black gram is common while at some places potato is taken after green manuring as in Patna, Jalandhar and both the systems have been found to be equally effective and beneficial. Under north Indian conditions growing green manure crop during summer has given better results. In N.E. Hills

Beans-potato or maize-potato and potato-cauliflower are common.

3. **Multiple Cropping Sequence:** Crop rotation of 400% cropping intensity like maize-Early potato-Late potato (seed crop)-minor millet (Cheena)/summer mung/cowpea/Urd is very popular in N.E. Plains zone. While selecting such rotations one has to very careful in selecting most suitable varieties of various crops. Maize-potato-onion/maize-potato-tomato or maize-potato-radish for seed tried in AICPIP (All India Coordinated Potato Improvement Project at Patna and Pantnagear) while at Chhindwara centre potato under intensive vegetable rotation had been found to be most remunerative. In Punjab potato-colocasia gave the best results.
4. **Integrated Inter-**relay Cropping System: At PAU, Ludhiana, three combinations were tried viz., potato-green gram, potato+wheat-green gram and potato-wheat-green gram. But potato+wheat-green gram gave the highest return. At Delhi potatp-wheat relay cropping gave better return than potato-wheat sequential cropping. At Ranchi also potato + wheat (2:2 row ratios) gave better return than potato-wheat sequential cropping. Number of trials was conducted in U.P. having sugarcane interplant between potato rows during autumn but there had been hardly any reduction in potato yield. In plateau again potato inter cropped between rows of sugarcane gave very high potato tuber yield. The best combination had been found to be maize-sugarcane+potato-Sugarcane.

Potato Varieties

For Plains

1. Early varieties
   a) **Kufri Chandramukhi:** It has replaced up-to-date and takes about 90 days to mature but it may be harvested even after 60 days. It is capable of yielding 120-150 q/ha when harvested earlier but it may yield as high as 250-350 q/ha when taken as main crop. It has attractive, oval and white tubers with good flavour but is susceptible to early blight.
   b) **Kufri Alankar:** It is of 90 days duration. The tubers are oval, flattened, white and attractive. The crop may be harvested within 60 days also with a tuber yield of 120-150 q/ha but when harvested after 90 days it produces as high as 350 q/ha. It has a good tolerance to early and late blight diseases.
   c) **Kufri Shakti:** It takes about 95-110 days to mature and is suitable for North Indian plains. The tubers are white, round and large in size. The stolons are short and tubers are placed near and around the main stem.
   d) **Kufri Hautej (G 2524):** It takes 90-100 days and yields 200-250 q/ha white large tubers of light yellow fresh. It may be grown on both plains and hills.
   e) **Kufri Bahar:** It takes 100-110 days and produces 250-300 q/ha tubers large, round while tubers. It is recommended for North Indian plains, Skins being very thin and weak is pealed off while transporting them.
   f) **Kufri Badshah:** it takes about 120 days and gives 250-300 q/ha tuber yield/ha. It has large, white, oval and smooth tubers. It is moderately resistant to early and late blight and can be grown on plains as well as hills.
   g) **Kufr Deva:** It matures within 110 days. Tubers are red and round. It has

short stolons and tubers are well laid around the main stem, hence harvesting is easy.

h) **Kufri Sheetman:** It is frost resistant variety suitable for Punjab, U.P., Rajasthan, Haryana, M.P., Bihar and West Bengal. It takes about 110 days to mature and it may be planted late due to its highest degree of resistance to frost. It may give 250 to 350 q/ha tuber yield. The tubers are oval, white, and large with attractive look.

i) **Kufri Chamatkar:** It has replaced Phulwa variety and is placed in medium late varieties. It takes about 110-120 days to mature and produces about 250-350 q/ha tubers. It produces uniform sized, shining, smooth, round, white tubers with fleet eyes and fine flavour. It is recommended for Punjab, Haryana, U.P. and M.P.

2. Medium and late varieties of Potato

j) **Kufri Sindhuri:** It is a prolific yielder which takes about 105 to 130 days to mature. The tubers are round, light red and medium sized. The haulms are erect with dark green leaves. The stems and leaf stalks are deeply pigmented. Under medium fertility it can produce about 250-350 q/ha tubers. It is resistant to early blight and can tolerate frost moderately.

k) **Kufri Red:** Selected from Dargiling Red Round another 120 days it has red round tubers with medium deep eyes and yellow flesh.

l) **Kufri Kissan:** It has medium larger white tubers and is recommended for plains of U.P., M.P. and Delhi.

Potato Varieties for Hills

1. **Kufri Jyoti:** A widely adaptable variety responsive to fertilizers. It is highly resistant to late blight and wart diseases. It is recommended for hills of H.P., Kumaon (U.P.) and Khasi (Assam). It bears white and oval tubers and is capable of producing about 180 q/ha tubers and takes about 110 to 120 days time.
2. **Kufr Jeevan:** It is recommended for H.P. hills and takes about 160 days to mature which gives about 175 q/ha tuber yield. The variety is resistant to late blight and wart. The tubers are white, oval and round.
3. **Kufri Naveen:** The tubers are attractive, oval and uniform. It yields about 200to 300 q/ha tubers. The variety is resistant to blight and rottage diseases and is recommended for Assam hills.
4. **Kufri Neelmani:** It takes 150 days to mature. It has white oval tubers with an excellent cooking quality and pleasant flavour. It can produce 250 to 400 q/ha tubers and is suitable for Nilgiris in South India.
5. **Kufri Moti:** Recommended for Nilgiris. The variety is resistant to blight and rest characters are similar to Kufri Neelmani.
6. **Kufr Lovkar:** It is photo-insensitive variety and matures within 70-80 days times. It is recommended for Deccan Plateau. The tubers are white, round with shallow eyes.
7. **Kufri Navtal:** It is an early maturing variety which takes about 70 to 80 days to mature and is moderately resistant to late blight. The tuber yield ranges between 150-200 q/ha. The tubers are white, oval with shallow eyes, light yellow flesh and waxy texture. It is rich in vitamin A and is good in cooking quality. It is recommended for Kumaon hills but it can be grown on plains also because of its frost resistance.
8. **Kufri Navjot:** A medium duration variety which takes about 100- 110 days in

plains and 120-130 days in hills. Tubers are white, oval having fleet eyes on them and they have dull white flesh and waxy texture. It may yield about 200 q/ha tubers. It may be grown on plains as well as on hills.

9. **Kufri Kundan:** It has oval and white tubers recommended for Northern hills which matures in 120 days.
10. **Kufri Kumar:** It produces large, white and oblong tubers and the crop matures in 150-160 days.

Hybrids

JH-222....................Yield 20-30 t/ha in 80-85 days.
JI-5857....................Yield 30-35 t/ha in 95-100 days.
JEX/C-166....................Yield 30-35 t/ha in 90-95 days.
PJ-376....................Yield 28-30 t/ha in 70-80 days.

True Potato Seed (TPS) Varieties

HPS. 1/113, C-3.

Being produced by private, companies, Centre for Technology Transfer, Bangalore and N.S.C., Bangalore.

Food Value

Potato contains carbohydrate 39.9 g, Calcium, 17.7 g, Iron 1.2 g, Phosphorus 70.7 g, Minerals 1.0 kg, Vitamin C 30 g, Vitamin B 2.4 g (Total of thiamine, riboflavin and nicotinic acid/ha/day). (50 years of Crop Science Research 1996 by ICAR, New Delhi.

SWEET POTATO (*IPOMEA BATATAS*)

Variety	:	H-42, Pusa sufed, Pusa sunehari, H-41, Pusa Lal.
Seed rate	:	50,000 vine cuttings of 25 cm length each.
Nursery	:	Keep plant vines in nursery after each harvest for next planting or plant second crop.
Land preparation	:	3-4 ploughings, add 20 kg Aldrin 5% dust and mix with soil, Prepare ridges.
Sowing/Planting time	:	June 15th (After 1st rain).
Spacing and sowing	:	75 cm row to row, 30 cm plant to plant, planting on ridges (30 cm high).
Manure and fertilizer	:	10 t/ha F.Y.M. or compost, 100 kg N+75 kg P_2O_5+100 kg K_2O/ha. Full dose of P_2O_5 and K_2O and ½ of Nitrogenous fertilizers be applied at planting and rest half after 1½ month followed by earthing up.
Interculture	:	Two weedings and and earthing, 1½ month and 2½ months after planting.
Plant protection	:	Spray 700 ml Nuvacron 40 EC or Thiodan 1000 ml in 500 lits of water to check defoliators.
Harvesting	:	October.

CASSAVA/TAPIOCA (*MANIHOT UTILISSIMA*)

Variety	:	H-226, H-97, H-2304, H-1687 - 10 months. H-165 - 7½ months.
Seed rate	:	10,000 numbers of 15 cm size cuttings.

Land preparation	:	3-4 ploughings. Apply 20 kg of Aldrin dust during ploughing. Dig pits, put P_2O_5 and K_2O and compost. Prepare mounds 30 cm high.
Planting time	:	June 15th (After first rain).
Spacing and planting	:	1 x 1 m, for long duration and 75 cm x 75 cm for short duration varieties. Plant the cuttings vertically on mounds inserting half in the soil. Plant on ridges if mounds are not done.
Manure and fertilizer	:	10 t/ha F.Y.M. or compost, 70 kg N+ 50 kg P_2O_5 + 75 kg K_2O/ha. Apply full dose of P_2O_5, K_2O and ½ N alongwith compost in the mounds, rest ½ N will be applied after 1½ months of planting.
Interculture	:	Weeding after 15 days and gap filling. Weeding, earthing and ½ N application after 1½ months. Disbudding of plants keeping two sticks per plant after 1½ months.
Plant protection	:	Spray 2.5 kg Blitox 50/Dithane Z-78 or 1.5 kg Dithane M-45 in 500 lits of water during September if leaf spot appears. Spray Rogor 40 EC @ 1 lit in 500 lit of water against cassava mosaic vactor. Rogue out cassava mosaic affected plants.
Harvesting	:	7½ months to 12 months. Remove rind, slice and dry to preserve for further use.

YAMS (*DIOSCOREA* SPP.)

Variety	:	Khumba or Hatikhojia type (*Dioscorea alata*)
Seed rate	:	10,000 pieces of 50 to 100 g weight.
Seed treatment	:	Treat the cut tuber with wood ash and keep under shade till sprouting starts. Benlate dusting prevents soft rot during sprouting.
Land preparation	:	3-4 ploughings. Dig pits 30 x 30 x 30 cm and prepare mounds. Apply fertilizer and compost.
Planting	:	15th June.
Spacing and planting	:	1 m either way from pit to pit. Plant the sprouted tuber in the mounds below 5 cm.
Manure and fertilizer	:	10 t/ha F.Y.M. or compost, 100 kg N + 50 kg P_2O_5 + 100 kg K_2O/ha. Apply full dose P_2O_5, K_2O and ½ N at planting and rest ½ N after 1½ month of planting.
Interculture	:	Two weeding and earthing up after 1½ and 3 months of planting. Give staking to crop immediately after sprouting.
Plant protection	:	Dust Carbaryl 5% @ 10 kg against hairy caterpillar damage. Spray Bavistin 0.02% against leaf spot disease.
Harvesting	:	January and February.

COCOYAM OR TARO OR ARUM (*COLOCASIA ESCULENTA*)

Varieties	:	Eddoe type – (Var. *antiquorum*) which has relatively small corm surrounded by large well developed cormels. Dasheen type – (Var. *esculanta*) which has a large central corm and fewer smaller, more compactly clustered cormels.
Seed rate	:	10 to 12 quintals per hectare. Arum is propagated vegetatively by using pieces or whole corms or cormels or stem cuttings which consist of the cormtip of about 0.6 cm or so, and the

		lower 15-25 of the petiole corms and cormels should weigh 50 to 80 g each. Germinate the piece before planting.
Land preparation	:	(a) Low land or wet culture – the field preparation resembles which is required for rice, with ploughing, discing and harrowing, to level the surface and provide a well puddle soil. Stem cuttings are normally used and planted by hand at a depth of 15-25 cm. (b) Upland culture – The field is ploughed, harrowed and then furrows are made (30 cm deep) and are spaced 45 cm a part. Pieces of corms and cormels are planted at a depth of 8-12 cm.
Planting time	:	Summer – February to March. Rainy season – June-July.
Spacing	:	In lowland culture, farmers plant as close as 30 cm x 30 cm. (1,08,900 plants/ha). Average spacing is about 60 x 60 cm.
Manure and Fertilizer	:	F.Y.M. or compost – 10-15 t/ha, 60 kg N. 40 kg P_2O_5 and 60 kg K_2O/ha. Apply ½ N and full P_2O_5 and ½ K_2O at planting time. Apply remaining N and K_2O as top dressing at one month after planting.
Interculture	:	Hoeing, weeding and earthing one month after planting and again one month after first interculture.
Plant protection	:	Leaf blight, Spray Dithane M-45 @ 1.5 kg/500 lit of water/ha
Harvesting	:	Arums are ready for harvesting when leaves begin to turn yellow and wither. It can be delayed for some weeks in dry weather without the corms deteriorating.

In addition to above five crops, three more starchy crops are also grown in India *i.e.*, (i) Elephant foot yam, (ii) Chinese potato (Coorka) and (iii) Arrow root (ararut).

CHAPTER 13

Production Technology of Crop: *Vegetables and Fruits*

(Ecological Requirements, Cultural Practices and Crop Production Management)

COWPEA

Variety	:	SEB-II (75 days duration). Swarna, FS-68, Asparagus bean, Local types. Pusa dofasali, Pusa Barsati, Pusa Komal.
Seed treatment	:	Treat the seeds with Agrosan GN or Ceresan @ 3 g/kg of seeds.
Seed rate	:	20 kg/ha
Land preparation	:	3-4 ploughings. Add all F.Y.M.
Sowing	:	June-July
Spacing and Sowing	:	45 cm row to row, 15 cm plant to plant. Sowing in line by hand.
Manure and fertilizer	:	10 t/ha F.Y.M. or Compost. 25 kg N + 50 kg P_2O_5 + 25 kg K_2O/ha. Apply ½ N and full P_2O_5 and K_2O dose at planting and ½ N after 20 days of planting.
Interculture	:	Weeding and hoeing followed by earthing up after 20 days of planting.
Plant protection	:	Spray the crop with Malathion 50 EC @ 1000 ml or dichlorvos @ 500 ml in 500 lits of water against pod borers and aphids.
Harvesting	:	August-September.

LADY'S FINGER OR BHINDI (OKRA) – FOR *KHARIF* SEASON

Variety	:	Pusa sawani, S-2, Pusa makhmali, Red wonder, and Perkins long green, Parbani kranti.
Seed reate	:	12 kg/ha.
Seed treatment	:	Treat the seeds with Thiram or Captan @ 3 g/kg of seeds.
Land preparation	:	2 to 3 ploughings, Mix 3/4th amount F.Y.M. at ploughing.
Sowing	:	3rd and 4th week of June.
Spacing and sowing	:	45 cm row to row. Sow in line by hand. Thinning should be done at 20 days at a spacing of 22.5cm.
Manure and fertilizer	:	10 t/ha F.Y.M. or Compost. 75 kg N + 50 kg P_2O_5 + 50 kg K_2O/ha. Mix full dose P_2O_5, K_2O and ½ N with ¼ th of F.Y.M. Apply in line and cover little soil. Apply the rest half of the nitrogen after 20 days.
Interculture	:	Hoeing, weeding ½ N fertilizer and earthing after 20 days of planting.
Plant protection	:	Spray 1.5 Cararyl 50% W.P. or Monocrotophos 36 SL 1000 ml or Malathion 50 EC 1000 ml in 500 lits of water against jassid, fruit borer, and red cotton bug. Dust 25 g sulfur to check mildew, if required. Tn fruiting stage Monocrotophos application should be avoided.
Harvesting	:	August to October.

LADY'S FINGER OR BHINDI (OKRA) – FOR *RABI* SEASON

Variety	:	Pusa Sawani, Parbhani Kranti, BO-1 and BO-2.
Pusa Sawani	:	Plant medium height, good yielder (75 q/ha), fruit medium size, green 5 ridges/fruit.
Prabhani Kranti	:	Plant tall, fruit long, green 5 ridges/fruit, 80-83 q/ha yield, resistant to Y.V.M. disease.
BO-1	:	Plant medium height, fruit long and bold, deep green, 7-8 ridges/fruit, very good yielder (105 to 110 q/ha). Highly tolerant to Y.V.M. disease,
BO-2	:	Plant tall, resistant to Y.V.M. disease, fruit long, green, tender, 5 ridges/fruit, very good yielder (95 q/ha).
Seed rate	:	10-12 kg/ha
Seed treatment	:	Treat the seeds with Thiram or Captan @ 3 g/kg of seeds.
Land preparation	:	3 ploughings, mix ¾ amount of FYM or Compost at planting.
Sowing time	:	Within 15th September and 1st week of February
Spacing	:	50 cm row to row and 25 cm from plant to plant within the row, sow twoseeds in one place in line by hand. Thinning should be done at 20 days after sowing.
Manure and fertilizer	:	10 t/ha FYM or Compost. 80:40:40 kg Npk/ha. Mix full dose P_2O_5, K_2O and ½ N with 1/4th of FYM or Compost. Apply in line and cover little soil. Rest half of the nitrogen apply after 20 days.
Interculture	:	Hoeing, weeding, ½ N fertilizer application and earthing up after 20 days of sowing.
Plant protection	:	Spray Monocrotophos 36 SL or Dimecron 375 ml/ha in 500 lits of water against jassid, fruit borer, red cotton bug. In fruiting stage Carbaryl 50 WP @ 2 kg/ha can be sprayed. Dust 25 kg Sulpher to checkpowdery mildew, if required.
Harvesting	:	November 1st week to 2nd week of December and 15th March to end of April. 75 to 100 quintals yield per hectare.

BRINJAL – FOR *KHARIF* SEASON

Variety	:	Pusa purple cluster, Pusa kranti, Pusa purple long, Black beauty and Penthi and Local good types, Anamalai muktakeshi (Rabi crop).
Seed rate	:	750 g/ha.
Seed treatment	:	Treatment of seed with Thiram @ 3 g/kg of seeds.
Nursery	:	Seedlings to be raised by 15th May in flat beds.
Land preparation	:	3-4 ploughings. FYM to be mixed.
Planting	:	June 15th to July 1st.
Spacing	:	75 cm row to row, 45 cm plant to plant depending on variety. Planting of 15 cm tall one month seedlings.
Manure and fertilizer	:	15 t/ha FYM or Compost. 125 kg N + 80 kg P_2O_5 +100 kg K_2O/ha. Full dose P_2O_5 and K_2O and ½ N at planting. Two weeding and hoeing at 20 days interval and apply rest half dose of N at 40 days and earth the plants in line.
Plant protection	:	Spray Dithane M-45 1500 g/Bavistin 500 g/Difoltan 1 kg in 500 lits of water against fruit rot disease and twig blight. Spray 1.5 kg Carbaryl 50% WP or Dichlorvos 500 ml/Malathion 50% EC @ 1 lit in 500 lits of water to check epilachna beetle and fruit and shoot borer. In fruiting stage Fenvalerate (20 EC) @ 500 ml/ha against the fruit and shoot borer may be applied twice at 20 days interval and subsequently Dichlorvos be used. Fenvalerate should not be

		applied more than twice during the growth stage of the crop. Two sprayings at 20 days and 40 days interval of planting to be given. Raise your own seedlings in beds sterilized by burning the soil. Apply Brassicol 75 WP @ 25 kg/ha to the soil against fungal wilt at planting time.
Harvesting	:	Harvesting from 15th August to November.

BRINJAL FOR *RABI* SEASON

Variety	:	Pusa Purple cluster, Pusa purple long, Pusa Kranti, ARU-2C, KT-4, Mukta Keshi, BB-1, BB-7, and BB-11 or any promising local type.
Pusa Purple Cluster	:	Plant tall, erect, compact, sturdy with purple pigmentation on stem, leave purple, nonspiny, fruits borne in clusters of 4-9. 10-12 cm long and deep purple in colour. Moderately resistant to bacterial wilt, good yielder (250-300 q/ha).
Pusa Purple Long	:	Plant dwarf, leaves light green, fruit long (20-25 cm), purple, glossy, tender, drop and touching the ground, 280 -300 q/ha of fruit yield.
Pusa Kranti	:	Plant medium tall, non-spiny, fruit uniform, thick, oblong 15-20 cm long, dark purple with shining green calyx and less seeded. Good yielder (270-280 q/ha).
ARU-2C	:	Plant medium tall, fruits are cylindrical, long, small, bright violetin colour, bearing in clusters (4 to 6), tolerant to bacterial wilt. Average yield 300 q/ha.
KT-4	:	Medium size plant, fruits born in cluster of 3 to 5 fruits. Fruits are cylindrical, purple colour which decreases with the fruit aging. Tolerant to bacterial wilt. Average yield 370 q/ha.
Mukta Keshi	:	Plant medium tall, fruit long, bold, deep purple, non-spinous, spongy, less seeded, yield 250-260 q/ha.
BB-1	:	Plant bushy and spreading type, Bacterial wilt resistant, spinous, fruit round, greenish purple, medium large, sugar content is more, very good cooking quality, good yielder (340-350 q/ha.
BB-7	:	Plant medium tall, fruit oblong, deep shinning purple, non-spinous, medium large, soft, less seeded, resistant to bacterial wilt, very good yielder (330 q/ha).
BB-11	:	Plant tall, fruit long, bold, non-spinous, deep shinning purple, medium large, soft, less seeded, resistant to bacterial wilt, very good yielder (370 q/ha).
Seed rate	:	750 g/ha.
Seed treatment	:	Treat the seeds with Thiram @ 3 g/kg of seeds.
Nursery	:	Seedlings to be raised 15th September to end of September in flat beds.
Land preparation	:	3-4 ploughings. FYM/Compost to be mixed.
Planting time	:	15th October to end of October.
Spacing and planting	:	60-75 cm row to row, 45-60 cm plant to plant depending on variety. Planting of 15 cm tall or one month old seedlings.
Manure and fertilizer	:	10-15 t/ha FYM or Compost, 120:50:50 kg NPK. Full dose P_2O_5 and ½ K_2O and 1/5th N at planting. Two weeding and hoeing at 20 days interval and apply 2/3rd N and ½ K_2O at 20 days and rest 2/5th at 40 days after planting and earthing up should be done each time after each top dressing.
Plant protection	:	Spray Dithane M-45, 1500 g/Bavistin 500 g/ Difolatan I kg in 500 lits of water against fruit rot disease and twig blight. Spray 1.5 kg Sevin 50% WP or Ekalux 25 EC 1 litre in 500 lits of water at 20 and 49 days after planting against the epilachna beetle. In fruiting stage Fenvalerate 20 EC @ 500 ml/ha be applied against the fruit and shoot borer twice at 20 days interval. Fenvalerate should

		not be applied more than twice during the growth period of the crop. Subsequently, Dichlorvos @ 500 ml/ha be applied.
Harvesting	:	Harvesting 15th December to end of April. After 7-10 days of planting apply Carbofuran 3G @ 30 kg/ha in the soil by placement method near the root zone and irrigate. On incidence epilachna beetle spray Carbaryl 50 WP @ 1.5 kg/ha and in the fruiting stage spray any Synthetic pyrethroid (Sumicidin Agrofen, Ripcord or Decis) twice at 20 days interval @ 1 ml/litre followed by Dichlorvos @ 1000 ml/ha. Synthetic Pyrethroid should not be sprayed more than 2 times.

PUMPKIN – FOR *KHARIF* SEASON

Variety	:	Arka suryamukhi, Arka chandan, Guamal local, Local type.
Seed treatment	:	Treat the seeds with Thiram or Captan @ 3g/kg of seeds.
Seed rate	:	6-8 kg/ha
Land preparation	:	3-4 ploughings. Add all FYM or Compost.
Sowing	:	June 15th (Just after 1st shower of rain).
Spacing and sowing	:	150 cm x 150 cm. Pit size 60 cm diameter.3-4 seeds per pit but later on thinned to 2 to 3 plants.
Manure and fertilizer	:	10 t/ha FYM/Compost. 75 kg N + 75 kg P_2O_5 + 75 kg K_2O/ha. Apply ½ N and full P_2O_5 and K_2O at planting time. Apply rest ½ N at one month stage.
Interculture	:	two intercultures, one at 30 days and another at 45 days stage.
Plant protection	:	Spray Carbaryl 50 W.P. @ 1.50 kg/ha in 500 lits of water. Apply bait spray against fruit fly at initial fruiting stage. Bait spray to be prepared with 20 cc Malathion, 200 g of molasses and 20 lits of water. Sometimes *kharif* crop is spoiled due to water stagnation. So provide better drainage. Other gourds like Cucumber, Ridge gourd, Snake gourd, Bottle gourd and Bitter gourd can be grown during *kharif* season. Seed rate is mostly 6 kg in all cases with a fertilizer dose of 50:35:30 kg NPK/ha. Others as per pumpkin.

PUMPKIN – FOR *RABI* SEASON

Variety	:	Arka Chandan, CO-1, CO-2, Baidyabati, Guamal, CM-14, Pusa Vishes and any local good variety.
Seed rate	:	5-6 kg seeds/ha
Seed treatment	:	2-3 g Thiram/Captan per kg of seeds.
Land preparation	:	Plough the field 3 to 4 times and prepare pits of size 45 x 45 x 45 cm and 1.8 m apart.
Sowing	:	Rabi season – November-December. Sow 5 to 6 seeds in each pit and allow 2 to 3 vines to train.
Manure and fertilizer	:	Apply 10 to 15 t/ha well rotted FYM/Compost at the time of planting. Incorporate 25 kg N, 30 kg P_2O_5 and 75 kg K_2O/ha before sowing. Second dose of 25 kg N/ha should be applied after 25 days of planting.
Interculture	:	Spray Ethrel 300 ppm, when the vine is at 5-6 leaf stage which encourages more number of female flowers. Fifteen per cent of the plant population should be left unsprayed. These unsprayed (Control) should remain in the midst of Ethrel sprayed plants.
Plant protection	:	Spray Kelthane 1-2 kg or Bavistin 0.15% to control powdery mildew atleast 2-3 times at an interval of 10-14 days. Aldrin 5% dust @ 25 kg/ha to be applied to the soil during land preparation

		and Carbaryl 0.15% concentration be sprayed when adult red pumpkin beetleds and fruit lies are noticed in numbers in the field.
Harvesting	:	200 to 250 q/ha.

SPINE GOURD (KANKOD)

Variety	:	There is no named variety and a great degree of variation is observed among seed population. However, better types means plants showing fruits above 2 kg yield per plant/season and tolerant to epilachna beetle are mostly preferred.
Sex ratio	:	Keep one male for 10 female plants.
Propagation	:	1. Tuberous root-collected from female. 2. Seeds are grown and unwanted population of male be removed. 3. Cutting treated with Seradix B and rooted cuttings planted.
Land preparation	:	1 to 2 plouhgings. Pil should be dug at 1 m distance of size 30 cm x 30 cm x 30 cm.
Planting time	:	1. June 1st fortnight for tuberous root. 2. Cutting 1st week of July.
Manure and fertilizer	:	2 kg of FYM/Compost per pit at planting time. Apply 60 kg N, 40 kg P_2O_5 and 40 kg K_2O/ha. Apply full P and K at planting and N in 3 splits at 15 days, 30 days and 45 days after planting.
Interculture	:	2-3 shallow hoeings weedings should be done one month after planting.
Staking	:	Staking at 1.5 -2 m height should be done. Put 3 bamboo small branches in a triangular position 30 cm from the centre.
Plant protection	:	Spraying with cararyl 50 WP @ 1.5 kg in 500 lits of water at every 15 days interval or whenever required depending on epilachna beetle incidence.
Special measures	:	1. Hand pollination during evening hours to ensure effective pollination at the end of the season. 2. Provide some shade by interplanting Dhaincha or Agasthi. 3. Water spraying during prolonged dry rainfree period to increase the yield.

KUNDRU (*COCCINIA INDICA*)

Variety	:	Any good genotype available in the locality.
Local		
Seed rate	:	16,000 to 18,000 pieces of stem cutting/ha. 12-15 cm long stem cuttings of pencil thickness having 5-6 leaves are taken and planted in basins of 60 cm diameter. 2-3 stem cuttings are planted in each basin.
Planting time	:	June-July and February-March.
Spacing and planting	:	1.5m x 1.2m. Vines are trailed on bowers. About 10% of cuttings should be male for effective pollination of female plants.
Manure and fertilizer	:	15 t/ha FYM or Compost. 60 kg N, 40 kg P_2O_5 and 40 kg K_2O/ha. Apply FYM/Compost, ½ N, full P_2O_5 and K_2O at planting. Apply rest ½ N two months after planting.
Interculture	:	Twice. One at 30 days after planting and another 60 days after planting.
Irrigation	:	During dry season apply irrigation at 4-5 days interval.
Plant protection	:	Powder mildew - Spray Karathane 0.3% at weekly interval. Downy mildew - Spray Dithane Z-78 at 0.2%. Aphids and jassids- Spray 0.1% Malathion.

AMARANTHUS – FOR *KHARIF* SEASON

Variety	:	CO-1, CO-2, Choti Choulai and Badi Choulai.
Seed rate	:	Direct seeding - 2 kg/ha Transplanting - 1 kg/ha.
Land preparation	:	4-5 ploughings (till fine tilth). Add FYM or Compost and prepare plots of suitable size.
Sowing	:	June-July.
Spacing	:	20 cm x 5 cm.
Manure and fertilizer	:	10 t/ha FYM/Compost. 50:50:20 kg NPK/ha. Basal application - 25 kg N, 50 kg P_2O_5 and 10 kg K_2O/ha Top dressing - 25 kg N, and 10 kg K_2O/ha
Tnterculture	:	First watering is done soon after broadcasting or transplanting. Subsequently the crop is gently irrigated once in 4-5 days.
Plant protection	:	Leaf webber is a serious pest of amaranthus and can be controlled by destroying larve manually. 0.3% Carbaryl 50 WP @ 1.5 kg or Dichlorvos @ 500ml/ha can check the attack of leaf webber and stem weevil. Blitox may be applied to check fungal diseases. Use of insecticide should be minimium.

AMARANTHUS – FOR *RABI* SEASON

Variety		
CO-1 (*A. deebices*)	:	Leaves – Dark green with ridged appearance. Stem - Dark green, round, succulent. Yield - 1.5 tonnes/ha.
CO-2 (*A. tricolour*):		Leaves - Green, Lanceolate, slightly elongated. Stem - Green, succulent. Yield - 2.80 tonnes/ha.
CO-3 (*A. tristi*):		Suitable for clipping of tender greens. Yield - 3 tonnes/ha.
Chhoti Chauli (*A. blitum*)	:	Suitable for leafy shoots. Leaves - Small. Green colour. Stem - Thin stemed, dwarf responds well to cutting. Yield - 3 tonnes/ha.
Badi Chauli (*A. tricolour*)	:	Suitable for leafy shoots. Leaves - Large and green. Stem - Thick, green-responds to cutting. Yield - 3 tonnes/ha.
Land preparation	:	The field is prepared to a fine tilth and beds of 2 m x 1.5 m size are usually made for growing the crop.
Seed rate	:	2 kg khada/ha 5-10 kg Kosala per hectare.
Sowing	:	In the prepared beds, the seeds are sown in rows, spaced at 20 cm, and later 10-15 days after sowing thinned to a spacing of 20 cm between plants, leaving one or two plants per hill.
Sowing time	:	For white and red Khada – June, February. For white and red Kosala sag - October – January.

Manure and fertilizer	:	Apply 10-15 t/ha FYM or Compost at the time of last ploughing with a basal application of 30 kg N, 50 kg P_2O_5 and 20 kg K_2O/ha.
Interculture	:	First watering is to be given soon after sowing the seeds and further irrigations are to be given once a week with a gentle flow of water. The field should be kept weed free by hoeing and weeding as and when necessary.
Plant protection	:	Diseases and pests are not a severe problem in amaranthus. Leaf webbing caterpillar and Amaranthus weevil pose serious problem to Amaranthus cultivation in certain years. Carbaryl @ 0.75 kg a.i.,/ha against the webber and 0.5 kg a.i./ha against the weevil can be sprayed. Minimum waiting period of 7 days between application of the insecticides and harvest should be maintained. The leaves should be thoroughly washed before cooking. Care should be taken to utilize the Amaranthus when it is succulent. Spray Dithan M-45 or any copper fungicides @ 1.25 kg per 500 lits of water to control blight. Spray 3 g of Carbaryl 50 WP per litre of water against incidence of leaf webber and caterpillars and wait for atleast 7 days for harvesting. To restrict white rust, spray Ridomil (0.15%) or Blitox (0.25%).
Harvesting	:	In green stage the plants are pulled as whole, washed and sent to the market as tender greens. The amaranthus are ready for harvest within 20 to 25 days. In A. tristis type a total of 10 clippings can be obtained with first clipping starting on the 20th day and thereafter at weekly interval.

TOMATO

Variety		
Punjab Chhuara	:	Early maturity. Large pear type fruit. Rich in vitamin-C. Resistant to blossom end rot and sun burnt injury. Yields about 400 quintals/ha.
Sioux	:	Plants dwarf spreading and determinate growth habit, medium to large size fruit, and uniform red colour on ripening of high yield.
Pusa Early Dwarf	:	Ripens early. Does well in both *kharif* and *rabi* season.
Pusa Ruby	:	Early variety. Plants are medium in height. Medium size fruit. Uniformly red when ripe.
Arka Saurabh	:	Multipurpose variety suited both for market and processing. Resistant to fruit crack. Yields about 35 t/ha.
Arka Vikas	:	Attractive red colour. Uniform ripening habit. Suitable for fresh market. Yield 35 t/ha.
LE-79	:	Wilt resistant variety. Yield 30 t/ha. Fruits are round.
B.T.-1	:	Bacterial wilt resistant, early maturity group, medium to small size fruit, very good yielder (425-450 quintals/ha.), oblong good keeping quality, pulpy.
B.T.-2	:	Bacterial wilt resistant, early maturity group, medium to small size fruit, 430-460 q/ha, round shape fruit pulpy, good keeping quality.
Hybrid	:	Mangala, Sheetal, Karnatak, Vaishali, Roopali.
Land preparation	:	Well drained high and medium lands are suitable for tomato cultivation. Land should be ploughed three to four times and leveled properly for furrow irrigation.
Seed treatment	:	Seed treatment should be done with Thiram (0.3%) or Captan (0.25%) or Diathan M-45 (0.25%).
Sowing	:	All determinate varieties can be planted early where as indeterminate varieties can be planted either in early, mid and late season depending on the variety. Pusa ruby and Pusa early dwarf are recommended for early planting. Seedlings

		of 3 weeks old are to be planted. Determinate types are to be planted 60 cm x 40 cm and indeterminate are to be planted with 60 cm x 60 cm spacing whereas F_1 hybrids are to be planted at 90 cm x 75 cm under staking. Seedlings are to be raised with 500 g of seeds for planting one hectare of area.
Manure and fertilizer	:	Apply well decomposed FYM/Compost @ 20-25 t/ha at the time of planting. Incorporate 60 kg N, 75 kg P_2O_5 and 60 kg K_2O/ha at the time of planting. Remaining quantity of Nitrogen 60 kg and 40 kg K_2O/ha should be applied one month after planting.
Interculture	:	Tomato produces good quality of fruits under staking. Staking is a must for tall F_1 hybrid varieties and all tall growing cultivars, 2 to 3 hoeing, weeding and earthing are to be given starting from 3rd week after planting upto 1½ month. Care should be taken to cause less injury to roots increases incidence of wilting. Depending on the varieties, tomato requires 6-8 irrigations. Lack of irrigation produces empty locule, light colour fruits.
Plant protection	:	Bacteria, fungal and nemastode wilts are important problems in tomato. Resistant types are not available. Wilt disease can be checked by soil ammendants and maintaining soil pH between 6-7. Boron and molybdenum deficiency cause low fruit set and cracking of fruit. Calcium deficiency causes blossom and fruit rot. Spray Endosulfan @ 1000 ml/ha at initial fruiting stage against fruit borer. In case the infestation continues in the fruit ripening stage spray Carbaryl 50 WP @ 2 kg/ha and maintain a minimum waiting period of 5 days for harvesting. For managing fungal blights spray with Copper oxychloride (0.2%) or Dithane M-45 (0.25%) or Carbendazim (0.15%). Grow resistant varieties – VT-3, VT-1.
Harvesting	:	Harvest fruits at turning stage (green to red) for better storage and marketing. Late planting produces sun scaled fruits with poor colour development.
Important practices	:	Stake the plants if possible for better fruits. Try to minimize injury to roots. In situ planting minimizes wilting. Rotate the crop regularly and always prefer new field where no solanaceous crops were grown previously.

PEA

Variety	:	Early badger, Bonevilla and Arkel, TP-3.
Seed treatment	:	Treat the seeds with Agrosan GN or Ceresan @ 3 g/kg of seeds.
Seed rate	:	75 to 100 kg /ha.
Land preparation	:	Plough the land 3 to 4 times to get good tilth. Level the land for better irrigation.
Sowing	:	September to November.
Spacing	:	30 to 45 cm row to row and 6 cm plant to plant.
Manure and fertilizer	:	Apply well decompose FYM/Compost @ 10 to 15 t/ha at the time of ploughing. Incorporate 15 kg N, 50 kg P_2O_5 and 50 kg K_2O/ha at the time of planting. Another 15 kg N should be applied after 20 days of planting.
Interculture	:	Hoeing, weeding, application of 15 kg N and earthing up should be done after 20 days of planting. Second earthing up should be after 15 days of first one. Irrigate the crop in furrows.
Plant protection	:	Powdery mildew and rust disease causes major loss in pea. Apply Sulpher

		dust 30 kg/ha or Calixin (0.1%). Apply Malathion 50 EC or Anthio 25 EC @ 750 ml in 500 lits of water.
Harvesting	:	60 to 70 quintals green pods/ha.

FRENCH BEAN

Variety	:	Contender, Pusa Parvati, Arka Kaal, Sel-2.
Seed treatment	:	Treat the seeds with Thiram @ 3 g/kg of seeds.
Seed rate	:	50-75 kg/ha.
Land preparation	:	Plough the land 3-4 times and level properly fpr better irrigation.
Sowing	:	September-October.
Spacing	:	30 to 45 cm row to row and 20 to 30 cm plant to plant.
Manure and fertilizer	:	Apply 10 t/ha of well decomposed FYM or compost per hectare at the time of final ploughing. Incorporate 25 kg N, 80 kg P_2O_5 and 80 kg K_2O/ha at the time of planting. Second dose of nitrogenous fertilizer @ 25 kg N/ha should be applied 20 days after planting.
Interculture	:	Hoeing, weeding, application of N fertilizer followed by earthing up should be taken up after 20 days of planting. Earthing up for second time may be taken up after 15 days of earthing.
Plant protection	:	Spray the crop with Malathion 50 EC or Anthio 25 EC @ 1000ml in 500 lits of water against pod borer and aphid, in the fruiting stage. To avoid leaf spots and blights spray with Blitox (0.25%) or Dithane M-45 (0.25%).
Harvesting	:	60 to 70 quintals/ha.

CABBAGE

Variety	:	Pride of India, Golden acre, Early drum head, Pusa Synthetic, Late varieties are Late drum head, Sri Ganesh Gol(F).
Golden Acre	:	Early variety, uniform solid, round heads,. Interior portion of the head is white and of excellent quality. Yield is about 220 q/ha.
Drum head	:	Uniform flat, solid small framed and short stalked head. Yield is about 400 q/ha.
Pride of India	:	Solid head. Matures in about 85 days and yields about 240 q/ha.
Pusa Synthetic	:	Medium maturity group, medium size head, compact, yield 350 q/ha.
Sri Ganesh Gol	:	Late maturity group, Compact head, 2.5 to 3 kg/ha weight, round, good quality, yield 400-450 q/ha.
Land preparation	:	Medium to high land should be preferred and 3-4 ploughings should be given to get a good tilth.
Seed	:	Seeds should be treated with Thiram (0.3%) + Plantomycin (100ppm) or with Captan. Hot water treatment (54°C) for 10 mts to check bacterial rot.
Planting	:	Raise early crop seedlings under polythene or raised beds. 500 g of seeds are sufficient for one hectare. 4 to 6 weeks seedlings should be planted at a spacing of 40-60 cm from row to row and 20 to 30 cm from plant to plant.
Manure and fertilizer	:	Apply 20 t/ha FYM or compost at the final ploughing. Application of 150 kg N, 50 kg P_2O_5 and 75 kg K_2O/ha gives the highest yield.
Interculture	:	Seedlings should be watered immediately after planting and thereafter irrigation at an interval of 15 days would be adequate. Weeding should be done frequently.

Plant protection	:	Against leaf webber apply Acephate 650 g or Carbaryl 50 W.P. @ 2.0 kg/ha and Malathion @ 1000 ml/ha against aphid. Waiting period of 21 days for Acephate and 7 days for other two compounds are maintained.
Harvesting	:	Harvest in properly matured stage. Irrigation to matured crop causes cracking of heads. Always harvest with few outer leaves to minimize the damage of heads during transit.
Yield	:	20 to 50 t/ha.
Important practices	:	1. Collect better seeds to prevent bolting. 2. Do not cultivate high chilling requirement varieties.

KNOLKHOL

Variety	:	Early – Early white, Early Vienna Late - Purple Vienna.
Land preparation	:	Deep, well pulverized and fine seed bed.
Seed treatment	:	Treat seeds with 2.5 g Captan/kg of seeds.
Raising of seedlings	:	Prepare raised seed bed of convenient length, 1 m width and 30-45 cm height. A nursery area of 100 m^2 is required for planting one hectare of land. Add adequate quantity of FYM or compost. Sow seeds of one cm depth. Drench the soil with 0.2% Captan solution to avaoid damping off. 10 days after germination spray on the seed bed 18 ml of Thiodan in 9 lits of water.
Seed rate	:	750 g of seeds/ha
Planting	:	Use 25-30 days old seedlings for planting.
Spacing	:	Row to row 45 cm and plant to plant 20-25 cm.
Manure and fertilizer	:	Apply 20 t/ha FYM or compost at final ploughing. 100 kg N, 50 kg P_2O_5 and 50 kg K_2O/ha is required. Apply ½ N, full P and K at planting and rest ½ N in 2 splits 25 days and 40 days after planting.
Interculture	:	One hoeing and weeding at 20 days after planting. Follow this with top dressing that earthing up and interculture.
Irrigation	:	Two hand watering for two days after planting. Afterwards 3 to 4 irrigations at weekly interval and then 10 to 12 days interval is needed. Weeding is to be frequently done.
Plant protection	:	Two weeks after planting, spray the crop with 1000ml Malathion against aphid in 500 lits of water in one hectare land.
Harvesting	:	Knolkhol is harvested when it is full grown and tender. The fibrous ones deteriorate the market value.
Yield	:	20 t/hectare.

CAULIFLOWER

Variety	:	
Snow ball	:	Medium size solid curds of attractive white colour. Required 110 to 120 days to harvest.
Pusa snow ball I	:	Straight upright leaves, covering the head very tightly. Solid, medium size, white curd. Requires 105 to 120 days to harvest, Yields about 220 quintals/ha.
Pusa snow ball-II	:	Late variety with good quality curd. Requires 110 to 135 days to harvest.

Pusa deepali	:	An inbred with uniform, medium-long, erect foliage with rounded tip, curds are white and deep and more number of plants can be accommodated per unit area. It is a self-branched variety.
Land preparation	:	Medium to high land should be preferred and 3-4 ploughings should be given to get a fine tilth.
Seed treatment	:	Seeds should be treated with Thiram (0.3%) + Streptocycline (0.15%) @ 3 g + 1.5 g; per kg of seeds.
Sowing	:	Raise early crop seedlings under polythene or thatch removal covers. Raise seedling from 500 g seeds in an area of 100 square metre for supplying seedlings to the area of one hectare of land. Seedlings of 3 weeks old should be planted. Frequent gap filling is required to maintain high plant population. Treat the seedlings with starter solution before planting. (Dissolve 4 kg 5:10:10 N, P_2O_5 and K_2O in 225 litres of water for preparing starter solution).
Manure and fertilizer	:	Incorporate FYM/Compost @ 20 t/ha during final ploughing. Apply 60 kg N, 60 kg P_2O_5 and 40 kg K_2O/ha at the time of planting and another 60 kg N and 20 kg K_2O one month after planting. Use of micronutrients likr sodium tetraborate (200 ppm concentration) will improve the curd quality.
Interculture	:	Light earthing, hoeing and weeding should be done for three times starting from 3rd week of planting, upto 1½ month after planting, 6-8 irrigations are given to the crop. When the curds attain maturity but delayed for marketing, blanching should be done with last leaves by covering the curds with rubber strap or straw. Exposure of matured or latterly formed curds to sunshine for long period produce bitter taste. Succession sowing should be done for better marketing.
Plant protection	:	Boron deficiency causes brown rot or red rot in curds and stems of cauliflower, Spray 50 kg of Borik acid/ha at a concentration of 0.3% to plants, two weeks after planting and another 2 weeks before curd formation. Molybdenum deficiency cause whiptail and blindness. Apply 2 kg of Sodium Molybdate/ha as spray or soil application. Leaf webber (Crocidoloneia), mustard saw fly, mustard aphid are important insects. Spray Carbaryl 50 WP @ 2 kg against the leaf webber and saw fly and spray Malathion 50EC @ 1000 ml/ha against the aphid in curd stage and wait for 7 days for harvestingin curd stage In pre-curd stage Endosulfan @ 1000 ml or Acephate @ 650 g/ha may be sprayed on observing pest incidence. To restrict blight and bacterial rots, spray Mancozeb (0.25%) atternatively with plantomycin (0.15%).
Important practices	:	1. Use good and healthy seeds. Use 3-4 weeks old seedlings. Blanch the curds. Go for succession sowing for easy marketing.

POINTED GOURD

Variety	:	Pointed gourd (Potal) varieties have not been assessed so far. Long tapering types are preferred and found fleshy and tasteful. Oval types crack during transit and on over maturity.
Land preparation	:	Early and medium rice land should be ploughed to get a fine tilth. Apply 10 t/ha FYM or Compost in pits of trenches. Row to row spacing should be one metre. Pits or long trenches should be dug in the row and roots planted at 50 cm

		distance.
Planting material	:	Pointed gourd strike roots at every node. Collect tuber roots along with swollen nodes and plant it as quickly as possible. Invariably, male roots are having rough nodular surface whereas female roots are smooth and slender.
Planting	:	Maintain 10 males in 90 female plants for better production. Planting should start by the end of October and continue up to January. Tip of the root should be on soil level and should be mulched with rice straw for better sprouting.
Staking	:	Yield can be doubled under staking. High well drained lands should be preferred for growing the crop under staking which can extend fruiting up to October and the crop can also be rationed. The crop should be grown on flat beds in medium land. Fruiting is earlier under flat method.
Manure and fertilizer	:	Apply 120 kg N, 80 kg P_2O_5 and 80 kg K_2O/ha. Apply 1/3rd N alongwith full P_2O_5 and ½ K_2O at planting. Remaining N should be applied in 2-3 splits at an interval of 20-25 days while ½ K_2O should be mixed and applied 3 months after planting.
Interculture	:	Hoeing, weeding and earthing should be done in early stage. Crop yields more fruits with frequent irrigation during summer.
Plant protection	:	Epilachana beetle and red pumpkin beetles are problem in early stage during February-April. So spray Carbaryl @ 0.75 kg ai/ha at flowering stage against the blister beetle eat away the flowers So dust Sevin 5% @ 25 kg/ha against the vine borer which is an important pest of pointed gourd which persists on the crop all round the year specially in western Orissa. The infestation can be reduced by: Planting insect free healthy roots. Spraying the infested vines with 0.5 kg/ha of Quinalphos or Phosphamidon or Endosulfan. In fruiting stage insecticides should be applied after plucking of the fruits, and a minimum waiting period of 21 days for Quinalphos and Endosulfan and 5 days for phosphamidon is to be maintained. To check the borer infestation in the ratoon crop, soil treatment with Aldrin dust *i.e.* 25 kg/ha should be made in the month of November-December. Vines also suffer for fungal wilt and leaf blight; sometimes even fruits are affected. To avoid this, Plants should be sprayed with Carbondazim 0.15% at 10 days intervals. Soil drenching with same compound should also be followed.
Harvesting	:	Fruits get ready for harvest by March, but in ratoon crop, by end of January.
Selection and preservation	:	The cultivators may select improved types from their fields. These types may be propagated through one or two node stem cuttings in pure sand. Rooted cuttings are planted in backyard to develop tubers and maintained there till next planting season. Male and female plants should be grown separately.
Important practices	:	Maintain male and female ratio 1:10. Stake the plants, if possible. Don't go for ratoon crops in coastal areas.

RADISH

Variety	:	Pusa Himani, Pusa Rashmi, Pusa Chetki, Japanese white, Bombay red. Pusa chetki can be grown in kharif season.
Seed rate	:	10-12 kg/ha.
Seed treatment	:	Treat the seeds with Thiram or Captan @ 3 g/kg of seeds.

Land preparation	:	3-4 ploughings. Apply FYM or Compost at ploughing.
Sowing time	:	September to November.
Spacing and sowing	:	30 cm from row to row. Proper thinning should be done at 18-20 days after sowing.
Manure and fertilizer	:	10 t/ha FYM or Compost. 50:50:75 kg NPK/ha. Mix full dose of P_2O_5, K_2O and ½ N and apply in line and cover little soil. Rest half of the nitrogen after 20 days.
Interculture	:	Hoeing, weeding, ½ N fertilizer applications and earthing up after 20 days of sowing.
Plant protection	:	On incidence of aphids and leaf webber in early growth period of the crop spray Methyl-Demeton 2.5 EC 1000 ml and Acephate 75 SP@ 650 g/ha respectively. In later growth stage of the crop Malathion 50 EC @ 1000ml /ha be sprayed if the pest incidence persists.
Harvesting	:	50-60 days after sowing. 150-200 q/ha yield depends on the variety.

CARROT

Variety	:	Pusa kesar, Halflong Nates, Coreless, Chantancy, Imperator, Streamines, Danvers.
Land preparation	:	Deep loose, Loamy sand. 3-4 ploughings are necessary.
Seed rate	:	5 to 6 kg/ha.
Seed treatment	:	Treat seeds with 3 g Captan/kg of seeds.
Sowing time	:	September to November.
Spacing	:	Row to row 30 cm plant to plant 10 cm.
Manure and fertilizer	:	10 t/ha FYM or Compost. 50:50:75 Kg NPK/ha. Mix all P_2O_5, K_2O and ½ N at basal application. Rest ½ N 30 days after sowing.
Interculture	:	One hoeing accompanied by weeding at 2 weeks after sowing. Follow with top dressing ½ N and follow up irrigation. Frequent weeding is to be done.
Irrigation	:	3-4 irrigations.
Plant protection	:	Two weeks after planting, spray over the crops @ 100 ml Malathion in 500 lits of water/ha.
Yield	:	20 tonnes/ha.

BEET

Variety	:	Crimson Globe, Detroit Dark Red.
Land preparation	:	Deep loose, Loamy soil with 3-4 ploughings.
Seed rate	:	5 to 6 kg/ha.
Seed treatment	:	3 g Captan/kg of seeds.
Sowing time	:	September to October.
Spacing	:	Row to row 30 cm, plant to plant 10 cm.
Manure and fertilizer	:	10 t/ha FYM or Compost. Beet has high boron requirement. 50:50:100 kg NPK/ha. All P_2O_5, K_2O and ½ N are applied as basal dose.
Interculture	:	One hoeing accompanied by weeding at 2 weeks after sowing. Then top dress the rest ½ N and then irrigate. Frequent weeding is necessary in beet crop.
Irrigation	:	8 to 10 irrigations are required.
Plant protection	:	To control beet leaf minor apply Malathion @ 1000 ml/ha. To control leaf spot apply Dithane M-45 2000 ml/ha.

Yield	:	25 tonnes/ha.

TURNIP

Variety	:	Purple top, White globe, Snow ball, Golden ball, Early Milan Red Top.
Seed treatment	:	3 g Captan/kg of seeds.
Seed rate	:	4 kg/ha.
Sowing time	:	September.
Spacing	:	Row to row 30 cm, plant to plant 10 cm.
Manure and fertilizer	:	10 t/ha FYM or Compost. 50 kg N, 50 kg P_2O_5 and 75 kg K_2O/ha.
Interculture	:	One hoeing accompanied by weeding.
Irrigation	:	7 to 8 irrigations are required.
Plant protection	:	Malathion @ 1000 ml in 500 of water/ha to control aphid.
Yield	:	20 tonnes/ha.

CHILLI

Variety	:	
Pusa Jwala	:	Resistant to virus disease. Attains maturity in 120 days. Fresh weight and dry weight 81 q and 15 q/ha respectively.
N.P.-46-A	:	Fruits are long, thin and green and punjent. Turns bright red in ripening stage. High yielder, 18 q/ha.
Lam-X-235	:	Attains maturity in 130 days with an yield of 92 q and 24 q/ha on fresh and dry weight basis respectively.
Pant C-1	:	Resistant to virus, high yielder. Good for green chilli.
BR- Red	:	Attains maturity in 130 days, suitable for dry dry chilli.
Sindhur	:	Attains maturity in 135-140 days. Only suitable for dry chilli, 18 q/ha as dry chilli.
Pusa Sadabahar	:	Disease resistant, multipurpose perennial chilli, 15-20 q/ha as dry chilli.
BC 21-2	:	High yielder. 22-24 q/ha as dry chilli. Very good for dry chilli.
Land preparation	:	High medium land having more organic matter is suitable for chilli. Plough the field for 3-4 tomes.
Seed treatment	:	Treat the seeds with Thiram (0.3%) or Bavistin may be used @ 1.5 g/litre of water for seed treatment. Soak seeds overnight in water for better germination.
Sowing	:	Sow 60 g of seeds in 100 sq m area and plant one month old seedlings. Plant the seedlings in 60 cm x 45 cm distance.
Manure and fertilizer	:	Incorporate 10 t/ha of well decomposed manure at the time of ploughing and apply 110 kg N, 70 kg P_2O_5 and 75 kg K_2O/ha. At the time of planting ½ N and full P and K should be applied in furrows above which, some manure are to be sprayed. Rest half N should be applied at 1½ month after planting.
Interculture	:	Three hoeing and weeding should be given to the crop. One earthing should be given at three week and another at top dressing. As most of the chilli crops are taken during the period from October to June, 10-12 irrigations are required. Water scarcity may cause fruit drop. Two sprays of Planofix (@ 50 ppm NAA or 5 ml Planofix/4.5 litre water) for better fruit setting one at the time of flowering and second spray 15-20 days after first spray. Chilli can be planted as an inter crop in Banana for which it should be planted at a spacing of 2.7 m x 1.2 m.

Plant protection	:	Thrip incidence cause considerable loss to the crop at all stages. Sprasy Monocrotophos or Demethoate 1000 ml/ha in 500 lits of water. In fruiting stage Malathion @ 1000 ml/ha be sprayed and a minimum period of 7 days should be maintained between application and harvesting of the fruits. Damping off is a disease at early stage and twig blights are seen in later stages. Spray any Coppor oxychloride preparation at 0.2% concentration or Dithane M-45 (0.25%). For dieback and anthracnose, spray Cuman-L @ 0.2% at 10-14 days intervals. Seed treatment should be done with 0.3% Thiram.
Harvesting	:	Chillies should be harvested in full ripening stage and should be dried at 130°F under full sunshine for 10-15 days. Under ripe fruits turn white on drying.
Important practices	:	Maintain proper plant population. Spray Planofix. Harvest chillies at full ripening stage.

CAPSICUM

Variety	:	California wander, Chinese giant, World Bitter, Yellow wander, Bharat (F1 hybrid).
Seed rate	:	750 g/ha.
Seed treatment	:	Treat the seeds with Thiram or Captan @ 3 g/kg of seeds.
Land preparation	:	3-4 ploughings, ¾ th FYM/Compost at ploughing.
Spwing time	:	September to November.
Spacing and planting	:	60 cm row to row and 30 cm plant to plant with the row. Plant 30-45 days old seedlings.
Manure and fertilizer	:	15 t/ha FYM or Compost. 120 kg N, 60 P_2O_5 and 120 kg K_2O/ha. Full dose P_2O_5 and ½ K_2O and 1/5th N at planting. Two weeding and hoeing at 20 days interval apply 2/5th and ½ K_2O at 20 days and 2/5th N at 40 days after planting and earthing up should be done each time after each to dressing.
Plant protection	:	Spray Dithane M-45, 1500 g/Bavistin 500 g in 500 lits of water to check aphids and thrips. In fruiting stage if the pests continue, Malathion 50 EC @ 1000 ml/ha should be applied.
Harvesting	:	Harvesting starts 90-100 days afyet sowing. 200-230 q/ha yield.

ONION

Variety	:	Important onion varieties are:
Pusa Red	:	Medium size bulb, flattish in shape purplish red, less pungent.
Pusa Ratnar	:	Bulb is large, round and deep red in colour, good storage quality, higher yield.
Early Grano	:	Globular bulb of yellow colour, mild pungency, suitable for salad purpose. Bulb attains full size in about 95 days after planting. High yielder.
N-53	:	Bright scarlet red colour bulbs. Suitable for planting in *kharif* season.
Arka Niketan	:	High pungency, very good storage life of 5 to 6 months. Yield 42 tonnes/ha.
Arka Pragati	:	Earliest variety. Yield 42 t/ha in 130 days.
Arka Kalyan	:	Suitable for planting in *kharif* season.
Agri found Dark Red	:	160-165 days, yield 32 t/ha, suitable for *kharif* season.
Land preparation	:	Medium land should be ploughed after the harvest of rice for 4-5 times till a fine tilth is obtained.

Seed treatment	:	Seeds should be treated with Thiram @ 3 g/kg of seeds or with Captan @ 2.5 g/kg of seeds.
Raising of seedlings	:	Raise seedlings in 3 m x 1.5 m nursery beds and incorporate 200 kg of well decomposed FYM or Compost per bed. Nursery of 0.05 ha with 10 kg of seeds will be sufficient for raising seedlings for planting one hectare of land. First week of October is considered to be the best time for sowing the seeds in the nursery.
Planting	:	Prepare 30 m x 3 m size beds and plant 1½ month old seedlings of 15-20 cm tall at a distance of 15 cm row to row and 10 cm plant to plant. Planting should be completed in the first fortnight of December.
Manure and fertilizer	:	Apply 20 t/ha of FYM/Compost. 120 kg N, 60 kg P_2O_5 and 100 kg K_2O/ha. FYM/Compost, 1/3rd of N, full P_2O_5 and 60 kg K_2O should be applied as basal dressing. The remaining quantity of nitrogen should be given in two split doses *i.e.*, 3 weeks and 6 weeks after planting. Another remaining dose of potash should be given at second top dressing alongwith second split of Nitrogen.
Interculture	:	Three weedings and light hoeing are required for the crop. In early stage, irrigate crop at short interval and the frequency of irrigation is increased with the increase in age o f the crop and decrease with maturity. Moisture stress decreases yield.
Plant protection	:	Spray Monocrotophos @ 1000 ml/ha against trips incidence. To check leaf blight/spot, spray the crop with Dithane M-45 @ 0.25% or Blitox @ 0.25% at 14 days interval.
Harvesting	:	Onion is matured in 4½ months. Harvesting should be done by top falling over, Onions are made bundles and dumped in the field for curing. Spraying of 250 ppm MH 15 days before harvesting suppresses sprouting
Important practices	:	Use 1½ - 2 months old seedlings. Use sulphate of potash as the source of K_2O. Harvest the crop at tops falling over.

GARLIC

Variety	:	Important garlic varieties are:
Fawari	:	Biggar bulbs, white colour, compact cloves.
Rajalle Gaddi		Medium sized bulb, high pungency.
T-56-41	:	Medium size bulb, white colour.
Jamnagar	:	Biggest bulbs, white colour, compact cloves.
Land preparation	:	Medium lands should be ploughed after the harvest of rice for 4-5 times till a fine tilth is obtained.
Seed treatment	:	Garlic cloves should be treated with tharam (0.3%) + Carbendazim (0.15%) of cloves.
Planting	:	The cloves should be dibbled as spacing of 20-30 cm between rows and 7.5 to 8 cm within the row.
Manure and fertilizer	:	Apply 20-25 t/ha of FYM or Compost at the time of field preparation and mix it thoroughly in the soils. Application of 120 kg N, 60 kg P_2O_5 and 120 kg K_2O/ha gave the highest yield. Half N, full P_2O_5 and ½ K_2O should be applied as basal dressing. The remaining quantity of N should be given in two split doses *i.e.* 3 weeks and 6 weeks after planting. Remaining dose of potash should be applied as second top dressing along with second split of Nitrogen.
Interculture	:	Three weedings and light hoeing at 20, 40 and 60 days after planting are required for the crop. In early stage irrigate the crop at short interval and the

		frequency of irrigation is increased with the increase in age of the crop and decrease with maturity. Moisture stress decreases yield.
Plant protection	:	Spray Monocrotophos or Dimethoate @ 1000 ml/ha against thrip incidence. To check fungal blights, spray Dithane M-45 (0.45%) or Blitox (0.25%) or Carbendazim (0.15%).
Harvesting	:	The crop is ready for harvesting when the tops turn yellowish and brownish and show signs of drying up and bend over. The bulbs begin to mature in about 4-5 months. The bulbs are lifted, cleaned and the leaves are tied at the top. The bulbs are dried for a week or so under shade. Then it is stored in the store room.
Yield	:	40 to 100 quintals/ha. Recovery of cloves from the bulbs ranges from 85 to 96 per cent.

GINGER

Soil	:	Can be grown well in sandy loam and red loam soil with with high organic matter content.
Climate	:	Grown well under warm and humid climatic conditions having rainfall ranging from 120 cm. Can be grown as an irrigated crop also.
Variety	:	Rio-De-Janerio, Narapatna, China, Maran, PGS-35 and PGS-19.
Seed rate	:	1000 kg rhizomes and each seed piece should have one or two sprouted eye.
Manure and fertilizer	:	20 t/ha of FYM or Compost. 75 kg N, 60 kg P_2O_5 and 100 kg K_2O/ha. Full P and ½ K at planting; ½ N and ½ K at 45 days after planting and rest half N at 75 days after planting should be applied.
Mulching	:	Green leaves or straw, first mulching just after planting -12 t, 2nd mulching 60 days after planting-5t, 3rd mulching 90 days after planting-5t/ha.
Planting time	:	May-June.
Spacing	:	15 cm x 20 cm. In case of sandy loam flat beds are prepared and clay soil raised beds of 12 cm height is preferred.
Interculture	:	Provide irrigation 6-8 days interval depending on soil and climatic condition. Keep the field free from weeds.
Plant protection	:	Insect-stem borer, leaf eating caterpillar, Desease- Leaf spot, soft rot and rhizome rot. Avoid ill drained soil and use disease free rhizomes. Treat the rhizome with Ceresan wet 2 g/litre of water before sowing. To control insects, spray 1000 ml Dimethoate in 500 lits of water and 2.5 kg or Copper oxychloride in 500 lits of water.
Yield	:	10-12 t/ha raw ginger.

TURMERIC

Soil	:	Can be grown in a variety of soils but sandy loam is always the best.
Variety	:	PTS-10, PTS-24, T-Sundar Armoor, Dugirala.
Seed rate	:	1700 kg per hectare. Good disease free rhizomes.
Manure and fertilizer	:	20 t/ha well decomposed FYM or compost. 30 kg N, 30 kg P_2O_5 and 60 kg K_2O/ha.
Basal dose	:	30 kg P_2O_5 and 30 kg K_2O at planting.
Top dressing	:	30 kg N at 30 days after planting. Rest 39 kg K_2O after 60 days of planting.
Mulching	:	Thick green leaves or straw, 15 t/ha applied twice, once at planting time and

		another 50 days after planting.
Planting	:	May-June. Treat the seeds with Bavistin and 5 g/1 kg of the rhizomes.
Spacing	:	15 x 30 cm. Plant well sprouted bits with bud upwards and at 7.5 cm depth.
Interculture	:	Provide irrigation 6-8 days intervals. Keep the field free from weeds whenever rhizome rot is seen. Grow the crop on raised beds.
Plant protection	:	Spray 1000 ml Malathion in 500 lits of water to check leaf spot and rhizome rot respectively and repeat the same at 2 weeks interval.
Yield	:	30 t/ha of rhizome with 20-25% dry marketable turmeric.

DRUM STICK

Variety	:	Jaffna, Chavakka Cheri Murungai, Local.
Soil	:	Sandy loam, Well drained
Planting distance	:	3-5 m either way.
Pit size	:	0.5 m x 0.5 m x 0.5 m.
Planting time	:	June-July and October.
Planting material	:	Limb cuttings or seeds. 1.0-1.3 m long and 14-16 cm circumferenceare planted in situ.
Manure and fertilizer	:	7-8 kg FYM/Compost per pit. Ammonium sulphate - 350 g, Single super Phosphate – 100 g, Muriate of Potash - 200 g per pit.
Time of application	:	June-July and October.
Irrigation	:	Watering is done until the plants are established.
Plant protection	:	Hairy caterpillars attack the plants. These can be controlled by spraying with fish oil resin soap.

BETELVINE

Betelvine (*Piper betle* L) is a perennial climber which is cultivated for its leaves. It is also important for its medical value in addition to normal use for chewing. The Cultivation of this crop comprises of over 400 hectares in Orissa, mainly, confined to the coastal belt of Balasore, Cuttack, Puri, Ganjam districts. Small pockets of its cultivation are also seen in the interior of Phulbani, Bolangir and Sambalpur districts. Cultivation of this crop is highly specialized which needs adequate skill, traditional ability and heavy investment. A betelvine garden once established is a perennial source of income providing much needed cash to the grower.

Method of cultivation: The crop thrives under tropical climate which provides a moderate temperature, shade and enough of humidity. In all the districts closed type of gardens are practiced under 'Baraj' conditions except in Ganjam where open 'Bada' type of cultivation is followed with live and bamboo standards. In closed 'Baraj' condition the vines are trailed on dead sticks of *Andropogon muricatus* (Inkad) and the top of the structure is covered with dead tops of the same plant for shade. In open (Bada) type of cultivation live plants of *Sesbania grandiflora* and *Leucaena glauca* are grown to provide natural shade by their top canopy of foliage. They also serve as standards for trailing of vines along with the bamboo stick in between.

Soil: This crop requires a well drained alluvium and sandy alluvium. Yet, it is being cultivated in clayey and sandy soil. Coastal sand dunes are also utilized in the sea-coast areas for its cultivation.

Variety: 'Bangla' is the main commercial variety grown in the coastal districts being variously named as 'Godi Bangla', 'Naua Bangla', 'Narasinghapuri', 'Jagannath' and 'Birikoli' etc. Deshi variety is grown in a small scale in the coastal belt under the name of 'Sanchi' and 'Alupatria'. This is exclusively grown in the interior districts being named as 'Kala Mahata' and 'Dhob Mahata'. A scented variety 'Bilhari' is also grown in small scale in these districts. 'Mithapan' is another variety mainly grown in the Madnapur district of West Bengal. Apart from these, there are varieties such as 'Maghai' (M.P.), 'Kapoori', 'Kalipatti' (Maharashtra), 'Mahoba', 'Bilhari' (Bihar and U.P.), 'Gachhipan', 'Angani'(Assam), 'Kalasakodi', 'Nottakodi' (A.P.) and 'Kutokodi', 'Kariballi'(Tamil Nadu) being cultivated in the country.

Land preparation: The soil should be well drained, pulverized by repeated ploughing and harrowing to obtain a fine tilth. Construction of 'Baraj' is essential before planting and shade is necessary to protect the seed vine from withering. In 'Bada' type of cultivation sowing 'Agasti' seed is done in lines in the month of June and seed vines are planted during September-October by which time 'Agasti' plants have attained a height of 1-1.5 m to provide sufficient shade to the seed vines. Before planting, ridges (Mandi) of 15 cm high are prepared with pulverized soil at an inter row distance of 1 m. The ridges are thoroughly drenched with water before planting.

Planting materials and method: Seed vine cuttings are obtained from apparently healthy and vigorously growing vines. Generally, 3 types of seed vines are used one leaf one node two leaves two nodes and tender terminals with 4-5 leaves. The variability is more in the later. Planting of seed vines is done twice in a year, *i.e.*, during September-October and February-March. The seed vines are planted at an inter plant distance of 15 -20 cm on the well drenched ridges being buried up to the first or second node keeping the attached leaves on ridges and mulched with damp straw to prevent senescence. The seed vines should be dipped in 0.5% Bordeaux mixture for 30 minutes before planting. Treatment with IAA and IBA (50 ppm) enhances early rooting of the planted vines. In open type of cultivation similar method is followed except that the seed vines are planted on ridges at the base of previously grown 'Agasti' plants. As 'Agasti' plants grow taller and taller, They are thinned out and after one year the shade plants are maintained at a distance of one metre. In such a fully developed garden the shade plants are kept at a distance of 2 metres and the vines in between them are supported by bamboo sticks.

After care: After a fortnight of planting young sprouts come out and when they are, 15-20 cm, in height the mulch is removed and top dressing is done with oil cake powder at the rate of 250 g per 'Aud' is the space in between two pairs of roof support measuring 2.5 m to 3 m in length). The growing sprouts are trailed on *Andropogon* sticks (Pakhudi) or on *Sesbania* plants as the case may be. After two months of planting the young vines are arranged alternatively in double row in the same line and trailed on standards at an inter-standard distance of 15 cm. The space in between the double row is known as 'Gampha' and such alternate arrangement of vines in a double row is called 'Pasha'. After completing 'Pasha' the vines are top dressed with oil cake powder at the rate of 500 g per 'Aud'.

Covering with fresh soil must be done one week after each top dressing after the fermentation of oil cake. Sufficient irrigation should be provided at the initial stage of planting to keep the soil always moist by sprinkling.

Manure and fertilizer: Well decomposed FYM or Compost should be given once in each year at the rate of 10 t/ha. Fertilizer at 150 kg N, 100 kg P_2O_5 and 50 kg K_2O/ha should be applied in 4-5 splits a year at a ratio of 50:50 oilcake and inorganic fertilizers. Mustard oil cake is mostly preferred to obtain good growth and leaves quality. However, others such as Groundnut, Sesame, Neem and Karanj oil cakes are in use as per need and availability. In open type of garden extra 50 kg of N and P may be given to nurish the shade plants.

Irrigation and drainage: Betelvine needs moist but well drained soil substractum for its good growth. During rainy season the garden should be made slanting in all directions keeping the mid space higher up to facilitate good drainage. During winter and summer vice-versa arrangement may be made to conserve moisture. Irrigation should be always optimum. During planting, sprinkling of water may be done 2-3 times a day and under established conditions irrigation may be given at an interval of 4-7 days depending on the type of soil. Tube well water is the best for irrigation. Tank water may be used after disinfection with commercial bleaching powder.

Plant protection: White fly, mites and aphids.

Occurrence and damage: Desap leaves causing retarded growth of plant and lower yield.

Control measures: Application of Methyldemeton 25 EC @ 750 ml or Dimethoate 30 EC @ 1000 ml/ha is suggested. Before 4 weeks of the plants coming to the leaf plucking stage, application of the above insecticides should be stopped. If incidence of the above pests are observed, Malathion 2 1000 ml or DDVP 2 500 ml/ha be applied.

Anthracnose and Leaf Spots

Symptoms of disease: It occurs from June to November or later. Irregular dark brown or black lesions on leaves which develop from the tip and margins, with age, entire leaves may be infected. Lesions are surrounded by a yellow halo. Anthracnose infection may be observed even on stem also, causing long black coloured streak, on nodes and internodes. Other leaf spots may also develop with variable sizes. In early stages they may be water soaked and light brown in colour. Later spots turn black and may have concentric rings with yellow halo.

Control measures: 1. Spraying with 1% Bordeaux Mixture + 200 ppm Streptomycin Sulphate once a month. 2. Planting disease free cuttings after dipping in antibiotic + fungicidal suspension. 3. Removal of dead vines from garden and their disposal away from the garden. 4. Restriction of surplus irrigation water.

Root/Foot Rot or Wilt Complex

Symptoms of disease: Lower portion of stem of Betelvine including root system is severely infected. It is discoloured to black colour, tissues shrink and may become hollow which can be easily pulled out from soil. White mycelial growth significantly is visible on lower stem region and roots. Brown coloured round mustard seed like bodies (scerotia) are visible on infested tissues. Plants may be wilted completely and die soon.

Control measures: 1. Soil drenching with 1% Bordeaux Mixture or with 1% Bordeaux Mixture + 0.2% Thiram. 2. Immediate removal and destruction of diseased vines. 3. Planting of healthy cuttings only. 4. Dipping of cuttings in 1% Bordeaux mixture + 200 ppm Streptomycin Sulphate before planting. 5. Avoid over irrigation of vines. 6. Destroy diseased betelvine tissues(leaves and dead vines) away from the gardens to avoid spread of inoculum. 7. When heavy infection on the foliage, spraying with Blue copper (0.25%) or Blitox (0.25%) or Bodeaux mixture (1%).

Harvesting, processing and packing: Generally, harvesting starts after 6 months of

planting. Plucking of leaves is done once in every fortnight or more depending upon the growth of vines and prevailing market conditions. Usually, after each fourth plucking the naked vines are lowered and buried in fresh soil which is followed by earthing up and top dressing. In open type of cultivation sometimes, the naked vines are twined up and hanged on the supporting 'Agasti' plants. On an average condition 60-80 lakh leaves may be harvested annually from one hectare of betelvine garden.

After harvest, the leaves are cleaned, graded according to their size, depetiolated and packed in bamboo baskets. The lots for local consumption are packed in bundles of 50 or 100 leaves. But for export to distant cities they are packed in 10,000 per basket being carefully wrapped with wet gunnies banana and 'Inkada' leaves. In some places the leaves are cured and bleached by coal heat and smoke before packing for better keeping quality during storage and transport.

FRUITS

PAPAYA

Variety	:	Coorg, Honey Dew, CO-1, CO-2, Pusa Delicious, Washington, Pusa Dwarf, Pusa Magesty.
Soil	:	Loamy soil, well drained.
Planting distance	:	2 m x 2 m.
Pit size	:	0.5 m x 0.5 m x 0.5 m.
Planting time	:	July-August.
Planting material	:	Seedlings.
Manure and fertilizer	:	Basal: FYM/Compost – 2 baskets, CAN – 200 g, SSP – 200 g and MOP – 200 g/ pit. (Per plant per year.).
		Top dressing: FYM/Compost – 2 baskets, CAN – 200 g, SSP – 200g and MOP – 150 g/pit.
		Time of application: June-July and October-November.
Interculture	:	One in two months.
Irrigation	:	During dry season once in a week.
Plant protection	:	Insect pests: Red spider mite – Sulphur dust.
		Diseases: Anthracnose – Dithane M-45 @ 3% for avoiding damping off seed treatment with Thiram (0.3%) or Captan (0.25%). Soil in nursery before seed sowing should be treated with 3-5% Formaldehyde.
		Papaya mosaic: Check insect vactor (aphids) by spraying Rogor or Metasystox @ 0.05%.
Important practices	:	1. Plant 2-3 seedlings per pit and remove the male plants as soon as their sex is known. However, to ensure pollination it is desirable to retain ten male plants for a group of hundred plants.
		Earthing up the soil around the tree trunk and adequate drainage facilities

		should be provided. Avoid deep ploughing as papaya is a shallow rooted crop. Plants should not be under shade of buildings or big trees.

BANANA (*MUSA SPP.*)

Variety	:	Champa, Robasta, Dwarf Cavendish, Manthan, Patkapura, Muguri and Bukuphata (Hilly variety)
Soil	:	Loamy, well drained.
Planting distance	:	2 m x 2 m.
Pit size	:	0.5 m x 0.5 m x 0.5 m.
Planting time	:	June-July.
Planting material	:	Healthy sword suckers (about 3 to 4 months old).
Manure and fertilizer per pit	:	Basal – FYM/Compost – 2 baskets., CAN (Calcium ammonium Nitrate) – 100 g, SSP (Single super phosphate) - 200 g and MOP (Muriate of potash) - 150 g. Top dressing after planting:

	60 days	90 days	120 days
CAN	100 g	150 g	-
SSP	200 g	-	150 g
MOP	150	-	150 g

Interculture	:	Twice in a year: 1. Pre-monsoon, 2. Post-monsoon.
Irrigation	:	During dry season once in a week.
Plant protection	:	
Insect pest	:	1. Aphid – Rogor 0.05% or Metasystox 0.05%. 2. Rhizome weevil – Phorate granules @ 1 g or Carbofuran granules @ 3 g/plant in the pit.
Diseases	:	1. Sigatoka leaf spot: Dithane M-45 @ 25% or Blitox (0.3%). 2. Anthracnose: Dithane M-45 @ 3% or Cuman L (0.25%). 3. Bunchy top: Metasystox 0.05% or Rogor 0.05%. 4. Panama wilt: Quick destruction of the affected plants if no chemical measure is recommended, soil drenching with 0.2% Bavistin.
Important practices	:	1. When the plants come to bearing stage some staking is provided to withstand from wind damage. 2. Remove the excess suckers from the base of the plant after 2 months of planting. When plant come to fruiting only one sucker is allowed to grow. Remove the male bud at 4 weeks of flowering. This encourages better fruiting and weight of bunches.

PINE APPLE

Variety	:	Kew, Queen.
Soil	:	Loamy, well drained.
Planting distance	:	

Plant population/ha	Plant to plant (cm)	Row to row (cm)	Trench to trench (cm)

		35,000 | 30 | 60 | 20
		43,500 | 30 | 60 | 90
		53,000 | 25 | 60 | 90
Pit size	:	10 cm deep hole or 20 cm deep trench.
Planting time	:	May-June and September-October.
Planting material	:	Healthy suckers of 300 g weight, Slips, Crowns.
Manure and fertilizer per plant	:	N - 12 g in 4 splits. P_2O_5 - 4 g in two splits K_2O - 12 g in two splits. FYM/Compost – 10 t/ha as basal. Time of application: First split 2 months after planting.
Interculture	:	Hoeing, weeding and earthing after each application of fertilizer.
Irrigation	:	During dry season once in 15-20 days.
Plant protection	:	Against thrips and slug caterpillar spray Monocrotophos 36 SL @ 1000 ml/ha.
Diseases	:	Heart rot or stem rot, Dip the suckers 1:1:3 – Bordeaux mixture before planting or in 0.15% Carbendazim, (Bavistin, Derosal, JK stein, Agrozim)
Important practice	:	In case the plants fail to flower, spray 10 ppm of NAA, when the plants have 45 leaves.

GUAVA

Variety	:	Allahabad Safeda, Lucknow-49, Chittidar.
Soil	:	Loamy, well drained.
Planting distance	:	6 m x 6 m.
Planting time	:	July-August.
Planting material	:	Grafts and gooties.
Manure and fertilizer	:	Basal: FYM/Compost – 2 baskets. CAN - 250 g. SSP - 200 g MOP - 100 g. Top dressing CAN - 250 g SSP - 200 g. MOP - 100 g.
Time of application	:	June-July and October.
Interculture	:	Twice in a year. 1. Pre-monsoon, 2. Post-monsoon.
Irrigation	:	During dry season, once in a week.
Plant protection	:	
Insects	:	1. Fruit fly: Sevin 2% + 1% molasses or Carbaryl 0.15% 2. Mealy bug: Methyl parathion at 0.05%, but fruiting stage 3. Malathion at 0.1% concentration be sprayed.
Disease	:	1. Fruit rot: Dithane Z-78 @ 2% or Blitox (o.25%) or Captan (0.25%).
Wilt	:	2. Bavistin 0.1% spray and soil drenching.
Canker	:	3. Dithane Z-78 @ 2% or Cuman-L (0.25%) or Blitox (0.25%).

CITRUS

Variety	:	Loose skinned - Nagpur Orange, Coorge OrangeMandarin), Khasi orange, Kinnow. Tight skinned - Mosambi, Malta (Mosambi), Sathgudi, Lemin, Eureka, Lisbon. Lime - Kagzi, Pant Lemon-1.
Soil	:	Loamy, alluvial, well drained.
Planting distance	:	Orange – 6 m x 6 m, Lime – 5 m x 5 m.
Pit size	:	0.75 m x 0.75 m x 0.75 m.
Planting material	:	Healthy budded plants, air layering (gooti).
Manure and fertilizer	:	Basal - FYM/Compost - 2 baskets CAN500 g. Neem oil cake – 200 g./plant/year. Top dressing - FYM/Compost – 3 baskets. CAN 500 g. SSP 500 g MOP 250 g.
Time of application	:	June and October in one to two doses.
Interculture	:	Twice in a year. Chemical weed control with Simazene @ 7.5-10 kg/ha. Weeding: 1. Pre-monsoon, 2. Post-monsoon.
Irrigation	:	During dry season, once in a week. Avoid over irrigation. Don't let the irrigation water in direct contact with the tree trunck. It can be prevented by earthing up soil around the tree.
Plant protection	:	
Insect pests	:	1. Caterpillar of lemon butterfly – Spray Carbaryl 50 WP @ 2.0 kg/ha after observing incidence of the larvae. 2. Aphids and 3. White fly - Spray Dimecron or Dimethoate at 0.03% concentration.
Disease	:	1. Fruit rot – Dithane Z-78 @ 2% or Carbendazim (0.15%). 2. Canker – Streptomycin 500ppm mixed with Blitox (0.25%). 3. Powdery mildew – Sulfex @ 0.3%. 4. Viruses/Mycoplasma – Spray Malathion.
Important practices	:	1. Regular foliar application of micro-nutrient like Zn, Cu, Mn as cocktail spray. 2. Prunning of dried twigs and dead branch after harvest. 3. Deep ploughing should be avoided. 4. Apply Bordeaux paste 60 cm from the base.

POMEGRANATE

Variety	:	Paper shell, Spanish ruby, Dholka, Ganesh Gole.
Soil	:	Loamy, well drained.
Planting distance	:	6 m x 6 m.
Pit size	:	0.75 m x 0.75 m x 0.75 m.
Planting time	:	June-July.
Planting material	:	Seedlings, Cuttings and air layering (gooti).
Manure and fertilizer	:	Basal - FYM/Compost - 2 baskets, CAN - 250 g. SSP - 500 g MOP - 250 g/plant.

Top dressing	:	FYM/Compost – 2 baskets. CAN - 500 g SSP - 500 g, MOP - 250 g/plamt.
Time of application	:	July-August and October-December (in two splits).
Interculture	:	Prunning and thinning frequently done in order to maintain shape size and number of branchesin a tree.
Irrigation	:	During dry season once in a week.
Plant protection	:	Caterpillar of Anar Butterfly – Spray Dimecron or Thiodan, bag the fruit immediately after set by polythene bag.

SAPOTA

Variety	:	Cricket ball, Kalipalti
Soil	:	Loamy, well drained.
Planting distance	:	7 m x 7 m.
Pit size	:	0.75 m x 0.75 m x 0.75 m.
Planting time	:	July-August.
Planting material	:	Healthy grafts and gooties.
Manure and fertilizer	:	Basal per plant/year – FYM/Compost - 2 baskets. Diammophos - 250 g, SSP 200 g. MOP 100 g.
Top dressing	:	FYM/Compost - 2 baskets Diammophos - 250 g, SSP - 200 g, MOP- 100 g./plant/year.
Time of application	:	June-July and November-December.
Interculture	:	Twice a year – 1. Per-monsoon, 2. Post-monsoon.
Irrigation	:	During dry season once in a week.
Plant protection	:	
Insect pests	:	White ants: Apply Sevidol granules in the soil. Sevin @ 2%. Stem borer: Monocrotophos at 0.05% concentration in the borer holes.
Diseases	:	Leaf spot - Dithane Z-78 @ 2%.

LITCHI

Variety	:	China, Calcutta, Muzafarpur
Soil	:	Loamy, well drained.
Planting distance	:	8 m x 8 m.
Pit size	:	0.75 m x 0.75 m x 0.75 m
Planting time	:	July-August.

Planting material	:	Gooties.
Manure and fertilizer	:	Basal per plant/year: FYM/Compost - 2 baskets. CAN - 250 g. SSP - 100 g. MOP - 150 g
Top dressing	:	CAN - 300 g, SSP - 100 g, MOP - 150 g/plant/year.
Time of application	:	June-July and October-November.
Interculture	:	Twice in a year. 1. Pre-monsoon, 2. Post-monsoon.
Irrigation	:	During dry season, once in a week.
Plant protection	:	Mite: Rogor @ 0.05%, Metasystox @ 0.05%. or Endosulfan at 0.07%. or Dicofol at 0.075%.

MANGO (*MANGIFERA INDICA*)

Variety	:	Banganpalle, Suvarnarekha, Latsundari, Totapuri, Malika, Amrapalli.
Soil	:	Loamy, well drained.
Planting distance	:	10 m x 10 m. For dwarf variety – Amrapalli - 2.5 m x 2.0 m.
Pit size	:	1 m x 1 m x 1 m.
Planting time	:	July-August.
Planting material	:	Healthy grafts (about one year old).
Manure and Fertilizer	:	Non Bearing Plant / Plant Bearing Plant / Tree

Basal	FYM/Compost	4 baskets	FYM/Compost	4 baskets
	N	75g	N	1 kg
	P_2O_3	110 g	P_2O_3	0.5 kg
	K_2O	55 g	K_2O	1 kg

Time of application	:	July
Interculture	:	Twice in a year. 1. Pre-monsoon, 2. Post-monsoon
Irrigation	:	During summer months once in a week.
Plant protection	:	
Insect pests	:	**Hopper** – Spray Endosulfan at 0.07%. **Mealy bug** - Provide greese band in the trunk. **Fruit fly** - Sevin 0.02% + 0.1% Molasses.
Diseases	:	**Powdery mildew** - Bavistin @ 0.1% or Calixin (0.1%) or Sulfex, at first flowering and subsequently spray at fortnight interval. **Anthraconse** – Blitox (0.25%) or Dithane M-45 (0.25%). **Die back** - Bordeaux mixture (5:5: 50) or Blitox (0.25%) or Dithane M-45 spray.

JACK FRUIT

Variety	:	Singapore, Cylone jack. or Local.
Soil	:	Loamy, well drained.
Planting distance	:	10 m x 10 m
Pit size	:	1 m x 1m x 1 m.
Planting time	:	July-August.
Planting material	:	Seeds and seedlings
Manure and fertilizer	:	Basal – FYM/Compost - 3-4 baskets, CAN - 250 g, SSP - 500 g, MOP - 150 g./plant.
Top dressing - FYM/Compost	:	3 baskets, CAN - 500 g, SSP - 500 g MOP - 500 g/plant.
Time of application	:	July and September.
Interculture	:	Twice in a year. 1. Pre-monsoon, 2. Post-monsoon.
Irrigation	:	During dry season, once in a fortnight.
Plant protection	:	Fruit rot – Use blitox or Dithane M-45.(0.25%).

CASHEWNUT

Variety	:	T.N. 40, Vengurla-36/3.
Soil	:	All types of soil.
Planting distance	:	8 m x 8 m.
Pit size	:	0.5 m x 0.5 m x 0.5 m
Planting time	:	July-August.
Planting material	:	Seedlings and grafts
Manure and fertilizer	:	FYM/Compost - 3-4 baskets. CAN - 250 g/plant SSP - 125 g,/plant MOP - 125 g/plant
Time of planting	:	July.
Interculture	:	Twice in a year – 1. Pre-monsoon, 2. Post-monsoon.
Irrigation	:	Not necessary.
Plant protection	:	Blossom blight – Spray 1% Bordeaux mixture. Stem borer - Spray or plug borer holes with 0.5% Monocrotophos.

COFFEE (*COFFEA ARABICA, COFFEA RUBUSTA*)

Soil and Climate	:	It requires rich forest soil and can be grown 1000-1500 m above M.S.L. Rainfall - 1600 mm. Shed- medium to light.
Soil	:	Deep, friable, porous, rich in organic matter.

Variety	:	Arabica – S-795, S-288, Cauvery, etc. Spacing: 1.5 m x 1.5 m.
Robusta	:	S-274, BR Series,
Spacing	:	3.0 m x 3.0 m. Grown in large scale for better performances.
Propagation	:	Through seeds – Seedlings of 1 year old is taken for planting.
Planting time	:	Disease free, healthy seedlings of 1 year to 1½ year age can be planted in June.
Weeding	:	The weeding is done whenever required.
Shade trees	:	Grivellia (Silver oak), Erithrina, Dapad are taken.
Thinning and pruning	:	Plant is trained for simple or multiple stem system after 75 cm height.
Manure and fertilizer	:	N.P.K. dose of 15:10:15kg/ha is applied in young stage after planting and than a dose of 40:30:40 kg/ha is followed twice in a year one before and one after monsoon. FYM or compost 10-15 t/ha may be applied as basal.
Irrigation	:	Must be given in November-January and February-April.
Plant protection	:	Disease: Leaf rust, black rot re common – Spraying of Bordeaux mixture 5% and other chemicals.
Insect pests	:	White stem borer, Short hole borer, Mealy bug are common pests, Control – Spraying with systematic insecticide.
Processing	:	Berries are harvested in ripening stage and depulping is done before processing after proper drying.

RUBBER (*HEVEA BRASILIENSIS*)

Climate	:	It is a good crop and can be grown up to an elevation of 500 m with well distributed rainfall not less than 200 cm and warm humid climate.
Soil	:	Well drained alluvial soil and loamy soil.
Variety	:	RRII-105, RRII-300, RRII-203, RRIM-600, etc.
Propagation	:	1. By clonal seeds. 2. By budded plants.
Planting distance	:	Density of planting in the case of budded plants 420-445 plants/ha and in the case of seedlings – 445-52 plants/ha.
Pit size	:	75 cm x 75 cm or 90 cm x 90 cm are necessary.
Mulching	:	Mulching should be provided at initial stage for establishment of plants.
Manure and fertilizer	:	Manuring is necessary in three stages: 1. At nursery, 2. Immature stage and 3. Matured stage.
At nursery stage	:	N:P:K -10:10:4 kg/ha
Immature stage	:	N:P:K-12:12:12 kg/ha.
Matured stage	:	N:P:K - 10:10:10 kg/ha.
Harvesting	:	Hevea latex is found in the latex vessels containing 30-40% of rubber. If is found from the bark, tapping is generally, done when the plant reaches a height 125 cm and girth is 50 cm. Best time of tapping is in March.
Plant protection	:	**Leaf fall:** Spray Bordeaux mixture. Powdery mildew Sulfur dusting, Scale insect, termite, Mite, Snail, Rat etc can be controlled by application of systemic insecticide.

OIL PALM (*ELAEIS GUENEENSIS*)

It is an oil yielding crop. The oil is edible. The oil content of the nuts varies uo to 27 to 30%.

Variety	:	Tenera, Dura and Pisifera are the varieties under palm. Tenera is a hybrid between Dura and Pisifera.
Nursery	:	Viability of seeds is 3-9 months. Ripen seeds are harvested under hot sun for two days and then socked in water for 5 days and sown in the nursery after maintaining moisture. Germination takes place within one month. Sometimes directly it is raised in polithene bags and seedlings of 2 months to 8 months old are ready for planting in main field.
Planting	:	Spacing 9 m is required and 140 plants per hectare is needed.
Planting time	:	May-June.
Manuring	:	

	N	P_2O_5	K_2O
	Gram/plant	Gram/plant	Gram/plant
1st year	400	200	400
2nd year	800	400	800
3rd year	1200	600	1200

		It is applied in two splits during May and September at a distance of 2 m diameter around the plant.
Harvesting	:	First harvest 3½ years – 4 years when the nuts are ripe and fall off.
Plant protection	:	
Disease	:	Anthracnose, spear rot, Bunch failure, are the major diseases found and can be controlled by spraying of Dithane M-45 or Captan at a strength of 200/100 tits of water.
Insect pests	:	Rhinoceros beetle, Red palm weevil, Birds etc. Application of Bordeaux paste and systemic insecticides.
Processing	:	Oil is extracted by hydraulic press or hand press after scraping the Exocarp and Mesocarp with steaming for 30-60 minutes.

BER

Variety	:	Gola, Banarasi Karaka, Umran, Kaithli, Seb.
Soil	:	It can be grown in all waste lands, even on slightly alkaline soils and rocky lands.
Planting distance	:	8-10 m
Pit size	:	1 m x 1 m x 1 m.
Planting time	:	June-August.
Planting material	:	Budded plants
Manuring	:	

Age of plant	FYM/Compost(t/ha)	Ammonium sulphate kg/tree
1st year	10	0.5
2nd year	15	1.0
3RD year	20	1.5
4TH year	25	2.0

Time of application	:	All the FYM/Compost and half of Ammonium sulphate is to be applied in June-July and rest half Ammonium sulphate as top dressing in October –November.
Interculture	:	It is essential to prun the ber tree to get a good crop. The best time is between 1st week of April to 1st week of May.
Irrigation	:	**Young plants** – At fortnight interval. Old trees - 20 days interval from October to

		March.
Plant protection	:	
Insect pests	:	**Fruit fly:** 1. Spray 500 ml Malathion and 5 kg Sugar or Gur solution in 500 lits of water. Repeat at weekly intervals. Mix 35 kg BHC 10% dust per hectare in the soil. Infected fallen fruits are buried deep in the ground or burnt.
Diseases	:	**Powdery mildew:** Spray 0.2% Karathane or 0.1% Benlate.
Harvesting	:	Use ladder to collect the fruits. Full grown trees can yield up to 200 kg.

KAGZI LIME

Variety	:	Kagzi, Kagzi Kalan, Malta lime.
Soil	:	Red loams are preferred. Shallow clay soils are not desirable. Water table should be 1 m below the surface. Good drainage arrangements are essential during rainy season.
Planting distance	:	3.5 m x 3.5 m
Pit size	:	1 m x 1 m x 1 m. Burn trash in pits. Fill the pit with mixture of tank silt or red earth + 20 kg of FYM/Compost + 3 kg of bone meal or 2 kg super phosphate.
Planting time	:	July-September.
Planting material		Healthy disease free budded tree.
Manure and fertilizer kg per pit	:	(see table below)

Year	FYM/Compost (t/ha)	N (kg/ha)	P (kg/ha)	K (kg/ha)
1st year	20	0.30	0.15	0.15
2nd year	25	0.60	0.30	0.30
3rd year	30	0.90	0.45	0.45
4th year	40	1.20	0.90	0.90
5th year & above	50	1.50	1.00	1.00

Up to a height of 60 cm, the trunk is pasted with Bordeaux paste.

To prevent chlorosis, micronutrient spray may be given as follows:

Zinc sulphate	2.25 kg
Copper sulphate	1.25 kg
Magnesium sulphate	0.90 kg
Ferrous sulphate	0.90 kg
Borax	0.45 kg
Lime	4.00 kg
Urea	4.50 kg/ha in 500 lits of water.

Time of application	:	The time of application may be January-February *i.e.*, when the flush leaves are expanding. First dose before January and rest after fruit set.
Interculture	:	The basin around the plant is dug and the weeds can be done after that green manuring may be practiced. Mulching with dry leaves conserves soil moisture. Water suckers should always should always be removed.
Irrigation	:	**In summer** – Irrigation at 10 to 15 days interval, while in winter irrigation after every 4 weeks.
Check pre-harvest fruit	:	Spray 2,4-D @ 10 ppm in August.

drop		
Most important	:	Buying the plants from a source certied as free from virus diseases should be taken. Production of nucellar seedlings in isolated areas may prove beneficial.
Harvesting	:	When harvesting is done fruits turn from pale green to yellow in colour. Immediately after rain, harvesting should not be done since the fruits will be inferior in quality. Degreening can be done using ethylene or by dipping in ethylene 1000 ppm solution.
Caution	:	Earthing up the soil around the trunk and adequate drainage should be provided. Avoid deep ploughing as Lime is a shallow rooted crop. Plant should not be under shade of building or big trees.

LEMONGRASS (*CYMBOPOGON FLEXUOSUS*)

Lemongrass is an important aromatic grass. The oil obtained by steam or hydro-distillation of leaves has strong lemon like 0dour and is used extensively in flavours, cosmetics perfumes, scenting of soaps and detergents.

Types	:	1. Pole purplish stemmed. 2. White stemmed.
Soil	:	Well drained sandy loam or travelly soils.
Planting distance	:	45 cm x 45 cm.
Planting time	:	June-July.
Planting material	:	Rooted slops.
Manure and fertilizer	:	(ha/year)

FYM/Compost	-	10 t/ha
N	-	100 kg
P_2O_3	-	100 kg
K_2O	-	50 kg

Time of application	:	All the P_2O_5 and K_2O should be applied as basal dose and N in 3-4 splits once during planting and subsequently after each cutting
Interculture	:	Hoeing and weeding at the early stage
Irrigation	:	During dry season once in a week for the healthy growth.
Plant protection	:	Insects: Termites – 10 kg Aldrin/ha.

PALMAROSA (*CYMBOPOGON MARTINI*)

Palmarosa is one of the important aromatic grasses cultivated in India. The whole herb on hydro-distillation yields resaceous oil rich in geraniol, which is used in cosmetic, perfumery soaps, detergents and flavouring tobacco.

Variety	:	RRL (B) 49, RRL (B) 60, RRL (B) 77, RRL (B) 77 E, IW -31245, Trishna, PRC-1.
Soil	:	Well drained sandy loam or gravelly soils.
Seed rate	:	2 kg seeds/ha.
Sowing time	:	May.
Planting distance	:	50 cm x 50 cm.

Planting material	:	Seedlings.
Manure & fertilizer	:	(ha/year)
Basal	:	

1st top dressing			2nd top dressing		
N	33 kg	N	33 kg	N	33 kg
P_2O_5	50 kg	P_2O_5	50 kg		-
K_2O	25 kg	K_2O	25 kg		-

Time of planting	:	Top dressing should be done after each cutting.
Interculture	:	Hoeing and weeding should be done once after 15 days of planting and subsequently at 1½ months interval.
Irrigation	:	During dry season once in a week for healthy growth
Plant protection	:	Insects: White ants – 10 kg Aldrin/ha.
Disease	:	Blitox spraying once during each cutting.

GLADIOLUS

Variety	:	There are hundreds of varieties. Some good large flowered types are Black Jack, Cherry Blossom, Friendship, Happy End, Hunting song, Charm Rose, Agnirekha, Apple blossom, Snow princes, Melody, Nazrana, Apsara, Arati, Sapna, Poonam, Meera and Shobha.
Planting time	:	September-October.
Propagation	:	Corms of atleast 4 cm diameter to be planted. The top shaped corms give better flower than the flat ones.
Selection of sites	:	Gladiolus is a sun loving plant. So the place where it is to be planted should be sunny.
Soil	:	Gladiolus can be grown in a wide range of soil from light sandy to clay loam which should be deep, well drained with a pH from 5.5 to 6.5.
Land preparation	:	The first ploughing is to be done to a depth of 30 cm, before two months of planting. About two to three weeks before planting second ploughing to be done followed by rolling for a fine tilth.
Planting	:	Gladiolus corms pass through a state of dormancy period extending upto three months after harvest. Only retard (*i.e.* dormancy broken) corms to be planted. Before planting, the brown scales covering the corms to be removed and corms to be soaked with 0,2% Captan or 0.1% Benlatefor 15 to 30 minutes. Before planting the corms are to be sprouted in wet sand. Planting should be done at a depth of 5-10 cm.
Planting distance	:	30 cm x 15 cm (row x plant).
Manure and fertilizer	:	Well decomposed FYM or Compost at the rate of 4-6 kg to be applied per metre in row before planting. If good manure is not available a green manure crop (*i.e.* sunhemp, cowpea) is to be grown and incorporated at least 2 months before planting. Besides that, NPK at the rate of 200 kg each per hectare should be applied. P and K shouls be applied as basal and N in 4 split doses such as 1. Before planting, 2. Side dressed after 3-4 leaves develop, 3. Side dressed at the time of spike emergence and 4. After the completion of flowering.
Interculture	:	Two earthings are to be done *i.e.*, one at 4th leaf stage and the other at spike emergenve stage. Regular weeding is to be done. Irrigation at an interval of 7-10 days is to be provided. Staking is to be provided along the rows at 2-3 m intervals on both sides.

Plant protection	:	**Disease** - Two soil borne diseases (*Stromatinia gladioli* and *Fusarium oxyiporum*) cause rotting of corms which can be avoided by treating the corms with fungicide before planting. **Pests** - Thrips are problematic which is to be controlled by spraying with Nuvacron (0.15%).
Harvesting	:	When the first floret (the lowermost) in the spike shows colour then the spikes are to be submerged in a bucket (up to 15 cm from the base) containing water. When the foliage turn yellow or brown the corms are to be harvested, treated with Captan or Benlate and stored in a dry room.
Yield	:	2,00,000 flower spikes per hectare.

HIGH DENSITY PLANTING IN BANANA

Variety	:	Robusta, Dwarf Cavendish
Soil	:	Well drained loamy soil.
Planting density	:	1.2 m x 1.2 m.
Pit size	:	0.5 m x 0.5 m x 0.5 m.
Planting time	:	October.
Planting material	:	Healthy sword suckers of 3-4 months old.
Manure and fertilizer	:	(Per pit) Basal - 10 kg FYM/Compost and 500 kg SSP. At 40 days - 350 g, CAN, 25 g SSP and 750 g, MOP. At 70 days - 350 g, CAN and 250 g SSP., At 110 days – 300 g, CAN and 200 g MOP.
Interculture	:	Earthing up should be done at 2-3 months stage and suckers should be removed whenever required.
Irrigation	:	During rainy season irrigation is to be given whenever needed. During dry seasons irrigation should be given once in a week.
Plant protection	:	Aphid - Rogor 0.05% or Metasystox 0.05%. Rhizome weevil – Phorate granules @ 1g or Carbofuran granules 3 g/plant in the pit. Bunchy top - Matasystox 0.05% or Rogor 0.05%.
Important practices	:	When plants come to bearing stage propping is provided to withstand damage from wind. Remove the male bud for high weight of bunches

INTERCROPPING IN COCONUT AND CASHEWNUT

Coconut and palm require about 10 years to attain normal bearing. In order to obtain some income from the farm, intercropping in coconut gardens may be adopted. In coconut garden, palms are spaced at 7.5 m x 7.5 m hence area is not effectively utilized. Pine apple, banana, elephant foot yam, groundnut, chillies, sweet potato, tapioca can be grown as intercrops and the income can be raised to 3-4 fold with effective management. Such systems also provide income round the year to the cultivator.

Til (sesame) recently, intercropping in cashew plantation received very little attention. However, there is good scope for cultivation of intercrops in cashew plantations. Once the canopy covers the area, there will be no scope for intercropping because completes shading of the interspaces. Legumes like horse gram, cowpea, and groundnut can be taken as intercrop. Casuarina, teak and Eucalyptus can be grown along with cashew. Many aromatic plants like palmarosa, lemongrass, citronella may be cultivated in large scale.

CHAPTER 14

Plant Protection

CHEMICAL CONTROL OF WEEDS

Weeds are the greatest enemies of the crop plants as they compete for everything the crop plant need. To raise a successful crop, it is therefore, essential to eliminate the weeds as early as possible. Certain chemicals have been evolved to achieve effective weed control in different crops. The use of certain tested chemicals for some of the important crops is furnished below.

RICE (*KHARIF*)

Upland Rice

Pre-emergence application of Thiobencarb @ 1.0 kg/ha (2 litres Saturn/ha) mixed with 1000 litres of water and should be sprayed in moist soil. OR Pre-emergence application of Butachlor at the same dose, time and method is also recommended.

Medium Land Rice

Direct sown: Use of Oxyfluorfan @ 0.05 kg/ha (212 ml Goal/ha) or Butachlor @ 1.5 kg/ha. (3 lit of Delchlor, Hitachlor, Dhanuchlor, J. K. Chlor) mixed with 1000 lit of water as pre-emergence application in moist soil is very effective.

Transplanted Rice: Use of Pendimethalin @ 1.0 kg/ha (Stomp @ 3.3 lits/ha) or Butachlor @ 1.5 kg/ha (3 lit of Trapp, Teer, Hiltachlor, J. K. Chlor) as spraying 4 days after transplanting (Water requirement – 1000 litres/ha) or sand mix application (Sand requirement – 50 kg/ha). Pur and dry sand should be chosen for herbicide application.

Low Land Rice

Direct sown: Use of Butachlor @ 1.5 kg/ha (3 lit of Delechlor, Dhanuchlor, J. K. Chlor etc.) or Pendimethalin @ 1.0 kg/ha (3.3 Stomp/ha) spray along with water requirement of 1000 lit/ha.

Transplanted Rice: Use of Butachlor @ 1.5 kg/ha or Pendimethalin @ 1.0 kg/ha applied as spray or sand mix as mentioned before. The herbicide should be applied in saturated soil.

RICE (*RABI*)

Hiltachlor/Teer/Delchlor/Butachlor @ 1.5 kg/ha (3 lit marketed product) in 1000 ml water/ha. 3-4 days after transplanting. OR Arozin @ 0.4 kg a.i./ha (3.3 lit marketed product) in 1000 lit water/ha. 3-4 days after transplanting. (These herbicides can be applied from as sand mix with 30-40 kg dry sands). Keep 1-2 cm standing water from 3rd day to get expected benefits from the herbicides. The sand-mix application should be done in crosswise manner to get higher weed control efficiency.

WHEAT

Apply Arelon @ 2 kg a.i./ha as pre-emergence one day after sowing mixed with

1000 lit of water/ha. OR Use Fernoxone @ 0.55 kg a.i./ha 15 to 20 days after sowing wheat with 1000 lit water/ha to kill only broadleaf weeds.

MAIZE

In sole maize, use of Atrataf @ 1 kg a.i./ha plus 1000 lit of water as pre-emergence spray is very effective in controlling weeds of maize. OR For maize + Cowpea intercropping, use stomp 1 kg a.i./ha as pre-emergence spray. OR. Application of Atrazin 1 kg/ha (2 kg of Atrataf) as pre-emergence spray controls predominant weeds of maize crop.

ARHAR

Use fluchloralin 1 kg/ha as PPL or Thiobencarb 1 kg/ha as pre-emergence spray to control the major weeds of arhar.

GREEN GRAM AND BLACK GRAM

1. Pre-emergence spray of Stomp @ 0.5 kg a.i./ha mixed with 1000 lit of water should be used to kill the weeds. Herbicide should be applied one day after sowing.
2. Saturn @ 0.75 to 1 kg a.i./ha may also be used with 1000 lit of water one day after sowing seeds. OR
3. Basalin @ 0.9 kg a.i./ha with 1000 lit of water should be applied one day before sowing of seeds. OR
4. Use Thiobencarb 0.75 kg/ha (1.5 lit of Saturn) of Butachlor 0.75 kg/ha (1.5 lit of Hiltachlor/Delchlor) or Pendimethalin 0.75 kg/ha (2.5 lit of Stomp) as pre-emergence spray *i.e.* one day after sowing.

GROUNDNUT

1. Pre-emergence spray of Stomp @ 1 kg a.i./ha is very effective against major weed flora of this crop. The herbicide is applied with 1000 lit of water one day after sowing groundnut. OR
2. Basalin @ 0.9 kg a.i./ha mixed with 1000 lit of water as pre-plant soil incorporation. After herbicide application care should be taken to mix the same thoroughly in the soil and then sow the seeds one day after herbicide application. OR
3. Saturn @ 0.75 kg a.i./ha or Butachlor 0.75 kg/ha also control grassy and some broad-leafed weeds. OR
4. Fluchloralin @ 1 kh/ha as pre-plant soil incorporation (Basalin 2 lit) is recommended for controlling weeds in groundnut or Pendimethalin @ 1 kh/ha (3.3 lit of Stomp) or Oxyfluorfan 0.075 kg/ha (319 ml of Goal/ha) be sprayed one day after sowing. Herbicides should be applied in moist soil for better results.

SESAME

1. Application of Basalin 1 kg a.i./ha as pre-plant soil incorporation controls the weeds of sesame (one day before sowing). OR
2. Fluchloralin @ 1 kg/ha in 1000 lit of water should be applied in moist soil and thoroughly mixed to get better control of weeds. This herbicide should be applied one day before sowing. Alachlor @ 1 kg/ha (Lasso 2 lit/ha) as pre-emergence spray mixed with 1000 lit of water also combact the weed.

MUSTARD

1. Spray Arelon @ 0.5 kg a.i./ha or Ronstar @ 0.5 kg/ha as pre-emergence spray one day after sowing mixed with 1000 lit of water.

JUTE

1. Fluchloralin @ 1 kg/ha (2.2 lit of Basalin) should be applied in moist soil as spray with water requirement of 1000 lit/ha one day before sowing and mixed thoroughly with soil. OR
2. Thiobencarb @ 0.75 kg/ha (Saturn 1.5 lit) should be applied one day after sowing with water requirement of 1000 lit.

SUGARCANE

1. Spray Atrataf @ 2 kg a.i./ha mixed with 1000 lit of water 2-3 days after planting sugarcane. OR
2. Apply 2,4-D Na-salt @ 2 kg/ha 10 days after planting sugarcane to control broad leaved weeds or use Atrazine 2 kg/ha (Atrataf) with 1000 lit of water as pre-emergence spray (1-2 days after sowing) for better control of weeds.

PRECAUTIONS

1. Atrazine is a wettable powder and is sparingly soluble in water. For this at first add a small quantity of water to the herbicide powder and make a fine paste. Add the remaining quantity of water to the paste, slowly with constant stirring than pour the prepared mixture to the tank for use. While deciding the day for spraying, ensure sunny weather, at least for 6 hours after spraying. Never spray during rainy or cloudy weather.
2. Use clean water for the spray.
3. Ensure adequate mixture in the field prior to spraying. This is more specially so far volatile compounds like (Fluchloralin).
4. Do not disturb the soil at least for a month in case of pre-emergence spray. Apply fertilizer if necessary with minimum soil disturbance during this period.
5. It is ideal to keep a separate sprayer is to be used for other purpose also, wash the sprayer thoroughly with large quantity of water and soap before such use.
6. Keep the herbicides out of reach of children.
7. Wash hands and legs with soap water after herbicide spray.
8. Use mark globes and eye protectors (goggles) at the time of spraying.
9. Use directed spray for post-emergent herbicides, like propanil etc.
10. Uniform spray in the field is necessary for ensuring efficient weed control.
11. If the same herbicide sprayer is to be used for other purpose also, wash the sprayer thoroughly with large quantity of water and soap before such use.

For better results adopt following measures. Dose may vary according to clay content of the soil.

1. Perfect leveling of land is needed.
2. Always spray the herbicide in moist soil.
3. Use only fresh water for herbicide mixture.
4. Stock solution with required herbicide has to be prepared first.
5. Use flat fan nozzle for herbicide spray.
6. Keep all the herbicide bottle in dark place and shake the container before use,
7. Spraying should be thoroughly uniform over the field.

SPRAY SOLUTION FORMULA FOR HERBICIDE REQUIREMENT

$$\text{Herbicide requirement in g or ml} = \frac{10 \times \text{area in sq m.} \times \text{rate in kg a.i./ha}}{\text{Active ingredient}}$$

CONTROL MEASURES FOR PARASITIC WEEDS

Orobanche: *Orobanche cernua*

Host crops: Brinjal, tomato, tobacco.

Cultural Method

1. Growing of trap crops such as sunflower, sesame, cotton, soybean and finger millet stimulates the germination of orobanche seed and these afterwards die in absence of host.
2. Adoption of crop rotation with non-host crops like red gram, horse gram, cowpea, gram, cotton and sesame minimizes the weed infestation.
3. Excess irrigation should be avoided to crops like tobacco, brinjal and tomato.
4. Growing of green manure crops like sunhemp and green gram before planting brinjal and tobacco also has been found useful to minimize orobanche infestation.

Chemical-directed Spray 10%

1. 10% copper sulphate solution to orobanche shoots control the parasite.
2. Pre-plant application of Dalapon @ 5.0 kg/ha Linuron @ 3 to 5 kg/ha and metabromuron @ 1-2 kh/ha is very effective in brinjal and tobacco.

Name of the Weed: *Cuscuta spp.*

Host plant – Niger.

Cultural Method

1. Use dodder (cuscuta) free clean seeds.
2. Thin the crop and destroy the infested plants at the early stage.
3. Remove the vegetative parts of the weed and destroy them completely by drying or burning.
4. Adopt crop rotation with crops like maize, bean, cowpea and cereals at least for 5 years.

Chemical Methods

1. Pre-emergence application of chloropropham (Furon-20 G) @ 4 kg/ha in moist soil 6 days after sowing niger. This should be mixed with wood ash or sand has to be broadcasted next day of sowing niger.
2. Pronamide (Kerb) @ 1.5 kg/ha 15 days after germination of *Cuscuta* or 2 kg/ha 20 days after germination.

Name of the Weed: *Striga spp.*

Host Crop - Sugarcane
Cultural Method

1. Eradicate the weed *Digitaria ciliaris*/ sanguinalis (Carb grass) completely in the areas striga spp is seen by using herbicide or by cultural operations. Grow trap crops such as soybean, cowpea, sunflower, groundnut and castor.

Chemical Method

1. Spray 2,4-D amine salt directly to striga @ 0.5 to 0.75 kg/ha two to three times during the crop growth period to destroy flushes of striga in its vegetative phase.

IMPORTANT INSECT PESTS OF CROPS AND THEIR CONTROL

RICE

Occurrence of Pests

- Among the five stem borer species seen in Orissa, the yellow stem borer (YSB) is dominant, followed by stripped stem borer (SSB), dark headed borer (DHB) and pink stem borer (PSB) in the state. YSB completes two broods in *kharif* and 3 broods in summer. In deep water rice (DWR), the relative abundance of YSB, SSB and PSB in grain maturity stage was 63.4, 18.3 and 18.3%, respectively. However, in the vegetative stage YSB is the sole species in deep water rice. In saline belts, stem borer and hispa are the major pests identified. In upland and medium land conditions, YSB and SSB over winter in equal proportions but SSB is dominant in medium and low land areas.
- The Hirakud Command area of Sambalpur, Orissa has been identified as the endemic pocket of rice gall midge. This pest, however, could be effectively suppressed by growing resistant varieties like Shakti, Phalguna and Vikram. Twelve species of grasses have been identified as alternate hosts of gall midge, In sambalpur district the gall midge tides over the off-season on ratoon rice under irrigated conditions.
- Among rice plant hoppers, the brown plant hopper (BPH) was considered the dominant species up to 1990s. The white backed plant hopper (WBPH) which was a minor pest attended major status in 1980s and now occurs in both *kharif* and summer seasons. For last ten years, leaffolders (LF) and case worm have attained major pest status. The green leafhopper (GLH) which was major pest in 1970s has changed its status to minor. Gundhi bug is continuing as a major pest in rainfed upland and medium land rice. The Surti caterpillar, Nisaga simplex, a pest of forest plants has migrated and is a major pest of medium land rice in 19 blocks of undivided Koraput and Kalahandi districts causing extensive defoliation.
- The yield loss assessment due to major insect pests of rice worked out is given in Table 14.1

TABLE 14.1: Insect damage intensity in rice

Sl.No.	Insect pest	Damage intensity	Yield loss (kg/ha)
1.	Stem borer	1% dead heart	32.4
	1% white earhead	62.5	
	1% increase in stem damage	0.01-0.33%	
2.	Gall midge	1% silver shoot	33.2
3.	Leaf folder	1% leaf area damage	
	Flag leaf	41.0	
	penultimate leaf	32.0	
	third leaf from top	29.0	
4.	Surti caterpillar	1 larva/m^2	29.6

- Through proper protection, loss can be avoided from 10 to 46% in *kharif*, 29 to 56% in deep water and 3 to 58% in *rabi* rice. Estimation of Economic Threshold (ETL) for major pests of rice has been worked out and presented below.

Sl. No.	Insect pests	ETL value
1.	Stem borer	9.70%dead heart (vegetative stage)
2.	Leaffolder	8.22% Leaf area damage (PI stage)
3.	Surti caterpillar	2 larvae/m^2

Host Plant Resistance

- Screening of large number of rice access for resistance to major insect pests of rice was done. Resystance due to antibiosis was observed in rice against YSB. Higher ratio of ash to silica imparted resistance to YSB. Rice varieties with higher silica content retarded growth and development of YSB larvae. Application of Boron-Zinc Oxytetrachloride complex induced resistance against YSB with 15% dead heart compared to 33.5% in control. Presence of aminoacids such as serine and clycine in growing primordial of rice tillers provided resistance to gall midge and stem borers in MTU 15 and Ptb 10 rice varieties. Unsuitability for feeding, lower fecundity, hatchability of eggs and honeydew excretions are the biological attributes exhibited by the BPH during feeding on resistant plant. Total free aminoacids and starch content are higher in susceptible plants.

Biological Control

- Research on biological control of YSB revealed that the egg parasitoid of YSB, *Tetrastichus schoenobii* behaved more like an egg predator and consumed eggs in succession but its natural occurrence in the field was low. The YSB eggs were also parasitized in the field by *Telenomous dignoides* (21 -29% eggs) and *Trichogramma japonicum.* Eight entomorphagous insects were recorded inside the rice silver shoot created by gall midge (GM) attack. In Orissa, the GM is attacked by the egg-larval parasitoid *Platygaster oryzae* and the purpal parasitoid *Neanastatus gracillius.* The BPH and WBPH of rice are chiefly predated upon by the wolf spider, *Lycosa pseudoannulata* and the mired bug, *Cyrtorrhinus lividipennis* in rice fields. Besides, the staphylinid beetle, *Paederus spp* and carabid, *Ophionea* spp have also been recorded as predators of nymphs and adults of planthoppers.

Botanical Pesticides

- The results on pest control with botanicals revealed that under low insect pest density application of *Azadirachta indica, Catharanthus roseus* and *Vitex negundo* extracts can be integrated with other control measures. Neem Seed Kernel Extract (NSKE) sprayed at 1% concentration with 0.1% teepol at boot leaf stage proved effective against rice SB. The neem derivatives such as NSKE, Welgrow and Repelin were found promising against rice GM. Neem oil and NSKE at 2% gave satisfactory control of rice leaffolder, stem borer, gall midge, GLH and found safe to predators such as spiders and mired bugs. Treatment of rice plants with various neem derivatives alone increased grain yield by 127-705 kg/ha as compared to neem oil plus monocrotophos treatment (1.0 t/ha). The use of phytoproducts resulted in delayed nymphal duration by 9.7 to 12.9% and reduction in nymphal survival by 49 to 84.3% and growth indices by 8.2 to 56.3% as compared to untreated control in case of plant hoppers.

Chemical Control

- Seed treatment with fipronil (Regent 5 FS) 100 g/ha was found effective in controlling stem borer in nursery bed and up to 42 days in the main field.
- Seedling root dip with 0.02% chlorpyriphos or isofenphos for 10 to 12 hours gave excellent control of SB and GM at the tillering stage. Treating the nursery beds with phorate, carbofuran, mephosfolan and diazinon granules 1 kg/ha at 10 days before uprooting seedlings protected the transplanted rice against the GM and SB for about 3 weeks.
- Carbofuran and phorate granules were found superior to other granular insecticides such as quinalphos, isofenphos, BPMC, MIPC, and disulfoton in controlling the sap sucking pests of rice viz., GLH, WBPH and thrips and gave 0.3 to 0.6 t/ha higher yield as compared to control.
- Soil application of phorate, carbofuran, BPMC, chlorpyriphos and carbosulfan G 1.5 kg/ha at 20 DAT temporarily inhibited rhizosphere microflora (fungi and bacteria) up to two weeks while rthofenprox at the same dose inhibited microflora for a week.
- Application of deltamethrin 0.012 kg/ha, quinalphos and chlorpyriphos each at 0.5 kg/ha induced resurgence of WBPH.
- Rice crop sprayed with the chitin synthesis inhibitor, buprofezin (0.1 kg/ha) against BPH caused 41% nymphal mortality 3 days after feeding.
- Sprayable formulations of many new insecticides have been field evaluated during last 10 years against major rice pests. Their bio-efficacy has been highlighted below in Table 14.2.

TABLE 14.2: Name and dose of new chemicals

Sl.No.	Chemical name Trade name evaluated	Dose	Rice pest controlled	
1	BPMC	Bipvin 1.8% D	30 kg/ha at 40 DAT	BPH, GLH.
2.	Imdacloprid	Confidor 200SL	20 g/ha at 20,30,40 DAT	GM, WBPH
3.	Fipronil	Regent 0.3G	75 g/ha	SM, LF.
4.	DDVP	Nuvan 76 EC	2 ml/lit	Gundhibug
5.	Abamectin	Vertimec 1.8 EC	2 ml/lit	Gundhibug
6.	Alpha-cypermethrin	Farsa 10 EC	30 g/ha	SB, LF, BPH,WBPH
7.	Lambda cyhalothrin	Karate 5 EC	12.5 g/ha	- do –
8.	Chlorpyriphos 16%+ Alphacypermethrin 1%	Ducord 17 EC	2 ml/lit	SB,LF.
9.	Monocrotophos 25% + Phosdom dichlorvos 11%	36 SL	600 ml/ha	SB,BPH,LF, GM, GLH, WBPH
10.	Monocrotophos 25%+ cypermethrin	3%Phoscpu 28 DC	800 ml/ha	- do –

Integrated Pest Management (IPM)

Until quite recently the control of insect pests was focused mainly on use of insecticides on calendar basis. This has resulted in suppression of parasitoid population, decrease in benefit cost ratio, pest resurgence, development of insecticide resistant strains of insects and environmental pollution. The minor pests are now gaining foothold due to indiscriminate use of pesticides. There are much concern about the persistence of harmful pesticide residues which poison the food grains of human and domestic animals. Thus, the emphasis is now on integrated pest management (IPM), where the use of insecticide is minimized by combining with other methods, including varietals resistance, cultural practices such as crop rotation and removing of crop residues to prevent pest survival, water management and biocontrol.

Judicious use of pesticides is recommended basing on the monitoring of pests and natural enemies and economic threshold levels of pests. Timely application of recommended dose of pesticides is the basic strategy of IPM. Pesticides need to be used only at the vulnerable stage of the pest (1) emergence of adults, just at the time of egg hatching, younger stage of pests, active stage of insects at specific site of the crop. Keeping in view the above basic requirements of IPM, the field worker can recommend use of insecticides depending on the situation of a particular location. The details of the cropwise pest management for crops are given below:

KHARIF RICE

- Transplant rice within second week of July for low pest incidence.
- Treat the seeds with fipronil 5 FS 1.25 g/kg to control stem borer.
- In gall midge endemic pockets grow tolerant varieties *e.g.* Sakti, Samalei, Bhuban, Nila, Sarasa, Khira, Udaya, Tara and Gouri.
- Seedling root dip in 0.02% chlorpyriphos for 12 hours before transplanting minimized the incidence of gall midge, stem borer and other sucking pests at the early growth stage.
- Apply optimum balanced dose of fertilizer *i.e.*, 60, 30, 30 kg N, P_2O_5, K_2O/ha.
- Monitor the crop at weekly intervals and apply insecticides when they cross economic threshold limit.
- Spraying may be done with quinalphos / chlorpyriphos / monocrotophos / ohosohamidon 0.5 kg/ha or newer insecticides (Table 14-2).
- If gall midge parasitization exceeds 25% then avoid insecticide application.
- In semi-deep water rice, spray monocrotophos 0.5 kg/ha once or twice between boot leaf and panicle emergence stage of control of yellow stem borer.

Summer Rice

- Collection and destruction of stem borer egg masses from nursery bed.
- Plough the rice field or burn the stubbles before mid January to restrict the carry over of SB to succeeding crop.
- Transplant rice before mid January to avoid severity of SB at heading stage.
- Use balance dose of fertilizer *i.e.*, 80, 40, 40 kg N, P_2O_5, K_2O/ha.
- In leaf folder endemic areas spray monocroptophos 0.5 kg /ha between 40 and 50 DAT or new insecticides (Table 14.2) and against SB at PI stage
- Avoid insecticide application when stem borer parasitization exceeds 30% and pest:defender ratio is 2:1.
- In BPH and WBPH endemic areas grow resistant varieties like Daya, Pratap, Lalat, Uday, Gouri and Bhuban.

- Foliar application of neem oil or NSKE 2% with 0.1% detergent at 20 and 70 DAT with an intermediate application of monocrotophos 0.5 kg/ha at 40 DAT is effective in controlling major insect-pests of rice.

Pulses

- Eight species of pod borers were recorded infesting various pulses. In pigeonpea, *Maruca testulalis* followed by *Nanaguna breviuscula* and *Helicoverpa armigera* are the dominant borers. *Nanaguna breviuscula* has been identified as a new borer species from the country. In chickpea, *Helicoverpa* is the only borer causing extensive damage to the crop. The intensity of damage is given in Table 14.3.

TABLE 14.3: Insect damage intensity in pulses

Crop	Insect pest	Damage intensity	Yield loss (%)
Pigeonpea	Pod borer	15% pod damage	
		(a) Kharif	0.73
		(b) Rabi	0.79
Green gram	Pod borer	1% pod damage	
		(a) Kharif	0.98
		(b) Rabi	1.34

- The ETL for pod borer complex in pigeonpea was found 2.7 larvae/plant or 6.5% damaged pods. In chickpea for gram pod borer the ETL was estimated to be 18 larvae/m^2 (mixed instars).
- In chickpea, increase in plant phenol content was associated with low borer incidence.
- The cattle egret was a major predator of *Helicoverpa armigera* in chickpea causing 15 to 51% reduction in the larval population. The Ichneumonid (*Campoletis chloridae*) and the Tachinid (*Carcelia illota*) parasitized up to 21 and 24% of late instar larvae of *H. armigera*, respectively during different months.
- For control of *Helicoverpa armigera* in pigeonpea under field conditions, spraying of NPV 250 LE (Larval equivalent)/ha or NPV 150 LE/ha + endosulfan (0.035%) was found effective resulting in 64% reduction in pod damage. Similarly, in green gram the whitefly, a vector of YMV was effectively controlled by spraying phosphamidon 375 ml/ha or endosulfan 1 lit/ha. The gram od borer in this crop was effectively controlled by spraying endosulfan 1 lit/ha, aphids by carbosulfan and other sucking pests by spraying carbosulfan/monocrotophos/malathion 1 lit/ha. In chick pea, *H. armigera* was successfully controlled by the biopesticide Btk (Halt) 1 kg/ha boric acid 0.1% or jaggery 0.5% added to Ha-NPV increased its efficiency.

Oilseeds

- The thrips (*Scirtothrips dorasalis* is the major pest of *rabi* groundnut causing 88% plant and leaf infestation. On the other hand, (*Aproaerema modicella*) appeared as a major pest in *kharif* groundnut resulting in 20 to 63% leaf infection. The incidence of bud necrosis disease transmitted by the groundnut thrips is increasing. The black aphid (*Aphis craccivora*) is also serious during *kharif* season.

- Late sown mustard is severely affected by the webber (*Crocidol mia binotalis*), aphid (*Lipaphis erysimi*) and sawfly (*Athalialugens proxima*). The mustard aphid is serious from December to February. One leaf webber larva/plant reduced 24.3 kg of seed yield/ha in mustard.
- In sesame, the leaf webber and capsule borer (*Antigastra catalaunalis*) is a serious pest in Orissa inflicting 80% avoidable loss. The economics threshold levels for various oilseeds were worked out and presented below.

Crop	Insect	ETL larvae/10 plants
Groundnut	Tobacco caterpillar (*Spodoptera litura*)	6.25 larvae/10 plants (15 DAS)
Sesame	Leaf webber	1.2% twig damage
Mustard	Leaf webber	2.46 larvae/plant

- As many as 17 species of aphidophagus insect predators have been recorded from Orissa. The lady bird (*Menochilus sexmaculatus*) consumed 244 aphids/day in grub stage. The adult male and female consumed 50 and 62 phids/day, respectively.
- Control with botanicals revealed that neem and karanj oil (0.4% effectively controlled the *Crocidolomia* population in mustard. Both the botanicals adversely affected the consumption, growth and development of mustard sawfly. Aqueous extracts of seven plants viz., custard apple, datura, Indian laurel, karanja, neem and rhizome extracts of sweet flag showed varied degree of antifeedant action in *Spodoptera litura*. The percentage reduction in leaf consumption was highest with NSKE (33%) followed by mandar (31%) and karanja (29%). These phyto-products prolonged larval period and exhibited abnormalities like difficulty in coming out of moulting skin, formation of larval-pupal intermediate and crippled adults.
- Among the synthetic chemicals, methyl demiton, menazon, Dimethoadate, monocrotophos and quinalphos 1 lit/ha were found effective against mustard aphid. For the mustard leaf webber, menazon, chlorpyriphos, fenvalerate, quinalphos, cypermethrin, deltamethrin and acephate at recommended doses were found effective. The combination insecticides Sherlone 29 EC (phosalone 24% + cypermethrin 5%) showed the highest efficacy against mustard leaf webber compared to endosulfan, neemazal, Biolep, Dipel, Delfin, lipronil, SINPV and Bioasp which were also in the order of effectiveness against the pest.
- In groundnut, application of carbofuran G 1 kh/ha reduced thrips damage and hence bud necrosis. In pre-*rabi* crop spraying of quinalphos of monocrotophos 0.4 kg/ha at 30 and 50 DAT effectively controlled major sucking pests. In summer groundnut, spraying once with dimethoate 0.4 kg/ha between 30 to 35 days proved effective against sucking pests.
- In sesame, quinalphos and phosalone 1 lit/ha was effective against the leaf webber and capsule borer. Besides, carbaryl 1 kg/ha and monocrotophos 0.5 kg/ha proved the best against this pest. Out of six combination insecticides evaluated against this pest, Nagata (ethion 40% + cypermethrin 5%) and Viraat (cypermethrin 3% + quinalphos 20%) proved superior to others in reducing the larval population, per cent twig, flower and pod infestation due to leaf webber and pod borer. Seed yield was the highest with both Nagata and Viraat at 450 g/ha and 230 g/ha, respectively.

- In hybrid sunflower, tobacco caterpillar (*Spodoptera litura*), capitulum borer, semi-looper, Bihar hairy caterpillar and jassids were the major insect pests recorded. Among the insecticides evaluated Spark 36 EC (deltamethrin 1% + triazophos 35%) 360 g/ha was found effective against Spodoptera.
- In soybean, whitefly, Bihar hairy caterpillar, *Hydylepta indica* and *H. diemenalis* were the major pests which could successfully be controlled by application of monocrotophos 1 lit/ha.

VEGETABLES

Brinjal

- In brinjal, fruit and shoot borer is the major pest. The yield loss assessment studies revealed 2.7% with every 1% increase in shoot infestation and a loss of 353 kg/ha with every 1% increase in fruit infestation. The avoidable loss due to the pest ranged from 43 to 64%. The ETL for the insect was worked out to be 4% shoot damage and 14% fruit damage.
- Application of micronutrient through seed treatment or foliar application, higher silica, crude fibre, total mineral contents and phenylalanine imparted resistance to fruit and shoot borer. Similarly, compact vascular bundle with lignified cells arranged in the stem attributed towards reduced shoot infestation.
- The presence of trichomes prevented oviposition by the moth as well as entry of newly hatched larvae into the fruit. Fruits having tight and semitight calyx, hard rind and more seeds confrred resistance to this pest. Brinjal epilachna beetles are parasitized by *Pediobius epilachnae* resulting in sufficient reduction in its population.
- Alternate application of triazophos and carbaryl 0.5 kg/ha as fpliar spray at fortnightly intervals commencing from 15 DAT gave excellent control of fruit and shoot borer and accrued the highest CBR 0f 1:9.

Tomato

- In tomato, the fruit borer (*Helicoverpa armigera*) is serious. Carbaryl 2kg/ha, Lindane and toxaphene 1 lit/ha were effective against the pest. Acephate 75 WP 600 WP 600 g/ha and cypermethrin 5 DF 60 g/ha reduced fruit damage by the borer, whereas the combination insecticide cypermethrin 5 DF + acephate 45 DF at 425 g/ha controlled the sucking pests of tomato vis-à-vis fruit borer.
- The American serpentine leaf miner (*Liriomyza trifoli*), anew pest, attacks 85 plants in Orissaincluding 4 pulses, 5 oil seeds, 25 vegetables, 25 vegetables, 23 ornamentals and 27 dicot weeds. Tomato is its favourite host. The avoidable loss in tomato was 37%. Insecticides like trizophos (Tarzan 40 EC), chlorpyriphos + cypermethrin (Anaconda 505) and phosalone 35EC) @ 0.05, 0.055 and 0.07%, respectively, were effective in controlling the pest.

Okra or Lady's Finger

- Newer insecticides like thiamethoxam (Actara 25 WG) and imidacloprid (Confider 200 SL) applied at 25 g/ha were superior to conventional pesticides like dimethoate or cypermethrin in controlling the sucking pests such as aphids and aphids. However, against the fruit borers and leafrollers, application of cypermethrin 100 g/ha, profenophos 500 g/ha and profennophos

+ cypermethrin (Rocket 44EC) at 440 g/ha proved effective. The borer infestation decreased to the extent of 77% in these treatments as compared to control.
- The okra leaf roller was successfully controlled by endosulfan 0.07% spray followed by endosulfan 0.035% + diflubenzuron (Dimilin 25 WP) 0.05% in the field.

Cabbage

- The avoidable loss due to cabbage pests was 55 to 75% in Orissa. The cabbage looper was successfully controlled by application of endosulfan 1 lit/ha or Dimilin 500 g/ha or carbaryl 2 kg/ha or acephate 500 ml/ha. The diamond back moth (*Plutella xylostella*), an International pest, is successfully controlled by anew insecticide chlorfenapyr (Rampage 10 SC) 75 and 100 g/ha. This gave 86 and 95% reduction of the pest up to 10 days of application, respectively. Next to it was cartap hydrochloride 500 g/ha.

Cauliflower

- In cauliflower, new combination insecticides like Spark and polytrin C 750 ml/ha applied at 55 and 75 DAT controlled major pests like aphids, tobacco caterpillar, leaf webber and diamond back moth with a record yield of 33 t/ha compared to 10.4 t/ha in control under field evaluation. These chemicals were toxic to predatory lady bird beetles and syrphid flies. But combined application of a biopesticide, Dipel + endosulfan (0.5 lit + 0.5 lit)/ha and neem product Multineem + endosulfan (1 lit + 0.5 lit)/ha controlled the above pests.

Chilli

- In chilli, the 'murda' disease caused by the thrips (*Scirtiothrips dorsalis*) and the yellow mite (*Polyphagotarsonemus latus*) is a serious disease. The avoidable loss due to these pests was worked out to be 54%. The ETL value for thrips was found to be 2 thrips/twig. Earlier, for control of chilli thrips carbaryl (2kg/ha), Sulphur, formothion and dimethoate (1 lit/ha) was recommended. New insecticides like ethion (Forsite 50 EC) 250g/ha sprayed at 15 days intervals on the thrips attaining ETL proved superior in controlling leaf-curl disease of chilli with 6,4% incidence and highet yield of green chilli (5 t/ha) compared to 84% leaf curl and 2.7 t green chilli yield/ha in untreated control. A new fermentation metabolite NACLFMOA at 20 and 25 ml/ha, abamectin(Vertimec 1.8% EC) and dicofol followed ethion in controlling leaf curl.

Cucurbits

- In cucurbits, fruitfly is a serious pest. Spaying of chtorpyriphos or endosulfan 1 lit/ha or dichlorvos 375 ml/ha proved effective. But controlling the fruitflies with baits is a good alternative, that avaoid insecticide application and killing of natural enemies. Providing ripe banana chips fortified with carbofuran granules (5 to 10 granules/chip) attracts and kills the adults. Besides, spraying acaphate 75 SP at 0.15% twice at vining and flowering stages of crop gave good control with highest CBR at 1:12.

Sugarcane

- A predatory phytoseiid mite (*Amblysius finlandicus*) was found attacking sugarcane mites in Puri district. Studies on the influence of abiotic factors on the predator and its prey, revealed that atmosphere of 21° C with 64 to 72% RH was most conducive for the predator. Investigation on biological control of sugarcane pests revealed that two egg paracitoids viz., *Ooencyrtus spp* and *Tetrastichus pyrillae*, a larval parasite(*Epiricania melanoleuca*) and a fungus (*Metarrhizium anisopliae*) attacked sugarcane pyrilla under natural conditions. Among these *E. melanoleuca* was the most predominant bio-control agent responsible for decreasing pest parasite ratio from 164:1 to 0.25:1 between August to October.

PLANTATION CROPS

Coconut

- Black headed caterpillar of coconut (*Oopisina arenosalla*) was controlled through insundative release of its larval parasitoid Bracon hebetor and Bracon brevicomis Studies on the biology of two other parasites of this pest viz., *Perisierola nephantidis* and *Brachymeria nephantidis* generated useful information regarding the influence of external factors on their sex ratio and reproductive potential under Orissa conditions.
- Studies on control of coconut eryophyid mite (*Aceria querreronis*) which has become a serious pest at present is continuing.

Cashew

- Fifty species of insect pests have been recorded from cashew and 22 species are of regular occurrence in Orissa. The stem and root borer, thrips (19 to 57% nut damage) and tea mosquito bug are important ones.
- Cleaning of the base of the trunk and removal of the affected trees effectively suppressed the stem and root borer of chewnut. Swabbing of tree trunks up to 1 m height width monocrotophos 0.2% after mechanical extraction of grub was found quite effective at early infestation. Besides, painting of the tree trunk (basal 1 m) with coaltar and kerosene (1:2) mixture twice a year *i.e.*, April-May and November to December is quite effective.

FIBRE CROP

Jute

- In jute, a few resistant varieties were identified against yellow mites. They were JRC 3120, JRC 2102, JRC 4145, JRC 9638 and C 1410.

Mesta

- Screening of varieties revealed that AR 71, AR 86 and ER 72 were found tolerant to mealy bug, the important pest of mesta.

Cotton

- Four boll worm species such as *Earias vitella* (Spotted boll worm), *Helicoverpa armigera* (American boll worm), *Pectinophora gossypiella* (Pink boll worm) and *Spodoptera litura* infested the cotton crop, the dominant being *E. vitella*. The damage due to boll worm complex ranged from 13 to 21% squares and flowers and 10 to 29% green bolls. The seed cotton yield loss was recorded to be 26 per cent.

- A new larval parasitoid *Rugadinae spp* was recorded on *E. vitella*.
- The IPM module developed for boll worm control comprised of: (1). Seed treatment with imidacloprid 7.5 g/ha, (2). Release of *Trichogramma chilonis* egg parasitoid 1.5 lakh/ha at 60 and 67 days after germination (DAG), (3). Application of NSKE (5%) at 40 and 55 DAG, (4). Spraying of endosulfan (0.07%) at 75 DAG and profenophos (0.05%) at 90 DAG, (5). Growing cowpea as inter crop, and (6). Installation of bird perches 20/ha. Through adoption of IPM schedule seed cotton yield of 1.56 t/ha and monetary return of Rs 9904/ha were obtained.

OTHER CROPS

Grain Amaranth

- Out of 9 major insect pests identified to attack grain amaranth, flea beetle and grasshopper were considered as serious pests inflicting 2.7 and 21% leaf infestation, respectively. Of the 14 popular grain amaranth varieties tested, IC-666 was least susceptible to the leaf eating caterpillar and leaf folder infestation. The variety R 104 was found tolerant to flee beetle and grasshopper.

Betelvine

- In betelvine, whitefly appeared to be the predominant pest. Out of 13 varieties evaluated Gandhi pan, Karpoori Telaku and Bainchi Godi were least infested by whitefly, blackfly and mealbug, respectively. Whitefly population was least in the month of September.

STORAGE ENTOMOLOGY

- The beetle (*Tenebrioides*) completed its life cycle in 78 days on whole wheat grains. At 15°C the adults survived for 114 days. The angoumois grain moth (AGM), *S. cerealella* completed its development in 34 days on wheat and it took longer time on rice and maize grains. Grains hardness imparted resistance to rice weevil.
- Grain characters like husk thickness, boldness and hardness with more silica and potash content in the husk conferred resistance against AGM. Groundnut varieties SG-84 and Smruti were found highly resistance. TAG-24, TG 26 were moderately resistant and JL 24, ICGS 44, TG 3 and AK 12-24 were proved susceptible to the groundnut seed beetle (*Caryedon serratus*). In sesame, the black seeded varieties (Krishna, ORM-14) were resistant to rice moth attack in stores, while the white seeded varieties (TKG-21, OMT-32) were susceptible. The green gram cultivars PDM 84-143 and PDM 84-155, black gram cultivars LB-20, and T-65 and pigeonpea cultivar T 6 exhibited resistant against pulse beetle. The thickness of the seed coat was positively correlated with resistance to pulse beetle.
- Application of 2 cm thick layer of rice bran on grain surface meant for consumption or sand for seed purpose protected the grains against major stored grain pests. Neem seed kernel powder mixed with grains at 5% provided protection to wheat and green gram against rice moth and pulse beetle for three months without impairing bad taste or smell.
- Dried begonia (*Vitax negundo*) leaf mixture with paddy exhibited a fairly high

level of adult mortality of AGM up to 15 days after application. Admixture of Karanj seed powder (2.5%), begonia or neem leaf powder (5%) with paddy grains protected from AGM attack. Similarly admixtures of custard apple or neem seed powder (5%) with wheat afforded protection against rice weevil for about 3 months,

- Dried citrus peel mixture at 3-5% checked rice weevil and AGM infestation for 1 month. Mixing of coconut or mustard oil at 200-500 ml/quintal of green gram or black gram provided good protection against the pulse beetle up to 3 months.
- Admixture of neem seed powder (1%) with pulse grains exhibited strong repellent action against pulse beetle. Neem oil (0.5%) also rendered satisfactory protection against the above pests.
- Another emerging pest, the groundnut seed beetles was controlled by admixture of neem leaf powder (2%) or turmeric powder (2.5%) followed by kendu leaf powder 2.5%). Even *Agreratum lantana* each at 2% or Econeem (0.5%) protected the groundnut pods from damage for 3 months. Impregnated jute bags with malathion (0.01%) or deltamethrin (0.009%) and coaltar lining gunny bags proved effective in long term storage of pods against seed beetle infestation.
- Sesame grains are protected from rice moth attack in stores up to 3 months by mixing Ageratum leaf powder or turmeric powder or custard apple seed powder @ 5% or oils like coconut and palmolin applied at 2.5%. Neem and karanj oil and their seed powder (2.5%) were found equally effective.
- Pre-harvest spraying of acephate, ethofenprox or azadirachtin at 0.05% kept rice grains free from AGM attack for a period of 3 months. Application of deltamethrin at 115 ml/m^2 on grain containers provided appreciable safely to wheat against Tribolium attack.
- Setting of pheromone (sex lure) traps in stores or godowns kept the AGM population under check.
- Eggs of red flour beetle were highly sensitive to gamma radiation. When eggs were exposed to 341 red gamma ray, the adult emergence sharply declined.
- Interspecific reaction revealed that in wheat grain *T. casetaneum* suppressed the population of *Rhizopertha dominica* due to its predatory activity. Infestation of grains by S. *Oryzae* and R. *dominica* accelerated the growth of *T. castaneum* and *Oryzaephilus surinamensis*.

PRODUCTIVE INSECTS

Honeybee

- Bee pollination increased by 11% in mustard, 55% in sesame, 33% in niger, 79% in sunflower, 65% in coriander and 64% in safflower. The increase in oil content in these crops ranged 2.2 to 3.9% in mustard, sunflower and niger, 5 bee colonies/ha is optimum to maximize yield through bee pollination.
- Among the natural enemies of bees the herpeto predator (Chameleon), king crow, bee hunter wasp were found predating upon foraging bees. Besides, wax moth attack was predominant in monsoon period causing 20% colony decline. In hilly areas bear is another important enemy of bee hives.
- Acarine disease and *Tropilaelaps clariae* mite are important enemies which can be controlled by formic acid fumigation. Five milliliters of formic acid

kept in small homoeopathic vials at the bottom of infested hives killed all mites without affecting the brood.

MANAGEMENT OF PLANT DISEASES

During *kharif* season due to high temperature humidity warmer temperature and intensive cultivation, some crops like rice severely suffer from a large number of infections plant diseases. These may be soils, seed or air borne in nature. As a consequence, heavy crop losses are encountered during the season. Unless it is an epidemic covering a large area, a sound approach to protect the crop by applying recent technology of integrated diseases management (IDM) may minimize the yield losses. For this the use of certified seeds, pre-treatment of seeds with effective fungicides, growing of tolerant/resistant varieties, application of judicious doses of basic (N.P.K.) soil nutrients, proper water management preventive schedules of spraying of crops with fungicides, maintenance of satisfactory sanitary conditions and follow up of suitable dates of sowing/transplanting of the crops are extremely necessary. Besides, these considerations it is also of utmost importance to procure the material from disease-free crops and regions. A large number of cases are known where in a region, incidence of a newly introduced diseases has become established and they threaten the high productivity of economically impart crops like vegetables, cereals, millets, pulses and oil seeds etc.

Recently enough information is available which shows to minimize some seed borne and air borne diseases by using selected natural products (developed from plant) and also by applying the principle of biological control. Wherever such varieties are known to show promise and depending upon their availability they could be used even of a limited scale to restrict the crop diseases.

Keeping in view the above a brief resume of symptoms and best possible management practices of some key diseases of important *kharif* and *rabi* crops are categorized in the following section. Regarding the recommendations made for use of fungicides, several have been listed so that if a particular product is not available, the alternative one may be used.

IMPORTANT DISEASES OF CROPS AND THEIR CONTROL

RICE

Udbatta Disease

- One of the pioneer research findings from the department of Plant Pathology, Orissa in the field of rice pathology was Udbatta disease of rice by *Ephelis oryzae*. This disease was first reported from Orissa by Dr. N. N. Mohanty, the then Lecturer-cum-State Mycologist during 1964. It was a serious problem in South Orissa particularly in Jeypore and Nabarangpur areas of undivided Koraput district. The flag leaves, the second and third leaves from the top and their corresponding leaf sheaths sowing lustrous, grayish-white appearance were reported. A number of collateral grass hosts for the pathogen were identified. Solar treatment of seeds before sowing and spraying of carbendazim at boot-leaf stage, have been recommended for the control of the disease.

False Smut

- This disease caused by *Ustilaginoidea virens* (= *Claviceps oryzae*) was a serious problem in late and super fine varieties of rice. Early sowing or transplanting of seedlings and spraying of Bordeaux mixture or with any copper fungicides twice at an interval of 10 days at pre-flowering stage were recommended for control of the disease.

Blast Disease

- High level silica in leaf reduced the blast incidence. A higher dose of nitrogen was reported to reduce silicate cells and therefore increased susceptibility to blast. Silicon content was high after 45 days of transplanting and at the time of emergence in resistant varieties (Table 14.4).

TABLE 14.4: Mean level of silicated cells in different varieties of rice

Variety*	15 days	30 days	45 days	At the time of emergence
Co-4	110.8	164.8	172.6	187.2
S-67	63.8	88.0	167.4	147.0
ADT-10	60.5	125.0	138.4	143.3
Co-13	52.8	94.0	130.6	118.4
Mean	72.0	117.9	153.2	148.9
CD (0.05)			21.584	

*Co-4 and S-67 were resistant and ADT-10 and Co-13 were susceptible.

Bacterial Blight

- It was first reported in Orissa from Sambalpur district during 1964 due to extensive cultivation of Taichung Native-1 and IR-8. Studies on serodiagnosis or the pathogen was taken up in the department of plant pathology for the first time in India during 1969-1977 and existence of a distinct serotype among 45 isolates from all over the world were reported. A new detection technique developed in the department on seed borne nature of X. oryzae pv. Oryzae was successfully demonstrated at the National Training Programme held at IARI, New Delhi during the year 2002.
- Over night soaking of seeds in 0.015% Streptocycline or 0.15% Plantomycin, draining out of water and spraying of Copper oxychloride (0.2%) along with Plantomycin (0.1%) or Streptocycline (0.01%) and balanced application of NPK could control the disease in large extent.

Sheath Blight

- Once regarded as a minor disease of rice, it is appeared for the first time in large scale during 1980-81) in the Hirakud command areas of Sambalpur district of Orissa. It spreads to other parts of the state in the passage of time. Rice cultivar OR 201-5 was found tolerant to sheath blight followed by OR 1206-25 and Mahanadi. Ragi, sugarcane, maize, jowar and bajra were found to be collateral hosts for the pathogen (*Rhizoctonia solani*) on which it perpetuated.
- Air temperature of 25°C to 30°C with RH of 80 to 85%, sunshine of 5 to 7 hours per day accompanied with moderate rainfall accelerated development under field conditions.
- Spraying of chemicals like propiconazole (Tilt) at 0.15% concentration or validamycin (Sheathmar) at 0.3% twice at 10 to 15 days intervals during

tillering stage could control the disease effectively. The rice varieties Swarna, Savitri and Mahalaxmi were found highly susceptible to this disease.

Sheath Rot

- Sheath rot of rice caused by *Sarocladium oryzae* appeared in epiphytotic form in coastal districts of Orissa during 1991-92 causing huge yield losses and again during post super-cyclone period of October 1999.
- Rice cultivars *i.e.*, Mahalaxmi, Kanchan, Savitri and Badami were tolerant to sheath rot, whereas Mahalaxmi and Savitri were susceptible to sheath blight. The disease was found less in early sown crops than late sown crops. Spraying the crop with carbendazim 50 WP (0.2%) or Mopsin 70 WP (0.2%) twice at the time of tillering and just before boot-leaf stage could control sheath rot disease.

Grain Discoloration

- The discolored grains became chaffy and their germination percentage, seedling vigour and protein contents were reduced. Association of a number of fungal flora such as *Curvularia lunata*, *Fusarium moniliforme*, *Trichoconis padwickii* and *Drechslera spp.* including one unidentified rice bacterium caused the disease. Varieties like Savitri, Manika, Swarna and Kanchan were found torerant.
- During the post-flood situation of last flood of September, 2003, minor leaf and grain spots of rice caused *Nigrospora oryzae* and stack burn disease caused by *Trichoconis padwickii* appeared in epiphytotic manner on rice crop covering a large area in Mahakalapara block of Kendrapara district.

PULSE PATHOLOGY

In order to enhance the yield of legume crops in Orissa, the state government sanctioned a scheme entitled "Rhizobial Inoculation of Legumes in Orissa" in 1964. The project work was initiated in the department. Rhizobial culture packs were prepared and were being regularly supplied to the farmers of different agricultural districts of the state for seed inoculation. The scheme was subsequently shifted to quality control laborotary of the state.

Pod rot of Black Gram and Green Gram

- During 1980-81 wet seasons, an epidemic on black gram and green gram was observed in several coastal districts of the state. Symptoms included discoloration and drying of leaves. Seeds in diseased pods were discolored, became small and malformed. Seed treatment with carbendazim mixed with Thiran was found to be most effective to check primary infection.

Other Diseases of Pulses

- Field screening of pigeonpea varieties against spot diseases indicated that Pusa-9, ICP-47452, ICP-3615 and ICP 71766 were resistant to leaf spot caused by *Alternaria tenuissima*. Wilt of black gram caused by *Fusarium spp.* was found to be effectively managed by seed treatment with bio-agent, "Bioderma" (*Trichoderma viride*) along with its application at the rate of 2.5 kg in 50 kg of FYM/ha during land preparation.

OIL SEED PATHOLOGY

Groundnut Collar Rot

- During 1966-69, collar rot of groundnut caused by *Aspergillus niger* and *A. flavus* was a serious problem for groundnut growers of the state because of prevailing high humidity and warm environmental conditions. Seed treatment with carbendazim (0.15%) mixed with Thiram (0.3%) and spraying the crop with mancozeb (0.3%) effectively controlled collar rot of groundnut in the field.

Groundnut Stem Rot and Tikka disease

- Stem rot of groundnut caused by *Sclerotium rolfsii* along with incidence of tikka disease was also one of the major set back in groundnut production of the state. Soil drenching of carbendazim (0.15%) along with Thiram (0.2%) was recommended for control of the disease. Spraying of carbendazim (0.15%) or mancozeb (0.38%) controlled the tikka disease.

Groundnut Shell Spot

- For the first time during 1985, maturing and harvested pods of groundnut from few groundnut cultivated fields particularly those with sandy to sandy loam soil, showed distinct black spots on the shells. Seeds from such pods were either aborted or partially shriveled. Sowing of seeds was observed to be one of the major reasons for gradual spread of the disease to large areas. Investigation on the association of micro-organisms as cause of the disease revealed the presence of *Rhizoctonia solani*, *Sclerotium rolfsii*, *Alteraria spp.*, a nematode and a bacterium (Erwinaa spp.). However, the primary incident of the disease and its nature has not yet been established.

Groundnut Rust

- Rust of groundnut which was not seen in India, has appeared during the early seventies and caused havoc in groundnut cultivation. The infected leaflets become chlorotic and blighted after appearance of brown coloured uredopustules on leaflets, petioles and stipules. In Orissa, the incidence of this disease was observed in late seventies appearing in severe form, more during rabi than kharif season and caused yield loss to the extent of 20 to 30% or more. The disease could be controlled by spraying Calixin (0.05%) for 2-3 times at an interval of 15 days commencing from one and half months of sowing.

Other Diseases of Groundnut

- Added to the above diseases in groundnut, severe out break of wilt caused by *Pseudomonas solanacearum* was recorded 1991-92 in areas of coastal Orissa as well as in western Orissa. Crop rotation with non-host like rice, sugarcane and maize was recommended wherever possible along with soil drenching of copper oxychloride (0.2%) mixed with Plantomycin (0.1%) for control of the disease.

Collar Rot of Mustard

- Collar rot of mustard by *Aspergilus spp.*, *Chatomium globosum*, *Fusarium spp.* and blight caused by *Alternaria brassicae* could be successfully controlled by spraying IAA (Indol Acetic Acid). CCC (Cycocel) and Zinc sulphate by way of inducing phytoalexin synthesis in host and there by inducing resistance to the host for respective diseases.

Sesame Diseases

- Leaf blight of sesame caused by *Alternaria sesami* and seedling wilt caused by *Macrophomina phaseolina* could be effectively controlled by treatment of seeds with Thiram/Deltan (3 g/kg) and spraying of mancozeb (o.3%).

VEGETABLE PATHOLOGY

Wilt of Solanaceous Vegetables

- Wilt complex of solanaceous crops (brinjal, potato, tomato and chilli) has been a serious problem for the farmers of Orissa. The management of this disease is difficult du e to involvement of fungal pathogens *Sclerotium solani* and bacterial pathogens, *Pseudomonas solanacearum* (= *Ralstonia solanacearum*) aling with root knot nematode (*Meloidogyne incognita*). An integrated approach for management of the disease was recommended to the farmers. It includes growing of resistant varieties (brinjal- Utkal Madhabi, Utkal Tarini; tomato – Utkal Pallavi, Utkal Dipti, Utkal Urbasi and chilli – Utkal Ava, Utkal Rashmi), seed treatment with Thiram (2g/kg), seedling treatment with carbendazim (0.15%) along with Plantomycin (0.1%), soil application of bio-agents in nursery soil, application of carbofuran (25 kg/ha), well decomposed FYM, lime (375 kg/ha, preferably paper mill sludge) and bio-agent (2.5 kg per 50 kg FYM/ha) during land preparation and drenching of basal soil with Copper oxychloride (0.2%) along with 0.1% Plantomycin twice at earthing and next pre-flowering stage.

Late Blight of Potato

- Late blight of potato caused by *Phytophthora infestans* has been a threat to potato growers of the state. The recommended control measures include selection of disease free seed tubers, wet seed tuber treatment with mancozeb (0.3%) mixed with 0.1% Plantomycin, application of bio-fungicide (Green Gayrd @ 1.25 kg FYM/ha) during land preparation and spraying of mancozeb (0.3%) or Cuman-L (0.2%) at least twice at 10 days intervals.

Potato Scab

- An out break of scab disease occurred in large areas of Cuttack during late seventies which spread to other areas through contaminated seed potato. The disease could be controlled by seed tuber treatment in 0.3% solution of mancozeb mixed with 0,1% Plantomycin. Green manuring with Dhaincha and application of mustard, groundnut, neem or karanj oil cake helped in reducing the disease. Application of Calcium ammonium nitrate favoured the disease development whereas; application of ammonium sulphate checked the disease incidence. Soil application of biofungicide (Bioderma or Niprot or Green Guard @ 2.5 kg + 50 kg FYM/ha) could successfully eradicate resting spores if applied during land preparation and again during earthing operation.

Powdery and Downy Mildews in Vegetables

- Powdery and downy mildew diseases infected several vegetable crops including various cucurbits and crucifers, field pea, coriander, cumin and channa (*Lathyrus sativus*). Spraying of dinocap (Karathane) 0.2% or Sulfex (0.5%) or Mopsin (0.2%) twice at an interval of 10 days during pre-flowering stage controlled powdery mildew. Spraying of Ridomil-MZ (0.2%) could effectively control the downy mildew.

Anthracnose and Die-back of Chilli

- It appeared in sever epiphytotic form in the state during 1969-70. The causal organism *Colletotrichum capsici* was observed to liberate a thermo-resistant and nonspecific toxin. The disease was found to be seed borne and could be controlled by treating the seeds with Ziram (0.25%) and spraying the crop 2 to 3 times with the same fungicide.

FRUIT PATHOLOGY

Powder Mildew of Mango

- Powdery mildew disease spreads rapidly and attack the mango flowers which become blighted and drop down. Very young fruits become shriveled, dry and fall down. Spraying of sulphur based fungicides like karathane (0.2%) twice on the blossom was effective in controlling the disease.

Anthracnose, Die-back and Twig blight of Mango

- Severe out break of anthracnose, twig blight and die-back of mango was observed in 1972-74. For blossom infection, 2 sprays of carbendazim (0.1%) at 15 days interval, for foliar infection, spray of opper oxychloride (0.3%) and for fruit infection, pre-harvest sprays with thiophenate methyl (Mopsin) at 0.1%, 15 days before harvest was recommended for management of the disease. Pruning (8 cm below the infection site) followed by spraying of Copper oxychloride (0.3%) and application of Copper oxychloride or cow dund paste at cut ends were recommended to die-back disease of mango caused by *Lasiodiplodia theobromae.*

Mango Malformation

- Mango malformation disease, a serious problem for mango plantations could be managed by application of 200 ppm NAA in the first week of October, regular removal of floral/vegetative malformed shoots, die-blossming of the early emerged panicles, establishment of separate “Mother Block” and propagation of varieties through malformation free scion shoots by verneer cleft grafting. Spraying of wettable sulphur mixed with monocrotophos and gum acacia (0.2 + 0.05 + 0.3%, respectively) @ 2% could effectively reduce the sooty mould disease intensity.

BIOLOGICAL CONTROL OF PLANT DISEASES

- Under integrated disease management programme, the Pathology department has undertaken studies on efficacy of plant products for control of various plant diseases. Neem oil and other neem products, bulb extract of garlic have been found effective in suppressing the growth of *Aspergillus niger*, *A. flavus* and *Sclerotium rolfsii* causing collar and stem rot of groundnut, respectively.
- Soil application of Bioderma(*Trichoderma viride*) along with FYM and neem cake could control stem rot of groundnut, panama wilt of banana, rot and wilt

of solanaceous crops and betelvine.

- *Gliocladium spp., Trichoderma spp. and Streptoverticillium spp.* have been evaluated with respect to their antagonistic effect against number of plant pathogenic fungi. Seed treatment with such antagonists proved successful in controlling seedling blight and damping off. Garlic extract was found effective against (*Erwinia carotovora* and *Pseudomonas solanacearum* causing rhizome rot of ginger.

MUSHROOM RESEARCH

- Cultivation of edible mushrooms like paddy straw mushroom (*Volvariella volvacea*), and oyster mushroom (*Pleurotus spp.*) had been one of the important activities of the department of Plant Pathology since early sixties. The techniques for pure culture, preparation of spawn and cultivation were standardized under the guidance of Dr. G. C. Rath, the then Associate Professor in the department of Plant Pathology, Orissa. Large scale production of spawn and production of various types of mushrooms were undertaken with the financial assistance of the Government of Orissa.
- Understanding the importance of mushroom production particularly as one of the sources of income by small and marginal farmers, the Centre for Tropical Mushroom Research and Training (CTMRT) was established in the department of Plant Pathology, OUAT, during 1991-1992 by the financial assistance of the state government due to efforts of Dr. A. K. Patra, Asst. Professor. The center, since then has undertaken research and training 0n various aspects 0f mushroom production and has commercialized the production of paddy straw mashroom (*V. volvacea*), oyster mushroom (*Pleurotus spp.*) and milky mushroom (*Calocybe indica*) in the entire state of Orissa as well as in the neighbouring states like West Bengal, Jharkhand, Chhatisgarh, Andhra Pradesh, Madhya Pradesh and Andaman and Nicobar Islands. Use of different types of crop residues as substrates were tried for production of different mushrooms, which revealed that paddy straw, stalks of sesame and sunflower after harvest used as substrates gave maximum productivity of *Volvariella*, *Pleurotus* and *Calocybe*, respectively. It was also found out that straw of ragi, bajra, jowar and maize; Chaff of mustard pods, maize cobs, and rice husk and sugarcane bag gases could be used effectively as substrates for *Pleurotus* and *Calocybe*.

NEMATODE MANAGEMENT

It is an establishment fact that although plant parasitic nematodes often act as limiting factors in agricultural production, the nematode damage to crops is mostly ignored or attributed to other causes such as soil fertility, deficient soil moisture or soil exhaustion. The role of this tiny thread like animals in agriculture has remained underestimated due to their soil-borne nature, microscopic size, hidden mode of life and nontypical and insidious feeding systems. Therefore, nematodes are one of the most confusing soil pest problems to identify, demonstrate and control. However, on an average, plant parasitic nematodes cause 10% (ranging from 5 to 15%) loss in major crop plants. Each soil will contain its own nematode fauna and in every place and in every crop there will be problems caused by plant parasitic nematodes which are mostly root feeders.

Symptoms of nematode disease on plants include poor growth chlorosis, yellowing, premature leaf drop, stunting patchiness, reduction in quality and quantity of produce dieback, incipient wilting and death of plants. The nematode affected plants produce symptoms which are more or less similar to conditions when the plants are deprived of properly functioning root systems. Moreover, the diseased plants lack vigour and have reduced ability to withstand drought, lack of nutrients and other adverse conditions. Moreover, roots damaged by nematodes are easily parasitized by fungi and bacteria resulting in root decay and causing various disease complexes including wilt. Intensity of damage depends on the number of individuals of particular parasitic species or several such species present in the unit volume of soil.

Various control measures have been successfully attempted against nematodes on different crops. But it has been found that the aim of merely killing or eradicating nematodes is neither economically feasible nor ecologically suitable. Therefore, it should be aimed at managing the crop losses due to nematodes to obtain maximum profit with the least disturbance to the ecological balance. The sole purpose of nematode management is to keep the nematode population sufficiently below the economic threshold limit or the injurious level. It can be achieved by various methods and practical nematode management technology consists of a multitude of cultural, physical, chemical, biological and genetic methods which can be categorized under broad groups such as soil, crop, pesticide and integrated management practices. The excepted practices are as follows.

GENERAL PRECAUTIONARY MEASURES

1. **Field Sanitation:** Follow regular intercultural operations to keep the land free from weeds and other noxious plants which serve as collateral hosts.
2. **Crop rotation:** Very advantageous where possibility exists for taking up alternative remunerative crop. However, the crop should be selected in due consultation with a Nematologist. Usually, a monocot followed up with a dicot or vice-versa is a good combination.
3. **Deep summer ploughing:** At least two deep summer ploughings during April and May will help to reduce the nematode population as well as other soil-borne organisms.
4. **Resistant Varieties:** Resistant varieties are available in some crops, which should be used.
5. **Follow integrated management:** Integrated management combining cultural, varietal and chemical practices should be followed.

NEMATODE PROBLEMS AND THEIR MANAGEMENT FIELD CROPS: RICE

White Tip Disease

(All types of land) Sun dry the seed for a fortnight during April-May or soak the seed for 15 mts in hot water at 52-55°C following a pre-soak in ordinary water for 12 hours before sowing.

Rice-Root-Nematode

(Medium and low land) (1) Treatment of nursery beds/fields with carbofuran @ 1.0 kg a.i./ha reduces nematode population and increases yield. Maintenance of soil fertility and general sanitation are very essential.

Rice-Root-Knot nematode (Nursery Problem)

Seedling root dip in 0.1% Carbisulfan 25 EC or Monocrotophos 40 EC for 12 hr, is very effective in reducing the intensity of disease.

PIGEON PEA

Root-knot and Reniform Nematode

Seed treatment with Carbofuran (75%WP) at 1.2% (w/w) gives better protection against all nematodes. Rhizobium treatment can be done one week after. Resistant varieties against root-knot and reniform nematode: P. 11-480, P. 11-62/1.

VEGETABLES

Brinjal and Chilli

Root-knot and Reniform Nematodes

Apply Carbofuran @ 0.32 g a.i./m^2 or Subuphos or Phorate @ 0.6 g a.i./m^2 of nursery sowing. Fresh neem cake @ 1 t/ha gives very good control of root-knot nematode and boots plant growth and yield under field condition.

DIRECT SEEDED CROPS

1. Small seeds like Okra Root knot and Reniform nematodes Cow pea - Pre-soaking seed treatment with Carbosulfan ST @ 3% w/w.
2. Large Seed like Gourds: Root knot and Reniform nematodes. Pre-soaking seed treatment with Carbosulfan ST 6% w/w.
3. Resistant variety against Root-knot Nematode.
   a) Brinjal: Ghatikia white, PBR-91-2
   b) Chilli : Pusa Jwala
   c) Cowpea : Swarna

Betel Vine

Root-knot Nematode

1. Use disease free vines.
2. Avoid sites used for vegetable cultivation within the preceding three years for new plantations.
3. In case of root-knot infection in standing crop, apply neem cake @ 2 t/ha in heavy soil or mustard cake @ 2 t/ha in light soil *i.e.* 2 kg/m^2.

NEMATODE RESISTANT CROP VARIETIES

Crop	Nematode species	Tolerant varieties/lines
Rice	*Meloidogyne graminicola*	Bhanja, Badami, Neela, OR-621-20-1
Green gram	*Meloidogyne incognita*	ML-62, T-44, M-13-7, ML-80, ML-70, Pusa 10-3
Black gram	*M. incognita*	Sarala, PAU 5, Pant-U-26, Ratnapur-1, UG-135, UG-301
	Rotylenchulus reniformis	Ratnapur-1, UG-201, UG-135, UG-301, JU-78/64.
Bengal gram	*Meloidogyne incognita*	P-44-1, P-116, P-235, ICC-999, GL-1002, GG-575, BG-268
Horse	*Meloidogyne incognita*	Kurubhai, BGM-1

Crop	Nematode species	Tolerant varieties/lines
gram		
	Rotylenchulus reniformis	K-42, Kurubhai, BGM-1
Tomato	*Meloidogyne incognita*	BT-2, BT-17, Radiant, Ronita, VFN-Bush, Patriot, Anahu, Sel 120, Azad Kirti, PB-NR-7, Piersol
	Rotylenchulus reniformis	Healani, Radiant, Sel 120, VFN-Bush, Piersol
	Meloidogyne incognita + Pseudomonas solanacearum	Marshal, Scl 120, Anahu
Brinjal	*M. incognita*	Ghatikia white, Gachha Baigan, Keonjar local (long), Vijaya hybrid, BB-48, PBR-91-2
	Rotylenchulus reniformis	LB-13, Gachha Baigan, ARU-1, T-3.
Chilli	*M. incognita*	Pusa Jwala, K-2, Sel-1, Sinduri, B-70A, TG-1, GA-506.
	Rotylenchulus reniformis	Pusa Jwala, K-2, CA-70A, GA-1366, G-82-21-3
Betelvine	*M. incognite*	Berhampuri, Awami-Pan, Dedhihata pochha, Bilhari, Halisahar

CHEMICAL CONTROL METHODS AGAINST PLANT PARASITIC NEMATODES

SN.	Crop	Nematodes	Chemical control
1.	Rice	*Hirschmanniella mucronata*	Carbofuran/Aldicarb 1 kg/ha in nursery + 6 hr root dip in 0.1% Aldicarb/Sulfone
		Meloidogyne graminicola	12 hrs seed soaking in 0.1% Carbosulfan 25 EC/monocrotophos 40 EC.
2.	Tomato	*M. incognita*	Application of Carbofuran 1 kg/ha + neem cake and urea 10 kg N/ha. Or 30 minutes root dip in 0.1% Carbofuran/Triazophos.
3.	Brinjal	*M. incognita*	Phorate/Carbofuran 0.5 kg/ha in nursery.
4.	Okra	*M. incognita*	Aldicarb/Carbofuran 1 kg/ha + neem cake and urea 10 kg N/ha
5.	Papaya	*M. incognita*	Neem cake 2.5 t/ha before planting.
6.	Pointed gourd	*M. incognita*	6 hr root cutting dip in 0.05% Carbofulsan.
7.	Ginger	*M. incognita*	Neem cake 1 t/ha before planting + Carbofuran 1 kg/ha 45 DAP.
8.	Banana	*M. incognita + Helicotylenchus multicinctus*	Paring and pralinage of rhizomes in 20 g Carbofuran/lit mud slurry.
9.	Betelvine	*M. incognite*	Carbofuran 1.5 kg/ha before planting or Neem cake 0.5 t/ha + Carbofuran 0.75 kg/ha at planting.

BIOLOGICAL CONTROL

Oviparasitic fungasus *Paecilomyces lilacinus* @ 8 kg of sterilized mustard cake showed promising result in reducing the root-knot nematode population in betelvine orchard. Under pot culture trial, VAM (*Glomus fasciculatum)* was found effective in reducing the root knot nematode population in blackgram.

CHAPTER 15

Crop Sequence and Research Findings

CROPPING SYSTEMS APPROACH

Cropping systems approach seeks to increase the benefits derived by crop production from available physical resources (such as rainfall, solar radiation, available irrigation and soil) that are not readily changed (invariants). Resource management concerns itself with the increase in quantity and quality of resources available for production (physical resources – variables). Thus agronomic research increases the resource use efficiency for a given crop, cropping system has cropping pattern (spatial or temporal combinations of crops in any one plot, generally time is limited to one year and a plot in defined as a contiguous area of crop planted in a homogenous manner during the defined period) as a variable with a view to use the available resources more efficiently.

Crop sequence or cropping pattern is a process of growing different crops in succession on a piece of land in a specific period of time with an object to get maximum profit from least investment without impairing the soil fertility.

PRINCIPLES OF CROP SEQUENCE

1. Crop with tap roots should be followed by those which have a fibrous root system. This helps in proper and uniform use of nutrients from the soil and the roots do not compete with each other for the uptake of nutrients.
2. The leguminous crop should be grown after non-leguminous crop because legumes fix atmospheric nitrogen into the soil and add more organic matter to the soil while on the other hand, non-legumes are fertility depleting crops.
3. More exhaustive crops should be followed by less exhaustive crops because crops like potato, sugarcane, maize etc, needs more inputs such as better tillage, more fertilizers, greater number of irrigations, more pesticides and better care than the crops like oil seeds and pulses etc. which need relatively less or little of the above mentioned inputs.
4. Selection of crops should be demand based *i.e.* the crops which are needed by the people of the area and by the family members should be chosen, so that the produce can easily be sold at higher price.
5. The selection of crops should be problem based *e.g.*
   a) On sloppy lands which are prone to soil erosion and alternative cropping of erosion promoting (erect growing crops like millet etc.) and erosion resisting crop like legumes, should be adopted
   b) Under dry farming or partial irrigated area the selection of crops should be such which can tolerate the drought, similarlby in low lying and flood prone areas, the crop should be such which can tolerate water stagnation *e.g.* paddy jute etc.
   c) The selection of crops should suit the farmers financial condition.
6. The crop selected should suit the soil and climatic condition.
7. The crop of the same family should not be grown in succession because they act like alternate hosts for insects, posts and disease pathogens. Thus the control of insects, pests and diseases becomes a perennial problem.

8. A part from this, different types of weeds found, are associated with various crops and selection of the same type of crop in the crop sequence encourage weed problems in the field.
9. An ideal crop rotation is one which provides maximum employment to the family and farm labour, the machines and equipments are efficiently used and all the agricultural operations are done timely.

ADVANTAGES OF CROP SEQUENCE

An ideal crop sequence has the following advantages.

1. Agricultural operations can be done timely for all the crops because of less competition. The supervision work also becomes easier.
2. Soil fertility is restored by fixing atmospheric Nitrogen, encouraging microbial activity, avoiding accumulation of toxins (HCN etc.) and maintaining physio-chemical properties of the soil. The soil may also be protected from erosion, salinity and acidity.
3. An ideal crop rotation helps in controlling insect pests and diseases. It also controls the weeds in the field.
4. Proper utilization of all the resources and inputs could be made by following crop sequence. The family and farm labour power, equipment and machines are well employed
5. The farmer gets a better price for his produce because of its higher demands in the locality or market.

CROPPING PATTERN IN COMMAND AREA

Cropping pattern is the yearly sequence of crops grown and the spatial arrangement of them and fallows in a given area. It is based on climate, soil, agricultural practice, availability of irrigation water, requirement of farmers for food, fodder, and cash and marketing facilities available.

The existing cropping pattern of a command area is the result of past and present decisions of individual farmers and farming communities. The decisions are usually based on experience, tradition, culture, personal preferences and available resources.

Factors Affecting Cropping Pattern of a Command Area

The factors governing cropping pattern in an irrigated area are as below

a) Climate
b) Soil characteristics
c) Hydrology
d) Water allowance and full supply days
e) Water and irrigation requirement of crop
f) Intensity of irrigation
g) Intensity of cropping and *kharif*, *rabi* and summer cropping ratio.
h) Farmers requirements for food, fodder and cash
i) Size of holding and family
j) Marketing and other physical facilities
k) Credit facilities

DEVELOPING A CROPPING PATTERN

Selection of a Crop

Crop planning involves selection or choice of crops from a large number of feasible crops (alternatives) which satisfy the planner under a host of site conditions (available resources, social needs, constraints in environment etc.). Crop planning in command areas with emphasis on maximum crop yield per unit of land or maximizing production per unit of water or both has limited relevance for decision making to allocate scarce water resources optimally since they focus on relative productivity of inputs without accounting for the equally important effects of relative input costs and output (farm produce) value in crop production.

Input

The most important trio of inputs-irrigation water, high yielding varieties and fertilizer nutrients on which crop yield depends have to be considered along with cost of inputs and value of farm produce (output) at market price for allocating water resources optimally. Field studies have to be conducted to determine the economic yield from crops to be grown in the command areas, at different irrigation frequencies and fertilizer nutrient levels, together called levels of management and there from the net benefit derivable from the farm produce at these management levels worked out be taking into account the cost of inputs and the value of a farm produce at market price besides the fixed cost of cultivation. Production functions with the water stored in the reservoir under different management levels as an independent variable and fertilizer nutrient as another independent variable with the crop yield as dependent variable should be developed. The net benefit obtainable for each crop under each management level is arrived at by taking into consideration the fixed costs of cultivation, the variable cost of inputs (including cost of irrigation water and application costs) and the value of output (farm products) at market price. Data is thus generated for several variables for the crops chosen to be grown in command area.

Linear Programming

Linear programming model can be used for maximization of net return for the project as the objective function. Water, land and cropped area are each treated as constraints. Maximization of net return to a command area is obtained by determining the optimum annual combination of cropping systems with varying levels of management and cultivated area under varying levels of water availability. The net returns obtainable under each management level are used as the coefficients and the area under each crop under each management level as variables in the equations. The combined effect of water needed and fertilizer nutrient level under each of the combinations are treated as one management level. A feasible and profitable cropping pattern has to be developed.

Water Availability

Reservoir water can be fully and efficiently used when it is adequately available. When there is shortage of irrigation water, the available water can be used to the best advantage of the project as a whole by suitably reducing the area under each crop and suggesting the optimum management levels for the crops grown or keeping the area constant and varying management levels for obtaining maximum returns for the project. Choosing crops with high water utilization efficiency and low water requirement and short growing season is the prime consideration in fixing the cropping pattern when water supply is limited.

Crop Calendar

Maximization of crop production in an irrigated area largely depends on properly designing the crop calendar in which timely sowing of crops and multiple cropping with short duration, high yielding variety of seeds is provided. Reasonable period of closure of canals for maintenance is essential. Timely sowing of crops, give higher yields due to better use of seasonal rainfall and sunshine and reduction in the incidence of pests and diseases. Decision on the most suitable crop calendars must be made at the early stage of project planning and design to attain the most efficient use of the available water supply and distribution system.

Procedure for Evolving Cropping Patterns

The following steps may be followed systematically to evolve crop patterns for an irrigation command area:

a) Data on meteorological parameters such as rainfall, temperature, radiation, humidity, wind, evaporation and humidity are first collected and crops are selected to suit the climatic conditions
b) A classified survey of land is undertaken and crops suitable for different classes of land are grouped.
c) A soil survey is made and soils are classified for crop growing and soil water management practices. Crops suitable for different soil conditions are then grouped.
d) The net areas that can be brought under cultivation are determined.
e) The areas that can be put under *kharif*, *rabi* and summer crops are worked out according to availability of water during the season.
f) Water and irrigation requirements of crops to be grown are calculated from the available data for the zone and subsequently the monthly requirement is worked out. Irrigation requirement is decided by deducting effective rainfall and the contribution from water table, if any, from the water requirement.
g) The area to be brought under each crop is decided as pre-*kharif*, *rabi* and summer cropping ratio.
h) Conveyance losses of water from distributory heads to fields are determined to work out the water allowance for fields. This is done by deducting the conveyance loss from the water allowance at the distributory head.
i) The intensity of irrigation is laid out according to availability of water at the field level.
j) Full supply days are worked out month wise and for the whole year in accordance with water needs of crops.
k) Requirements of farmers for different crops and area to be put under them should be assessed for modification of the model crop pattern for individual farmers according to the size of this holding
l) Crops that can be profitably grown under irrigation in the area for better economic return are assessed.
m) Cropping sequence, and mixed, catch or relay cropping that can be followed is decided.

The procedure as stated in the foregoing lines gives a general guideline to formulate a cropping pattern for an area. All the factors governing the cropping patterns should be considered together.

CROPPING PATTERN FOR IRRIGATED AND RAINFED AREAS IN ORISSA

A set of cropping pattern for different situations are indicated in the enclosed statements. The crop patterns that are proposed to be taken up in a given situation should be decided basing on the availability of water, soil type, fertility level of the soil, marketability of the producer and other social economic factors. While choosing the crops on the above criteria, it is essential to keep in view the following points as well.

1. No two crops sharing in common, and same pests and diseases, should be grown in succession.
2. One crop should be deep rooted and the other with rather shallow root system so that different layers of the soil can be trapped for nutrients.
3. Attention to the restoration of soil fertility through the cultivation of leguminous crop may preferably find place in the rotation.

Statement – 1

Major/Medium Irrigation Project, (Canal flow)

	1st Crop	2nd Crop	3rd Crop
High land	H.Y. Paddy	Mustard	Pulses
	-do-	Pulses	G. Nut
	-do-	G. Nut	Pulses
	-do-	Potato & Veg.	Vegetable
	-do-	Cotton	Cotton
Medium Land	Sugarcane	Sugarcane	Sugarcane
	H.Y. Paddy	Wheat	Til(Sesame)
	-do-	Mustard	Pulses
	-do-	Pulses	G. Nut
	-do-	Potato	Til (Sesame)
	-do-	Cotton	Cotton
	Sugarcane	Sugarcane	Sugarcane
Low Land	Jute	Paddy	Mung
	Jute	Paddy	Mung
	Paddy	Under sown crop of black gram or Field pea.	H.Y. Paddy

Statement – II

Cropping Sequence for Dugwell Areas

	1st crop	2nd crop	3rd crop
High land	Veg/Ragi	Vegetable	Vegetable
	G. Nut/Maize	Potato/Veg./Onion/Chillies	Mung/Biri/Til (Sesame)
	Sugarcane	Sugarcane	Sugarcane

	1st crop	2nd crop	3rd crop
High land	Veg./Ragi	Vegetable	Vegetable
	G. Nut. /	Potato/Veg./Onion/Chillies	Mung/Biri/Til (Sesame)

	1st crop	2nd crop	3rd crop
	Maize		
	Sugarcane	Sugar cane (Inter crop of Mung first 21/2 months)	Sugarcane
Medium land	Paddy	Potato/Wheat	Vegetable/Pulses/Til
	Sugarcane	Sugarcane	Sugarcane
	Medium Cotton	Hybrid Cotton	Inter crop of Mung in Cotton during 1st 2½ months
Low land	Rice	Pulses/Groundnut	

Statement – III

Rainfed Condition

	1st crop	2nd crop
Upland (Unbunded)	Ragi	Horse gram
	Maize	Niger
	Groundnut	
	Vegetable	
	Arhar/Castor	
Upland (Terraced)	Early H.Y. Paddy	Niger/Mustard/Mung/Biri/Cowpea.
Medium Land	Medium H.Y. Paddy	Mustard/Safflower/Gram/Linseed
Low Land	Long duration Paddy	Green gram/Field pea (Under sown crop)

MAJOR CROP SEQUENCES FOLLOWED IN DIFFERENT AGRO-CLIMATIC ZONES OF ORISSA STATE.

Agroclimatic zone	Crop sequences		
	High land	Medium land	Low land
1	2	3	4
North Western Plateau	Rainfed / dry farming		
	Sole crops		
	Groundnut, Arhar (Red gram), Castor, Maize, Jowar, Niger, Green gram, Black gram		
	Inter cropping		
	Arhar + Groundnut / Green gram / Black gram / Rice / Ragi		
	Sequence cropping		
	Miner millet – Horse gram / Green gram / Niger / Mustard.	Rice-Mustard / Lentil / Safflower / Green gram / Bengal gram / Horse gram	Rice-linseed / Lentil / Bengal gram / Field pea
	Maize / Jowar / Rice-Horse gram / Green gram / Bengal gram mustard.	Rice-field pea as paira crop	
	Irrigated		
	Maize-potato / Mustard-til	Rice-green gram-	Rice-rice

Agroclimatic zone	Crop sequences		
	High land	Medium land	Low land
1	2	3	4
	(sesame) / Vegetables.	groundnut.	
	Groundnut / Rice-mustard / Wheat / Maize-Green gram / Vegetables.	Rice-Mustard / Potato / Wheat-til / Green gram / Black gram.	Jute-rice-Green gram / Black gram
	Rice-potato / groundnut / vegetables / til / green gram / vegetables.	Rice-groundnut-green gram / black gram	
North Central Plateau	Rainfed / Dry farming		
	Sole crop (Single crop in *kharif*)		
	Rice / maize / green gram / gulji / arhar / ragi / black gram / til / groundnut		
	Sequence cropping		
	Rice-fallow	Rice-fallow	Rice-fallow
	Rice-horse gram	Rice-Bengal gram /	Rice-linseed
	Fallow-niger	Green gram /	Rice – kheshari /
	Maize-mustard	black gram	mung (Paira crop)
	Fallow-horse gram / groundnut (In limited area)		
	Irrigated		
	Rice-vegetables / mustard / wheat / potato	Rice-wheat-green gram	Rice-rice
		Rice-potato-til	Rice-green gram
		Rice-mustard-rice	Jute-rice-gram
			Green gram / lady's finger
North Eastern Coastal Plain	Rainfed /dry farming		
	Rice / arhar / groundnut / jute / maize / green gram / black gram / rice / maize / groundnut-horsegram / green gram / blackgram	Rice-groundnut / black gram / green gram / mustard / Jute-rice / groundnut – mustard.	Rice – Fallow / Jute-rice / Jute-mustard / rmung / til. (Sesame)
	Irrigated		
	Arhar-mustard-vegetables	Rice-potato-til	Rice-rice
	Groundnut-mustard-til	Rice-groundnut / Wheat mustard-til / mung / black gram	Jute-rice-rice.
	Rice-potato-til / mung / black gram	Jute-rice-rice	
	Rice-wheat-mung / til / vegetables	Jute-wheat-veg. / gram	
East and South-	Rainfed / dry farming		

Agroclimatic zone	Crop sequences		
	High land	**Medium land**	**Low land**
1	**2**	**3**	**4**
Eastern coastal plain			
	Rice / jute / arhar / pulses / veg / minor millets / musta	Jute-rice-pulses	Rice-pulse
	Groundnut / til-fallow	Jute-groundnut-g. nut/Til	
	Rice-horse gram / mustard / til (sesame)	Mung / green gram	
	Jute-mustard / groundnut / pulses	Rice-groundnut	
	Groundnut – horse gram		
	Ragi – horse gram / mung		
	Irrigated		
	Rice-potato-til / mung / b. gram	Rice-rice	Jute-rice-rice
	Rice-wheat-til/pulses	Jute-rice-rice	Rice-rice
	Rice-groundnut-mung / veg.	Rice-mung-rice	Rice-mung/
	Pulses / groundnut-wheat / potato-til	Rice-mustard/	mustard-rice
	Ragi – vegetables.	mung-rice/mung	
North Eastern Ghat	Rainfed / Dry farming		
	Arhar / groundnut / Turmeric / cowpea / minor millets / til fallow	Rice-potato-til / vegetables	Rice-pulses mustard
	Arhar + groundnut / mung / maize	Rice-garden	Rice-green gram / gram pea
	Rice-fallow	pea-cowpea	
	Minor millets / rice-til / horse gram / mustard	Runner bean – mustard potato	
	Maize / jowar / ginger / niger / black gram-fallow		
	Irrigated		
	Ground nut / til-horse gram / black gram	Rice-cowpea / wheat / mustard-groundnut	Rice-rice
	Ragi-horse gram/pulses	Maize-potato-vegetables.	Rice-potato til/pulses
	Mesta / maize / rice-niger / mustard / linseed / h. gram	Rice-gram-groundnut – vegetables.	Jute-rice-rice. Rice-mustard-rice
Eastern Ghat	Rainfed / Dry farming		
	Soybean-mustard	Rice / minor-millets / mesta-pea / til.	Rice-pea / linseed.
	Minor millets-niger	Maize / mesta-linseed / mustard	Rice fallow.
	Potato-niger / mung	Cotton-mung / black gram	Rice – mung.
	Irrigated		
	Maize / rice / ragi-potato / wheat-	Rice-wheat	Rice-potato-til / veg.

Agroclimatic zone	Crop sequences		
	High land	Medium land	Low land
1	2	3	4
	mung / cowpea / vegetables		
		Rice / ragi-potato / groundnut / soybean / mustard.	Rice – rice.
		Ginger / turmeric-mung.	Mesta-rice.
South Western Ghat	Rainfed / Dry farming		
	Mesta / maize / ragi / veg. / til / cowpea-fallow.	Rice / maize / mung / horse gram / ragi-fallow.	Rice-fallow.
	Irrigated		
	Rice-mung / Groundnut / potato	Rice-potato-til/	Rice-rice
		Rice-groundnut-til.	Rice-mustard-
		Rice-wheat-mung.	Rice.
		Rice-gulji-horse gram	Rice-groundnut-til
Western Undulating	Rainfed / Dry farming		
	Rice / groundnut / ragi-fallow.	Rice / groundnut / ragi / jowar / maize / arhar-fallow.	Rice-horse gram / mung / pulses.
		Gulji-jowar/mustard	
	Irrigated		
	Rice-vegetables	Rice-rice	Rice-rice.
	Rice-sugarcane.	Rice-wheat / vegetables.	Rice-sugar cane.
	Rice-wheat / pulses / mustard.	Rice-potato-til.	Rice-rice-rice
Western Central Table Land	Rainfed / Dry farming		
	Arhar / pulses / minor	Rice-house gram / castor.	Rice-rice / safflower / mustard / Mung / Black gram.
	millets / rice / ground nut / cotton-fallow.	Maize / ragi-horse gram / Mung.	
	Irrigated		
	Groundnut-wheat-Pulses.	Rice-wheat-mung.	Rice-rice.
	Cotton-groundnut.	Rice-mustard-rice / ground nut	Rice-wheat / mustard / groundnut / mung-til / pulses.
	Groundnut-mustard – Til.	Rice-rice.	Rice-rice-rice.
	Pulses-mustard-veg.	Rice-cotton	
	Rice-groundnut-pulses		
	Rice-potato-til (Sesame)		
Mid Central Table Land	Rainfed / Dry farming		
	Arhar / groundnut / til / rice /	Arhar / groundnut / black gram / mung / rice / minor	Rice-mung / black

Agroclimatic zone	Crop sequences		
	High land	Medium land	Low land
1	2	3	4
	pulses – fallow	millets-fallow. Rice-horse gram. Arhar + groundnut-fallow.	gram / khesari. Rice-mustard.
	Irrigated Rice-vegetables. Rice-mung/horse gram. Groundnut-mustard-vegetables	Rice-wheat-mung Rice-fieldpea-mung. Rice-mustard / groundnut – vegetables. Rice-potato-til.	Rice-rice. Rice-rice-rice.

MULTIPLE CROPPING SYSTEMS IN HIRAKUD COMMAND

Panda, *et. al.*, 1972 recommended after thorough study that generally, rice-rice system was the practice in the Hirakud Command area. During the period from 1970 to 1987, sincere efforts were made to introduce new cropping systems such as rice-mustard-rice, rice-potato-til (sesame), rice-potato-rice, rice-wheat-mung (green gram), and groundnut-potato-summer vegetables.

Rice-mustard-rice was made the most popular cropping system in 1.8 lakh hectares of Hirakud command area comprising of 14 blocks of old Sambalpur and Bolangir districts. The system could be made beneficial to the farmers as the disease and pest load over years were minimized.

As an instant case, rice-radish-potato-maize + green gram-lady's finger/cowpea system could prove 600 per cent cropping intensity over the recommended practice (100%) under crop intensification programme in irrigated villages at Kujapalli and Baijamunda, Sadar Block, Sambalpur. The increasing benefit of cropping intensity also attracted the attention of other farmers and the system is in practice in other command areas which could change the economic standard of the farmers.

Further, Panda, 1972 also recommended that five rotations such as rice-tomato-rice-green gram, rice-radish-potato-maize +green gram-lady's finger/cowpea, rice-potato-rice-green gram, rice-potato-rice and rice-radish-wheat-green gram were found to be the most profitable ones.

For the maintenance of soil productivity, it is desired to include a leguminous crop in the crop rotation. Cultivation of green gram in summer months from April to June is becoming popular with farmers as this enables them to use land and water resources which other wise remain idle during the period. Pusa Baisakhi, being a short duration and a drought resistant crop can fit into multiple cropping programmes aiming at maximum food production per unit area per unit time.

CROPPING IN ABERRANT WEATHER CONDITIONS

Average rainfall of Orissa is 1498 mm (1901-1973) and 1358 mm (1901-1983). The reduction of rainfall over years is more conspicuous in inland districts than in coastal ones. Number of rainy days is all decreasing. The coefficient of variation is widening more at the beginning and end of monsoon. All over analysis is a posteriori and can not be forecast what will happen next year/ in such cases contingent planning is more important. The following are the aberrant situation for which contingent planning is necessary. The usual pattern of rainfall in Orissa is that the South-west monsoon breaks in second week of June and ceases in second week of October. July and August are the heavy rainfall months. The question of aberration comes if monsoon does not fallow the normal pattern. Delayed or early setting up of monsoon, early ceasing, and mild drought are all aberrant situations. But farmers follow the tradition and complete the sowing up of crops.

If monsoon does not set in till last week of June, the following may be amended:

Upland

1. Sowing of rice in att, bhattas, dungar, padar, or high lands should not be done.
2. Sow non-rice crops like maize, ragi, jowar, arhar, castor, groundnut, or sesame as per command of the areas, sow the crops in lines so that interculture and/or cropping becomes easier.
3. Sow short duration rice varieties after late setting of monsoon, less than 85 days.

Medium Land

1. Sow rice varieties of 15-20 days shorter in duration than usual duration varieties for these lands.
2. If sowing was already done, there was germination, and no rain, there is seed mortality.
3. If mortality is more than 60%, resow the crop. If less, follow the usual operation, adjust population by khelua or clonal propagation.
4. Maintaining a high density pattern at one corner of the land to get seedlings for gap filling.
5. In lower topography lands, broad cast seeds at the rate of 15-20 kg/ha in wet soil up to 15th, July in inland districts.

Sow short duration varieties of ragi, maize, jowar, groundnut, and arhar. Sweet potato can be a good remunerative crop.

1. If rain ceases, after mid July-August.
   a) Not sufficient rain for beusaning.
   b) Don't wait for beusaning, weed the crop follow other packages or practices as usual.
   c) In uplands, weed the crop, earth up. Reduce N top dressing.
2. If drought occurs in mid August- mid September.

Upland

1. If drought occurs in mid August-mid September, sow arhar/castor in between lines and dibble in broadcast crop.
2. If searching and parching of plants could not be advanced and no hope of getting the rice crop, plough the land. Start early *rabi* sowing.

Medium Land

1. Weed the crop.
2. No late beusaning.
3. Held up all water.

Late June Flood

Damage to bua, rotting and washing away, resow after recession of flood. Sow arhar in main lands. In low lands, transplant rice.

July Flood

If the crop is completely damaged, transplant the crops in direct seeded area, give on beusaning, do early khelua and maintain crop stand. Follow usual package fertilizer. August and September (flood) – Transplant old seedlings of 60-65 days, maintain population by clonal propagation. Loss due to late September flood can not be recouped. After flow, put the land to black gram, gram by direct sowing or by broadcasting the crop.

October Rainfall.

Good for *rabi* crops. More of uplands can be put to oil seeds (mustard, linseed, groundnut) and pulses (Horse gram, green gram and black gram) and vegetables.

November Heavy Rainfall

Will damage gram, vegetables. But the crop can not be replanted with vegetables. Mustard fields can be resown with coriander and linseed. More area in medium land can be put black gram and gram and coriander after ploughing and usual land preparation. Experience is the best guide.

SALIENT RESEARCH FINDINGS ON CROP PRODUCTION

- Under soil fertility and fertilizer management study in Khunta block of Mayurbhanj district, rice (var. Bhoi) yield to the tune of 39 t/ha was recorded under recommended fertilizer use along with farm yard manure in medium land situations. The yield of rice (var. Prachi) was 47q/ha is low land condition of the same block. Under upland condition, rice variety Vandana performed well compared to local varieties tested.
- Application of sulphur and foliar spraying of 0.25 per cent borax solution to rice crop enhanced the grain filling by 9 to 13 per cent over control.
- Rice variety ZHU-XI when intercropped with green gram variety PDM 54 in rainfed uplands of Phulbani district, it becomes profitable with yields of 1.6 t. rice and 0.3 t. green gram per hectare. Under rainfed medium lands, rice-green gram sequence is the best in Kalahandi district, whereas rice-vegetable (cauliflower) becomes the most remunerative (net profit Rs 18,880/ha) in irrigated uplands of the same district.
- Medium duration rice variety (Naveen/Lalat/Swarna) can suitably be grown after harvest of jute crop (Baladev/Jagadev) in Kendrapara area to obtain 3.6 to 4.0 t. of rice yield per hectare.
- The major intercropping systems recommended for the rainfed uplands of the state arhar+groundnut, arhar+rice/ragi, maize+cowpea both for grain and fodder. Nearly 2.00 lakh hectares have been put under arhar+groundnut/rice intercropping system. Fertilizer application to arhar+rice or maize+cowpea on the basis of area occupied by each crop is remunerative
- Rice-field pea-sesame or rice-potato-sesame are more profitable (Rs.29,170/- to Rs.47,189/ha) than the rice-rice system in the irrigated commands. Rice-groundnut/mustard/green gram/Bengal gram in two crop sequence and rice-cabbage-lady's finger (Rs. 42,600/ha), rice-tomato-lady's finger/cowpea (Rs.

38,820/ha) and rice-mustard-groundnut in three crop sequence are found promising for irrigated areas.

- Rice-fish farming is suitable for the coastal tract. There is a feasibility of putting 25000 ha of land under the system.
- In rice-groundnut cropping system application of the recommended dose of fertilizer (RD)+lime(0.25 LR) to kharif rice and RD+gypsum (250 kg/ha)+boron (13 kg borax/ha) soil application once in two years maximized the yield of rice(5.8 t/ha) and groundnut (3.0 t/ha). All P should be applied to rice as a mixture of SSP and RP(1:1) required for both the crops.
- Improved methods of crop management gave 41% higher return than farmer's method in rice-sunflower and 28% in rice-green gram sequences.
- Soil application of biofertilizers (Azotobactor+PSM+KSM) with 75% of RD produced 2.3 t/ha of rice grain under rainfed upland conditions during kharif season.
- For direct seeded medium and low land rice, apply full P as basal, and N and K in 3 splits – 50% at khelua(Clone splitting stage), 25% 15 days after and 25% at panicle initiation stage.
- Application of 0.3% borax as foliar spray in mustard maximized the seed yield (1.8 t/ha) and oil yield (0.7 t/ha).
- Groundnut intercropped with teak found promising. The net return was Rs. 4,000/ha and benefit-cost ratio was 1:4.
- Rajmash, a non-traditional pulse, produced a seed yield of 2.1 t/ha with 6 irrigations and 80 kg N/ha. The water requirement was 500 mm and water use efficiency was 46 kg seed/ha-cm.
- Niger should be sown during the first fortnight of August for doubling the yield. It is an important non-monetary input for the tribal farmers.
- Yam can be grown in rainfed uplands without staking at a spacing of 100 cm x 50 cm. It conserves soil and water and yields up to 17 t/ha. It can impart food security to the tribal farmers.
- For mulching in turmeric and ginger, mustard and gram stalks, mango and jackfruit leaves can be used in place of sal leaves (*Shorea robusta*). Arhar and cluster bean may be intercropped with ginger.
- Patchouli (*Pogostemon patchouli* Hook. L.) yields an essential oil of commerce used in perfumery. The average of yield is 40 kg/ha. Application of irrigation to oat at IW/CPE of 1.0 (8 - 10 irrigations) with 200 kg N/ha produced 164 kg oil/ha. The water requirement was 1075 mm.
- Cereal+legume intercropping of maize and cowpea in 2:2 ratio recorded the highest green forage yield (37 t/ha) with net return of Rs. 15,678/- and benefit-cost ratio of 3.41.
- Application of Azotobacter with 80 kg N/ha to sorghum increased the forage yield and saved 20 kg N/ha.
- Oat, a non-traditional crop, produced 25 t/ha of green forage with 75% recommended dose plus 40 kg S/ha under Bhubaneswar condition
- Among different cropping sequences, rice-maize-green gram gave maximum net profit of Rs 50,564/ha with B:C of 2.45. Other promising systems are rice-potato-okra, rice-cauliflower-rice.
- Seed inoculation with liquid bio-fertilizer (Rhizobium and PSB) resulted in 10 per cent increase in the yield of black gram.
- Integrated Nutrient Management in Jute-rice system gives fibre yield of 22.8

q/ha and rice yield of 39 q/ha.

- The rice varieties ZHU 11-26 (29.5 q/ha, 85 days) and Vandana (30.3 q/ha, 94 days) were more suitable for cultivation in rainfed uplands.
- The promising varieties of black gram suitable for growing in rainfed uplands during *kharif* season are OBG 23 (11.37 q/ha, 69 days), B 3-8-8(10.79 q/ha, 69 days), OBG 15 (10.6 q/ha, 66 days), LBG 645 (10.51 q/ha, 85 days) and Pant U 30 (10.5 q/ha, 72 days).
- OG 52-1, JL 24, TAG-26 and IGGS 11 with average yield of 16.0 to 17.0 q/ha, are the promising groundnut varieties suitable for cultivation in rainfed uplands during *kharif* season.
- Mixed cropping of groundnut and pigeonpea in 6:2 ratio and rice+pigeonpea in 5:2 ratio covering two hectares each were demonstrated for risk management and better resource utilization during kharif season and found to be suitable for watershed.
- Chemical weed control with Butachlor @ 1 kg a.i,/ha at the post-emergence stage along with one mechanical weeding at 40 days after sowing, is found the most economical and beneficial among all the weed control measures employed in rice.

RESEARCH FINDINGS OF OUAT: AGRONOMY CULTURAL PRACTICES

Rice

- The optimum date of planting for hybrid rice was 15th July with an average yield of 5 t/ha. The yield decreased by 10 and 22% when planted on 30th July and 15th August, respectively. It gave high yield when planted at 20 cm x 15 cm spacing with 1 or 2 seedlings/hill.
- Ratooning of transplanted *kharif* rice cv. Savitri when fertilized with 80,40,40 kg of N, P_2O_5, K_2O/ha produced the highest grain yield of 4.72 t/ha with net return per rupee invested of Rs 2.25.

Variety	Grain yield (t/ha)	Return/rupee invested
Savitri	4.72	2.25
Manika	4.11	1.60
Mahalaxmi	3.25	1.01
CD (0.05)	0.22	-

- Under rainfed situations, *Beushaning* of direct seeded rice was equally productive (4.2 t/ha) but more remunerative (1.94 benefit cost, B.C.) than transplanting (4.4t/ha, 1.87).

Treatment	Grain yield (t/ha)	B:C
Broadcasting + *Beushaning*	4.25	1.94
Broadcasting + Weedicide + Hand weeding	3.65	1.70
Transplanting	1.36	1.87
CD (0.05)	0.34	

- For 10 days delay in planting from mid July to 10th August, the yield reduction was 11, 14 and 23% respectively in *kharif* rice.
- Under post flood situations, sprouted seeds of short duration rice varieties like Nilagiri, Pathara and Parijat could be sown up to end of August to obtain

a yield of 2 t/ha.

- Long duration rice varieties like CR 1014 and Jajati were found suitable for late planting up to mid September with realizable yield of 3 t/ha. Planting of 60 days old seedling of CR 1014 was better than 30 or 45 days old seedlings under the late planting conditions with yield advantage of 35 and 15%, respectively.

Wheat

- Wheat sowing should not be delayed beyond November under any circumstances under Bhubaneswar condition.
- Unidirectional sowing at 15 cm interval (2.44 t/ha with B:C of 1.40) with respect to yield and economics.

Maize

- Sweet corn variety Madhuri sown at 60 cm x 20 cm in *kharif* season produced the highest green cob yield of 4.11 t/ha with B:C of 4.52 as compared to 45 cm x 2 cm or 60 cm x 30 cm.

Ragi

- Ragi yield decreased by 14 and 38% from 1.75 t/ha when sowing was shifted to shifted to third and fourth week of June from 1st week, respectively.

Pulses

- Recommended spacing for ricebean during *kharif* season was 30 cm x 10 cm (0.84 t/ha) with an yield advantage of 19 and 57% as compared to 45 cm x 10 cm and 60 cm x 10 cm spacing, respectively.

Oilseeds

- Pod yield of groundnut deceased by 20 and 33% when sowing was delayed to third and fourth week of June from 1st week (1.24 t/ha), respectively.
- Sowing of *kharif* groundnut in flat bed (1.22 t/ha) was as good as sowing behind the plough (1.14 t/ha) but was superior to sowing behind marker (0.95 t/ha) or with groundnut planter (0.81 t/ha).

Sowing method	Pod yield (t/ha)
Behind plough	1.14
Behind marker	0.95
Flat bed	1.22
Groundnut planter	0.81
CD (0.05)	0.22

- Recommended row spacing for toria was found to be 20 to 25 cm and mustard was 30 cm.
- Toria var Parbati produced 18 and 50% higher yield than Anuradha and TS 29, respectively.
- Optimum time of sowing for toria (cv Parbati) was found to be 15 October. Early (5 October) or late (25 October, 5 November) sowing reduced yield by 14, 10 and 43%, respectively, as compared to 15 October sowing (0.36 t/ha.
- Hybrid sunflower sown at 60 cm x 30 cm produced maximum seed yield of

1.97 t/ha as compared to 45 cm x 30 cm and 75 cm x30 cm spacing, The recommended spacing for improved sunflower cv. Morden was 45 cm x 20 cm.

- Optimum date of sowing for niger under Bhubaneswar conditions was 15 July (0.37 1/ha). Sowing beyond this date reduced the yield by 6, 11, 23, 33 and 39% for every fortnight delay up to 1 October.
- Niger planted at 25 cm x 10 cm gave 18% higher yield than 30 cm x 10 cm (0.27 l/ha)

WATER MANAGEMENT

Rice

- Application of 12 irrigations at weekly intervals produced maximum grain yield (4.2 t/ha) of summer rice cv. Sarathi. Water table depth of 20 cm was better than 30 cm.
- Irrigating summer rice one day after disappearance of ponded water (DADPW) saved about 32% (33 cm) of water with an yield of 4 t/ha as compared to submergence (5± 2 cm) during tillering and reproductive stages.

Treatment	Yield (t/ha)	Water requirement(cm)
One DADPW	4.11	122
4 DADPW	3.60	94
Submergence (5±2 cm) at tillering fb one DADPW	3.84	135
Submergence at reproductive fb one DADPW	4.23	141
Submergence at tillering and reproductive stage Fb one DADPW	4.30	155

DADPW = Days after disappearance of ponded water. fb = followed by

- During *kharif* season keeping the field moist for 45 days after transplanting and submergence thereafter till dough stage in dwarf rice produced same yield as that of continuous shallow submergence (5±2 cm). Maintaining 3 to 5 cm standing water from *Beushaning* to ripening resulted in maximum grain yield of 3.7 t/ha.
- Upland rice sown behind the plough (4% KCl treated seeds) in mid June with the onset of monsoon and supplied with full dose of FYM and split application of 40 kg N/ha escaped terminal drought and produced higher grain yield. Rice cv ZHU XI-26 produced 20 and 30% higher yield than Kalinga-III and Vandana, respectively, in unbunded uplands.

Wheat

- Wheat performed the best when irrigated at 10 days intervals with shallow depth of irrigation (4 cm) at the initial stage and 5 cm thereafter at IW/CPE ratio of 0.9 to 1.2. The critical stages in order were crown root initiation (CRI), flowering, milk, heading, dough and jointing.

Maize

- Moisture stress during seedling, early growth stage, tasseling and milk stages decreased the yield of *rabi* maize by 41,32, 50 and 50%, respectively as compared to non stressed crop (6.32 t/ha).

Stress at	Grain yield (t/ha)	Water requirement (cm)	Water Use Efficiency (WUE) (kg/ha-cm)
Seedling stage	3.76	33	99
Early growth stage	4.31	37	118
Tasseling stage	3.31	35	88
Milk stage	3.16	33	93
No stress	6.32	43	148

- Mulching with 20 micron black polyethylene sheet and irrigation at 50% available soil moisture (ASM) reduced the water requirement of summer maize by 33% (25 cm).

Pulses

- Withholding irrigation at vegetative, flowering and seed filling stages reduced the yield of summer green gram by 13, 14 and 16%, respectively, from maximum seed yield of 1.0 t/ha obtained with irrigation at all the stages of growth or at 0.6 IW/CPE ratio.
- Application of 4 irrigations at 0.6 IW/CPE ratio produced 31% higher yield of black gram than 2 irrigations applied at 0.3 ratio (0.57 t/ha).
- Rajmash required 5 irrigations at IW/CPE ratio of 1.2 which produced the grain yield of 1.52 t/ha with B:C ratio of 1.95 and consumptive use (CU) of water of 41 cm.

Oilseeds

- From a number of studies it was revealed that maximum pod yield of *rabi*/summer groundnut was obtained by irrigating the crop at 0.8 to 0.9 IW/CPE ratio. Reduction in pod yield due to stress at flower initiation, peg penetration, pod initiation and pod development stage was 11, 18, 15 and 8% respectively.
- Yield response of summer groundnut to irrigation was linear at deep water table (110±10 cm) but was quadratic in shallow (50±10 cm) and medium deep (90±10 cm) water table conditions. The water requirement of the crop was 42 cm with 9 irrigations.

Depth of water table	Pod yield (t/ha)	Water requirement (cm)	Optimum No. of Irrigations
50±10 cm	1.43	40	8
90±10 cm	1.97	42	9
110±10 cm	2.03	42	12

- Optimum yield (0.89 t/ha) of sesame was obtained through application of 6 irrigations at IW/CPE ratio of 0.1. Application of only one irrigation at vegetative, flowering, bud initiation and pod development stage reduced the yield by 34, 28, 20 and 18%, respectively, as compared to the optimum yield.
- Application of 2 irrigations to toria (cv. M 27) at pre-flowering and maximum pod filling stage with Jalshakti (5 kg/ha) applied in line recorded 59% increase in yield under residential moisture conditions.
- Stress at 4 leaf, button, flowering and grain filling stages of *rabi* sunflower reduced yield by 8, 17, 25 and 17%, respectively, as compared to normal irrigated crop (1,61 t/ha). Summer sunflower(NSFH 2) required 7 irrigations at 1.2 IW/CPE ratio to produce maximum yield.

Stress at	Seed yield (t/ha)	% reduction	Consumptive use (cm)
Four leaf stage (30 DAS)	1.48	8	18
Botton stage (45 DAS)	1.33	17	21
Flowering (60 DAS)	1.20	25	20
Grain filling (80 DAS)	1.33	17	21
No stress	1.61	-	25

- *Rabi* castor (cv. Aruna) produced maximum bean yield of 0.9 t/ha with 7 irrigations at 0.75 IW/CPE.

IW/CPE	No. of irrigations	Castor bean (yield t/ha)	Consumptive use (cm)	WUE (kg/ha-cm)
0.50	5	0.83	47	17.7
0.75	7	0.90	57	15.8
1.00	9	0.80	69	11.7

WUE = Water use efficiency.

Cotton

- Plastic mulch reduced irrigation requirement of cotton during *kharif* season and saved 12 cm of water.

Jute

- The crop required 30 cm of water in 6 irrigations to produce maximum fibre yield of 1.8 t/ha.

Potato

- Potato required 9 irrigations of 3 cm depth each at IW/CPE ratio of 1.1 and produced maximum tuber yield of 19.6 t/ha. Mulching with black polyethylene increased the proportions of A grade tubers and decreased water requirement by 13 cm.

Patchouli

- Patchouli, an important essential oil yielding plant, required 8 irrigations at IW/CPE ratio of 1.0 to produce 120 kg oil/ha with CU of 81 cm.

Pigeonpea Based Intercropping System

- Consumptive use of pigeonpea based intercropping system was less than that of combined CU of individual sole crops. The effective rainfall was maximum (40%) in rice grown alone or in combination with pigeonpea.

Treatment	Pigeon pea Equivalent yield (t/ha)	Effective rain fall (cm)	CU (cm)	Land equivalent
Pigeon pea + Rice	1.77	55	61	1.27
Pigeon pea + Ragi	1.24	53	60	1.05
Pigeon pea + Groundnut	1.98	53	60	1.23
Pigeon pea + Green gram	2.24	51	57	1.71
Pigeon pea + Black gram	1.98	51	58	1.40

Treatment	Pigeon pea Equivalent yield (t/ha)	Effective rain fall (cm)	CU (cm)	Land equivalent
Pigeon pea	1.94	54	60	-
Rice	0.60	55	39	0
Ragi	0.74	40	35	-
Groundnut	1.23	44	40	-
Green gram	0.64	30	25	-
Black gram	0.93	40	34	-

SOIL FERTILITY MANAGEMENT

Rice

- Early rice variety Neela responded upto 60 kgN/ha and produced 72% higher grain yield (2.23 t/ha) than control.
- Late duration rice varieties CR 292-5258 and CR 1009 responded up to 120 kgN/ha in presence of 60 kg each 0f P_2O_5 and K_2O/ha during *kharif* season.
- Application of Mossouri rock phosphate (MRP) and SSP each at 30 kg P_2O_5/ha was as good as 60 kg P_2O_5/ha as SSP but was superior to 60 kg P_2O_5 as MRP alone.
- Rice hybrid NPH 4507 produced desirable yield of 6.86 t/ha by the application of 120, 60, 60 kg N, P_2O_5. K_2O/ha which was at par with that of 150, 75, 75 kg N, P_2O_5, K_2O/ha dose.

Fertility level (NPK kg/ha)	Grain yield (t/ha)				
	Hybrid			HYV	
	NPH	Pro-Agro	RH	Lalat	Mean
	4507	6201	204		
60,30,30	5.86	5.19	5.00	4.42	5.12
90,45,45	5.95	5.49	5.76	4.95	5.54
120,60,60	6.86	5.91	6.75	5.17	6.17
150,75,75	6.97	6.01	6.81	5.27	6.22
		NPK	Variety		
CD (0.05)		0.30	0.62		

- Combined application of 90 kg N and 25 kg $ZnSO_4$/ha to *kharif* rice produced 4.01 t grains/ha which was about 8% higher than N alone. It also increased the N uptake by 14% and Zn uptake by 50% as compared to N alone. Soil application of Zn was superior to foliar application.

Treatment	Grain yield(t/ha)	N Uptake(kg/ha)	Zn Uptake(kg/ha)
N alone	3.73	69.4	0.34
N + Zn	4.01	79.1	0.51
Soil application (Zn)	4.04	81.1	0.52
Foliar application(Zn)	3.97	72.4	0.54
CD(0.05)	0.50	0.46	0.007

- Combined application of 60,30,30 kg N, P_2O_5, K_2O/ha and 20 kg $ZnSO_4$/ha to rice (cv. Lalat) resulted in maximum grain yield of 4.88 t/ha.

- Soil application of Zn @ 2.5 kg/ha with 20 kg elemental sulphur produced same yield (4.16 t/ha) as that of 5 kg Zn+ 5 t FYM/ha.
- Application of 60 kg N/ha in two splits either through USG(Urea Super Granule), GCU(Gypsum Coated Urea), NCU (Neem Coated Urea) or MRPU (Mossouri Rock Phosphate Urea) was superior to prilled urea with an yield advantage of 69,63,62 and 53%, respectively.

Treatment	Grain yield (t/ha)	Return/rupee invested(Rs)
Urea	2.20	1.30
USG	3.72	2.02
GCU	3.58	1.91
NCU	3.57	1.92
MRPU	3.37	1.78
CD(0.05)	0.27	-

- Basal application of neem coated urea (NCU) at 30 kg N/ha was as good as split application of prilled urea at 30 kg N/ha with regards to yield. The nitrogen use efficiency (NUE) was the highest with application of NCU.
- Application of 75% RD (60, 30, 30 kg N, P_2O_5, K_2O/ha) + Azotobacter + PSM + KSM in upland rice produced 2.32 t grains/ha and was at par with 75% RD + Azotobacter + PSM.

Treatment	Grain yield (t/ha)	Return/rupee invested (Rs)
RD (60:30:30 kg NPK/ha)	1.97	1.72
75% RD	1.87	1.32
75% RD + Azotobacter + PSM	2.31	2.68
75% RD + Azotobacter + Azosprillium + PSM + KSM	2.32	1.84
CD (0.05)	0.35	-

RD = Recommended Dose.

- Application of FYM 10 t/ha or Azolla 1.0 t/ha with 75% of RD (60,30,30 kg N, P_2O_5, K_2O/ha) resulted in grain yield of 5.92 and 5.75 t/ha, respectively, which were at par.
- Application of FYM 5 t/ha with dhaincha and 60, 30, 30 kg N, P_2O_5, K_2O/ha produced the highest grain yield of 5.6 t/ha and returnper rupee invested (2.84) in cv. CR 1018 including 76% yield advantage over recommended dose.

Wheat

- Application of 60 kg N/ha in two equal splits (sowing + crown root initiation, CRI) produced same yield (1.82 t/ha) as that of 120 kg N/ha (1.89 t/ha).
- Under late sown conditions at Bhubaneswar application of 40,29,20 kg NPK/ha to Sonalika wheat resulted in the highest grain yield of 0.82 t/ha was at par with that of 80,40,40 kg NPK/ha (0.75 t/ha)

Maize

- Application of 80 kg N/ha in 3 equal splits (Basal, 25 DAS and 40 DAS) recorded maximum gree cob yield (5.47 t/ha) of sweet corn cv. Madhuri.

- Soil as well as foliar application of Zn and B with RD recorded 21% higher yield of maize than NPK alone.

Treatment	Grain yield (t/ha)	B:C ratio
RD (100,50,50 kg NPK/ha)	2.98	1.29
RD + 20 kg $ZnSO_4$ + 1 kg B/ha (Soil)	3.59	1.43
RD + 0.5% $ZnSO_4$ + 0.25% Borax (Foliar)	3.61	1.45
CD (0.05)	0.16	-

Pulses

- Yield of green gram, black gram and horse gram increased due to application of lime (2.5 t/ha) and 10 kg phosphate in acid lateritic soils. Response was better to lime than that of phosphate.
- Seed yield of green gram increased with increasing level of P_2O_5 and optimum yield was obtained with 40 kg P_2O_5/ha (1.04 t/ha).
- Application of 40 kg P_2O_5/ha to black gram produced 0.74 t/ha of grains which was at par with that of 60 kg P_2O_5/ha and 29% higher yield than no phosphate application.
- Application of 60 kg P_2O_5/ha to pigeonpea resulted in maximum yield of 1.33 t/ha which was at par with that of 30 and 90 kg P_2O_5/ha.
- Rice bean responded up to 40 kg P_2O_5/ha (0.670 t/ha). Residual effect of P was observed up to 60 kg P_2O_5/ha in subsequent *rabi* crops such as tomato, groundnut, maize, wheat, ragi and mustard.
- The yield of Rajmash increased with increasing level of N. The highest yield of 1.58 t/ha was recorded with 90 kg N/ha which was 83% higher than control.

Oilseeds

- *Rabi* groundnut responded up to application of 40 kg K_2O with an yield advantage of 20% over control.
- TMV-2 variety of groundnut responded up to 60 kg P_2O_5/ha and limes 1.0 t/ha which increased the pod yield and oil content.
- Combined application of SSP (40 kg P_2O_5/ha) and Gypsum (250 kg/ha) to *rabi* groundnut resulted in maximum pod yield of 2.89 t/ha which was 36% higher than control.
- Optimum dose of fertilizer for sunflower variety Morden was found to be 60 kg N and 40 kg P_2O_5/ha with seed of 0, 92 t/ha.
- Hybrid sunflower gave maximum yield by application of 80, 40, 40 kg N, P_2O_5, K_2O/ha.
- Application of 20 kg elemental sulphur/ha to mustard increased the seed yield, oil content and B: C ratio followed by 40 kg S/ha through gypsum.
- Sesame variety Kalika produced the highest yield (0.82 t/ha) with minimum oil content of 49.50% at 45 kg N/ha in an irrigated sandy loam soil during *rabi* season.
- Application of 5 t poultry manure with 30, 15, and 15 kg N. P_2O_5, K_2O/ha to sesame produced maximum seed yield of 0.898 t/ha and B: C ratio of 1.83.

Forage Crops

- Cowpea (cv. EC 4216) produced the highest green fodder yield (GFY) of 20.8

t/ha in two cut system with 20, 40 kg N, P_2O_5/ha.

- Oat (cv. Kent) at closer spacing of 15 cm with 120 kg N/ha in one cut system at 50% flowering stage produced maximum green fodder (30.1 t/ha), crude protein (0.64 t/ha) yield with the highest B:C ratio (5.39) and energy efficiency (4.54)
- Use of biofertilizer (*Azospirillum*) with 100 kg N/ha in fodder maize (Ganga 5) produced the highest GFY of 29.9 t/ha with B:C of 2.57. Combined application of *Azotobacter* and *Azospirillum* increased the GFY by 22 per cent.
- Inoculation of *Azospirillum* with 80 kg N/ha produced 8% higher GFY in oats with B: C ratio of 1.66 while inoculation with 80 kg N/ha recorded the highest GFY of 27.0 t/ha and B:C of 3.24 in fodder sorghum.
- Cereal + legume mixture consisting of maize (HGT 3/African tall) and cowpea (EC 4216) in 2:2 row ratio recorded the highest forage yield (47.9 to 50.8 t/ha) with B: C ratio of 2.13 to 2.30.
- As a sole crop, cowpea (Russian giant) produced GFY of 21.22 t/ha. When it was grown with Teosinte in 2:1 ratio it produced maximum GFY of 25.79 t/ha with the highest B: C of 2.23.
- Hybrid napier and Guinea grass were suitable in the alleys of *Acacia mangium* and *Gmelina arborea*, respectively, whereas Thin napier was suitable in the alleys of all the three trees species including *Dalbergia sissoo*. The B: C ratio was the highest in thin napier (1.80) followed by Guinea and Hybrid napier.

Farming System

- In a pond based farming system, crop enterprise gave the highest employment of 304 man days and net income of Rs 24,595/ha. This was followed by mushroom production which provided 180 man days employment and net income of Rs 12,856. Composite pisciculture provided an attractive net income of Rs 16,603 with employment generation of 31 man days in term of return per rupee invested, apiculture gave the highest (Rs 7.94) followed by pisciculture (Rs 5.46) and crop production (Rs 3.05).

A pond based farming system model on 1.25 ha farmland at Bhubaneswar.

Components	Employment generation (man days)	Total expenditure	Net return (Rs)	Return/rupee invested (Rs)
Field crops	98	3315	5638	2.70
Multistoried cropping	88	3831	9089	3.37
Pomology	18	900	1466	2.63
Olericulture	96	3812	8302	3.18
Floriculture	4	25	100	1.80
Pisciculture	31	3722	16,603	5.46
Poultry	23	9240	981	1.11
Duckery	23	5387	713	1.13
Mushroom	180	18184	12856	1.70
Apiary	1	170	1180	7.94
Biogas	11	600	1431	3.38
Total	**573**	**49,286**	**58,380**	**2.18**

Cropping System

Sequential Cropping

- Application of rock phosphate and single super phosphate in 1:1 ratio to rice in a rice-green gram cropping system produced the highest yield of rice (3.43 t/ha) and green gram.
- Among the new emerging rice based cropping systems, rice-tomato gave the highest net return of Rs 65,184/ha and B:C of 4.47.
- Under normal planting, agronomical practices for optimizing production of rice-rice system include – timely planting (21 July) of rice variety Pratap (140 days) in wet season with 15 cm x 10 cm spacing and 80, 40, 40 kg N, P_2O_5, K_2O/ha followed by timely planting (17 January) of rice cv, Lalat (126 days) at the same spacing and fertilizer dose in dry season. It produced a total yield of 8.7 t/ha of grain.
- Under delayed planting (20 August) of *kharif* rice cv. Mahalaxmi (150 days) at 15 cm x 10 cm spacing and 100,50,50 kg NPK/ha followed by timely planting (17 January) of Lalat during dry season at 15 cm x 10 cm spacing with 80,40,40 kg NPK/ha produced 8.5 t/ha.
- Hybrid rice (Pro Agro 6201) responded up to 120 kg N, 60 kg P_2O_5 and 80 kg K_2O/ha and produced the highest grain yield of 6.4 t/ha during *rabi* season. The residual effect on succeeding *kharif* rice (Lalat) was the highest with the application of 120 kg N and 120 kg K_2O/ha to the *rabi* rice.
- Among various rice bean based systems, riceben-tomato produced the highest ricebean equivalent yield of 3.73 t/ha with net return of Rs 17,367 and B:C of 1.70.
- Application of additional amount of 20 kg N/ha with rice straw @ 5 t/ha to *kharif* as well as *rabi* rice increased grain yield by 25% and 35%, respectively, as compared to recommended dose of N.
- In rainfed upland, timely sowing of rice (cv. ZHU-XI-26) in the third week of June followed by horse gram in the last week of September produced 3.5 t/ha of rice and 1.6 t/ha of horse gram.

Intercropping

- Application of 100% RD of fertilizer to both the crops on area basis in pigeonpea + rice (2:5) intercropping system produced the highest pigeonpea equivalent yield of 1.10 t/ha *i.e.*, 0.78 t/ha of pigeonpea and 1.29 t/ha of rice grain. This could fetch a net return of Rs 9150/ha with B: C of 1.88.
- Intercropping of with rice in 2:8 ratio could fetch a net return of Rs 5368/ha having an yield advantage of 59%.
- Wheat + mustard intercropping under irrigated condition produced the highest wheat equivalent yield of 3.48 t/ha *i.e.*, 2.22 t of wheat and 0.47 t of mustard/ha.
- Application of 100% fertilizer to individual crops in maize + cowpea (2:2) intercropping system was found remunerative which resulted in maximum maize equivalent yield of 86.4 thousands cobs/ha, the highest LER (2.27) and B:C ratio(2.43).

Weed Control

Upland Rice

- Use of 20% extra seed produced 19% more grain than the recommended seed rate. The net return and B:C ratio were also increased due to high weed control efficiency(WCE) (64%).
- Interculture at 20 days after sowing (DAS) effectively controlled weeds (65% WCE) and produced maximum grain yield (2.42 t/ha), net return (Rs 3643) and B:C (1.33).
- Pre-emergence application of butachlore 1 kg/ha day following sowing in moist soil produced maximum grain yield (2.54 t/ha) which was 36% higher than unweeded control. It also registered the highest net profit (Rs 4780/ha), B:C (1.42) and WCE (75%).

Direct Seeded Medium Land Rice

- *Beushening* with *khelua* at 30-35 days after germination (DAG) followed by two hand weedings (HW) at 40 and 60 DAG have the highest grain yield of 4.0 t/ha which was 20% higher than no weeding. The WCE was 89% with the former.
- Application of butachlor 1.0 kg/ha at 2 DAS in moist soil produced the maximum grain yield (4.9 t/ha) which was 75% higher than weedy check.

Transplanted Rice

- Application of Rice Guard (anilophos 0.375 kg + ethoxysulfuron 0.01 kg/ha) or cyclosulfamuron (10WP) 30 g/ha at 7 days after transplanting (DAT) produced 3.95 and 5.24 t grains/ha, respectively, with corresponding WCE of 86 and 77%.
- Pre-emergence application of oxyfluorfen 0.5 kg /ha at 1 DAT supplemented with use of rotary weeder at 45 DAT produced the maximum grain yield of 3.62 t/ha with WCE of 78 per cent.

Wheat

- Pre-emergence application of pendimethalin 1 kg/ha day following sowing supplemented with one HW at crown root initiation stage recorded the highest grain yield of 2.79 t/ha which was 59% higher than the weedy check. It is also recorded the highest net return of Rs 4731/ha and WCE of 97 per cent.

Maize

- Pre-emergence application of atrazine or 2,4-D Na salt @ 1 kg day following sowing in moist soil produced grain yield of 4.28 and 5.10 t/ha, respectively, with WCE of 93% and 86% and B:C of 2.5 and 3.4.

Ragi

- Application of thiobencarb 0.75 kg/ha at 4 DAT doubled the yield (2.49 t/ha) with the highest WCE of 69% and B:C of 1.5.

Groundnut

- Maximum pod yield (2.24 t/ha) was obtained with pre-emergence application of imazapic 120 g/ha. The WCE was 61% and B:C ratio was 1.84.

Sesame

- Pre-emergence application of alachlor at 1 DAS produced the highest yield and increased the yield by 5 times with WCE of 66 per cent.

Ricebean

- Pre-emergence application of metochlor 1 kg/ha at 1 DAS increased the seed yield (0.53 t/ha) by 70% over weedy check and WCE was 96 per cent.

Cotton

- Post-emergence application of paraquat 0.2 kg/ha at 30 DAS supplemented with one HW at 60 DAS recorded the highest seed cotton yield (1.49 t/ha) which was 149% higher than weedy check. It also gave the maximum net return (Rs 20,704/ha), B:C (2.41) and WCE (82%).

Jute

- Preplant incorporation 1 kg/ha at 1 day before sowing DBS) with one HW at 20 DAS recorded the highest fibre yield of 2.80 t/ha which was 248% more than weekly check. The WCE was 70 per cent.

Onion

- Application of oxyfluorfen 0.03 kg/ha at 6 DAT supplemented with one HW at 30 DAT registered the maximum bulb yield (12.76 t/ha) which was 312% higher than the weedy check. It also recorded the highest WCE (69%), net return (Rs 2740/ha and B: C (3.5).

Tomato

- Application of metribuzin 0.5 kg or napropamide 1.5 kg/ha at 3 DAT enhanced the yield from 14.40 to 26.39 t/ha with WCE of 84%, maximum net return of Rs 38,832/ha and B:C of 1.43.

Cropping System

- Preplant incorporation of fluchloralin 1 kg/ha at DAS with one HW at 20 DAS in jute and 15 DAS in groundnut and black gram gave the highest fibre yield (2.80 t/ha), pod yield (2.63 t/ha) and seed yield (0.68 t/ha) of jute, groundnut and black gram, respectively in jute-groundnut-black gram cropping system. It also recorded the maximum net return of Rs 17,569/ha.

PLANT BREEDING AND GENETICS

The department continues to play a pivotal role in development and release of new crop varieties. The founder member of the department were Dr. G. V. Chalam and Dr, P. D. Dixit who had contributed substantially to crop improvement, particularly in rice, the premier crop of Orissa state. Dr. Dixit has the credit of improving upon a number of local land races through pure line selection.

Achievement in Rice Breeding

Improvement of Traditional Tall Rice

Research on improvement of tall *indica* rices yielded more than 30 varieties developed through pure line selection. Some of these varieties have maintained their popularity either as cultivars or as parents in hybridization programmes till today.

- T 141 and T 90: Still grown in sizeable lowland for their wide adaptability and better quality grains, often used as parents in high yielding rice breeding programme.
- T 1242: Maintains its popularity as a lowland cultivar for its grain quality and

pigmented leaf base.

- BAM 6: Popular lowland variety widely grown in Ganjam district.
- FR 13 A and FR 43 B: Considered to be the top donors both at national and international levels for flood/Submergence tolerance and are being used extensively as parents in breeding for flood resistance.

Breeding for High Yielding Varieties (HYV)

Breeding for high yielding rice varieties in Orissa was initiated in the mid 1960s by Prof H. K. Mohanty and associates and the efforts during last 40 years led to release of as many as 44 rice varieties for commercial cultivation in Orissa. Characteristics of the rice varieties released so far are given in Table 15.1.

- Jagannath, a short height mutant developed by X-ray irradiation of T 141, is photosensitive, moderately responsive to nitrogen, late in maturity duration and has adaptability to lowland. This variety is a unique achievement of the department which was the leading lowland variety during 1970-80 and had revolutionized the lowland rice breeding programme in the country as an important parental source.
- Parijat and Khandagiri have spread to neighbouring states like West Bengal and Chhatishgarh.
- One of our lowland cultures OR 142-99 has been released in Cambodia under the popular name Santepheap-3 and is one of the leading varieties of that country.
- The variety Rambha has been identified as suitable for Myanmar and Sarathi for irrigated lands in Thailand, Malaysia, Indonesia, Vietam, Bangaladesh and Egypt.
- Recently, one of our lowland culture OR 1128-7-1 was released in Tamil Nadu as ADT 44 suitable for growing in Samba season.

Ongoing Rice Breeding Programme

In the current breeding programme special emphasis has been given to develop rice varieties possessing:

- Adaptability to lowlands particularly semi-deep water situation.
- Tolerance to flood, waterlogging and salinity.
- Tolerance to drought in uplands and early seedling drought in rainfed lowland where the crop is direct seeded.
- Genetic yield enhancement in short grain indigenous aromatic rices for export and economic development of rural poor.

Achievements in Non-rice Crops

- Two varieties of wheat, 4 in ragi, one each in green gram, black gram and niger (Alasi 1), 3 in sesame were released for cultivation in the state (Table 15.2).

TABLE 15.1: High yielding rice varieties (1969-2002)

Sl. No.	Name of the variety	Culture/IET	Cross combination	Days to maturity	Major features	Yield (kg/ha)		Year of release
						Average	Potential	
Varieties for uplands								
1.	Parijat	OR 34-16, IET 2684	TN 1/TKM 6	95	Semi dwarf, hull straw colour, MS grains, White kernel	3790	6940	1976
2.	Suphala	OR 45-61-23	T 141/TN 1	95	Semi dwarf, hull straw colour, SS grains, white kernel	3880	7223	1976
3.	Subhadra	DR 92, IET 4097	TN1/SR 26B	85	Semi dwarf, hull straw colour, Apiculus pigmented, MB grains, white kernel	2800	4570	1980
4.	Keshari	OR 63-252, IET 6215	Kumar/Jagannath	100	Semi dwarf, golden colour hull, MS grains, white kernel	3500	6330	1980
5.	Rudra	OR 165-97-15	Partijat / IET 3225	85	Semi dwarf, hull straw colour, MB grains, white kernel	2300	4200	1985
6.	Shankar	OR 165-86-12	Parijat/IET 3225	85	Semi-dwarf, hull straw colour, MB grains, white kernel	2400	4800	1985
7.	Pathara	OR 88-23 IET 6639	Hema Suphala	95	Semi dwarf, hull straw colour, MB grains, white kernel	3370	6560	1985
8.	Badami	OR 164-5 IET 7259	Suphala/ Annapurna	95	Semi dwarf, hull and kernel golden colour with MB grains	3500	7560	1992
9.	Nilagiri	OR 163-104 IET 10393	Suphala/DZ 192	95	Semi dwarf, hull straw colour, MB grains, white kernel	3150	6920	1992
10.	Khandagiri	OR 811-2, IET 10396	Parijat/IR 13429-94	95	Semi dwarf, hull straw colour, MS grains, White kernel	3220	6140	1992
11.	Ganteswari	OR 377-85-6 IET 9802	IR 2061-628/N22	95	Semi dwarf, hull golden colour, MB grains, dull red kernel	3296	7430	1992
12.	Udayagiri	OR 752-38-1 IET 12136	IRAT 138 / IR 13543-66	95	Semi dwarf, hull golden colour, MB grains, dull red kernel	3550	5600	1999
13.	Lalitagiri	OR 1045-1-38-1	IRAT 138/ IR 13543-66	95	Semi dwarf, hull golden colour, MB grains,	2975	6750	1999

Sl. No.	Name of the variety	Culture/IET	Cross combination	Days to maturity	Major features	Yield (kg/ha)		Year of release
						Average	Potential	
		IET 12136			white kernel			
Varieties for Medium lands								
14.	Kumar	OR 10-112	T 90/IR 8	120	Semi dwarf, hull straw colour, MS grains, white kernel	3850	6330	1971
15.	Data	OR 131-13-13 IET 7430	Kumar/ CR 57-49	125	Semi dwarf, hull straw colour, MB grains, whit ekernel	4100	5830	1985
16.	Sarathi	OR 79-21 IET 5975	T 90/IR 8/W 1263	120	Semi dwarf, hull straw colour. MB grains, white kernel	4000	5000	1985
17.	Ananga	OR 131-5-8 IET 7433	Kumar/ CR 57-49	125	Semi dwarf, hull straw colour, MB grains, white kernel	4399	8073	1988
18.	Shrabani	OR 367-SP-11 IET 9946	Mahsuri/IR 30	125	Intermediate height, golden colour hull, MS grains, white kernel	4082	7870	1988
19.	Hema	OR 5-11	T 90/IR 8	135	Semi dwarf, hull straw colour, LB grains, white kernel	4000	7500	1971
20.	Rajeswari	CR 10-193	T 90 / IR 8	135	Semi dwarf, straw coloured hull, MB grains, white kernel	4000	7000	1971
21.	Jajati	OR 47-2 IET 7284	Rajeswari / T 141	135	Weakly photosensitive, intermediate height hull straw colour, MS Grains, white kernel	4000	7000	1980
22.	Pratap	OR 131-3-1 IET 6661	Kumar / CR 5749	135	Semi dwarf, hull straw colour, MB grains, white kernel, apiculur pigmented	4000	7500	1985
23.	Gouri	OR 148-1-44 IET 7428	Rajeswari / Vikram	135	Semi dwarf, hull straw colour, MB grains, white kernel	4500	8500	1985
24.	Bhubana	OR 447-20 IET 7804	OR 158-5 / Rasi	140	Semi dwarf, hull straw colour, MB grains, white kernel	4320	8967	1985
25.	Birupa	CR 253-2 IET 8620	Adt. 27 / IR 8 / Annapurna	135	Semi dwarf, straw coloured hull, MB grains, white kernel	4250	6850	1992
26.	Bhanja	OR 443-80-4 IET 10451	IR 36 / Hema / Vihram	140	Semi dwarf, straw coloured hull, MB grains, white kernel	4030	6830	1992

Sl. No.	Name of the variety	Culture/IET	Cross combination	Days to maturity	Major features	Yield (kg/ha)		Year of release
						Average	Potential	
27.	Samanta	OR 487-30-3 IET 10451	T 90 / IR 8 / Vikram / Siam 29 / Mahsuri	140	Semi dwarf, straw coloured hull, MB grains, white kernel	4220	6890	1992
28.	Sebati	OR 776-SSD, 26 IET 11786	Daya / IR 36	125	Semi dwarf, straw coloured hull, MS grains, white kernel	4120	6290	1999
29.	Bhoi	CR 987-13 IET 12443	Gouri / RP 825-45-1-3	125	Semi dwarf, straw coloured hull, MB grains, white kernel	3890	6240	1999
30.	Konark	OR 1143-230 IET 12734	Lalat / OR 135-3-4	125	Semi dwarf, straw coloured hull, MS grains, white kernel	4573	8580	1999
31.	Surendra	OR 447-20-P IET 12815	OR 158-5 / Rasi	135	Semi dwarf, straw coloured hull, MB grains, apiculus pigmented, white kernel	4710	7680	1999
32.	Gajapati	CR 820-29 IET 13251	OR 136-3 / 13429-196-1-20	130	Semi dwarf, Straw coloured hull, MS grains, white kernel	4432	7060	1999
33.	Kharavela	OR 815-3 IET 13253	Daya/IR 13240 108-2-2-3	125	Semi dwarf, straw coloured hull, MS grains, white kernel	4406	8108	1999
Varieties for Lowlands								
34.	Jagannath	BBS 873	Mutant of T-141	150	Strong photosensitive, semi-dwarf, hull straw colour, MS grains, white kernel	5000	7500	1969
35.	Mahalaxmi	OR 621-6 IET 10048	Pankaj / Mahsuri	155	Semi dwarf, hull straw colour, MB grains, white kernel	4070	6300	1992
36.	Manuka	OR 624-46 IET 9737	CR 1010 / OBS 677	155	Semi dwarf, hull straw colour, MB grains, white kernel, Panicle weight type	4700	6300	1992
37.	Urbashi	OR 645-18 IET 10811	Rajeswari / Jajati	145	Tall, hull straw colour, MB grains, white kernel, panicle weight type	4050	6170	1992
38.	Rambha	OR 143-7 IET 7975	Pankaj / W 1263	160	Tall, photosensitive, resistant to int. water depth, MB grains, White kernel	4260	5500	1986
39.	Kanchan	OR 609-15 IET 10016	Jajati / Mahsuri	160	Tall, photosensitive, resistant to int. water depth, MS grains, White kernel	4010	5500	1992

Sl. No.	Name of the variety	Culture/IET	Cross combination	Days to maturity	Major features	Yield (kg/ha)		Year of release
						Average	Potential	
40.	Santepheed	OR 142-99	Pankaj / Sigadis	155	Int. ht, stiff straw, weekly photo sensitive, MB grains, hull straw colour, white kernel	4500	7000	1992
41.	Prachi	OR 888-4-30 IET 12786	IR 9764-45-2-2 / CR 149-3-2	155	Semi dwarf, hull straw colour, MB grains, white kernel, panicle weight type, photosensitive	4270	8333	1999
42.	Ramachandi	OR 912-10-190	IR 1749-45-2-2 / Jagannath	155	Semi dwarf, hull straw colour, MB grains, white kernel, panicle weight type, photosensitive	4151	6046	1999
43.	Mahanadi	OR 1301-13 IET 13356	IR 19661-131-1-3-1 / Savitri	150	Semi dwarf, hull straw colour, MB grains, white kernel, panicle weight type, photosensitive	4421	6983	1999
44.	Indravati	OR 1128-7-S1 IET 13396	IR 56 / OR 142-99	150	Semi dwarf, hull straw colour, MB grains, white kernel, panicle weight type, photosensitive	4071	7026	1999
45.	Jagabandhu	OR 1206-25-1 IET 14800	Savitri / IR 4819 Sel. / IR 27301 Sel	150	Intermediate height, hull straw colour MB grains, white kernel, photosensitive	4415	6838	2002

TABLE 15.2: High yielding non-rice crop

Sl. No.	Name of the variety	Culture No / IET No.	Cross combination	Days to maturity	Major features	Yield kg/ha		Year of release
						Average	Potential	
WHEAT								
1.	Sagarika	OW 6	NP 798 / Kalvanisona	95-110	HT 75-90 cm, spreading at early stage of growth subsequently becoming erect, Stiff straw, semi compact ear head, white bold grain	2800	4400	1983
2.	Utkalika	OW 801-3	HB 1742 / FAO 106/68 R	95-110	Ht 80-100 cm compact tillers, stiff straw, compact ear heads	2700	4100	1983

Sl. No.	Name of the variety	Culture No / IET No.	Cross combination	Days to maturity	Major features	Yield kg/ha		Year of release
						Average	Potential	
					with high spikelet density, medium bold white grains			
RAGI								
1.	Dibyasinha	T 20	Mutant of AAKP 7	90-95	Tall plant ht, moderate tillering habit, finger length 5.5 to 6 cm, small spherical brown coloured grains, ear head slight top curved with white pericarp	2050	2640	1971
2.	Nilachal	B4-10-56	Mutant of IE-642	101-105	Tall plant ht, few productive tillers, light brown and medium bold grains, panicle length 5.8 cm with top incurved ear heads	2340	3080	1985
3.	Suvra	OUAT-2 Mut-18	Mutant of C 0-9	100-105	White grain with white flour, suitable for *kharif* and *rabi*, moderate resistance to blast, Earworm.	2240	3220	1998
4.	Bhairabi	BM 9-1	Mutant of Bhudamandia	105-110	Grain suitable for *kharif* and *rabi*, resistance to logging, blast, stem borer and earhead caterpillar	2550	3680	1999
BLACK GRAM								
1.	Sarala	B 12-4	T 9 mutant	75-80 (*kharif*) 85-90 (*rabi*)	Short plant ht, photoinsensitive, compact in habit with larger number of branches, dark green seeds, medium size and light black, protein content 24%, tolerate	850	1270	1980

Sl. No.	Name of the variety	Culture No / IET No.	Cross combination	Days to maturity	Major features	Yield kg/ha		Year of release
						Average	Potential	
					to YMV			
GREEN GRAM								
1.	Dhauli	TT 9-E	Local CV / T 51	60-65 *kharif*	Intermediate plant ht, photo insensitive, compact branching, seed colour shining green protein content 26%, tolerant to YMV, Cercospora and powdery mildew	750	1180	1980
SESAME								
1.	Binayak	Sel-14	A pure line selection from local collection	85-90 (*kharif*) 80-85 (*rabi*)	Less compact branching, capsules glabrous, seed colour light brown, oil content 48.5%	500	850	1972
2.	Kanak	BS 6-1	Vinayak / T 4	85-90 (*kharif*) 80-85 (*rabi*)	Compact branching, capsules pubescent, seed colour grey, oil content 47%	600	950	1972
3.	Kahika	BM 3-7	Mutant of Vinayak	90-95 (*kharif*) 80-85 (*rabi*)	Compact branching, capsules glabrous, seed colour brown, oil content 48.7%	700	1000	1980

CHAPTER 16

Integrated Farming Systems (IFS) Under Agroecosystem

INTEGRATED FARMING

Integrated Farming is defined as biologically integrated farming system which integrates natural resources and regulation mechanisms into farming activities to achieve maximum replacement of off-farm inputs; secures sustainable production of high quality food and other products through ecologically preferred technologies; sustain farm income; eliminates or reduces sources of present environment pollutions generated by agriculture, and sustains the multiple function of agriculture (IOBC, 1993). Thus integrated farming systems must be built on the sound foundation of ecological science. It emphasizes a holistic approach. Such an approach is essential because agriculture has a vital role to play that is much wider than the production of crops, including providing diverse, attractive landscapes and encouraging biodiversity, and conserving wild life. Sustainable development in agriculture must include integrated farming systems with efficient soil, water, crop and pest management practices, which are environmentally friendly and cost effective. The future agricultural system should reorient from the single commodity system to food diversification approach for sustaining food production and income. Integrated farming systems, therefore, assume greater importance for sound management of farm resources to enhance farm productivity which will reduce environmental degradation and improve the quality of life of resource poor farmers and to maintain agricultural sustainability. The aim of the integrated farming system can be achieved by:

- efficient recycling of farm and animal wastes;
- minimizing the nutrient losses;
- maximizing the nutrient use efficiency;
- following efficient cropping systems and crop rotations; and
- complementary combination of farm enterprises

THE ROLE OF IFS IN CROP-LIVESTOCK INTERACTION

A review of Mc In tire and Gryseels (1987) reveals that crop-livestock interactions take place in two main systems. First, there is interaction without full integration, defined here as a segregated system. Farmers own animals, and herders might grow crops, but the integration is not complete because power and manure are not fully used by either group. This type of interaction is dominant in the West African Sahel and Savannah.

The second is complete crop-livestock interaction, defined here as an integrated system. The cultivator owns and manages animals, uses draft power, feeds crop residues or fodder crops and recycles manure for soil fertility or for fuel. This type of interaction is dominant in the Ethiopian and Malgache high lands and in parts of Zimbabwe, Malawi, and Senegal.

It also reveals that the goals of Farming systems research in the study of crop-livestock interactions are to increase the direct physical contribution of animals to crop production via inputs of power and manure; to increase the indirect contribution of one sector to the other via cash income; and to increase the direct physical contribution of crop production to animal production, especially from crop residues. For each goal, there are different strategies according to the current degree of integration and according to the constraints to crop and animal production.

INTEGRATED FSR ACTIVITY AND MODELS

Integration of various agricultural enterprises *viz.*, cropping, animal husbandry, fishery, forestry etc. have great potentialities in the agricultural economy. These enterprises not only supplement the income of the farmer but also help in increasing the family labour employment.

The integrated farming system approach introduces a change in the farming techniques for maximum production in the cropping pattern and takes care of optimal utilization of resources. The farm wastes are better recycled for productive purposes in the integrated system. A judicious mix of agricultural enterprises like dairy, poultry, piggery, fishery, sericulture etc. suited to the given agro-climatic conditions and socio-economic status of the farmers would bring prosperity. Farm as a unit is to be considered and planned for effective integration of all enterprises.

A review of an article by Obias (1986) reveals that the research and development team of Maya Farms, an integrated livestock farm and meat processing enterprise on the outskirts of Metro Manila, Philippines, developed a number of integrated farming systems, in which the biogas plant provided a vital link between crop and livestock farm, between coconut plantation and cattle raising, between pig and poultry production and between livestock and fish farming.

In an integrated crop and livestock farm, a demonstration model for 1.2 ha developed at Maya Farms, had one hectare under crops and the remaining area for piggery unit. Manure from pigs and water buffalo (a draft animal for land preparation), processed in a bio-gas plant, which supplied the farm house with enough fuel for cooking, ironing clothes, lighting and running a small gas refrigerator. The sludge was sufficient to fertilize the one hectare field, producing one crop of rice during the wet season and two crops of corn during the dry season. The corn and the rice bran were used to supplement the feed for the pigs and the rice straw and corn stover were fed to the water buffalo. The farmer was able to sell five or six porkers every month.

In the integrated coconut and cattle production program, the farm was able to increase the nut production to triple times by utilizing the sludge from the biogas plant and through the irrigation and fertilization of the forage crops intercropped in the coconut plantation while the cattle bring additional profit.

Obias (1986) also indicated the possibility of integrating piggery and poultry so that the biogas produced over and above the requirement of pig farm is sufficient to cover the biogas deficiency in the poultry operation. Moreover he attributes that by combining pig production with poultry raising, chicken manure helps in increasing the consistency of pig manure slurry for the biogas unit. It has been identified that a 500-sow piggery would produce enough excess biogas to support a 12,000 layer operation reared on the farm itself.

In another attempt it was reported that by combining pig and fish, the pig manure supplements the fertilizer need of the fish pond. To avoid the direct effect of pig manure on fish through infection of pathogens and the tainted taste of fish, the processing of manure through a biogas plant is being effectively attempted. An additional advantage in the latter is that the system produces a fuel gas before the waste is used for the fish pond. Thus biogas can serve as fuel for the farm house and the piggery.

Obias (1986) concludes that recycling of farm wastes though biogas plant relieves the farmer of the problem of waste disposal and environmental pollution. According to him it also provides some relief for problems caused by the high cost of three major farm inputs – fertilizer, feed and fuel. By using integrated farming systems that optimize the utilization of farm wastes, the farmer reduces his production costs and increases his productivity.

Mehla *et. al.* (1986) reported that in Northern India, with the triple cropping intensity of fodder crops and recycling of feed nutrients in a milch animals and biogas system, from one hectare of land, it is possible to produce 82.63 GJ of energy and 906.11 kg of protein from milk; 263.93 kg of N; 148.93 kg of P_2O_5 and 90.94 kg of K_2O fertilizer from biogas slurry and 2897 kilowatts of power per annum for farm operations from biogas using the Kirloskar biogas engine. Traditional systems of wheat and rice cultivation can produce 93.87 GJ of energy and 654.25 kg of protein from food grain per hectare per annum. The recycling of feed nutrients in the available straws of rice and wheat crop can yield an additional 16.31 GJ of energy and 166.67 kg of protein from milk; 81.23 kg of N; 56.00 kg of P_2O_5 and 25.14 kg K_2O fertilizer from biogas slurry and 1097 kilowatts of power per hectare annually from biogas.

A model as constructed by Mehla *et. al.* (1986) to integrate the process of recycling of feed nutrients through a crop, animal, biogas system is given in Fig. 16.1. The model was constructed on the basis of the following observations.

1. There is a linear relationship between availability of feed and fodder inputs and milk production system with a provision of horizontal growth of herd size in relation to feed and fodder inputs (Mishra, 1975). This observation holds true under Indian conditions.
2. During anaerobic digestion of animal waste (dung) in a biogas system, the carbon content of organic matter is degraded into carbon dioxide and methane (biogas). There is a linear relationship between degradation of the carbon content of organic matter and biogas production (Prakash, 1976).
3. The biogas energy can be burnt to produce energy in the form of heat and can be used as fuel to generate power through a biogas engine in which 80% of diesel oil is replaced by biogas.
4. The residues available from the biogas system contain organic manure and can be used for replenishing nutrients in the soil.
5. In the model, the flow of uncontrolled inputs and losses are not accounted.

INTEGRATED APPROACH IN WETLAND SITUATION

A model was developed at TNAU, Coimbatore for 0.4 ha (one acre) farm considering small and marginal farmers of Tamil Nadu. The components integrated in the rice based farming system are (a) cropping (b) fish culture and (c) poultry. While programming for cropping system, due consideration was given to crops under cultivation in the locality, the crop rotation that is being followed, and water requirement for fish culture included. An area of 0.36 ha was allotted for cropping and remaining 0.04 ha for fish rearing.

Rice Based Cropping with Poultry-Cum-Fish Culture

In wetland, the normal cropping programme followed is rice-rice-green manure/pulse. But in the integrated approach, a modification was made in the existing cropping pattern by including crops like maize, groundnut and gingelly. The cropping intensity increased to 300 per cent as against 240 per cent recorded earlier.

The polyculture fingerlings are preferred for this system as the growth rate within a year is found more economical than monoculture. The fingerlings were released in the pond as per the recommended stocking density of 7500 numbers per hectare of ponded water; accordingly 300 fishes in the following ratio were released in the pond.

Grass-carp (grass and algal feeder)	30 (1)
Rohu (column feeder)	60 (2)
Mirgal / common-carp (bottom feeder)	90 (3)
Silvercarp / catla (surface feeder)	120 (4)
Total	**300**

No separate feed was supplied to the pond, instead, then poultry droppings from the poultry shed placed well above the farm-pond formed the feeding material.

It has been quantified that 500 layer chicks are sufficient to provide required quantum of excreta to meet the feed requirement of 7500 polyculture fingerlings in a hectare of pounded water. As such 20 Bapkok chicks were used in the proposed program for egg yield and to meet the feed requirement of fingerlings released in the 0.04 ha (10 cent) pond. The fingerlings which solely depend on the poultry excreta grew at the rate of one kg per year.

In this integrated system, the waste products of rice formed 33 per cent of the poultry feed and the feed cost could be reduced by 57.5 per cent by substituting rice grain, maize, oil cakes of crop component and the Lucerne grown around the fish pond. In commercial poultry farm, cost of production per egg will be around 42 paise. By way of reducing the feed cost, the cost of production per egg under integrated farming system was reduced to 19 paise.

Thus, the net income under the integrated farming system was Rs. 15,145 / acre (0.4 ha) while Rs. 9,010 acre (0.4 ha) was obtained form the cropping system, rice-rice-green manure. The employment opportunity generated due to integration was 73 man-days and 246 woman-days while in pure cropping it was 68 man-days and 180 women-days (Rangasamy *et. al.*, 1988).

The planning strategy for lab our surplus economy in the rural sector have the objectives of maximization of employment opportunities (Thiagarajan, 1977) However, various programs launched and efforts made in the country in the past have not been adequate enough to absorb the growing and existing manpower. It has been stated that about 300 million lived in the rural areas. If the rural poor consisting of the landless laborers, marginal and small farmers to be helped to get better income, there has to be diversification of lab our use through the creation of employment opportunities so that the growing surplus lab our force could be absorbed in the villages (Swaminathan, 1981). This is possible by integrated faming system approach.

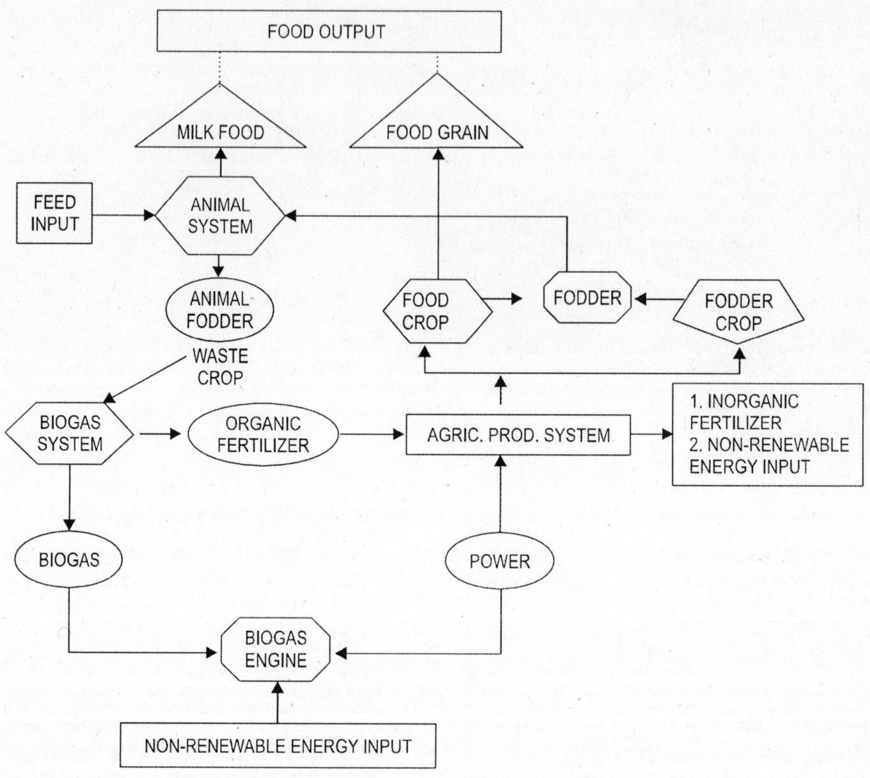

FIG. 16.1: Model for recycling feed nutrients in crop, animal and biogas systems. (Mehla et. al 1986)

Crop-Poultry-Fishery in Cauvery Delta

A model was developed at Tamil Nadu Rice Research Institute, Aduthural with poultry-cum-fish culture as a component in mixed farming. The study was conducted in a holding size of 2 hectares. In one hectare conventional cropping (rice-rice-rice fallow pulse) as practiced by farmers was followed. In another holding, modified cropping (rice-rice-rice fallow cotton/maize) was followed in 0.96 ha and an area of 0.04 ha was allocated for fish pond over which the poultry shed was constructed. The poultry droppings collected in the pond were used by fish as feed material.

In the third season, rice fallow pulses fetched a net profit of Rs. 373 in the existing farmers' practice. But in the modified cropping system, a net profit of Rs. 6,660 and Rs. 729 were obtained by introducing cotton and maize respectively instead of rice fallow pulses. The income from the poultry sub system was Rs. 581. Per day income was worked out to Rs. 3.81 for taking five months rearing period. On an average, one bird voided about 110 g of droppings per day (on net weight basis) and about 794 kg fresh droppings were recycled in the fish ponds. On moisture free basis about 260 kg of droppings were recycled in the fishponds. The results of the experiments showed that higher fish production of 4503 kg/ha in 6 and half months could be achieved in Cauvery delta by recycling poultry droppings in fishponds.

A net income of Rs. 8311 was obtained from conventional method of cropping and a net income of Rs. 15,009 was obtained by the modified cropping system. A net profit of Rs. 2201 was obtained from poultry-cum-fish culture unit. Totally an additional income of Rs. 8,899 was obtained from the modified cropping linked with poultry-cum-fish culture over the farmer's method of cropping. The additional employment generated from the system was 123.9 man-days.

Application of poultry manure in combination with inorganic fertilizer to rice-fallow-cotton recorded the highest seed cotton yield (19.97 q/ha) with a net income of Rs. 10,903/ha (Govindan, 1988).

Cropping with Dairy

Two holdings of one hectare each was taken for the study purpose, rice-rice-rice fallow pulse was the normal cropping system followed in Tanjore delta. In other holding, modified cropping system was follow ed with rice-rice-rice fallow black gram, rice fallow cotton and green grass for 3 number of milch animals.

A net return of Rs. 8,422 was obtained from farmers method of cropping and Rs. 10,912 from modified cropping system. Net return from dairy enterprise was Rs. 8,986. The additional net income of Rs. 11,477 was obtained from the dairy based mixed farming over the existing farmer's method of cropping.

Inclusion of dairy unit generated an additional employment of 190 man-days. The total additional employment generated by dairy based mixed farming was 309 man-days over the practiced cropping system. Similar findings have been reported by Pracer *et. al.* (1983) and Rai and Singh (1982). The farm yard manure obtained from dairy unit was applied to rice-fallow-cotton in combination with inorganic fertilizers. The treatment recorded the highest seed cotton yield of 17.6 q/ha and a net return of Rs. 8,573/ha (Rajkumar, 1988).

Cropping with Goat Rearing

Field experiments were conducted at farmer's holding in Kabisthalam village at Tanjore district of Tamil Nadu to study the feasibility of rearing Tallichery goats under deep litter system (5 does and 1 buck). The conventional method of cropping rice-rice-rice fallow black gram was compared with modified cropping rice-rice-rice fallow cotton, para grass. The goat manure collected under deep litter system was recycled as a source of nutrient to rice fallow cotton. The annual net income obtained from one hectare of conventional cropping and modified cropping were Rs. 7,654 and Rs. 15,409, respectively. The goat unit resulted in a net return of Rs. 4,330. Totally the additional income generated by introduction of mixed farming with goat rearing was Rs. 12,084.

The farmer's method of cropping generated 305 man-days in a year while 413.5 man-days were generated with mixed farming. The additional employment generated was 108.5 man-days. For landless laborers, rearing of goats and sheep may provide continuous employment throughout the year. This result adds to the findings of Rastogi (1976), Faroda *et. al.* (1978) and Singh *et. al.* (1986)

Application of goat manure in combination with inorganic fertilizer (50:50) gave the highest seed cotton yield of 18.89 q/ha (Santhanagopalakrishnan, 1988)

Cropping with Aquaculture

The farmers depend largely on cultivable crops for their income. They could enhance their farm receipts by adopting integrated fish farming methods such as cropping-cum-fish culture and livestock-cum-fish culture. These practices are already in vogue in several countries of Southeast Asia from the beginning of 19th century although it is yet to be popularized in countries like India.

The integration of cropping with fishery includes rice-fish farming; banana-fish farming and coconut fish farming. Rice-fish farming especially plays an important role in the rural economy as fish culture levels itself well to labour intensive farming operations.

Rice-fish Culture

The different types of exploitation of rice field fisheries are as follows:

Captural Systems

In this system, there is no stocking of fish fingerlings in rice fields. On the other hand, wild fish population reproduced in the flooded rice fields are harvested at the end of rice growing season. In captural systems, sumps of about 40-50 m^2 size and 2 m deep are dug in the lower region of a group of fields. Species of Channa, Clarias, and Anabas are cultivated in such systems.

Cultural Systems

As in fishponds, fish fingerlings are stocked in these systems at a known density depending on the fertility of the systems. The practices involved in cultural systems may be ''concurrent culture'' in which fish and rice are simultaneously grown and harvested or 'rotational culture' in which rice and fish are grown alternatively.

In concurrent culture system, fish controls weeds, mollusks and harmful insects of rice and helps in securing higher productivity of rice. However, this practice limits the use of modern agricultural techniques like chemical fertilizers, insecticides and herbicides.

In rotational culture system, insect icides and herbicides could be used for better rice production. Normally after one or two weeks of rice harvest, the field is prepared for fish rearing. Bunds of 50 cm height are formed around each field and ditches of 50 cm width and 30 cm depth are dug around the field. Simple bamboo pipe or screens are kept at the inlet and outlet to prevent passage of fish and detritus during the culture period. Common carp is known to be very ideal for this system.

Integrated fish farming with agriculture and livestock is certainly profitable venture as the risks are less in these practices than in either rice culture or livestock. In many farm complexes, excess livestock wastes could be diverted to either biogas plants or to fishponds, thereby fuller and profitable utilization of these could result especially through fish culture ponds.

INTEGRATED FARMING SYSTEMS FOR IRRIGATED AREAS

A model on integrated farming system to suit the farmers in irrigated area of Coimbatore was developed. An area of 1.0 ha was taken for the study with the following enterprise mix.

Cropping with Dairy, Biogas and Siliviculture

The normal cropping pattern followed in Coimbatore district of Tamil Nadu in garden lands is finger millet-cotton-sorghum. In the integrated approach, the same cropping pattern was slightly modified by inclusion of intercropping such as green gram in cotton, cowpea in sorghum and sunflower as border crop in finger millet (0.75 ha). The perennial grass fodder (0.15 ha) and perennial legume fodder, lucerne (0.05 ha) were also raised in the holding.

Three graded jersey cross bred milch cows with 2 calves were included under dairy component. For effective recycling of farm and animal waste, a biogas unit 2 m^2 capacity was installed for the production of fuel, light and enriched manure. Sixty kg of cow dung expected out of 5 animals is sufficient enough to produce m^3 of gas everyday which is equivalent to 1.5 litres kerosene.

Two hundred subabool trees were planted all along the boundary of the field for fodder and timber production.

The total grain production in the normal cropping component was 8230 kg (grain + fodder) valued at Rs. 18,184/ha while 9830 kg grain valued at Rs. 26.300/ha, was obtained due to integration of dairy and biogas. This system facilitated continuous flow of income to the farmer throughout the year besides generating additional employment.

INTEGRATED FARMING SYSTEM FOR RAINFED AREAS

One hectare of dry land was selected for the study involving crops, goat rearing and silvipasture (Rangasamy, 1988).

Cropping with Goat and Silvipasture

Generally in dry land, farmers raise only fodder sorghum yielding 3 to 10 tonnes of fodder/ha depending on the rains received. In the integrated approach, the cropping pattern was modified by including both sorghum grain crop (0.2 ha) and sorghum fodder crop (0.2 ha). To meet the fodder requirement of goats, 0.2 ha of land was raised with subabool and *Cencrus ciliaris*, a pasture grass as an intercrop. Trees like *Acacia senegal* and *Prosopis cineraria* were raised in o.20 ha.

Tellichery goats (1 buck + 5 does) were advantageously integrated into this dry farming situation. The sorghum grain crop with 1830 kg grain and fodder valued at Rs. 1660 / -, 240 kg of goat meat valued at Rs. 6000 /- and manure 3000 kg valued at Rs. 750 /- were obtained in the integrated approach with total value of Rs. 8410 /- while 12000 kg of fodder valued at Rs. 6000 /- only was obtained in the conventional system of cropping alone. Thus, an additional income of Rs. 2410 /ha was obtained by integration. This system had generated additional 54 man-days employment over the conventional system (Rangasamy, 1988).

During last few years, some studies conducted on profitability of various farming systems indicated mixed farming's as assured high income enterprises. (Faroda *et. al.*, 1984).

Integrated Aquaculture – Agriculture Systems

Small ponds on a mixed enterprise can improve the whole farm productivity and profitability as well as produce fish. Farm ponds could at best be an alternative to the large scale, intensive, specialized fish farms. The scientists at the International Center for Living Aquatic Resources Management (ICLARM) in the Philippines conducted a series of experiments in farmer's holdings in Africa. They have found that wide adoption of small ponds could result in sizable increase in farm fish production in Africa, in addition to facilitating increases in other produces such as vegetables, fruits and condiments from pond dikes and margins and adjacent garden and from pond bed itself in dry periods. This type of integrated farming system practices are spread to a number of other Asian countries. *e.g.*, the Mayan Chinampa system of water channels and raised beds for plants. The small farm ponds make as bio-digesters of waste, providers of fertile soil and water and help in site improvement that permits new operations for rotating and linking enterprises for risk management.

A set of six 'traditional' ponds were integrated with vegetable plots, maize gardens, banana groves, leucaena species, hedge rows and rice fields. The small ponds (50 sq.m) were undrainable and filled with runoff and underground infiltration. Waste materials from the other enterprises were used as pond inputs. Fish were periodically harvested and no special efforts made to control predation from birds and others. Pond muds and water were used to fertilize the vegetable plots. In studies using these traditional ponds, sizable yields of mustard, cabbage, hybrid maize and fish were obtained. The total productive area of 639 sq. m (250 sq. m of ponds and 139 sq. m of vegetables and maize) yielded an equivalent of 43 tonnes of mustard- cabbage, 42,000 cods and 454 kg fish/ha. Various combinations of farming enterprises with aquaculture have proved to be rewarding.

INTEGRATED FARMING SYSTEM FOR HILL REGIONS

The North Eastern Hill Regions of India are unique in their physiographic characteristics of land, varied agro climatic conditions, land tenure system and cultivation practices from the rest of the country. The region falls under high rainfall zone and is characterized by hilly terrain with wide variation in elevation from 100 m to 5,500 m. Majority of the farmers in the region are maintaining dairy cattle and the major source of green fodder comes from lopping of the fodder trees and locally available grasses. Crops and cropping patterns of such land uses will differ depending on the type of enterprise such as milk, beef, mutton, wool, pork and poultry production. Combination of cultivated varieties of perennial legumes, grasses, shrubs and trees can extend availability of green fodder upto February at low altitude, thereby shortening the requirement of conserved fodder for lean season feeding. Land upto 100% slope having soil depth greater than 1 m can be used for agri-horti-silvipastoral system. This system has the potentiality of maintaining 1.18 livestock unit/ha of land use.

INTEGRATED FARMING SYSTEM FOR ISLAND ECOSYSTEMS

Integrated farming system models have been developed for the Andaman and Nicobar Islands. Major part of Andaman and Nicobar Islands comes under medium slope category. The topography is mainly characterized by undulating sloppy lands. Present cropping system in this region is raising of fruits trees such as banana, papaya, mango, guava, pineapple, sapota, pomegranate, lemon, lime, sweet orange, mandarin, jack fruit, custard apple etc. Among the plantation and spice crops, coconut, arecanut, cashewnut, black pepper, nutmeg and clove are mainly grown. Rice is predominantly cultivated during the rainy season (May to December) and field crops such as green gram, black gram, cowpea, sesame, groundnut, vegetables, etc. are grown in the dry season on a very limited scale in rice fallows. Animal husbandry is in its infant stage. Fishery resources are very rich but remain unexploited.

Some of the possible integrated farming system (Shyam Singh and Gangwar, 1989) considered to be useful for the socio-agro climatic conditions of Andaman and Nicobar Islands are discussed below:

1. **Coconut-cum-fodder-cum-milch cattle:** Mixed farming by raising fodder grass in coconut has been found profitable. Raising grasses such as hybrid

Napier alone or with leguminous fodder crops like Stylosanthes in the coconut garden can support 4-5 dairy animals. The animals supply large quantity of cattle manure which, when applied to coconut garden, improves the soil fertility considerably.

2. **Coconut-cum-fodder-cum-fish or prawn culture:** This system is suitable for tidal marshy area. The coconut palms are grown on platforms raised upto the highest tidal level and fish and prawn in channels between the coconut plantations. On platform with coconut trees, fodder crops like Guatemala grass, thin Napier and guinea grass could be raised successfully with an average forage yield of 290, 190 and 184 q/ha respectively under high salinity conditions.
3. **Coconut-cum-fish culture in salt affected lands:** Coconut grows successfully even in salt-affected paddy fields if the fields are within the approach of brackish water. The bunds of fields at water entrance gate should be raised to maintain the required level of water. Field bunds should also be raised as per requirement. In high tides, water will be exchanged and thus fish rearing can be taken up together with coconut in dry period. In rainy season, salinity-tolerant rice varieties could be cultivated in the same field with coconut and fish culture.
4. **Fruits-cum-fodder-cum-milch cattle:** The space available between the fruit trees can be utilized for growing fodder crops such as cowpea, rice bean, green gram and black gram. The fodder could be used for feeding the cattle and cattle manure could be applied to the fruit trees and fodder crops.

CHAPTER 17

Farming Systems Rescarch (FSR) For Different Situations

FSR FOR DEGRADED LANDS

A chain of soil conservation demonstration and training centres was set up during the 1950s to develop soil and water conservation systems to increase productivity and reduce erosion from the degraded lands of rainfed areas. These centres evolved methods for the mechanical measurement of erosion control, and developed effective land use systems based on the best engineering, agronomic, and soil and water management practices. They also trained staff in soil conservation, although their research was limited in scope and concept. In 1967 these centres were transferred to ICAR and one of the leading centres, at Dehra Dun, was converted into a Central Soil and Water Conservation Research and Training Institute with seven regional stations. The research achievements of these centres over the 15 years from 1956 to 1971 have been reviewed by Tejwani *et. al.* (1975). They showed that although the research undertaken was on multidisciplinary lines, research on alternative land use systems and integrated farming systems based on the watershed concept became more prominent after 1971, but have yet to make a significant contribution to real farm situations. The data on national average yields for the period from 1971-84, compared with the average and highest yields from National Demonstration plots is given in Table 17.1.

TABLE 17.1: National average yields 1971-84 (t/ha), compared with the average and highest observed yields from National Demonstration plots

Crop	National average yield	National demonstration average	Highest yields from national average demonstrations
Paddy	1.8	5.1	11.5
Wheat	1.5	3.7	7.6
Maize	1.1	3.3	7.5
Sorghum	0.6	3.7	7.1
Pearl millet	0.5	2.6	4.8

The watershed management approach has been effective in improving the productivity of land by reducing erosion though the introduction of appropriate land use systems, but it has proved difficult to obtain the cooperation of small farmers who own small parcels of land in the watershed. However, when the social aspects have been effectively managed, as in Sukhomajri and Nada villages in Haryana, the equitable distribution of the benefits of watershed management has ensured the acceptance of the system even by small farmers (Mittla Gurmel Singh, 1990).

FSR FOR WASTELANDS

Cultivable Wastelands

The land, which is capable or has potential for the development of vegetative cover and not being used due to different constraints of varying degrees, such as erosion, water logging, salinity etc. The lands can further classified into:

a) Lands, which could be reclaimed with reasonable investment on inputs and labour:
   i) Gullied and degraded land.
   ii) Scrublands with rough pastures ---- slope and elevation
   iii) categories (undulating stony graveled wastelands).
   iv) Surface waterlogged land (high water table, temporary flooding).
   v) Salt affected land (alkaline).
   vi) Shifting cultivation area.
   vii) Degraded forest land.
   viii) Sands-Coastal lands, semi-desert lands.
   ix) Mining and industrial wastelands.

b) Land that could be improved with higher investments of inputs and labour;
   i) Marshes of Permanent nature.
   ii) Desert sands.
   iii) Ravenous land.

Uncultivable Wasteland

These lands cannot be developed for vegetative cover, for instance barren rocky areas and snow-covered glacial areas. These are classified into:

a) Rocky outcrop or stony waste.
b) Snow or ice covered areas.
c) Salt pans and salt marshes.

PASTURE LANDS

Reseeding Methods

Natural succession of desired grass species is a very slow process in arid regions. Reseeding of pasturelands with perennial grass species suiting to the region is the only solution for upgrading the degraded lands. The best sowing time is end of June or after the first effective showers. Sowing of seed of *Cenchrus cliaris* at the Central Research Farm, Jodhpur in dry soil in the last week of June gave 36 per cent higher forage yield over post monsoon sowing in the month of July.

Grass-Legume Mixture

Performances of *Stylosanthus hamata* appeared to be promising in association with *Dichanthium annulatum* in rainfall zone of more than 400 mm. The total productivity of grass-legume mixture of D. *annulatum* + *S. hamata* was highest (42.0 q/ha) which was 26 per cent higher over the control. *Prosophis cineraia* is an important drought hardy-to-feed free, which grows in pasturelands without any detrimental, effect on grass grown in association with it. Silvipastural studies conducted with *Acacia tortilis*, *Azadirachta indica*, *Albizzia lebbek* and *Holoptelia integrifolia* and four grasses namely *Cenchrus ciliaris*, *C. settigerus*, *Dichanthium annulatum* and *Panicum antidotale* proved to give better economic return.

Sand Dune Stabilization

A check-board or parallel hedge system can be adopted depending upon the direction of the wind. If the prevailing wind blows from one direction or in the opposite direction then mechanical barriers can be established in parallel lines. In western Rajasthan locally available brushwood, like *Leptadenia pyrotechnoia, Zizyphus mauritiana, Lelligonum, Polygonoides, Lasiurus sindicus, Panicum turgid and Helianthus mina* have been used by cutting and burying them vertically downwards in lines 2-3m apart. In areas where wind blows from different directions a check board system (2-3 m^2) has been found to be very effective in arresting the movement of sand.

Species Suitable

1. *Prosophis chinensis.*
2. *Prosophis cineraria.*
3. *Acacia tortilis.*
4. *A. raqdiana.*
5. *Zizyphus mauritiana.*
6. *Calliconum polygonoides.*
7. *Crotalaria burbia.*
8. *Aerva javenica.*
9. *Laiurus sindicus.*
10. *Panicum turgidum.*

It was found that sand dune afforestation with *Prosophis chinensis* yielded fuel-wood of about 38 tonnes/ha after 10 years.

FARMING SYSTEM FOR ARID LANDS

Studies conducted at CAZRI Jodhpur, revealed that growing of pearl millet as single crop after 7-9 months fallow period proved to be most productive (17.2 q/ha). Years receiving more than 800 mm rainfall with an extended rainy season favour double cropping. An analysis of 80 years rainfall data revealed that 20 years *i.e.* one out of four were such which experienced more than 500 mm of rainfall with an extended monsoon and showed promise of double cropping. It was found that pearl millet followed by mustard gave the highest total productivity and gross monetary returns. Among the intercropping systems tried, planting of one row of pearl millet in between paired rows of mung bean or cluster bean resulted in highest productivity.

Horticultural Crops in Arid Land IFS

Criteria for selecting fruit crops are:

1. Fruit crops chosen must have flowering and fruit development synchronous with moisture availability.
2. They must have deep and strong root system to pierce the hard pan and to extract moisture from deeper layers of soils.
3. They should be tolerant to salinity in soil and saline irrigation water.

Fruit crops like ber Zizipus mauritiana, Pomegranate (Punica granatum), custard apple (Annona squamosa) gonad (Cordia dichotoma), karonda (Carissa carandas), amla (Emblica officinalis), and sapota (Arches sapota) and guava (Psidium sativa) are suitable.

Tree Farming or Wood Lots

In the semi-arid areas (500-1150 mm of rainfall), tree farming as a component of the farming system is a sure way of reasonable productivity and income. The tree species found suitable are:

1. *Acacia abide.*
2. *A. militia.*
3. *A. nilotica var. cupressiformis.*
4. *Annona squamosa.*
5. *Albizia labbek.*
6. *A. indica.*
7. *Casuarina equisetifolia.*
8. *Dalbergia sisso.*
9. *Gliricidia sepium.*
10. *Grevillea robusta.*
11. *Leucaena labusta.*

Range Management

Improved grass cultivars with high yield potential and good quality forage are now available. Of the 5 grasses *Cenchrus ciliaris*, *C. setigerus*, *Chloris gayana*, *Panicum antidotale* and *Urochloa mosambicansis* tried for three years under semi-arid conditions at Hyderabad. *Cenchrus setigerus* yielded the highest amount of fodder. *Stylosanthes hamata*, *S. scabra* and *S. viscose* also performed well.

Contingency Planning

The following aberrations can be observed in many situations on analysis of rainfall:

a) Early onset of monsoon.
b) Late onset of monsoon.
c) Early recession of monsoon.
d) Breaks in monsoon.

Crop selection should be done in such a manner to suit the above situations.

Ley Farming

This system fits well where moderate rainfall (500-700 mm) is received but the soil is deep (770 mm). Studies revealed that inclusion of *stylosanthes hamata* as a pasture legume in the rotation with castor or sorghum improved the yield of sorghum in succeeding style. Grain yield of sorghum can be increased and sustained at a higher level without any application of nitrogen under dry lands in ley-farming system.

Wasteland Development

Soil is basic resource for all agricultural activities. However, over time as biotic pressure is built up beyond the carrying capacity of this resource, there is misuse or over use of crop land. On account of disturbed hormony of natural environment, there are increasing instances of soil degradation. The reasons for land degradation may be due to inherent or imposed disabilities or both such as location, environment, chemical and biological properties and even suffer from management conditions. The explosive increase of population, increasing desire of man to exploit marginal and sub-marginal lands, pressure from industry for its raw material needs, defective land use and improper irrigation management emerged as the main factors causing land degradation. At some places, degradation has reached disastrous proportions and if not corrected in time, will have a devastating effect on our future population.

Failure in the restoration of degraded areas and increase in the population may reduce the available land area for productive use. Due to environmental destruction, the weaker section, particularly the rural poor and tribes, are affected mostly because their life revolves around constant search for food, fodder, fuel and water. As ecological conditions worsen, their survival becomes increasingly difficult.

Late Prime Minister of India, Sri Rajiv Gandhi, in his broadcast address to the nation on 5 th January, 1985 said 'Continuing deforestation ha brought us face to face with a major ecological and socio-economic crisis. The trend must be halted. I propose immediately to set up a National Wasteland Development Board with the objective of bringing 5.0 million ha of land every year under fuel wood and fodder plantations'. A Technical Task Group was constituted by the Planning Commission and National Wasteland Development Board. The primary task was comprehensive estimation of wastelands in the country and to propose standardized definitions for various categories of wastelands.

According to dictionary meaning, wasteland is the area which is unfit for use. The lands which are unproductive and least utilized are termed as **wastelands**. This does not mean that they are unfit for use. Due to neglect or degradation, these lands are not utilized to their full potential. The Wasteland Survey and Reclamation Committee (1961) defined wastelands as those lands which are either not available for cultivation or left out as fallows and cultivable wastes.

According to the National Wasteland Development Board, wastelands are defined as the lands which are currently lying unutilized due to different constraints. These include culturable and unculturable watelands. The wastelands are also termed as **degraded Lands**

Classification of Wastelands

Out of 329 million ha of total geographical area in India, about 158 million ha comes under different kinds of land degradation. Wastelands are broadly grouped into culturable and unculturable wastelands.

Culturable Wastelands

The lands are capable or have the potential for the development of vegetative cover. However, they are not being used due to different constraints of varying degrees such as erosion, water-logging, salinization etc. Most of these lands can be put to some productive use after proper treatment and reclamation measures. Actually these lands are problematic lands instead of wastelands.

Unculturable Wastelands

The lands that can not be developed for vegetative cover are called unculturable wastelands *e.g.*, barren rocky areas, snow covered glcial areas.

Classification of wastelands based on National Commission on Agriculture (1976), Ministry of Agriculture and Irrigation, Government of India is given in Table 17.2.

TABLE 17.2: Classification of wastelands

Category	Spread
Raverine land, Coastal sandy areas, High altitude and steep sloping areas	Coastal region, interior mountainous regions, steep slopes in Himalayas
Salt affected areas	Semi-arid and Indo-Gangetic plains and other semi-arid areas in Peninsular region.
Waterlogged areas	Low-lying areas in high rainfall regions and irrigated areas.
Ravine land	Gujarat, Rajasthan, Madhya Pradesh and Uttar Pradesh
Shifting cultivation	North-Eastern region.
Barren hill ridges and rock outcrops	Distributed in various states of the country.

National Wasteland Development Board (1987) classified as follows:

Utilizable/Culturable Wastelands

1. Eroded lands include gullied and/ ravenous land
2. Undulating upland
3. Waterlogged land
4. Salt-affected areas include saline and alkali soils
5. Shifting cultivation areas
6. Degraded forest lands
7. Degraded non-forest plantation
8. Sand dunes, sand bar and coastal sand
9. Minespoils and industrial wastelands
10. Strip lands
11. Pasture and grazing lands

Unculturable Wastelands

1. Barren rock/stony wastes/sheet rock areas
2. Steep sloppy areas
3. Snow covered and/ glacial area

MANAGEMENT OF WASTELANDS

The characteristics of the wastelands and their development vary with the categories. Agro-forestry system plays greater role in the development of culturable wastelands.

Raverine Land

Rivers originating from Himalayas have twin actions of erosion and deposition. Depending on the course of rivers and speed of flow, heavy load of sand and silt is carried by the gushing streams. They deposit large quantities of sediments on way to the sea. Gradually, land masses are formed which remain for some part of the season atleast above the flood level. Raverine lands are called as '*khadar* 'in North India, '*diara*' in Eastern Uttar Pradesh and '*char*'in West Bengal. In Punjab and Haryana, they are called '*Rao*' and 'cho' which are most common in branched or multiple channel rivers. Development of such land is mainly based on adoption of suitable soil conservation measures. Planting live hedge is most beneficial in controlling erosion and large deposition of sediments.

Coastal Sandy Areas

India has a long coastal line about 5600 km. Wind erosion is more serious and extensive, especially during monsoon seasons. There is blowing and shifting of sands during the monsoon and cyclonic periods. This causes heavy damage to the standing crops in the neighbouring areas. The problem is acute in Eastern coast of Orissa and Western coast of Keral. In such areas, afforestation is necessary as it acts as a moderator of the effect of cyclone. All along the coastal areas, wide forest belt is to be created for moderating the effects of cyclones. Casuarina is an important tree suitable for coastal belts as it can stand high degree of salinity and moisture stress. It also acts with least attention. Casuarina plantation on the sandy foreshore should be backed by plantation of other adaptable tree species further inland. These plantations can be maintained upto 10 metres width along the coast. The inner areas can be planted with cashew trees interspaced with coconut.

The species such as *Salvadora persica*, *S. oleoides*, *Atriplex nummularia* and *Juncus regid* are suitable for saline coastal regions. *Simmondsia chinensis* (Jojoba), *Simaroba spp* and *Prosopis spp* are highly adopted for sand dune areas. Grasses viz., *Panicum maximum*, *Cenchrus spp*, *chloris spp*, *Spinifex squamosus*, *Sporobolus spp* and *Cynodon dactylon* are most suited.

High Altitude and Steep Sloppy Areas

In high altitude region of Himalayas, the soils suffer from excessive run off during monsoon, causing floods and there is moisture deficiency during the dry period. Indiscriminate grazing by cattle, especially sheep and goat including the nomadic herds, are the major causes for soil erosion in hilly areas. Hedge row planting is required under hilly terrain. Nitrogen fixing tree species and shrubs suitable for agro-forestry may be planted at close spacing. The space between trees and within the rows may be utilized for growing hedges with perennial grasses. Tree species such as *Acaia modest*, *A. catechu*, *Prosopis juliflora*, *Populus alba*, *P. nigra*, *P. ciliate* and *P. candicuas* are best suited. In Nilgiris with land slope of above 33% perennial vegetation preferably with silvipasture and tea plantation is recommended.

Salt-Affected Soils

Salt-affected soils may be classified as saline soils and alkali soils. Saline soils are predominant in arid, semi-arid and coastal areas. The alkali (sodic) soils occur in the Indo-Gangetic plain and parts of Madhya Pradesh. In peninsular semi-arid tracts, sodic soils occur where the soils are irrigated injudiciously. Most of these soils have high productivity potential due to abundant nutrient reserve and ideal moisture retaining and release characteristics. Classification based on the United States Salinity Laboratory (USDA) and Soil Science Society of America (SSSA) is given in Table 17.3 and 17.4)

TABLE 17.3: Classification of salt affected soils (USDA)

Class of soil	EC (mmhos/cm)	ESP	pH
Saline	> 4	<15	< 8.5
Saline-alkali	> 4	> 15	> 8.5
Non saline-alkali	< 4	> 15	8.5 – 10.0

TABLE 17.4: Classification of salt affected soils (SSSA)

Class of soil	EC (mmhos/cm)	SAR	pH
Saline	> 2	-	-
Saline – sodic	> 2	> 15	> 6.5
Sodic	-	> 15	-

Salt Affected Soils of India

In India, about 7.0 m ha of otherwise productive land is salt affected and a large part of it has got out of cultivation. The problem is acute in semi-arid and sub-humid IndoGangetic plains. The extent salt affected areas in India is given in Table 17.5.

TABLE 17.5: Salt affected soils in India

	Group	States	Approximate area (m ha)
1	Coastal salt affected soils		
	Coastal salt affected soils of arid region	Gujarat	0.714
	Deltaic coastal salt affected Soils of the humid region	Orissa, Andhra Pradesh Tamil Nadu, West Bengal	1.394
	Acid salt affected soils	Kerala	0.016
2	Salt affected soils of the medium and deep black soil region	Karnatak, Madhya Pradesh Andhra Pradesh, Maharashtra	1.420
3	Salt affected soils and semi- arid regions	Gujarat, Rajasthan, Punjab Haryana, Uttar Pradesh	1.000
4	Alkali soils of Indo-Gangetic plain.	Haryana, Punjab, Uttar Pradesh, Bihar, Madhya Pradesh	2.500

Characteristics of Salt Affected Soils

Saline Soils

Soils containing excess amount of soluble salts are called saline soils. The predominant soluble salts are chlorides and sulphates of sodium, magnesium and calcium. These soils sometimes contain appreciable amount of nitrate. Presence of excess amount of soluble salts induce flocculation of the clay fraction and hence the physical physical properties of the saline soils are generally good. Plant growth is affected by the osmotic effect of excess soluble salts. Accumulation of specific ions such as B, Cl and Na in plants to toxic levels may often affect plant growth adversely.

Alkali Soils

Alkali soils have high Exchangeable Sodium Percentage (ESP) of above 15 which is generally considered as the limit between the normal and alkali soils. The predominant salts are carbonates and bicarbonates of sodium. The soils are also termed sodic soils. In the moist situation, there are black patches due to the dissolution of organic matter and subsequent evaporation. Surface of the soil becomes very hard during drying and turns muddy with the receipt of rain. The soils are completely disbursed due to high ESP and hence water stagnation occurs in the field. The bulk density of the sodic soils is normally (1.5-1.85 g/cm^3) and hence root penetration is hampered. The tilth of the sodic soils are poor because of high pH and ESP. Seed germination and root growth are affected due to the presence of hard surface layer. Plant growth is adversely affected due to the toxic quantities of sodium or deficiency of other essential plant nutrients like Ca and Zn.

Sources of Excess Salt in the Soil

1. The salts get deposit due to impeded drainage, as the soil may have developed from the parent material having excess salts originating from weathering.
2. Irrigation water may contain salt and by continuous use of such water, salt accumulates and is deposited in large areas.
3. Salt accumulates on the surface due to capillary movement from the ground close to the soil surface.
4. Excess salt accumulation occurs due to high sub-soil water table, aridity and poor drainage.
5. Back water flow or intrusion of sea water also causes sodicity in the coastal areas

Management of Salt Affected Soils

Management of salt affected soils depends on the climate, soil characteristics, including nature of the soils, available water resources and associated features in different regions.

Reclamation

Reclamation of sodic soils basically requires replacing excess sodium by calcium on the soil exchangeable complex. This is done through application of amendment and leaching out of the exchanged Na from the root zone. Gypsum ($CaSO_4$-$2H_2O$) is the cheapest and the most commonly used amendment for reclamation. Gypsum dissolves to a greater extent in sodic soils due to high ESP. The solubility is also found to increase with the fineness of the particle. The quantity of gypsum to be applied will depend upon the pH and texture. Gypsum @ 15 t/ha may be required to reclaim the affected soil. Combined application of gypsum and organic like green manuring (daincha), is found to be better than inorganic amendments alone.

Leaching with good quality water is necessary to leach down or remove the excess salt at the surface and the sodium released from the exchange complex. Proper drainage is important to keep the salt below the root zone.

Agro-forestry for Salt-affected Areas

Utilization of problematic soils for cultivation of forest trees is a desirable alternative since the crop production on the salt affected soil is uneconomical. In places where the ground water is saline and scarce in quantity as in arid and semi-arid regions, agro-forestry could be practiced. Afforestation and pasture development are the two ways of managing the salt affected lands when other measures are considered uneconomical. Putting these lands under multipurpose tree species will fulfil the twin objectives of biological amelioration and harvesting biomass to meet the fuel, fodder and timber requirements. In general plantations for fuel wood are rated better for saline soils than timber wood species.

- **Choice of tree species** The following tree species are recommended for sodic soils: Prosophis chinensis, P. juliflora, Acacia nilotica, A. catechu, A. auriculiformis, A. tortilis, A. arabica, A. repens, A. aneura, Tamarix articulate,

Casuarina equisetifolia, Eucalyptus hybrid, Albezia lebbeck, Azadirachta indica, Terminalia arjuna, Dalbergia sisso, Salbadora persica, Leucaena leucocephala.

- **Grass species:** Grasses can also be grown in saline/sodic soils for reclamation and economic returns. Karnal grass- Diplachne fusca, Rhodes grass- Chloris gayana, Bermuda grass- Cynodon dactylon, Para Grass-Brachiaria mutica, Blue panic- Panicum species, Hybrid Napier- Pennisetum purpureum, Guinea grass- Penicum maximum, Kallar grass- Leptochloa fusca, Alkali sacaton- Sporobolus spp
- **Fruit trees:** Among the fruit trees, sapota is found to be the most promising. Date palm is highly salt tolerant. Guava, ber, pomegranate and custard apple are also suitable fruit trees for salt affected areas

Management of Trees in Problem Soils

Tree species can be grown successfully on the sodic soils by treating the soil of the planting pit with gypsum (3-4 kg/pit), FYM (25 kg/pit) and fertilizer application. This practice gives better response than replacing the sodic soil of the pit with good soil. The pH value considerably decreases throughout the profile soon after the application of amendment. Incorporation of gypsum and farm yard manure in the soil of the planting pit also helps in checking subsequent deterioration of the soil as a result of salt movement from the surrounding salt affected areas.

Auger Hole Planting

This technique has been developed for planting trees in alkali soils. In this method, 100-150 cm deep and 20-30 cm diameter auger holes are made with tractor operated auger and trees are planted. It has been found that height, girth, growth and fuel wood production of *Prosopis* were more with auger hole planting.

Sub-surface Planting

In this method, saplings are planted in the trenches or furrows of 30 cm deep and these trenches are also used for irrigation. A maximum salt builds up in the surface soil, planting saplings in the furrows is found highly suitable fpr saline soils.

Waterlogged and Marshy Lands

In the waterlogged areas, stagnation of water occurs for most part of the year because of imperfect drainage. In the coastal regions, water logging may be caused by intrusion of sea water in the lands below mean sea level as found in the parts of West Coast. Large extent of fields in the canal irrigated areas become unproductive due to water logging. The Central Board of Irrigation defined water logging as an area where the water table rises to an extent that the soil pores in the root zone of the crop becomes saturated resulting in restriction of the normal circulation of air, decline in the level of oxygen and increase in the level of carbon dioxide.

Marshy lands are permanently or periodically inundated by water and characterized by vegetation which include grasses and weeds.

Proper drainage is important to reclaim these areas. Tree species suitable for the waterlogged areas are: *Eucalyptus robusta, Syzygium cumini, Salix spp.,*

Populus nigra, Terminalia arjuna, and Acacia nilotica. In marshy places, trees species best adopted are: *Baringtonia spp. Bischofia javania, Eucalyptus robusta, E. rudis, Lagerstroemia flosreginea* and *Casuarina equisetifolia.*

Gullied and Ravenous Land

Gully erosion is described as 'Cancer of the land" and is considered as the severest form of erosion. A narrow deep gash formed as a result of localized and concentrated run off by running water is called gully. Very deep (>9 m) and wide gullies with steep sides are known as 'Ravine'. All gullies and ravines contribute considerably to soil erosion. The ravines are characterized by the absence of vegetation of any type. Run off is maximum, slopes are steep and gullies are of varying depth, few metres to 100 metres or more. The ravines are classifieds as shallow, medium and deep based on the gully depth, bed width and area. Ravenous lands are mainly confined to the states of Uttar Pradesh, Madhya Pradesh, Raasthan and Gujarat along the banks of Yamuna, the Chambal, the Mahi and Sabarmati and tributaries. In addition to the existing areas, about 8 thousand ha are added every year due to faulty management practices.

Afforestation

Small and narrow gullies are put under agro-forestry. Silvipasture is suitable for medium and shallow gullies.

The deep and narrow gullies are best controlled by putting them to permanent vegetation. Deep gullies are utilized as woodlands by planting suitable tree species. Gully heads and sides are stabilized by planting grass species. Reducing the slopes of gullies and diverting the surface flow are essential pre-requisites for any successful afforestation programme. Constructing gully plugs in the bed of main and branch gully is important to conserve soil and moisture. Suitable species for afforestation of gully and ravenous land are given below:

- **Trees and Shrubs:** Acacia nilotica, Acacia catechu, Azadirachta indica, Albizia lebbeck, Dendrocalamus stratus, Dalbergia sissoo, Prosopis chilensis syn, P. juliflora, Eucalyptus spp.
- **Life hedges:** Impomea carnea, Agave Americana, Arundo donax, Vitex negundo.
- **Grasses and legumes:** Cenchrus ciliaris, Pennisetum peiecellatum, Saccharum spontaneum, Dicanthium annulatum, Cynodon dactylon, Heteropogon contortus, Stylosanthes species.

Shifting Cultivation

Shifting cultivation is a primitive cultivation technique and is a transition between food gathering and food production. Shifting cultivation is dominant in the tropical and sub-tropical high land regions of the world, having high rainfall, steep slopes and shallow soils. It involves 10% of population and 30% of exploitable land of our planet. In India, this system is prevalent in several parts of the North-Eastern Hill regions of Arunachal Pradesh, Meghalaya, Mizoram, Manipur, Nagaland and Tripura. It is also prevalent in some parts of Assam, Orissa, Madhya Pradesh, Andhra Pradesh and Bihar. Shifting cultivation is popularly known as '*Jhuming*' in the North Eastern zone of India. It is called as '*Khallu*' or 'Kurwa'in Bihar, '*Dahiya*' or '*Podo*' in Orissa, Madhya Pradesh and Andhra Pradesh. Shifting cultivation is also known as '*Slash and burn*' or '*Swidden*' cultivation. It is estimated to cover an area of 4.36 million ha. Shifting cultivation is defined as the practice of cultivating clearings scattered in the reservoir of natural vegetation (forest, grass, woodland) and abandoning them as soon as the soil is exhausted and this includes the practice of shifting homesteads in order to follow the cultivator's search for new fertile area.

The practice of shifting cultivation consists of three phases: Land clearing, Cultivation phase and Restoration phase.

Land Clearing

Forest vegetation is felled down at various successional stages on the hill slope during winter months of December and January. The slash are then allowed to dry and burnt during February to April for nutrient enrichment, moderation of pH and charcoal. Fields are cleared of boulders and stones from surface in the process on clearing and piled up forming bunds on slope edges and field boundary.

Cultivation Phase

Minimum tillage is generally adopted and practiced on contour, using hand tools and bullock drawn implements. Hill slopes under shifting cultivation are mostly cultivated during kharif season, with pure or mixed crops. The crops include Indian millet, finger millet, upland rice, pigeon pea, black gram, rice bean, niger, horse gram, etc. This cycle of cultivation varies from 3 to 5 years. The crop yield sharply declines in the successive years.

Restoration Phase

A short period of cropping generally alternated with **a** long period of fallow ranging **5** to 50 years. Fields are allowed for natural restoration.

Impact of Shifting Cultivation

Burning of vegetation in shifting cultivation alters the plant nutrient supply from organic to inorganic form. Some of the nutrients, C, N and S are lost by volatilization while burning the slash. A sizable quantity of nutrients, particularly K, Ca and Mg are lost before the onset of the first rain through blown off ash due to string winds in the drier period. During the short '*Jhum*' cycle, the soil fertility deteriorates to a very low level.

During the fallow period, soil fertility is lost due to erosion and other causes. Wrong tillage practices in shifting cultivation promote run off and soil erosion. Average soil loss due to shifting cultivation has been recorded to be 40.9 t/ha in Meghalaya. It is reported that erosion rates on 50-60% slope are 147 t of soil loss/ha under first year of shifting cultivation and 170 t/ha in the second year.

Under '*Jhum*' cultivation, some of the weed species, *Eupatorium odoratum* at lower elevation and F. *adenophorum* at higher elevation, become dominant. Shifting cultivation causes large scale damage to the forests and results in deforestation and denudation of hilly slopes. During '*Jhum*' cycle, large number of edible vegetations are cut and burnt. Shifting cultivation and deforestation have been identified as major constraints in developing sustainable crop production system in the North-Eastern region. About 19.8% of the total geographical areaof the region has become degraded or wastelands.

Management of Shifting Cultivation

It is necessary to integrate conservation measures and improved farming to stabilize production at higher level. These measures may help in slowing down resource degradation and accelerating aggravation in the system.

Clearing and Cultivation Management

The land use and farm planning should depend upon the hill land classification. Area should be treated at terrace and inter-terrace level with mechanical and vegetative conservation measures. Improved vegetative cover, land configuration management and alley farming will form the base for sustained production achieving better soil and moisture conservation and improved nutrient recycling. The terrace should be protected against erosion. Contour bunds, bench terraces and half moon terraces are the conservation measures that are required for the treatment of land. Such land uses are expected to retain about 90% of the annual rainfall with negligible soil loss. Plantation of vegetative barrier could be ideal for both erosion control and improved nutrient cycling. It is possible to get higher production of finger millet by 80% under natural barrier.

Restoration Management

Lack of restoration management results in slow build up of soil fertility normally it takes 5 years to build up vegetation and 8-10 years to restore fertility status. Retained fallow vegetation, planted vegetative barriers and integration of cultivation to restoration phase by perennial pigeon pea are successful in Orissa.

Agro-forestry Systems for Shifting Cultivation

Suitable farming systems in shifting cultivation include forestry, silvipasture/horticulture and agriculture. Three tier systems have been suggested by the ICAR centre, Shillong. In this system, trees are recommended at the hill top, horticultural crops in the middle portion of the hill slopes and opening of terraces at the lower portion for cultivation of field crops. The forestry systems are primarily based on the production of fruits (pineapple, jack fruit, tamarind, etc.), nuts (arecanut) and betel vine. These should be integrated with woody perennials which supply fuel and timber for domestic use.

Broom grass+schema systems are commonly practiced on most of the degraded boulder lands with high slopes and shallow soil depth. The silvipastural system comprising *Alnus nepalensis*, pine apple and forage crops like Guinea grass (*Panicum maximum*) or Nandi grass (*Setaria sphacelata*) coupled with Stylo (*Stylosanthes gyanensis*) could be the sustainable agro-forestry system for hills with 30-60% slope in Meghalaya.

Lower areas of hilly slope are terraced and utilized for raising field crops according to the requirement of the farmers. Crops such as rice, maize, finger millet, mustard sesame, black gram and root crops like ginger and turmeric may be grown. The terraces may be stabilized with mechanical measures and planted with fodder grasses or legumes on risers. Fodder grasses/legumes such as *Stylosanthes hamata*, *S. quyanensis*, *Clitoria ternatia*, *Setaria sphacelata* and thin napier are suitable for terrace rises.

Taungya System

'Taugya' is a Burmese word meaning hill cultivation. In this system, government lands are temporarily allotted to the cultivators. They are allowed to plant forest trees species and to raise agricultural crops in between the rows of forest tree species simultaneously for about 3-5 years. Thereafter, cultivators are shifted to another patch of land.

This system was practiced as early as 1856 in Burma (Myanmar) in establishing teak plantation. Regular 'Taungya' cultivation as a system of raising forest plantations was started in North Bengal in 1911. In India, it is now followed in Kerala, West Bengal, Uttar Pradesh, Tamil Nadu, Andhra Pradesh, Orissa, Karnataka and North Eastern Hill regions. This system provides proper use of land and maximizes production. It also facilitates employment opportunity to unemploy rural people. Important tree species and field crops suitable for 'Taungya' system are given in Table 17.6

TABLE 17.6. Important trees and agricultural crops for 'Taungya' system

Trees	Crops
West Bengal	
Tectona grandis	Paddy, maize, millets
Shoerea	Turmeric, ginger
Schima wallichii	Pineapple
Cryptomeria japonica	Hemp
Michelia spp.	
Uttar Pradesh	
Shoerea robusta	Paddy, maize, sorghum
Tectona grandis	Wheat, barley
Acacia catechu	Pigeon pea, soybean
Dalbergia sissoo	Rape seed
Eucalyptus spp	
Populus spp.	
Kerala	
Eucalyptus spp.	Paddy
Tectona grandis	Tapioca
Bombax ceiba	Ginger, turmeric

Sand Dune Management

The soils of the arid zone are mostly sandy to sandy loam with excessive permeability. In India, about 22.7% of the total geographical areas is occupied by sandy soils mainly in Rajasthan, Gujarat, Punjab and Haryana. The major characteristics of these regions are severe wind erosion, high temperature, devoid of Perennial River and sparse vegetation. Most of these areas is covered by drifting or semi-established sand dunes.

The movement of sand dunes occurs particularly during dry rainless period with high wind velocity. This seriously affects the houses, buildings, farmlands, roads, railways, canals and water courses. Sand dunes shift from one place to another due to blowing wind at high speed in regions devoid of natural vegetation. The problem is more serious in the western part of the desert in India where the vegetation is negligible and sand dune concentration is maximum. They need immediate attention for their stabilization. Some of the dunes are semi-stabilized but their crest and flanks remain acute because of cultivation activity on them or due to improper management.

Afforestation

The afforestation option varies with respect to degree of aridity, wind velocity and availability of different materials. Shifting sand dunes are managed with the following methods:

1. Protection against biotic interferences
2. Mulch barrier in parallel strips
3. Planting suitable trees, grasses and shrubs to act as wind breaks

The plant species chosen should have deeply buried rhizomes or prostrate growth with stem anchored by adventitious fibrous roots for rehabilitation of sand dunes. Suitable species for afforestation of sand dunes are listed in Table 17.7. Among the tree species *Prosopis cineraria* is considered as the king of trees in desert areas because of its natural speed in this area. However, under plantation condition, some initial care of the tree is required during the first year planting. *Dalbergia sissoo* is found more capable of establishing itself in desert ecosystem. Among the grass species, *Cenchrus ciliaris* performs better.

TABLE 17.7: Species suitable for sand dunes

Trees	Bushes	Grasses
Prosopis cineraria	*Clotalaria burhia*	*Cenchurs ciliaris*
P. chilensis	*Calligonum polygonoides*	*Lasiurus sindicus*
Acacia tortilis	*Aerva javanica*	*Panicum turgidum*
Acacia radiana	*Ziziphus* spp	
Ziziphus mauritiana		
Erianthus munja		
Dalbergia sissoo		

Wind Breaks

- Tall plants of the following species may be used as wind breaks for sand dune management.
- Dalbergia sissoo, Albizia lebbeck, Acacia nilotica, Tamarix spp., Parkinsonia aculeate.

Barren Rock Hill Ridges and Rock out Crops

The rock exposure for varying lithology is often barren and devoid of soil cover and vegetation. There will be low canopy coverage at the top of the hill primarily due to poor/restricted growth as there is no soil in the hills. Extensive rock outcrops exist in the Deccan Plateau, Eastern and Western ghats, the Aravallis and Vindhyan ranges. They are generally termed as skeletal areas were the soils are very shallow, coarse, poor, eroded and degraded. The selection of species depends on the rainfall and other climatic factors of the areas. The species suitable for barren rock hill ridges and rock out crops are given in Table 17.8.

TABLE 17.8. Species suitable for barren rock hill ridges and rock outcrops

Trees species	Shrubs	Grasses
Areas of low to medium rainfall		
Eucalyptus tereticornis	*Carrisa spinarium*	*Cenchrus ciliaris*
Dendrocalamus strctus	*Dicrostachys cinerea*	*C. setigerus*
Melia azadarach	*Indigofera articulate*	*Dicanthium annulatum*
Albizia lebbeck	*Inga dulsis*	*Panicum antidotale*
Acacia catechu	*Pennisetum pedicellatum*	

Trees species	Shrubs	Grasses
Hardwickia binata		
Ailanthus excelsa		
Cassia siamea		
Prosopis chilensis		
Dalbergia sissoo		
Area of high elevation upto 2000 m		
Robinia pseudoacacia		
Populus ciliate		
Pinus roxburghii		

CHAPTER 18

Enterprises of Farming System

In the integrated farming system, it is always emphasized to combine cropping with other enterprises/activities. Many enterprises are available and these include cattle maintenance, sheep or goat rearing, poultry, rabbit rearing, bee keeping, etc., any one or more can be combined with the cropping.

CATTLE MAINTENANCE

Rearing of cattle in India is carried out under a variety of adverse climatic and environmental conditions. The cattle are broadly classified into three groups.

1. **Draft breeds:** The bullocks of these breeds are good draft animals, but the cows are poor milkers. *e.g.* Nagore, Hallikar, Kangeyam, Mali.
2. **Dairy breeds:** The cows are high milk yielder but the bullocks are of poor draft quality. *e.g.*, Sahiwal, Sindhi, Gir.
3. **Dual purpose:** The cows are fairly good milkers and the bullocks are with good draft work capacity. *e.g.*, Hariana, Ongole and Kankrej.

Exotic breeds: The exotic breeds are high milk yielder, *e.g.*, Jersey, Holstein-Friesian, Aryshire, Brown Swiss and Guernsey.

Buffaloes

Important dairy breeds of buffalo are Murrah, Nili Ravi (Which has its home tract in Pakistan), Mehsana, Suti, Zafarabadi, Godavari and Bhadawari. Of these Godavari has been evolved through crossing coastal buffaloes in coastal regions of Andhra Pradesh with Murrah.

Housing: Each cow required 12 to 18 sq. m space and the buffaloes need 12 to 15 sq. m. It is important to provide good ventilation and an open shed of housing is always preferable. Dairy building should be located at an elevated place to facilitate easy drainage. The floor should be rough and have gradient of 2.5cm for every 25cm length.

Breeding and maintenance: The cows remain in milk for 9-10 months, the average calving interval being 16-18 months. A cow does not require more than 6-8 weeks of dry period. From the economic point of view, cow should ordinarily be bred during the second and third months after calving. In weak animals and heavy milkers, 1 or 2 months may delay breeding. Cattle animals come to heat in more or less regular cycles of about 21 days, which lasts for about a day. The best time to serve a cow is during the last stage of heat. If artificially, inseminated, a second insemination may be given within 6-8 hours after the first insemination. The gestation period varies with individual cows and breeds and normally it is about 280 days.

In the case of buffaloes, the lactation period last for 7-9 months. She-buffalo comes to heat every 21-23 days. The gestation period is 310 days. Calf rearing is very important in the case of buffalo maintenance. Since they require abundant water, wallowing is required. Regular deworming is needed for buffalo maintenance.

Under Indian condition, cattle commonly mature at the age of about three years. This period can, however, be reduced by six months under well managed herd.

Feeding: Cattle feed generally, contains fibrous, coarse, low nutrient straw material, called roughage and concentrates.

Roughages: Dairy cattle are efficient user of the roughages and convert large quantities of relatively inexpensive roughage into milk. Roughages are basic for cattle ration and include legumes, non-legume hays, straw and silage of legume and grasses.

Concentrates: Grains and byproducts of grains and oilseeds constitute the concentrates. They are extensively used in dairy cattle ration. These include cereals (maize, sorghum, oats, and barley), cotton seeds, industrial waste (bran of wheat, rice, gram husk) and cakes of oilseeds (groundnut, sesame, rape-seed, soybean, linseed).

Vitamins and mineral mixtures: It is advisable to feed a supplement containing vitamins A and D. Mineral mixtures containing salt, Ca and P should also be provided in the ration.

The ration per animal per day normally includes concentrate @ 1 kg for 2 litres of milk yield, green fodder (20-30kg), straw (5-7kg) and water (32 litres).

Sheep and Goat Rearing

The system of sheep and goat rearing in India is different from that adopted in the developed countries. In India, smaller units are mostly maintained. In the developed countries, however, it is practised in large scale in fenced areas.

Goat Rearing

In India, activity of goat rearing is sustained in different kinds of environments, including dry, hot, wet and cold, high mountains or low lying plains. The activity is also associated with different systems such as crop or animal-based, pastoral or sedentary, single animal or mixed herd, small or large scale. Goat is mainly reared for meat, milk, hide and skin. Goat meat is the preferred meat in India. A goat on hoof (live goat) fetches a better price than a sheep on hoof.

Housing: Goats can be maintained under stall-fed conditions. Successful goat rearing depends on the selection of site. Goats do not thrive on marshy or swampy ground. Goats are to be provided with a dry, comfortable, safe and secure place, free from worms and affording protection from excessive heat and inclement weather. The kids are kept under large inverted baskets until they are old enough to run along with their mothers. Males and females are generally kept together. The space requirement for a goat is 4.5 to 5.4 sq.m.

Breeding and maintenance: Goat matures in about 6-7 month. Breeding is allowed for buck at one year and doe after 10 months of age. Gestation period is 145-155 days. It gives birth to 1-3 Kitts per time. Numbers of evings are three per 2 years. The kids can be weaned after 30-40 days. Mother can be allowed for mating 45-60 days after eveing. Once in five years, the buck can be changed to avoid deterioration due to inbreeding. When the young ones attain a body weight of about 25-30 kg in about nine months, they can be sold.

Feeding: The per head requirement of nutrients in respect of goats is relatively low. Hence, they are suitable for resource poor small farmers with marginal grazing lands. They are essentially browsers and eat plants, which any other animals won't touch. Goats eat 4-5 time of their body weight. Since the profit depends on weight addition, adequate proteins and calorie should be giving to goats. They eat more of tree fodder and hence 40-50% of green fodder should contain tree leaf fodder the rest with other grass species. Goats should be fed with concentrates of maize, horse gram, groundnut cake, and fish meal and wheat bran. Common salt and vitamin mixtures should also be added. Abundant clean fresh water should be made available to the goats. Water should be changed every morning and evening. Fresh water is required for digestion, blood circulation and removal of waste form the body. Water is also required for regulation of the body temperature.

Sheep rearing: Sheep are well adapted to many areas, excellent cleaners and make use of much of the waste feed. They consume large quantities of roughage, converting a relatively cheap food into a good cash product. Housing need not be elaborate or expensive. However, to protect the flock from predatory animals, the height of the fencing should be raised to two metres.

Breeds of Indian sheep: There are three types of sheep in India based on the geographical division of the country.

1. **The temperate Himalayan region:** Gurez, Karanah, Bhakarwal, Gaddi, Rampur- Bushiar.
2. **Dry western region:** Lohi, Bikaneri, Marwari, Kutchi, Kathiawari
3. **Southern region:** Deccani, Nellore, Bellary, Mandya, Bandur.

Breeding and maintenance: One ram can be maintained for 40-50 ewes. Rams are liable to fight when two or more of them are put in a pen. Unlike other farm animals, ewes in general do not come in heat at regular intervals throughout the year but are seasonal in this respect. The duration of the heat period will range from 1-3 days and 75% of the ewes remain in heat for 21-39 hours. The optimum time of service is towards the end of heat period. Average heat interval is 18 days during the breeding season. The gestation period will vary from 142-152 days with an average of 147 days. A normal ram is in full vigour for breeding from the age of 2 and half –5 years. Sheep grow fully at two years of age when the ewe is ready for breeding. Under average range condition, may be expected to produce about five crops of lambs.

Feeding: A sheep requires about 1-2 kg of leguminous hay per day depending on the age of sheep and its body weight. Proteins may be supplied through concentrates such as groundnut cake, sesame cake or safflower cake when the pastures are poor in legumes or when scarcity conditions prevail. Normally 110-225 g of cake is sufficient to maintain an average sheep in good condition. Feeding a mixture of common salt, ground limestone and sterilized bone meal in equal parts is required to alleviate deficiency of minerals in the feed.

POULTRY REARING

Poultry is one of the fastest growing food industries in the world. Poultry meat accounts for about 27% of the total meat consumed world wide, and its consumption is growing at an average of 5% annually. Poultry industry in India is relatively a new agricultural industry. Till 1950, it was considered a back yard profession in India. In the sixties, the growth rate of egg production was about 10% and it increased to 25% in the seventies. The growth rate came down to 7-8% by 1990 due to price-rise in poultry feed. By 2000, the total egg production may reach upto 5000 crores. Broiler production is increasing at the rate of 15% per year. It was 31 million in 1981 and increased upto 300 million in 1995 (Singh, 1997). Nearly 330 thousand tones of broiler meat are currently produced. The average global consumption is 120 eggs per person per year and in India, it is only 32-33 eggs per capita year. As per the nutritional recommendation, the per capita consumption is estimated at 180 eggs/year and 9 kg meat / year.

Breeds: Specific poultry stocks for egg and broiler production are available. A majority of the stocks used for egg production are crosses involving the strains or inbred lines of white Leghorn. To a limited extent, other breeds like Rhode Island red, California Grey and Australop are used. Heavy breeds such as white Plymouth Rock, white Cornish and New Hampshire are used for cross-bred broiler chickens. Hence, it is essential to consider the strain within the breed at the time of purchase. Several commercial poultry breeders are selling day old chicks in India. It is best to start with the day old chicks.

Housing: Adequate space should be provided for the birds. Floor area of about 0.3-0.4 m^2 per bird is required for heavy breeds. The house should have good ventilation and reasonably cool in summer and warm during winter. It should be located on well-drained ground, safe from flood waters.

Feed: The feed conversion efficiency of the bird is far superior to other animals. About 60-70% of the total expenditure on poultry farming is spent on the poultry feed. Hence, use of cheap and efficient ration will give maximum profit. Ration should be balanced containing carbohydrates, fats, minerals and vitamins. Some of the common feed stuffs used for making poultry ration in India are:

- ***Cereals:*** Maize, barley, oats, wheat, pearl millet, sorghum, rice-broken.
- ***Cakes/meal:*** Oil cakes, maize-gluten-meal, meat meal, blood meal.
- ***Minerals/salt:*** Limestone, Oyster shell, salt, manganese.

From the day old to 4 weeks of age, birds are fed on starter ration and thereafter finisher ration, which contains more energy and 18-20% protein. Feed may be given 2-3 times. In addition to the food stuffs, antibiotics and drugs may also be added to the poultry ration. Laying hens are provided with oyster shell or ground limestone. Riboflavin is particularly needed.

Maintenance: The chicks must be vaccinated against Ranikhet diseases with F1 Strain vaccine within the first 6-7 days of age. One drop of vaccine may be administered in the eye and nostril. When chicks get the optimum body weight of 1.0-1.5 kg around six weeks, they can be marketed for broiler. Hens may be retained for one year for production *i.e.*, upto the age of about 1 and half years. After that they are disposed off for table purpose. It may not be economical to keep the hens beyond 1 and half years since egg production would get reduced. One hen is capable of laying 180-230 eggs in a year starting from the sixth month. In addition, a laying hen produces about 230 g of fresh droppings (75% moisture) daily.

Duck Rearing

Ducks account for about 7% of the poultry population in India. They are popular in coastal and water logged states like West Bengal, Orissa, Andhra Pradesh, Tamil Nadu, Kerala, Tripura and Jammu and Kashmir. Dusks are predominantly of indigenous type and reared for egg production on natural foraging. They have a production potential of about 130-140 eggs/bird/year. Ducks are quite hardy, more easily brooded and resistant to common avian diseases. In places like marshy river side, wetland barren moors where chicken or any other type of stock do not flourish, duck farming can be a better alternative.

Breeds: The important Indian breeds are Sylhet Mete and Nageswari which are mostly found in the Eastern region of the country. Their annual production is 150 eggs/bird/ year. Improved breeds for egg and meat production are available. Khaki Campbell and Indian Runner are the most popular breeds for egg laying. Khaki Campbell has a production potential of 300 eggs/bird/year. Indian Runner is the second best producer. White Pekin, Muscovy and Aylesbury are known for meat production. White Pekin is the most popular duck in the world. It is fast growing and has low feed consumption with fine quality of meat. It attains about 3 kg of body weight in 40 days. Indigenous types, however, still continue to dominate in duck farming. Desi ducks are robust, well adapted to local conditions and free of diseases.

Housing: Ducks prefer to stay outside day and night even during winter or rains. In mild climate, it is possible to raise ducks without artificial shelter. A light fence of at least 1.2 m high enclosing the yard is enough to stop any predators. One nest of size 0.3 x 0.3 x 0.45 m to every 3 ducks is sufficient. In case of laying birds, a mating ration of 1 drake: 6-7 ducks and in meat type 1:4-5 is allowed. The duck house should be well ventilated, dry, leak and rat proof. The roof may be of thatched or asbestos sheeted. A water channel of 0.5 m wide and 0.20 m deep is constructed at the far end on both sides parallel to the night shelter in the rearing or layer house.

Feeding: Ducks normally require lesser attention. They supplement their feed by foraging, eating fallen grains in harvested paddy fields, small fishes and other aquatic materials in lakes and ponds. However, for intensive rearing, pellet feeding may be given. Ducks prefer wet mash due to difficulties in swallowing the dry mash. Hence, ducks should never have access to feed without water. During the first 8 weeks, birds should always have an access to feed. Later on they may be fed twice a day in the morning and late afternoon.

Maintenance: The general management of ducks is similar to that of the chickens. The incubation period is 28 days. A broody duck or hen may be used for small scale hatching and incubator for large scale hatching. During the early part of the life, newly hatched ducklings require warm temperature under the natural or village conditions. A duck or broody hen can take care of 10-15 ducklings. Artificial brooding may be resorted for large number of ducklings. High egg-laying strains of duck come into production at 16-18 weeks of age.

Ducks are resistant to common avian diseases. Some of the common diseases in duck are plague, duck plague, duck virus hepatitis, duck cholera and aflatoxicosis.

Turkey Rearing

Turkey is a robust bird and be reared in humid tropics. It actively feeds on a variety of crop residues and insects in the farm and hence can be raised with much ease.

Physical features: There are a few peculiar features in the physical appearance of a turkey. Male has black beard attached to the upper breast regions. This beard is small in females of white turkeys and absent in those of coloured varieties. A fleshy protuberance called snood is present near the base of the beak in both males and females. The head bears caruncles and a large flap of skin below the chin is known as dewlap. Broad Breasted Bronze breed are ordinarily seen in India. Norfolks and the Cambridge are the other important breeds.

Housing: Housing for turkeys may be of range system, confinement or a combination of both. There is a growing trend towards confinement housing. Breeds require a floor space of 0.4 m^2/bird in confinement and 32.5 m^2/bird in the range rearing.

Feeding: Turkeys have much higher growth rate compared to chickens. Hence they have relatively higher feed and nutrient requirement. Turkey can be fed with all types of dry feed in well types of dry feed comprising pellets, wet mash, grain and green feed in well balanced combination. They also need plenty of clean water.

Maintenance: Mating ratio in natural flock is 14 hens: 1 tom for small and 12:1 for medium sized turkeys. The birds should be care fully segregated when they reach the mating stage. Mating takes place during summer. Particular attention should be bestowed in regulating mating. The males should be kept away from their mates in the morning and selectively mated with other birds by turn. Mating in the evening increases the number of fertile eggs.

Generally turkey lays eggs on alternate days from six months onwards. In a year, they lay about 150 eggs. The bird attains a weight of 5-8 kg in a year. The birds should not be allowed to fatten too much as it would make them vulnerable for diseases. Changing the breeding stock periodically would be gainful for the farmers.

PIGEONS FOR MEAT

Domestic pigeon (*Columba livia)* belonging to the sub-family Columbia can be reared for its meat. Females are smaller than the males and finer in the head. They hold their tails higher than males and their pelvic bones are spaced wide apart. The males are large, aggressive and make a loud cooling noise. They strut around females during the breeding season. There is difference in the growth rate among the males and females. The former grows faster than the latter. Thus, even at an early stage, the gender difference can be easily made based on their body weight.

Feeding: Adults feed the young ones (squabs) with 'pigeon milk', a fluid that is secreted in their crop. The pigeon milk looks like thick custard and has high protein content. Pigeons require water for drinking and for bathing to remove external parasites. A pair of breeds drinks about 500 ml of water per day.

Pigeons are not fussy eaters, but do best on diets having whole grains. Nutrient requirement of pigeons is crude protein (13.5%), carbohydrates (65%), crude fibre (3.5%). The daily feed intake of an adult pair is about 120 g.

Maintenance: Each pair normally produces 14 squabs in a year. The incubation period of the eggs is 18 days. Both partners build their nest and take turns to sit on the eggs. The female starts laying eggs again when the squabs are two weeks old. The young ones become independent in 35 days. Squabs of 30-35 days age are best suited for meat purpose since recovery is high with the body weight of 350 g. in addition to the meat, pigeon droppings (25 g per bird per day) can be used as fish feed or enriched manure.

Japanese Quail

Quail belongs to class Aves, family *phasianidae* and genus *Coturnir.* Quails, initially as experimental birds, have become popular for their eggs and meat. Efforts are being made by Central Avian Research Institute, lzatnagar, to develop the guinea fowl for meat production. Males are lighter than females and are identified by the cinnamon coloured feather on the upper throat and lower breast region. Female is similar to male in colour except that the feathers on the throat and upper breast are long pointed and much lighter cinnamon in colour.

Maintenance: Quails need small space and are highly suitable for cage rearing. Therefore, it is possible to rear them even in urban areas where space is a major constraint. They are socially and commercially more adaptable due to their small body size, shorter generation interval, highly prolific nature and simplicity in rearing and are more economical. Females lay eggs as early as at the age of 5 weeks. The average age of the first egg production is 40 days. They are in full production by 50 days. They have a capacity to lay 300 or more eggs each per annum. The shell of a typical quail's egg is a mixture of greenish black and brownish speckles or patches of varying sizes. An egg weighs about 10-12 g on an average and is comparatively thin and delicate in nature. Albumen content is 65 to 59%, yolk 28-32% and shell 12-14%. The yolk content in relation to albumen is more in a quail's egg than in a chicken's egg. Special care should be given for collecting and handling the egg.

The rapid growth rate of the Japanese quail renders it an ideal broiler. At six weeks, quails attain a body weight of 125-150 g with a feed efficiency of 3.0. Japanese quails are fairly resistant to most of the common diseases affecting chicken. However, they are sensitive to abrupt environmental changes during first and second week of their life.

PIGGERY

Pigs are maintained for the production of pork. They are fed with inedible feeds, forages, certain grain byproducts obtained from mills, meat byproducts, damaged feeds and garbage. Most of these feeds are either not edible or not very palatable to human beings. The pig grows fast and is a prolific breeder, farrowing 10 to 12 pigs at a time. It is capable of producing two litters per year under good management conditions. The carcass return is high at 65-70% of the live weight.

Breeds: Imported breeds of Large White Yorkshire and Landrace are being used widely. Yorkshire is the most extensively used exotic breed in India. It is a prolific breed having good carcass quality, growth rate and conversion ability. For a small breeding farm or unit, the selection of a herd boar is extremely important. A good boar weighs 90 kg in about 5-6 months and is strong on feet and legs. The mother of the pig to be selected should have large litters of eight piglets or more.

Housing: Housing should provide maximum comfort to pigs so that their growth is optimum. There should not be dampness, dampness, draft and over heating. Locally available materials can be used for housing. One pig requires about 2.7 sq. m with a wall of 1.2 m height. Eight boars can be kept in 2.7-4.5 sq. m area with 2.4-6.0 sq. m open space.

Feeding: Feed plays an important role in successful pig production. Pigs are the most rapidly growing livestock and suffer more from nutritional deficiencies than the ruminants. Protein, carbohydrates, fats, minerals, vitamins and ample good water form a complete diet for pig. Pigs have a simple stomach; therefore, they must be fed with maximum of concentrates and minimum of roughage. The main ingredients of swine ration are cereals and millets and their byproducts. The fibre content in swine ration should be very low (around 5-6%) for better feed utilization efficiency. Mixed ration should also contain 0.5% of added salt. Swine requires comparatively higher percentage of Ca and P than do cattle or sheep. When pigs are maintained with agricultural/kitchen waste or fish and slaughter house waste, the cost of production remains low. On an average, the consumption of feed is 3.5% of total weight. Feed allowance is calculated as 2.5 – 3.0 kg/100 kg body weight + @ 0.25 kg feed per piglet with the lactating mothers.

Management: As a general rule, well developed gilts weighing about 100 kg, when 12-14 months old, may be used for breeding. The body weight is more important than age at breeding. Sows with low body weight show higher rate of fetal and per-weaning mortality and have been proved as mothers with poor nursing ability. The gestation period is on an average 114 days. Litter size at birth may be 1-16 numbers with the body weight of 1-25 kg. Normal period between birth of piglets is 10-20 minutes. Time taken for the whole process of farrowing ranges from 1.5 to 4.0 hours. Sows are weaned after 40 days. It is advantageous to cull the sow after fifth or sixth litter in a commercial herd. Weaned sows come into heat in 3-10 days after weaning and may be allowed to breed. The boar-sow ratio should be 1:15. It is profitable to raise two litters from each sow each year.

Mortality in piglets is an important cause of heavy economic loss leading to failure of pig industry. In general, one fourth of piglets farrowed die before they are weaned. Another one tenth are also categorized into stunt or unprofitable group due to disease or parasite infection. Thus about 60-65% of the piglets farrowed perform as healthy piglets at slaughter age. Death rate is high during farrowing and the first week after farrowing. The farrowing season also determines mortality rate. Mortality in new born piglets is maximum when farrowing takes place in acute cold or hot climate. Therefore, mating should be planned in such a way that farrowing could be avoided in such seasons.

RABBIT REARING

Rabbit (*Oryctologus cuniculles*) is reared all over the world in various climatic conditions. Rabbit rearing in India is of recent origin, though hunting of wild rabbits for meat is not uncommon. Rabbits can easily be reared with relatively less concentrate feed and with very high reproduction rates. Rabbit meat does not have any religious taboo, unlike pork and meat from cow.

Breeds: Rabbits can be classified into two categories as meat and wool producing ones. Important meat breeds are New Zealand White, Soviet Chinchilla and Kandinrex. Breeds for wool are Angora, Polish Bolamina, Dan and Flirorida.

Housing: Rabbits can be kept in hutches outside or in cages inside. Cages should have door (30 x 30 cm) with welded wire mesh at floor and on sides. The area required by a rabbit depends on the body weight. For one kg weight, 1.8 sq. m area is needed. Angora rabbits cannot be kept at high temperature and should be restricted to hills. Breeding bucks and adult does are housed single. Castrated male can be kept in group. Young rabbits also live in group till the age of 4 months and thereafter they are separated. Rabbit keeping requires calm and quite areas. Sudden noise can cause death of rabbits. Disturbance by dogs, cats and humans should be kept to the minimum.

Feeding: Rabbit is a non-ruminant herbivorous animal having a simple stomach. Like any other animal, rabbit requires protein, carbohydrates, fats, minerals and vitamins as its feed. Rabbits digest the fibre fraction very poorly and a diet having 5-7% crude fibre is adequate. Alfalfa and other forages are well utilized by rabbits. The feed ingredients may constitute maize (20%), alfalfa (25%), wheat bran (40%), groundnut cake (12%), mineral mixture (2%) and salt (1%). The feed mixture should be given twice a day, in the morning and evening. Other grasses and vegetables can be given intermittently in small quantities. Water should always be provided.

Maintenance: Rabbits of 5-6 months age are ready for breeding and the body weight should not be less than 2.0 kg. The female rabbit is taken to the cage of male for mating. At the time of mating, the male produces a particular noise. After that the female rabbit is taken out of the cage of the male. Early morning is the best time for mating. Rabbits mostly breed from March to July.

Gestation period is 28-32 days. The eyes of the young ones (pups) open around two weeks and at 3 weeks, they feed themselves. During this period, the pups should not be often handled. Around four weeks, the young ones are separated from the mother. After two weeks of weaning, the female can be allowed for mating.

In an year, it gives birth five times, each time producing about six numbers. Mortality in the kindlings is the most important component that determines the economics of rabbit rearing. In general, mortality in adult rabbit is very low ranging from 3-5%. However, it ranges 10-15% in kindlings up to the age of 30-42 days. Mating during extreme hot and cold season causes high mortality of kindlings.

BEE KEEPING

Bee keeping is one of the most important agro-based industries which does not require any raw material from the artisan like other industries. Nectar and pollen from flowers are the raw materials which are available in plenty in nature. Bee keeping can even be started with a single colony.

Species: There are two species, *Apis cerana indica and A. mellifera* are complementary to each other but have different adaptations. *A. cerana* is better acclimatized to higher altitudes of the Himalayan region. *A. mellifera* is more profitable in the plains.

Management: The bee- keeper should be familiar with the source of nectar and pollen within his locality. The most important sources of nectar and pollen are maize, great millet, bulrush, sunflower and palm. The beginner should start with 2 and not more than 5 colonies. A minimum of 2 colonies is recommended because in the event of some mishap, such as the loss of the queen occurring in one, advantage may be taken with the other.

The hive consists of floor- board, brood chamber, super chamber, top cover, inner cover, frames and entrance rod. These parts can easily be separated. The hive may be double walled or single walled. The single walled hive is light and cheap.

The most suitable time for commencing bee keeping in a locality is the arrival of the swarming season. Swarming is a natural tendency of bees to divide their colonies under conditions that are generally favourable for the survival of both parent colony and the swarm. This occurs during the late spring or early summer.

Honey collection: Honey should have good quality to meet the national and international standards. Qualities such as aroma, colour, consistency and floral sources are important. Proper honey straining and processing are needed to improve the quality of the produce. Honey varies in the proportion of its constituents owing to the differences in the nectar produced by different plants. The nectar collected by bees is processed and placed in comb cell for ripening. During the ripening, sucrose is converted into glucose and fructose by an enzyme called invertase which is added to it by the bees. Honey is an excellent energy food with an average of about 3500 calories per kg. It is directly absorbed into the human blood stream, requiring no digestion.

AQUACULTURE

Ponds serve various useful purposes, viz., domestic requirement of water, supplementary irrigation source to adjoining crop fields and pisciculture. With the traditional management, farmers obtain hardly 300-400 kg of wild and culture fish per ha annually. However, poly-fish culture with the stocking density of 7500 fingerlings and supplementary feeding will boost the total biomass production.

Pond: The depth of the pond should be 1.5-2.0 m. This depth will help for effective photosynthesis and temperature maintenance for the growth of zoo and phytoplanktons. Clay soils have higher water retention capacity and hence are best suited for fish rearing. Pond water should have appropriate proportion of nutrients, phosphate (0.2- 0.4 ppm), and dissolved oxygen (5.0-7.0 ppm). Water should be slightly alkaline (pH 7.5-8.5). If the pH is less than 6.5, it can be adjusted with the addition of lime at an interval of 2-3 days. Higher pH (> 8.5) can be reduced with the addition of gypsum. Application of fresh dung also reduces high pH in the water.

Soil of the pond should be tested for N and P content. If the nutrient content is less, nitrogenous fertilizers like ammonium sulphate and phoshatic fertilizers like super phosphate can be added. Organic manures such as FYM and poultry droppings may also be applied to promote the growth of phyto and zooplanktons.

Species of Fish

1. Among the Indian major carps, Catla (*Catla catla)* is the fast growing fish. It consumes a lot of vegetation and decomposing higher plants. It is mainly a surface and column feeder.
2. Rohu (*Labeo rohita)* is a column feeder and feeds on growing plants, decomposing vegetation, large colonial algae, zooplanktons and detritus to a small extent.
3. Calbasu (*Labeo calbasu)* is a bottom feeder on detritus. Mrigal (*Cirrhimus mrigale)* is also a bottom feeder, taking detritus to a large content, diatoms, and filamentous and other algae plants. Common carp (*Cyprinus carpio)* is a bottom feeder and omnivorous.
4. Silver carp (*Hypophthalmichlthys molitrix)* is mainly a surface and phytoplankton-feeder and also feeds on micro plants.
5. Grass carp (*Cyernus carpia)* is a spcialized feeder on aquatic plants, cut-grass and other vegetable matter. It is also a fast growing exotic fish.

Poly fish culture: The phytophagous fish (Catla, Rohu and Mrigal) can be combined with omnivorous (Common carp), plankton-feed (Silver Carp), mud-eaters (Mrial and Calbasu) in a composite fish culture system.

Management: The fish are to be nourished with supplementary feeding with rice bran and oilseed cakes. This will enable faster growth and better yield. Each variety of carps could be stocked to 500 fingerlings with the total of 5000-8000 per ha. This stocking will enable to get a maximum yield of 2000 to 5000 kg/ha of fish annually.

SERICULTURE

Sericulture is defined as a practice of combining mulberry cultivation, silk worm rearing and silk reeling. Sericulture is a recognized practice in India. The total area under mulberry is 240 thousand ha in the country. It plays an important role in socio-economic development of rural poor in some areas.

The climatic conditions in India are favourable for luxuriant growth of mulberry and rearing of silkworms throughout the year. The temperature in Karnataka state, major silk producing state in India, ranges from 21.2 to 30°C. Climatic conditions in Kashmir are favourable to rear silk worm from May to October.

Moriculture: Cultivation of mulberry plants is called as 'moriculture'. There are about 20 species of mulberry, of which four are commonly cultivated. They are *Morus balba, M. indica, M. latifolia.* The crop can yield well for 12 years, after which they are pulled out and fresh planting is done. Yield of mulberry leaves is 30-40 t/ha/year.

Silk Worm Rearing

There are four types of silk worm:

1. Mulberry silk worm – *Bombyx mori*
2. Eri silk worm – *Philosamia ricini*
3. Tassar silk worm – *Antheraea mylitta*
4. Muga silk worm – Antheraea *assami*

Rearing and maintenance: The fertilized moth is covered with an inverted funnel or cellule and eggs are allowed to be laid over a card board. Parasites may be removed by brushing the egg masses with a fine brush. This will also enable to obtain a uniform hatch. In a bamboo tray rice husk is spread. Tender chopped mulberry leaves are added to the tray. The hatched out larvae are transferred to the leaves. It is important to change the leaves every 2-3 hours during the first 2-3 days.

The cocoon is constructed with a single reel thread of silk. If the moths are allowed to emerge from the cocoons, the silk thread is cut into pieces. Hence, the pupa are killed 2-3 days before the emergence of moth and processed. The cocoons required for further rearing are kept separately and moths are allowed to emerge from them.

MUSHROOM CULTIVATION

Mushroom is an edible fungi with great diversity in shape, size and colour. Essentially mushroom is a vegetable that is cultivated in protected farms in a highly sanitized atmosphere. Just like other vegetable, mushroom contains 90% moisture with high in quality protein. Mushrooms are fairly good sources of vitamin C and B complex. The protein have 60-70% digestibility and contain all essential amino acids. It is also rich source of minerals like Ca, P, K, and Cu. They contain less of fat and CHO and are considered good for diabetic and blood pressure patients.

Varieties

1. Oyster mushroom – *Pleurotus sp.* (var CO 1, CO 2)
2. Paddy straw mushroom – *Volvariella volvacea*
3. White button mushroom – *Agaricus gisporus* (var A-11, Horst u3.)

Method of Production

Oyster mushroom: Take fresh paddy straw and cut into small pieces of 3-5 cm length. Soak them in water for 4-6 hours and then boil for half an hour. Drain the water and dry the straw in shade till it is neither too dry nor wet. Take polythene bags of 60 x 30cm size and make two holes of one cm diameter in the centre of the bag such that they face opposite sides. Tie the bottom of the bag with a thread to make a flat bottom. Fill the bag with paddy straw to 10 cm height. Then inoculate with the spawn. Likewise prepare 4-5 layers of straw and spawn alternatively. The last layer ends up in straw of 10 cm height. Keep this in a spawn running room maintained at a temperature of about 22-28°C and with RH 85-90%. After 15-20 days when the spawn running is completed, cut open the polythene bag and take it to cropping room and allow it to grow for 7 days and harvest the mushroom. Yield: 0.5-1.0 kg/bed.

Paddy straw mushroom: Cut the straw into long pieces of 60-90 cm and soak in water for 12 hours and sterilize 15 minutes. Arrange the straw in bundles. Lay the moistened straw bundles on the slightly raised concrete floor or on wooden platform in layers of four bundles width. Spawn or seed the beds simultaneously in each layer either by broadcasting or placing the grain spawn at different spots. Sprinkle grain dhal over each layer on the spawn. Don't spawn below the topmost layer. Maintain it at 30-35*C. Harvesting is ready after 25-30 days. Yield: 1-1.5 kg/bed.

European mushroom: It requires a complex method of preparing compost which is used as a subtract for mushroom production. Spawning is done by three methods. viz., surface spawning, layer spawning and trough spawning. Fill the trays with compost and do spawning. After spawning, compost is pressed hard to make it compact. The trays are arranged in the cropping room in tiers and cover with newspaper sheet sprayed with 2% formalin. The temperature of 20-25°C and RH of 90-95% should be maintained. After spawn running is completed in 15-20 days and do casing. Pin heads appear within 10-15 days after casing. Cropping continues for 60-75 days. Mushrooms can be harvested at button stage. Yield ranges from 6-7 kg/m^2

BIOGAS PLANT

Biogas is a clean, unpolluted and cheap source of energy, which can be obtained by a simple mechanism and little investment. The gas is generated from the cow dung during anaerobic decomposition. Biogas generation is a complex biochemical process. The celluloses material is broken down to methane and carbon-di-oxide by different groups of microorganisms. It can be used for cooking purpose, burning lamps, running pumps, etc.

Types of Biogas Plant

1. **Float dome type:** Different models are available in this category, *e.g.*, KVIC vertical and horizontal, Pragati Model, Ganesh model.
2. **Fixed dome type**: The gas plant is dome shape under-ground construction. The masonry gas holder is an integral part of the digester called dome. The gas produced in the digester is collected in dome at vertical pressure by displacement of slurry in inlet and outlet. The entire construction is made of bricks and cement. The models available in this category are Janata and Deen-Bandhu.

The selection of a particular type depends on technical, climatological, geographical and economic factors.

Size: The size of the bio-gas plant is decided by the number of family members and the availability of cattle. One cubic metre capacity plant will need two to three animals and 25 kg of dung. The gas produced will meet the requirement of a family of 4-6 members. It would suffice to have a 2 cubic metre plant to cater to the need of a family of 6-10 members.

Site selection and management: The site should be close to the kitchen or the place of use. It will reduce the cost of gas distribution system. It should also be nearer to the cattle shed to reduce the cost of transportation of cattle dung. Land should be levelled and slightly above the ground level to avoid inflow or run-off of water. Plant should get clear sunshine during most part of the day.

Generation of dung has a direct bearing on the quantity of gas generated. The amount of gas production is considerable higher in summer followed by rainy and winter seasons. Gas production would be maximum at a temperature between 30 to 35°C. If the ambient temperature falls below 10°C, gas production is reduced drastically.

Bio-gas slurry: Slurry is obtained after the production of bio-gas. It is an enriched manure. Another positive aspect of this manure is that even after weeks of exposure to the atmosphere, the slurry does not attract fleas and worms.

INTEGRATION OF ENTERPRISES

In agriculture, crop husbandry is the main activity. The income obtained from cropping is hardly sufficient to sustain the farm family throughout the year. Assured regular cash flow is possible when cropping is combined with other enterprises. Judicious combination of enterprise, keeping in view of the environmental conditions of a locality will pay greater dividends. At the same time, it will also promote effective recycling of residues/wastes.

Choice of enterprise: The basic points that are to be considered while choosing appropriate enterprise in Integrated farming system (IFS) are:

- Soil and climatic features of an area/locality
- Resource availability with the farmer
- Present level of utilization of resources
- Return/income from the existing farming system
- Economics of proposed integrated farming system
- Farmer's managerial skill
- Social customs prevailing in the locality

Enterprise Integration

Livestock is the best complementary enterprise with cropping, especially during the adverse years. Installation of a biogas plant in crop-livestock system will make use of the wastes, at the same time provides valuable manure and gas for cooking and lighting. In a wetland farm, there are greater avenues for fishery, duck farming and buffalo rearing. Utilizing the rice straw, mushroom production can be started. Under irrigated conditions (garden lands), inclusion of sericulture, poultry and piggery along with arable crop production is an accepted practice. The poultry component in this system can make use of the grains produced in the farm as feed. Pigs are the unique components that can be reared with the wastes which are unfit for human consumption. In rain fed farming, sheep and goat rearing form an integral part of the landscape. Sericulture can be introduced in rain fed farming, provided the climatic conditions permit it. Agro-forestry (silviculture and silvi-horticulture) are the other activities which can be included under dry land conditions. In the integrated system, selection of enterprise should be on the cardinal principle that there should be minimal competition and maximum complementary effect among the enterprises.

CHAPTER 19

Integrated Farming Systems (IFS) Models under Agroecosystem

IMPORTANCE OF INTEGRATED FARMING SYSTEMS IN THE TROPICS

Population in the most tropical countries is increasing (Subbian and Palaniappan, 1992 and Jayanthi, 1996) warranting to produce more food from the limited area available for cultivation. Compounding with this, the agricultural systems have encountered numerous ecological and environment constraints. The goal of sustainable agriculture can be achieved by seeking the optimal use of internal production inputs in a way that provide acceptable levels of sustainable crop-productivity and live stock production resulting in economically profitable return. It would be possible to reach this level with proportionately less use of inputs through integrated farming systems. One choice of components in the farming systems and their management must ensure that the interaction between components should be the most complementary with least competitiveness. Integrated farming system approach is not only a reliable way of obtaining fairly high productivity, but also a concept of ecological soundness leading to sustainable agriculture (Swaminathan, 1987).

Integration of Sub-Systems in Farming System

The various enterprises that could be included in the farming system are crops, dairy, poultry, goat rearing, fishery, sericulture, agroforestry, horticulture, mushroom cultivation etc. Thus it deals with the whole farm approach to minimize risk and increase the production and profit with better utilization of wastes and residues. It may be possible to reach the same level of yield with proportionately less input in the integrated farming and the yield would be more sustainable because the waste of one enterprise becomes the input of another, leaving almost no waste to pollute the environment or to degrade the resource base. To put this concept into practice efficiently, it is necessary to study linkages and complementarities of different enterprises in various farming system. The knowledge of linkages and complementarities will help to develop farming system (integrated farming) in which the waste of one enterprise is more efficiently used as an input in another within the system.

Generally, there are 3 different farming situations, *viz.*, dry land, garden land and wetland farming systems. The improved cropping systems, which form the major component of farming system were best verified and evaluated in farmers holdings at various locations in different agroclimatic conditions. Depending on the ecosystem, the enterprise mix will vary.

Wetland system	:	e.g. Rice-based system + fish culture + duck/ poultry.
Garden land system	:	*e.g.* Garden land crops with forage + livestock/poultry (Crops with fodder, mulberry + dairy+ silkworm rearing).
Dry land system	:	*e.g.* Grain/fodder crops + sheep / goat rearing + agroforestry (tree crops/agri-silvi/ agrihorti).

Arable farming results in narrow profit or less margin to the farmers, warranting development and adoption of integrated farming systems with various combinations of components. Integration of milch cow with cropping resulted in higher net returns through maintaining high yield of crop and animal than cropping alone (Arora, 1980; Siddeswaran and Palaniappan, 1990, and Chinnusamy and Purushothaman,1992). Increased net returns due to the introduction of poultry-cum-diary based farming systems over cropping alone were reported by Sirohi *et ai.*, (1980) Mohamed Ali *et. al.* (1984) and Govindan (1988). Higher net return could be obtained with duck rearing in the integrated farming systems (Sharma *et. al.*, 1985). Higher net income with rice + poultry + fish + mushroom culture under low land rice conditions (Rangaswamy *et. al.*, 1990 and Govindan *et. al.*, 1990), dairy + biogas + silviculture under garden land conditions (Rangaswamy *et. al.*, 1990 and Siddeswaran and Palaniappan, 1999) and cropping + goat rearing under rainfed situations (Rangaswamy *et. al.*,1990) were reported. Rice + fish + azolla integrated farming system was found to be the best for maximum net and per day returns over conventional cropping under lowland rice fields in Tamil Nadu, India (Balusamy and Shanmugasundaram, 1994 and Balasubramanian, 1994). Integrated farming systems involving dairy cattle became more relevant under existing ecological and socioeconomic conditions in the tropics and would provide an increased quantity of manure for cropping (Swaminathan, 1975). Vikraman Nair *et. al.* (1979) and Sundaresan (1979) found that there was little loss of nutrients from the cattle manure in integrated farming systems. Also by recycling the wastes of dairy cow and poultry birds, the value of nutrients added to the soil could be increased (Mohamed Ali *et. al.*, 1984). Rice - poultry - fish farming system was found to increase the nutrient uptake of rice and soil organic matter content (Tiwari, 1993 and Lightfoot *et. al.* 1992). Intensive integrated farming systems involving duckery-cum-fish culture provided excellent opportunity to organic recycling and farmers' dependency on market purchased inputs was also reduced (Mishra and Pasupalan, 1994). Integration of enterprises like cattle rearing, fishery, poultry, goat rearing/sericulture and mushroom cultivation with cropping could provide proper recycling of residues for deriving maximum compatibility and replenishment of organic matter.(Rangasamy *et. al.* 1995).

FARMING SYSTEM MODELS FOR THE TROPICS

Development of ecologically stable and environmentally sound integrated farming system model for different eco-region to replace or improve the traditional farming systems would maximize the productivity and profitability of the farming systems. Different farming system models developed for varied agro-ecological region are given in Table 19.1.

TABLE 19.1: Integrated farming system models for different agro-ecological regions

S.N.	Farming system model	Region	Suggested by
1.	Composite fish culture	India	Jayaseelan *et. al.,* (1988)
2.	Duck cum fish farmnig	Hangkong	Sin (1980), Woyanarovich (1980)
		Taiwan	Chen and Li (1980)
		India	Sharma *et. al.,* (1985)
3.	Pig cum fish farming	India	Jhingranand Sharma, (1980)
		India	Paulrai (1984)

S.N.	Farming system model	Region	Suggested by
4.	Poultry cum fish farming	India	Sharma *et. al.*, (1985)
		India	Kumar and Singh (1984)
		India	Govindan (1988)
		Indonesia	Djajadiredja *et. al.*, (1980)
		Malaysia	Tan and Khoo (1980)
		Malaysia	Delmando (1984)
5.	Cropping cum Mushroom	India	Rao (1984)
		India	Loganathan (1989)
		India	Rangasamy and Jayanthi (1984)
		India	Vangriensven (1987)
6.	Rice + Pig + Fish + Duck + Azolla farming	Philippines	Govina (1990)
7.	Rice + Fish + Poultry farming	Tamil Nadu	Govindan *et. al.*, (1990)
8.	Rice + Fish + Poultry + Mushroom Farming	India	Rangasamy (1992)
9.	Rice + Fish + Azolla Farming	India	Shanmugasundaram, S. *et. al.*, (1992)
10.	Rice + Fish + Dairy Farming	India	Veerabudran (1994)
11.	Crop + Dairy + Biogas + Silviculture	India	Siddeswar and Palaniappan (1990)
12.	Crop + Livestock Farming	Himachal Pradesh (India)	Sharma *et. al.*, (1991)
13.	Rice + Fish + Shrimp + Vegetable Crops	Vietnam	Lighfoot and Minnick (1990)

Farming systems research came into prominence as a part of agricultural research in the past decades with a main stimulus of growing realization that most farmers are economically rational (Simmonds, 1986). The meaning of FSR varies differently to different authors and to varied agro-ecological situations, but the objectives of the farming systems, in general, are conversing to the development of suitable location specific farm technologies to raise and sustain the total farm productivity in terms of food, fodder, fuel and to meet the felt needs of the farmers within the circle of this agro-eco-socio-political favorities and constraints (Gomez, 1988 and Norman and Colinson, 1985). FSR has been applied to a variety of activities and viewed the farm in a holistic manner and considered interactions in the system (Byerlee et al., 1982). The FSR approach is directed at small farmers in marginal areas who operate complex systems characterized by important interactions between enterprises (Gilbert et al., 1980; Maxwell, 1984; Simmonds, 1985 and Collinson, 1988).

FSR was considered to be very important in providing scientific approach for problem identification and technology development aimed at improving agricultural production systems and human welfare through increased but sustainable agricultural productivity (Plucknett *et. al.*, 1987 and Spencer, 1991). John (1991) viewed FSR as an concept open to multiple interpretations existing for the numerous local varieties of farming systems activities. FSR is a research method designed to understand the farmers' priorities, strategies and resource allocation decisions, with a view to identify and develop technologies useful for each farm environment of varied agro-climatic conditions (Kanwar *et. al.*, 1992; and Pandey *et. al.*, 1992).

Composite Fish Culture

Fish culture in inland waters as an economic and paying proposition or trade has been carried on to a great extent in Tamil Nadu, Andhra Pradesh and Kamataka. A pond fish culture was given with the innovation of polyculture or composite fish culture technique in 1970's. The technology developed under the composite fish culture programme has put India on the threshold of aquapulsion. Though this system of farming with different species of carp together was practiced in Bengal even earlier, the availability of the Chinese carps (silver carp and grass carp) and common carp for farming from midseventies changed the potentiality of pond fish production. The principle involved in this culture is fuller utilization of fishpond in a body of water. Three exotic fishes viz. grass carp, silver carp and common carp along with the three indigenous varieties viz. catla, rohu and mrigal are grown in the same pond.

The feeding habits of these fishes are follows:

Catla	:	Surface (microscopic animal or zooplankton) feeder
Silvercarp	:	Surface (microscopic plants or phytoplankton) feeder
Rohu	:	Column feeder
Grass crap	:	Aquatic vegetation feeder
Common crap	:	Omnivorous bottom feeder

Integrated Fish Farming

Integrated farming approach can be adopted to boost food production by clubbing aquaculture with agriculture, dairy, ducks, poultry, pigs and horticulture. This system combines aquaculture, animal husbandry and agriculture and forms-self-reliant integrated farming. This system recycles the organic wastes such as weeds and crops, by-products from fields, wastes from the farm itself. In this system, substances that may otherwise be considered as waste are utilized to increase fish production. Pond bottom humus is used as fertilizer for crops grown adjacent to ponds or directly on their dykes. Live stock manures (Pig, duck, goose, chicken) are used as pond fertilizers for plankton production. Agricultural by-products from wire processing *i.e.* wheat and rice, coconut or groundnut cake are also put into the ponds as fish feed. Recently anaerobic digesters are being used in integrated fish farms to produce methane gas as an energy source from agricultural wastes and manures. The nitrogen rich slurry residues are then used to fertilizer ponds. Even with the great variety of pond inputs, the limited quantity of locally available nutrients is the central factor restricting pond area expansion where integrated fish farming is, presently practiced. There are various systems of integrated fish farming and the most common are given below:

1. Paddy cum fish culture.
2. Fish cum duck farming.
3. Paddy-fish-azolla farming.
4. Paddy-fish-vegetable farming.
5. Crop husbandary and fish culture (pond system).
6. Fish cum poultry farming.
7. Fish cum pig farming.
8. Fish cum sericulture.

9. Fish cum horticulture.
10. Fish cum biogas plants.
11. Crop-Livestock-fish farming.

Integrated fish farming is a diversified and Co-ordinated way to farming or producing agricultural items and livestock in the fish farm with fish as the main product. The item produced are to be used either as source of feeds or fertilizer source of additional income or both. Judicious and wise integration of these items in a fish farm promotes the full utilization of its land area and recycling of wastes and by-products; minimize the operational expenses in feeds and fertilizer; improves the living condition of workers due to the increase of income and maintains a balanced ecosystem.

Paddy-Cum-Fish Culture

This farming system 6f growing fish in the paddy fields is commonly practiced in the coastal areas in the States of Kerala and West Bengal. However, the paddy-curn-fish culture in the freshwater paddy fields is rare. This production system has great scope in the State of West Bengal, Assam, Bihar, Uttar Pradesh, Orissa and Andhra Pradesh where only a single crop using the deep water paddy varieties is cultivated. Investigations carried out by CIFRI have indicated that the freshwater fishes can be cultivated in the specifically designed paddy fields with trenches or pools constructed within the paddy plots so that two crops of paddy (tall variety of rice in kharif and high yielding dwarf variety of rice in Rabi) and a single crop of fish, covering both the seasons can be raised. The trenches can be classified as under:

1. Perimeter type.
2. Central pond type.
3. Lateral trench type.

The depth in the trench may vary between 1 m and 1.5 m. Composite fish culture can be undertaken in these trenches with fingerlings of catla, roghu, mrigal and common carp. Deep-water paddy varieties are usually selected for_cultivation during kharif season. Enough care must be taken to avoid the spraying of pesticides during the cultivation. In order to overcome this problem of pesticides affecting the fish, integrated pest management may be adopted. Pesticides like neem-based pesticides, carbonates and selective organophosphates may be used in the low doses so that the fishes are not affected. When rice is harvested during November-December, the water recedes from the rice field and the fishes take shelter in the trenches. With the onset of the *rabi* season, the rice field is once again prepared and the seedlings of the high yielding variety of paddy such as Ratna, Jaya, etc., are transplanted. Before the application of the pesticides, an earthen ridge is erected along the periphery of the paddy plot to prevent the drainage of the pesticides into the trenches with the fishes. In this system of integrated farming, a fish yield of 700-1000 kg/ha/year is expected, giving a financial return of 20 per cent over mono-cropping of rice.

Duck-Cum-Fish Farming

The integration of ducks in fish culture is economically advantageous to fish farmers. In this type of farming, the increase in production of both the animals may be met and at the same time the input cost on fish culture operations can be reduced considerably. In duck fish integration the ducks are sources of animal protein in addition to fish. In this system of farming, an ecological balance is maintained and sufficient variety of products is obtained to meet farmer's family needs in terms of food and income. Duck-fish culture is highly renunerative since ducks could be a good addition where the turn over is quick and both fish and duck are really acceptable by a majority of the population. A series of experimental trials have helped to evolve a scientifically sound and economically viable system for integrating freshwater aquaculture with duck farming, with actual benefits to both. The droppings of the ducks can be recycled for manuring the fish ponds @ 10 to 15t/ha/year. The dropping can either be collected from the duck house or the duck houses can be constructed above the water surface of the pond itself. A total of 200320 ducks (about 1 kg in size) are sufficient to provide enough quantity of manure for fertilizing a hectare of water area of the pond. In this system, a production of 3000-4500 kg/ha/year has been achieved, besides the income derived through the sale of duck meat and eggs.

Rice-Fish-Azolla Farming

A field study was conducted at low lying wetlands of Lower Bhavani Project (LBP) Area, Agricultural Research Station, Bhavanisagar, Tamil Nadu with a view of evaluating the feasibility and economics of rice-fish-azolla farming. Experiment was conducted with two rice crop seasons (kharif and rabi) planted with rice varieties ADT 36 and ADT 38 with a total duration of eight months with one fish crop harvest. *Azolla microphylla* was applied at the rate of 2.0 t/ha as fish feed cum N fixer five days after transplanting. Calotropis, a wasteland weed, was incorporated (12.5 t/ha) as green leaf manure, 10 days before transplanting. Fertilizers were applied at the recommended level (100:50:50 kg NPK/ha). Irrigation was scheduled to maintain 5 cm depth of water throughout the crop growth period. Need based plant protection measures were taken with 5 per cent neem seed kernel extract. Fingerlings of catla, rohu and mrigal were stocked in equal ratio with a total of 3000/ha, 15 days after transplanting. Banana pesudostem with cowdung (1:1) mixture and rice bran were fed at the rate of 5 per cent of the fish body weight. Field trenches were provided with 1.0 m depth and 1.0 m width occupying 10 per cent of rice area to shelter the fish (Shanmugasundarm and Balusamy, 1993).

The results revealed that in low-lying wetlands the most profitable farming system was rice-rice-azolla-calotropis + fish farming. There was a slight decline in rice yield under rice-rice-fish farming due to the presence of fish trenches occupying 10 per cent of the area. However, the yield reduction was made to good when azolla and calotropis were introduced. The soil fertility status indicated that in general there was an increase in the available nutrient contents after completion of cycle. Even after heavy depletion, through crop removal, the nutrient status could be replenished by the addition of organic residues. This might be due to decomposition of azolla, fish manure and fish feed in the system.

Rice-Fish- Vegetable Farming

For effective restoration and recycling of organic wastes, an experiment was started to integrate rice-vegetable and fish culture in wetlands at Agricultural Research Station, Bhavanisagar, Tamil Nadu. The objective is to maximize the return per unit area and also efficient recycling of crop residues along with fish culture. The experiments consisted of rice alone, rice +rice + fish, rice-rice-vegetable + fish farming. The experiment was continued for two rice seasons with a total duration of eight months. The study area was 0.25 ha/ system. Field trenches (1m deep x 1 m wide) were provided for fish shelter. Catla, rohu and mrigal were stocked 15 days after transplanting at an equal ratio of 3000 fingerlings/ha. Vegetable crops viz., bitter gourd, ribbed gourd, beans etc. were raised along the raised bund of fish trenches. The results indicated that rice-rice-vegetable + fish culture was more remunerative and it was possible to get 2 rice crops, 1 vegetable crop in eight months in rice fields under two season rice.

Crop Husbandry and Fish Culture (Pond System)

By integrating the fish culture with rice, increased returns could be obtained. Further, the nutrients leached out from rice fields could be effectively utilized for fish culture. With a view to study the feasibility of rice-cum-fish culture under low-lying wetlands, an experiment was conducted at Agricultural Research Station, Bhavanisagar, Tamil Nadu. In one acre holding, conventional cropping system rice-rice-black gram practiced by the farmers was followed. In another one acre plot, integrated farming system (rice-rice-maize + fish culture) was practiced. An area of 0.85 acre for crop husbandry and 0.15 acre was allotted for fish culture. Fish pond size of 600 m^2 area was dug to a depth of 2.0 m. The fish fingerlings of catla, mrigal and common carp were stocked in the fishpond to a density of 3000 fingerlings/ha. The excess water in the paddy fields was drained once in four days from 10 days after transplanting of rice. After draining, irrigation was given up to 5 cm depth each time. Whenever fertilizer was applied as top dressing, the subsequent drainage was done six days after fertilizer application. The same practice was followed for plant protection chemical application. Supplementary feeding were given @ 5 per cent of body weight of fishes.

The comparative economics showed that rice-rice maize + fish culture recorded maximum net return and it increased by Rs. 6,383 and Rs. 5,795/ha over conventional cropping of rice-rice-black gram during 1991-1992 and 1992-93, respectively. Drained water increased the yield of fishes and this might be due to increased nutrients available in the drained water from the pond as revealed by increased plankton population.

Fish-Cum-Poultry Farming

As in the case of duck farming, the fish ponds can be manured with the litter from the poultry pens @ 50 kg/ha/ day every morning. About 500-600 birds can provide adequate quantity of litter for fertilizing a hectare of water area of the fish pond. A net financial return on variable cost @ 54 per cent is achieved through the sale of fish, chicken meat and eggs.

Fish-Cum-Pig Farming

The pig sites can be constructed on the embankments of the fish pond. The pig manure (faeces and urine) can be drained directly into the pond. Besides serving as rich manure, the pig excreta is directly utilized by fish as feed, as it contain about 70 per cent digestible food for the fish. Hence, in this system of integrated farming, no supplementary feed or pond fertilization is required, thus reducing the input cost to about 60 per cent for the fish farming. About 30-40 pigs are sufficient to provide adequate quantity of excreta required for manuring a hectare of water area of the fish pond. A high fish yield of 6000-7000 kg/ha/year is generally obtained. A financial return of 52 per cent on the variable cost is achieved through the sale of fish and pig meat.

Fish-Cum-Sericulture

In this system of integrating the traditional method of raising fish and silkworm together, high yield of both is ensured. The mulberry trees are grown on the bonds of the fish ponds. The droppings of the silkworms are fed to the fish as grass carp feeds directly on the chrysalides and the droppings. The excreta of the fish decompose in the bottom of the pond into ammonia, phosphorus and potassium and other nutrients, which are essential for the growth of plants. Hence, this organically rich bottom soil of the fish pond is utilized for fertilizing the mulberry trees. Thus, the fish and the plants sustain their yield through mutual benefit. Besides, the silkworm pupae and their cake after extraction of the oil can be utilized as the supplementary feed for the fish.

Fish-Cum-Horticulture

It is common to come across raising of some horticultural crops on the embankments of the fish ponds. Fruit trees like papaya, banana, etc. can be cultivated on the bunds of the pond for making the integrated farming system economically more viable. The leaves of the banana can be utilized as supplementary feed for the grass carp.

Fish-Cum-Biogas Plants

Biogas plants are usually located as a component in the integrated farming system. The slurry from the biogas plant is rich in humus and nutrients in available form helps in the production of adequate quantity of Zooplankton, when manured in the fishpond. The water manured with the biogas slurry is less acidic when compared to the ponds manured with raw cow dung.

Integrated Crop-Livestock-Fish Farming

Integrating crops, livestock and fish in smallholder farming systems has many ecological and economic advantages. Such systems are conducive to nature conservation as they promote habitat stability and diversity for wildlife living on the farms and in adjacent areas. As these integrated systems optimize the use of on-farm and adjacent resources, they encourage habitat conservation rather than destruction. Such systems are productive and profitable because they utilize waste as inputs in other enterprises within the farm, because the fishes are highly nutritious and valuable traditional food. They use microenvironments within a farm system, which add to farm productivity as security (Chambers, 1990). The following bio-resource flow models, prepared by farmers in Vietnam an Malawi, illustrate how farm enterprises are integrated to exploit different land and water resources and to substitute for external inputs. Making these models helps farmers transform their farming systems by clarifying how existing enterprises can be better integrated and how new enterprises can be better fitted into the farming system.

In Vietnam, the bio-resource flow models show how water from the trenches is used to irrigate vegetables, grown in a mulch of rice straw, on dikes made from trench mud. Immediately after trenching, chicken and cattle manure is put into the rice-field trenches to promote phytoplankton blooms for the fish and shrimp to feed on. Although the shrimps' diet is primarily natural, they are also fed during their first two months with farm-grown by-products, such as germinated rice grain, cassava flour, rice bran, coconut and groundnut oilcake, and trash from the irrigation canals. Mango and eucalyptus branches are also put in the trenches to keep out cattle and poachers, and to provide an undisturbed habitat, which shrimps need. Such integration reduces the need for external inputs. As the shrimps and fish eat rice weeds, weeding expenses can be reduced by 30 per cent with no detrimental effect on rice production, because animal manure and fish faeces fertilize the paddy.

In Malawi, bio-resource flow models show how fish ponds are integrated with other farm enterprises. The pond water is fertilized with rotten fruits of guava, papaya and avocado, and manure from cattle, sheep and goats. Leaves of leucaena, pumpkin, wild vegetables and cowpea are also put into the ponds to feed the fish, and coarse maize bran is used as a fish food when available and affordable. Apart from fish, other outputs from the ponds are fertile sediments for the vegetable gardens and water for irrigation. The main purchased input on these farms is maize bran which can be supplemented and partly substituted with yam leaves and napier grass.

IFS MODELS

In an agricultural country like India, the average landholding is very small. There is no scope for increase in farm size. The population is steadily increasing without any possibility of increase in land area. The income from cropping of an average farmer is hardly sufficient to be assured of regular income for a decent living. He has to think of other enterprises, which will complement his cropping activity. This is made possible through optimal crop and livestock mix consistent with the farm resources immediately available. Farming includes crop, livestock, poultry, fishery and tree culture. A judicious mix of one or more with cropping complements, the cropping enterprise through effective recycling of residue/waste. A carefully chosen and planned farming system, keeping in view the soil and environment conditions, will be helpful to the users in greater dividend. Farm as a unit is to be considered and planned for effective integration.

Integrated Farming System

Experience through case studies and experiment conducted at Tamil Nadu Agricultural Uuniversity, Coimbatore, revealed that by integration of complementary enterprises like livestock, fishery, agroforestry etc., additional income and generation of additional employment potential could be achieved. Hence models of integrating different components were formulated for different land types viz., dry land, garden land and wetland. The models of the different farming systems are as follows.

Model 1 (Wetland Farming System)

A model was developed for 0.40 ha (1.00 acre) considering small and marginal farmers of Tamil Nadu State (Rangasamy *et. al.*, 1988 and Ranagasamy, 1992). While programme for the cropping system, due consideration has been given to the crops under cultivation in the locality, the crop rotation that is followed, and water requirement for the fishpond. The details of each system integrated are furnished below:

Components	Allocation of land
I. Cropping	0.36 ha
II. Fisheries	0.04 ha fish pond
III. Poultry (Poultry shed is placed over the pond)	

Components	Allocation of land
Total	0.40 ha

Cropping Programme

Rice (IR-50) (May-August)	-	Rice (IR-20) (Aug. – Dec.)	-	Maize (CO 1) (Jan. – April)	0.16 ha
Rice (IR-50)	-	Rice (IR-20)	-	Groundnut (CO-2)	0.10 ha.
Maize (CO 1)	-	Rice (IR-20)	-	Gingelly (TMV-3)	0.10 ha

Fisheries

The number of fingerlings used for the proposed programme has been arrived at based on the recommended stocking density of 7500 numbers per ha of ponded water. Accordingly 300 numbers at the ratio noted against each of the following varieties were released in the pond.

Silver carp/Catla	:	120 (4)
Rohu (Column feeder)	:	60 (2)
Mrigal/Common carp (Bottom feeder)	:	90 (3)
Grass carp (Grass feeder)	:	30 (1)

Poultry

Five hundred layer chicks are sufficient enough to void required quantum to excreta to meet the feed requirement of 7500 polyculture fingerlings in a hectare of ponded water. Accordingly twenty papkok chicks were used in the study.

Control: The normal crop rotation that is being followed in this tract under low land condition is also included in the programme to compare with income under integrated farming system. The details of which is furnished below:

Rice(IR-50) (May-AUG.)	-	Rice (IR-20) (Aug. – Dec.)	-	Green gram (CO-3) (Jan. - March)	0.20 ha
Rice (IR-50)	-	Rice (IR-20)	-	Sunnhemp	0.40 ha

In rice cultivation, even with best management practice we are getting about 5 to 7 per cent ill filled or partly filled grains. This waste product of rice when it is hulled becomes a substitute of 35 per cent of poultry feed. The poultry feed cost can be reduced by about 57.5 per cent by substituting the ill filled/partly filled rice grain, maize grain, oil cakes of crop component, the lucerne grown around the fish pond and fish waste anticipated through disposal of fish. The reduction in feed cost by way of effective recycling the products of crop component, leads to reduction in the cost of production per egg. In commercial poultry farm, cost of production per egg is around 42 to 44 paise. By way of reducing the feed cost, the cost of production per egg is reduced to 19 paise, thus leading to fingerlings which solely depend on the poultry excreta grows at the rate of one kilogram per year yielding about 250 to 300 kg/year, which on marketing will be a bonus.

The growth of fingerlings was solely due to the dropping of the birds kept over the fishpond. The poultry droppings contain protein 30 per cent and fat 2.5 per cent (Table 19.2). The poultry droppings were found to enhance the nutritive value when compared to periodical fertilization and supplemental feedings of fish with groundnut cake and rice bran. Mortality of birds was almost nil till the end of the program. This may be due to the cool environment created by the fishpond. In rice cultivation even with best management practice about 5-7 per cent ill filled or partly filled grains are obtained. This waste product of rice when it is pooled becomes 35 per cent of poultry feed can be' reduced to certain extent. By converting the groundnut and sesame seed into its by product viz., oil and cake, the cake can be made free of cost to the poultry feed component and thus can reduce the cost of the poultry feed. The poultry feed cost can be reduced by about 57.5 per cent by substituting the ill filled or partly filled grain, maize grain of the farm; oil cakes of the crop component and the alfalfa grown around the fish pond. The reduction in feed cost by way of effective recycling the products of crop components, leads to reduction in cost of production of egg. It works to Rs. 0.19 as against Rs. 0.40 in the conventional poultry farm Gayanthi *et ai.*, 1994).

TABLE 19.2: Nutritive value of poultry manure (Dry matter basis)

Name of the constituent	Percentage
Protein (Crude)	30.00
Fat	2.50
Crude Fibre	13.00
Nitrogen free extract	30.50
Total ash	24.00
Minerals	
Calcium	4.60
Phosphorus	3.00
Sodium	0.40
Potassium	1.37
Copper (ppm)	61.00
Zinc (ppm)	325.00
Manganese (ppm)	291.00

The economics worked out for the system as a whole for all the five years reveal that the mean net income under integrated farming system (IFS) and conventional system were Rs. 34067/ha/yr and Rs. 17586/ha/yr respectively. The net profit was Rs. 93.33/ha/day and it was 93.7 per cent higher than conventional system. The employment opportunity in IFS was enhanced to the tune of 102 per cent in mandays over the conventional system. The result of the integrated farming system program was compared with conventional crop activity. The income obtained from the integrated farming system program was compared with conventional farming system 59.20 per cent was from cropping, 8.30 per cent from poultry, 7.50 per cent from fisheries and 25.00 per cent from mushroom. The additional net income realized from the IFS was Rs. 14126 over conventional system. The additional employment generated through farming system over the conventional system was 627 days/ha/yr.

Model II (Garden Land Farming System)

A model on integrated farming system to suit the small farmers under upland condition was developed. The following are the components of the model.

Cropping : 1.00 ha 11 A Vinayak Complex

Ragi + Sunflower – Cotton + Maize + Greengram/Cowpea	0.56 ha
Summer cotton + Maize + Wheat + Greengram / Cowpea /Sunflower / Radish / Beetroot / Bengalgram / Bellary onion	0.19 ha
Perennial grass fodder (CO-1)	0.15 ha
Perennial legume fodder Lucerne)	0.05 ha
Farm shed	0.05 ha

Dairy

Three number of graded jersey cross breed milch animals with two calves.

Biogas Plant

For effective recycling of farm and animal waste and for fuel, light and enriched manure.

Farm Silviculture

Two hundred Subabul trees along the boundary of field for fodder and timber.

A model on integrated input management to suit the small farmers under garden land conditions was studied during 1988 to 1993 at the garden lands of TNAU, Coimbatore, Tamil Nadu (Premsekar and Rangasamy, 1994). The model integrating the components of cropping, dairy, spawn production, biogas and silviculture were contemplated in. this study. (Table 19.3). The receipt, expenditures and net income of 1988 to 1993 in Integrated Farming System (IFS) and conventional cropping system (CCS) was Rs. 34584 and Rs. 13496/ha respectively. IFS thus provides an additional mean net income of Rs. 20638/ha over a period of five years by way of effective recycling. The employment opportunity in the new integrated model with proposed components was enhanced to the tune 770 man days per year over CCS. The results of five years study revealed that by way of integrating agricultural allied enterprises with crop activity, the standard of living of the farmer, income per unit area, employment opportunity for the family etc. enhanced over the CCS (Table 19.4 and 19.5).

TABLE 19.3: Treatment details of integrated farming system and conventional cropping system

Treatment details				Area (ha)
I. Integrated farming system (IFS)				
Cropping	Cotton + Greem gram (Aug.-Feb.)	Maize + Fodder cowpea (Feb.-may)	Bellary onion (May-Aug.)	0.56
	Wheat + Sunflower	Maize + Fodder cowpea	Summer cotton	0.19
	Beet root	Bengal gram	Radish	
	Grass bajra Napier (CO 1)			0.15
	Luceme			0.05
	150 trees of Lucerne and *Leucanea lecucocephala* planted on bunds			
	Farm shed			0.05
Dairy unit	3 Jersey cows + 2 calves			0.05
Spawn and mushroom production capacity 1.5 to 2 kg/day				
II. Conventional cropping system (CCS)				
	Cotton	Sorghum	Ragi	0.20

Treatment details				Area (ha)
	(Aug.-Feb.)	(Feb.-May)	(May-Aug.)	

TABLE 19.4: Receipt, expenditure and net income for five years (1988-89 to 1992-93) in garden land IFS

Years	Receipt (Rs./ha/yr)		Expenditure (Rs./ha/yr)		Net income (Rs./ha/yr)	
	CCS	IFS	CCS	IFS	CCS	IFS
1988-89	42701	79862	23887	34547	27161	10660
1989-90	33977	70252	13910	26400	36276	12490
1990-91	41987	77832	14620	26740	35655	12420
1991-92	28589	45707	10338	37845	17117	27507
1992-93	25665	101105	6975	35615	75440	28640
Mean	34584	74952	13946	32220	40369	18343

Net income in IFS (Rs, /ha/yr) = 20638
Additional net income in IFS (Rs./ha/yr) = 5054

TABLE 19.5: Contribution from different components in IFS (Garden land)

Details	Crop	Dairy	Biogas	Silviculture	Spawn	
Receipt in Rs./ha/yr	19488	33728	5247	5996	104493	
Expenditure (Rs./ha/yr)	8380	14503	2256	2578	4512	
Net income (Rs./ha/yr)	11108	19255	2991	3416	5980	
Percentage to the total	26%	45%	7%	8%	14%	Labour consumed
Man days/ha/yr	668	274	91	15	162	

Model III (Dry land Farming System)

Problems and Potential in Dry Farming

Dry land farmers are poverty stricken and their resource base is poor. They do not have dependable infrastructure to support their family and livestock. 10 to 40 per cent of rainfall is unutilized and along with it, an estimated 8.4 million tones of NPK are lost. Soil erosion has been and will continue to be a problem. Fodder shortage is another problem and because of this, there is death of draft power, which, in turn, leads to difficulty in timely completion of field operations. Dry lands are impoverished which is further compounded by poor rainfall with respect to quantity and distribution (Sivasankaran, 1994).

Despite this fragile ecosystem, scope to improve the productivity is vast. That means, there is lot of hidden yield potential. Experiments go to show that the yield of rainfed cereals could be increased 6-8 times over the farmers' practice. In marginal lands of SAT, 6 t/ha of *Stylosanthus hamata,* 3 t/ha of *Cenchrus ciliaris* (all in dry state) besides 5 t/ha of fuel wood could be obtained with the incorporation of scientific management practices in silvipasture. The yield of ber could be 5O kg/tree/annum. So much is the potential in dry lands. But unfortunately in dry farming, technologies cannot be uniformly recommended because of wider variability of rainfall from place to place.

Crop Intensification

Crop intensification can be by and large, in space dimension (inter cropping) only barring areas where there is bimodal distribution of rainfall. Intercropping is different from the mixed cropping. Depending upon the rainfall, the inclusion of intercrop aids to achieve yield stability, and under favourable rainfall, increases to yield/income per unit area.

Alternative Land Use Systems

Dry farming is at best risky. If marginal and sub-marginal lands as per land capability classification are earmarked for appropriate alternative land use systems such as silvipasture, dry land horticulture etc dry farming may be less risky. Such alternatives could rescue the farmers from reaching disaster level of income in the event of crop failures due to poor monsoon. A hint at the economics of dry land use system. While the B : C ratio with annual field crops never goes beyond 2.0, it is always more than that in dry land horticulture (Korwar *et. al.*, 1991). Similarly inclusion of subsidiary enterprises like goat or sheep rearing not only rescues the farmers in terms of crop failures but also maximizes the income for poverty stricken dry land farmers, That is why, goats and sheep are considered 'walking banks' of the farmers. Findings of an ongoing IFS experiment at Regional Research Station, Aruppukottai, Tamil Nadu, are given below (Table 19.6).

TABLE 19.6: Economics of IFS (Rs./four acres

Particulars	Expenses	Income	Net income	Additional returns from four acres
1989-90				
Control (Crop alone)	4050	9576	5526	
IFS*	9277	16966	7889	2163
1990-91				
Control	5500	9340	2840	
IFS*	5664	12060	6390	2556
1991-92				
Control	13752	16098	2346	
IFS*	22428	29981	7552	5206

IFS*: Annual crops + Tellicherry goat * fruits trees (all in 4 acres)

Dairy Project

Studies show that dairy based mixed farming practiced in marginal farmers holdings with 3 milch animals was more renumerative than other mixed farming practices which gave an annual mean net income of Rs. 48000/ha at Coimbatore, Tamil Nadu. In Paiyur, Tamil Nadu, in a 0.8 ha farm, dairy based farming gave the maximum return (Rs. 4,872/ha) and employment (518 man days) per year. Four dairy animals can be maintained in one hectare of land in such a way that two give milk throughout the year. By keeping 4 milch animals, it was clear that the benefit cost ratio of 1.61 and employment generation of 274 man days could be obtained (Chelliah and Chandrasekaran, 1989).

Biogas-Project

Based on the number of dairy animals available, a biogas plant can be constructed near the farm shed itself. The dung collected from 3-4 milk animals is enough to generate 2 m^3 capacity of biogas. This can be used for the preparation of gruel to the dairy unit, lighting two lamps in the farm-stand, mushroom and spawn production.

Spawn and Mushroom Project

From the cropping component, sorghum grain could be obtained as per the schedule, which is the basic material for spawn production. From the mother spawn, we can inoculate the sorghum seeds to produce spawn bottles by recycling the biogas plant. Mushroom cultivation is a viable proposition in the mixed farming system and the production technology is also simple. Small and marginal farm families with thatched shed and less expenditure can do the mushroom cultivation (Chelliah and Chandrasekaran, 1989). *Pleurotus sp.*, the commonly grown edible type of mushroom, can be grown easily. Paddy straw is cheap and a good substrate for cultivation of oyster mushroom. It is clear that in a small project with a production of 3 kg of mushroom the farm family can get a net income of Rs. 6,400/ha and a per day income of Rs. 35/-.

Improved Technology Available

Efforts to develop improved technologies for stepping up the production of dry lands are being made since long in the country, as a result the technologies capable of increasing productivity and income appreciably have been generated. A critical examination of different technologies reveals that the dry land technology should consist of soil and water conservation, crop management, efficient cropping system and Integrated Farming System (Rajendra, 1994).

Existing Cropping System

The normal cropping programme of the tract is sorghum and cowpea and was raised for grain purpose during Northeast monsoon period.

Improved Cropping System

In this system, field crops and both tree and forage crops were included. All the improved technologies available for ensuring a successful crop production and income of the farm in dry lands were applied.

Crop Components

(i) Sorghum + Cowpea (grain)	:	0.20 ha
(ii) Sorghum + Cowpea (fodder)	:	0.20 ha
(iii) Subabul + Cenchrus ciliaris	:	0.20 ha
(iv) Acacia senegal - tree fodder	:	0.20 ha
(v) Prosophis cineraria - tree fodder	:	0.20 ha

Goat Rearing

Six Tellichery goats (5 female and 1 male)

Technology Integrated

Taking advantage of the rain received during March and April, summer ploughing was resorted for better absorption of rainwater and eradicates the perennial weeds. Since the slope of the land is less than one per cent and the rainfall is comparatively low, broad bed furrows 1.5 m interval was formed to increase the intake of rainwater. Much care was taken in selecting the improved crop varieties. The selected varieties have high yield potential, high harvest index, flexibility in sowing time; better response to management, compatibility for crop mixture and resistance to pests and diseases. The sorghum seeds were treated with 10 per cent potassium dihydrogen phosphate and the cowpea seeds with zinc sulphate solution for drought tolerance.

These field crops were sown during third week of September, (Pre-monsoon sowing) before the onset of North East Monsoon. Optimum seed rate and depth of sowing were ensured for getting sufficient population. Application of organic manure in the form of enriched farmyard manure (750 kg/ha) and composted coir pith waste and tank silt would augment the moisture holding capacity of the soil. Sorghum seeds were treated with Azospirillium and cowpea. with Rhizobium to increase the root growth to absorb moisture from deeper layers of soil at times of soil drought. Foliar spraying of antitranspirants like Kaolin and also potassium chloride (0.5 per cent) at critical stages when crops suffered due to acute moisture reduced the yield loss by 10-15 per cent.

Tree Farming

The tree crops subabul, *Acacia senegal* and *Prosophis cineraria* were planted for fodder, timber and fuel. These perennial fodder tree crops are highly drought resistant hardy plants. In deep vertisol, fruit trees like pomegranate, sapota and guava were planted. Annual crops were raised in the interspaces of the perennial crops for about 4 or 5 years.

Water Harvesting

In the rainfed area, about 50-60 per cent of the total rainfall is received during a short period of 3 months. About 50 per cent of the rain goes as run off. Having farm pond could collect this water. A farm pond was dug in an area of 7-8 cents (280-320 m^2), which could store about 240 m^3 ($1m^3$ = 1000 lit) of water during monsoon period. The water so collected was used for pot watering the tree crops at least during the critical stages. The water could be stored in the farm pond for about 30-35 days under Coimbatore condition. For grown up trees, slopy basins can be formed so as to collect rainwater at the root zones.

Goat Rearing

Tellicherry goats was integrated with the improved cropping under dry land in Coimbatore. These goats are medium size animals with average weight of 23-25 kg capable of yielding 1.5 lit. milk per day. The goats yield more meat, exhibiting grazing habit similar to sheep. They thrive on several types of grasses and leaves and fairly resistant to diseases. Initially five female and one male goats were maintained and in course of time they multiplied to 20:1 in a area of one hectare land. These goats were stall fed under deep litter system.

Organic Recycling

Organic waste was effectively recycled in Integrated Farming System. The goats were fed with annual crops like (sorghum and cowpea) and perennial fodder trees (*Acacia senegal* arid *Prosopis cineraria*) at the rate of 3 kg of dry fodder, 3 kg of green fodder and 100 gm of concentrate per adult animals. The animals were grown in deep litter system and the manure so collected was applied to the field. We could get an average of 3 tones of organic manure valued Rs. 750/- to Rs. 1,000, which is an additional income by way of recycling of the organic waste.

Economics

The system of integrating crops, goat and trees recorded increased net income compared to the conventional cropping system in all the five years of experimentation. The above mentioned IFS gave a mean net income of Rs. 5666/ha/yr where as the conventional cropping system gave a net income of Rs. 1919/ha/yr. (Rajendran, 1994). This resulted in an additional net income of Rs.

3747/ha/yr. The per day net profit worked out was Rs. 15.52/ha in the Integrated Farming System whereas it was only Rs. 5.26 under conventional cropping system. The average net income from IFS under crop and animal component were Rs. 3085 and Rs. 2584 respectively. In IFS, 54.4 per cent of net income was obtained from crop component and remaining 45.6 per cent was from the animal component.

The average income from the crop component in IFS exceeded the average income of crop in the conventional cropping system. This was due to inclusion of various crops and trees and improvement in soil fertility due to effective recycling. This also provides employment opportunity to family labour round the year. The average labour requirement for maintenance of goat works out to 52 man days per year and it was 101 man day /ha/yr for the crop component under integrated farming system. The total employment generated under IFS was 153 man days/ha/yr whereas it was 40 days/ha/yr under conventional cropping system thus generating an additional employment opportunity of 113 man days/ha/yr.

The different farming systems developed independently, without any consideration for utilizing the wastes arising out of the industrial, agricultural and other activities of the communities are not viable. Hence, in course of time, these wastes have assumed staggering dimensions. Thus detailed investigations were undertaken for recycling these wastes more profitably under the integrated farming systems. The recent scientific advances have pave the way to blend our traditional practices of integrating various farming systems according to the present day needs. The adoption of this farming system brings down the operational cost considerably and increases the yield per unit by recycling the wastes or the byproducts of the various farming systems.

A Model Study at TNAU, Coimbatore, Tamil Nadu

One hectare of dry land was selected in Tamil Nadu Agricultural University, Coimbatore. The mean annual rainfall of this tract is 648 mm. The mean maximum temperature ranges from 29°C to 35°C with the minimum temperature range of 17°C to 23°C. The soil type is clay loam. Six goats (5 female and one male) were taken up for the study with the following cropping programme.

Cropping System

IFS

1. Sorghum + Cowpea (grain)	:	0.20 ha
2. Sorghum + Cowpea (fodder)	:	0.20 ha
3. Subabul + *Cenchrus cilliaris* grass	:	0.20 ha
4. *Acacia Senegal* (tree fodder)	:	0.20 ha
5. Prosophis cineraria (tree fodder)	:	0.20 ha

Control

Sorghum + Cowpea (grain)	**:**	**0.20 ha**

Feed for the Goat

The goats were allowed in the developed pasture viz., *Cenchrus cilliaris* grass + *Clitoria ternata* (leguminous plant) grown in the pasture land. Dry fodder was fed to the animal as stall feeding at 2 kg/day.

Crop Yield

The straw and green fodder yield from the experimental area was recorded and income calculated. The animal yield was recorded in terms of meat, the weight gained by the six animals for one year. The results of the one year experiment reveals that the net income under Integrated Farming System may be increased by increasing the herd size/ha that can be fed with the biomass produced from one hectare of land, besides increasing the employment opportunity round the year.

Model IV (Coconut based Integrated Farming System)

Coconut is mainly grown as a pure crop under rainfed conditions. In India, ten million people depend on directly or indirectly on coconut cultivation, processing and marketing. Being a perennial crop, it starts yielding from fifth year extending up to sixty years providing insufficient farm income to sustain even small families. One of the feasible way to step up production and to enhance the farm income is to grow compatible intercrops in association with coconut. In addition, part of the garden can be entitled to grow fodder crops, which in turn helps in rearing milch animals, generating conventional energy and satisfies the family requirement.

Coconut utilizes only 40 per cent of the solar radiation and 23 per cent of the soil space and remaining area is found wasteful and can be efficiently tapped through companion crops. An endeavor was undertaken in the existing coconut plantation aged six years planted with normal spacing 7.5 x 7.5 m at a density of 180 palms/ha to assess the suitability in intercrops. Out of one acre taken for study, 60 cents were alloted to crop components and remaining 40 cents left for rearing milch animals.

Crop Components

I tier	:	Coconut
II tier	:	Fruit crops (Papaya, anona, pomegranate, banana)
Vegetables	:	Bitter grourd, ribbed gourd, lab lab
III tier	:	Spices and condiments: ginger, turmeric
IV tier	:	Coriander, mentha (Pudhina).
V tier	:	Grass: Bajra Napier
Trellis	:	Betel vine and pepper

Fertilizers. were applied as per recommendations of each crops. Irrigations were scheduled once in a week except betel vine which was irrigated in alternate days. Water requirement of coconut (120 litres/day) was apportioned and shared by all crops, monitored through a water meter. Income derived is furnished in Table 19.7. The results revealed that growing compatible intercrops in coconut realized the income of Rs. 9,477.25/ha for a period of ten months over the pure coconut plantation. Additionally there was an employment generation of 400 man days.

TABLE 19.7: Economics of coconut based farming system

Sl. No.	Crop	Quantity (kg)	Amount (Rs.)
1.	Ribbed gourd	42	126.00
2.	Bitter gourd	15	60.66
3.	Coriander	-	615.75
4.	Ginger	10.2	307.50
5.	Turmeric	806	6448.00
6.	Grass	1320	1320.00

Sl. No.	Crop	Quantity (kg)	Amount (Rs.)
7.	Sunnhemp	600	600.00
Total			**9477.25**

Integrated Nutrient Management (INM) through IFS

India would need to produce an additional food grains of 100 and 160 mt. by 2030 and 2050 AD respectively, to feed its projected population. More than 85 million, out of 105 million operational holdings in India are with less than 2 ha farm size, necessiating intensive cultivation involving the modern concept of time and space by adopting heavy dose of chemical fertilizers leading to accumulation of chemical residues in the soil (Swaminathan, 1990). As agriculture becomes intensive, soil toxicity and deficiencies of various kinds threaten the sustainable production, yields start declining in high production areas, if sustainability concepts are to be integrated in different farming situations. One of the main objectives in IFS is recycling of produces/wastes among the components involved. The end product of such recycled materials is used as organic manure for crops. In this venture, some of the products are improved in their quality on recycling and thus the program provides double benefit to the farmer. A judicious mix of one or more enterprises with cropping complements each enterprise through effective recycling of waste/residues and in addition, when carefully chosen keeping in view the soil and environmental conditions.

Integrated farming system approaches are not only a reliable way of obtaining fairly high productivity with substantial fertilizer economy, but also a concept of ecological soundness leading to sustainable agriculture. The need for system approach of nutrient supply to the system as a whole to increase the fertilizer use efficiency in economizing the costly mineral fertilizers by accounting residual effect of applied fertilizers and preceding crops was greatly emphasized (Rangasamy, 1994).

Crop and Animal Waste

The productivity of livestock mainly depends on the availability of quality feed and fodder in requisite quantity. It is observed that most of the livestock population in the country in underfed. The residues of different agricultural crops constitute the major source of fodder for livestock. In India, till late 1980's the main source of livestock feed consisted of conventional agricultural residues like rice straw, wheat, sorghum and maize stalk, sugarcane trash and remains from pulses. Based on the grain:straw ratio, approximately 334.5 million tones of agricultural crop residues are available in India.

The area under green fodder is not increasing to meet the increased demand. The area under green fodder in the country is estimated to be approximately 4 per cent of the total cultivated area. Since the farmers feel that animal power is one of the costliest inputs, they switched over to mechanical power for most of the operations. Moreover, the animals meant for milch and meat are replaced by cross-bred and improved varieties, which need to be supplemented with quality concentrated feeds. Thus, the estimated total production of crop residues in 2020 AD to the tune of 447.0 m t is to be effectively utilized. The details of the crop residues in India as given by FAO (1985) projected to 2020 AD. are furnished in Table 19.8.

TABLE 19.8: Projected availability of crop residues in 2020 A.D

Crop	Quantity of residue (m.t.)	
	FAO (1985)	Projected for 2020 A.D.
Rice	118.9	160.0
Wheat	57.5	80.0
Maize	21.0	35.0
Millets	40.0	56.0
Sorghum	41.0	52.0
Sugarcane	43.0	64.0
Total	**321.0**	**447.0**

In addition to crop residues, there is a possibility for collecting the following quantities of animal voids in India (FAO, 1985).

Cow dung	:	750 m.t.
Buffalo dung	:	250 m.t.
Voids of small ruminants	:	130 m.t.
Total	:	**1130 m.t.**

At present, the voids are being utilized for fuel and as FYM/compost. The same objectives can very well be attained with boosted value through recycling is found through experimentation. When recycled through biogas unit, there is good possibility to improve the organic source of nutrients apart from generation of fuel energy, when fossil fuel is getting depleted. Apart from the major nutrient, there is good amount of increased availability of secondary and trace elements required for the excellent performance of crop communities. By virtue of adoption of one of the modem agricultural technology viz., integrated farming system (IFS) in the farm activity, there is a possibility of improving untapped potential of each and every produce by recycling with dual benefits.

The possibility of improving the quality of products/waste materials of different components involved in the integrated farming system trials were conducted at Coimbatore for lowland, upland with supplemental irrigation and rainfed land through effective recycling and integrated nutrient supply system and their results are presented.

Lowland Farming

In the lowlands IFS, cropping, poultry, fishery and mushroom enterprises are involved with a view to recycle the residue and byproducts of one component over the other. The trial was conducted for a period of five year (1992-1996) in an area of 0.4 ha (1.0 acre) considering the average holding size of marginal and small category farmers. In one acre farm, 90 cents were assigned for crop activity and the remaining 10 cents allotted to fish pond. Twenty fowls sheltered over 10 cents of fish pond to feed 400 polyculture fingerlings gave about 700 kg poultry droppings in a cropping year. This could yield about 33.7, 21.4, 10.1 kg of N, P and K respectively. In IFS, at the end of one year after the harvest of grown up fish about 4500 kg of silt was obtained from the pond with a nutrient content of 3.52, 1.38 and 1.06 per cent, N, P and K respectively. This could relieve the burden of applying equal quantity of inorganic fertilizer to the crop component of the IFS. Even after accounting the nutrient value of the voids of fowls as manure, additional benefit of 124.7 kg, 40.7 kg and 37.6 kg of N, P_2O_5 and K_2O respectively, can be achieved by recycling (Table 19.9) (Rangaswamy, 1996).

TABLE 19.9: Nutrient value of recycled poultry manure

Particulars	Content (per cent)			Nutrient added (kg)		
	N	P	K	N	P	K
Raw poultry manure	4.81	3.06	1.44	33.7	21.4	10.1
Settled silt of the pond in which poultry dropping is used as source of fish feed.	3.52	1.38	1.06	158.4	62.1	47.7
Additional benefit through recycling				124.7	40.7	37.6

Similarly, inclusion of mushroom with the production capacity of 2 kg/ day as one of the components in IFS utilizes about 1800 kg of paddy straw and could yield about 2340 kg of mushroom spent after the harvest of edible mushroom at the end of one year. The enhancement in weight in the mushroom spent is due to unharvested mycelial growth. The nutrient value of mushroom spent in IFS is given in Table 19.10.

TABLE 19.10: Nutrient value of mushroom spent in IFS

Particulars	Content (per cent)			Nutrient added (kg)		
	N	P	K	N	P	K
Paddy straw	0.62	0.24	1.72	11.2	4.3	31.0
mushroom spent	0.71	0.29	1.74	16.6	6.8	34.6
Additional benefit through recycling				5.4	2.5	3.6

Field experiments were conducted at Tamil Nadu Agricultural University, Coimbatore to identify and optimize the component linkage for lowlands of Tamil Nadu and to sustain the productivity through effective recycling of wastes from the linked components associated in the system during September, 1993-August, 1995. The components of integrated farming systems in lowlands involve crop + fish + mushroom, crop + poultry + fish + mushroom and crop + pigeon + fish + mushroom. The efficiency of the component linkages was evaluated predominantly on the basis of productivity, its income and employment generation with the possibility of utilizing recycled organic wastes as nutrient to enrich the soil fertility.

The results on productivity, economics, employment generation, energy budgeting and resource recycling revealed the advantage of appropriate component linkage for lowland integrated farming systems (Jayanthi, 1995). Experiment on enterprises linkage for lowland farming systems revealed that rice-soybean-sunflower and rice-sesame-maize cropping systems each in 0.18 ha with pigeon (40 pairs), fish (400 polyculture fingerlings in 0.04 ha of ponded water) and mushroom (2 kg/ day) was the best in obtaining higher rice grain equivalents than the conventional cropping system of rice-green gram-maize (0.20 ha) and rice-sunnhemp-maize (0.20 ha) cropping systems (Model. I). Cropping + pigeon + fish + mushroom integration earned the highest gross and net return with better per day return and benefit: cost ratio. Integration of cropping with pigeon + fish + mushroom generated the highest employment of 798 man days/ha/year with 1.16 man days from the allied enterprises linked with equi-distribution through out the year. Integration of poultry + fish + mushroom + cropping applied with recycled poultry manure sustained the productivity of soil through the addition of bio-resource residue with better NPK nutrient supply potential (Model 2).

The growth of fish cultures included was comparable among artificial feeding and the treatment with poultry and pigeon dropping feeding. Fish production was higher with artificial feeding than with other two methods (Model 3). However, the net returns and benefit: cost ratio were higher in the fish fed with poultry dropping followed by pigeon dropping.

Rice-Fish-Azolla Farming System

Field experiment was conducted at Agricultural Research Station, Bhavanisagar, Tamil Nadu to develop an integrated N management practice for rice-fish-azolla farming in wetland. Farming systems consisted of rice+rice+fish and rice-rice+azolla+fish and two levels of nitrogen (100 and 75 per cent of recommended) with and without green leaf manure (*Sesbania rostrata*) applications. In the rice-fish system, rice and fish crop were raised together (synchronous system) in rice field. Field trenches were provided with 1.0 m depth and 1.5 m width occupying 10 per cent of the rice area, for sheltering the fish. *Azolia microphylia* was grown in rice field throughout the cropping period. The dual culture method of growing azolla with rice has gained widespread adaptability, because standing water is available in rice field from seedling to panicle maturing in lowland rice fields and is effectively used as biofertilizer for rice. Azolla cultivation in rice field can improve the fish food organisms. Fish culture in the decomposition of organic matter and promote release of nutrient from soil. The excreta of fish directly fertilize the water in rice fields leading to increase in utilizable source of N to the rice crop. Integration of allied components like azolla + fish with rice in lowland farming could provide wider scope for bio-resource recycling.

Rice-rice + azolla + fish farming with 75 per cent recommended N as well as incorporation of green leaf manure resulted in higher productivity with increased economic returns and improved the soil fertility through recycling of organic residues. The quantum of organic residue addition and N added through recycling were higher in rice-rice + azolla + fish farming with *Sesbania rostrata* incorporation (Table 19.11). The unutilized fish feed, decayed azolla and fish excreta settled at the fish trench bottom had a higher nutrient-value, which can be recycled to enrich the soil (Balusamy, 1996).

TABLE 19.11: Organic residue recycling and N addition through residues

Systems	Organic residue added (kg/ha/yr)			Total N addition (kg/ha/yr)
	Crop residue	Fish trench residue	Azolla biomass	
Rice-Rice	3620	-	-	19.85
Rice-Rice + Azolla + Fish	3860	952	49800	136.29

Farming in Upland with Supplemental Irrigation

Bearing in mind the advantage of recycling of cowdung, a waste from the animal component (3 adult + 2 calves), biogas unit was linked in the IFS of upland with supplemental irrigation instead of directly utilizing it as FYM. It gave a good proportion of methane gas, a fuel supplement, apart from its enhanced manurial value of the slurry that comes out of the biogas chamber. Moreover, by this recycling, some of the weed seeds present in the raw cowdung also get killed during digestion process, thus improving the quality of the slurry over its raw materials used. The quality improvement by way of recycling the cowdung through biogas chamber was studied by analyzing the NPK content of slurry and FYM prepared utilizing the cowdung from the system. The total quantity secured from the unit over the period of 365 days has been taken on equal weight basis and the analyzed data is furnished in the Table 19.12 and Table 19.13. The net income and employment generated through this system were Rs. 20,702/ha/year and 620 additional man days/ha/year respectively. Recycling of cowdung also led to the production of 730 m^3 of biogas with the possibility of enchancing the nutrient value of NPK to the tune of 44.5 kg, 65.9 kg and 28 kg respectively in a year. Trace elements like Fe, Mn, Zn and Cu are also present in an enhanced level over FYM (Rangasamy and Premsekhar, 1994)

TABLE 19.12: Nutrient content in biogas slurry and FYM

Particulars	Nutrient content						
	Major element (per cent)			Trace element (ppm)			
	N	P	K	Fe	Mn	Zn	Cu
Biogas slurry	1.43	1.21	1.01	4200	550	150	52
Farm Yard Manure	0.94	0.56	0.72	4000	490	100	45

TABLE 19.13: Nutrient saving by recycling (in kg)

S.No.	Particulars	N	P	K	Fe	Mn	Zn	Cu
1.	Quantum of nutrient in slurry obtained (11.0 t on dry weight basis)	157.3	133.1	144.4	46.2	6.05	1.65	0.57
2.	Quantum of nutrient in FYM obtained (11.6 t) in an year	112.8	67.2	86.4	44.0	5.39	1.10	0.49
3.	Additional nutrient realized by way of recycling	44.5	65.9	28.0	2.2	0.66	0.55	0.08

Rainfed Farming

The model integrating Tellichery goat as a component with crop activity in rainfed condition has been tried for a period of five years. The data collected on the availability of organic source of crop nutrients is furnished in Table 19.14. The organic manure like litter from the goat unit can readily be used for soil application, and thus will help in enriching the soil. Goat droppings are found to be a good energy source, which can also be linked with biogas unit before it is utilized as manure. This will generate good value of gas (22 kg of goat dropping will generate one cubic meter of gas as against 30 kg of cattle dung) as well as enhance nutrient availability. Thus, through recycling in the farming systems approach, the potential of each produce can be exploited to a greater extent.

TABLE 19.14: Nutrient gain from 20+1 productive

Tellichery goats unit under deep litter system	
1. No. of kidding / year	1.5
2. No. of kids / kidding	1.5
3. No. of kids / year / adult	2.25
4. No. of kids from 20 adults / year	45
5. Mean wt. Of dropping recorded / day	Adult Kid
(excludes the pellets dropped during open grazing)	900 g 110 g

6. dropping received in an year	6800 + 1800 = 8600 kg
7. Deep litter waste obtained / year	8600 + 2600 = 11200 kg
8. Wt. of coir waste used for the stall / year	2600 kg
9. Analytical value of deep litter was obtained from the unit (per cent)	N: 1.79, P: 0.95, K: 0.82
10. Total nutrient available from the goat unit / year (kg)	200.5 106.4 91.8

The experimental evidences indicated above, dearly show that the recycled organic residues/wastes of crops and allied activities could supplement the chemical fertilizers. Moreover, this will also provide opportunity to reduce the environmental pollution by way of burning the fossil fuel in the process of manufacturing the chemicals. The soil health problem by way of dumping more and more quantities of inorganic fertilizers can also be solved to greater extent by proper utilization of organic sources as indicated above under different field situations. This programme, if properly implemented, by involving small and marginal farmers who occupy more than 83 per cent of the total holdings in this country, will certainly pave way for increasing the productivity, profitability and also will help in sustaining the potential of the production base viz., soil.

CHAPTER 20

Achievements and IFS Models in Orissa

FINDINGS OF FSR UNITS IN ORISSA

Albeitlate, FSR has attracted the attention of the scientists' world wide. This is mainly useful for small and marginal farmers with limited resources.

FSR is now in progress in all the agricultural universities in India with the financial assistance of Ford foundation/IRRI/IFAD in Bihar, West Bengal, Orissa, Tamil Nadu and by voluntary organizations like Ramakrishna Mission (RKM).

The FSR/E programme was started in Orissa during 1987-88. It was undertaken through on-farm and on-station programme by OUAT, Bhubaneswar funded by Ford Foundation. On-farm Farming System Research units were established in districts of Kalahandi, Phulbani, Balasore and Puri. Besides on-station FSR/E unit was established in Agronomy research farm O.U.A.T., Bhubaneswar to meet the research, extension and teaching needs.

After getting financial support from Ford Foundation in August, 1989 under the Farming Systems Research/Extension (FSR/E) project, work was intensified in the remotest part of Orissa where improved agricultural technology was practically not available. From 1989 to 1992 work was undertaken at larger scale in Cuttack (alluvial soils), Balasore/Mayurbhanj, (P-deficient red and yellow mixed soils), Kalahandi (light textured and black soils), and Dhenkanal and Phulbani (lateritic soils) districts of Orissa. During 1992-93 FSR/E work was continued in Balasore and Mayurbhanj districts to confirm and extend the finding of 1991-92. The experiments were also initiated at the news sites (Khuntuni and Bhaipur) of Cuttack district during the years 1993-95 (wet seasons). Besides this, two more new sites in Cuttack and Dhenkanal districts representing mainly lateritic soils were also taken up. In addition to these sites, some scattered work in different villages of Orissa (on demand) was also undertaken under the project. The experimental findings of these said units of FSR Orissa are presented in the table 20.1 to 20.6

A model on-station unit was established in the year 1991-92 at the University Head quarter under the Department of Agronomy for research, teaching and demonstration purpose, On-Station Farming System Research (FSR) at the Orissa University of Agriculture and Technology, Bhubaneswar has served as model for the coastal track of Orissa.

The basic objectives of this demonstration is to generate maximum income and employment by combining different enterprises such as Field crops, Multistoried cropping, Pomology, Oleariculture, Multi-layer pisciculture, Dairy, Duckery, Poultry, Apiary, Mushroom production, Floriculture, Agroforestry, Bio-gas generation, Commercial nursery by recycling of produces and by-products.

TABLE 20.1: Income-expenditure statements of beneficiaries under on-farm farming system research at Ranital Centre 1992-93 (rainfed)

Component	Marginal farmer (0.48 ha) (One farm family)	
	Expenditure (Rs.)	Income (Rs.)
Field crops	980	1519
Vegetable	3492	6629
Fruits	550	2132
Mushroom	657	960
Animal Husbandry (Including poultry)	10976	13220
Fishery	440	962
Total	17085	25422
Net return (Rs.)	8337	
Family labour earning (Rs.)	4500	
Total Income (Rs.)	12837	
Return per rupee invested	1.48	

TABLE 20.2: Income-expenditure statements of beneficiaries of on-farm farming system research at Ranital Centre 1992-93 (Irrigated)

Component	Small farmer (2.0 ha) (One farm family)		Marginal farmer (0.98 ha) (One farm family)	
	Expenditure (Rs.)	Income (Rs.)	Expenditure (Rs.)	Income (Rs.)
Field crops	11,972	20,580	4,911	7,620
Vegetables	11,301	16,336	3,445	5,285
Fruits	170	450	570	4,200
Mushroom	264	525	155	360
Animal Husbandry	8,900	11,959	7,200	10,330
Fishery	605	1,300	3,450	7,000
Apiary	-	-	1,250	2,200
Total	33,212	51,130	20,981	36,995
Net return (Rs.)	17,918	16,014		
Family labour earning (Rs.)	14,530	9,580		
Total income (Rs.)	32,448	25,594		
Return per rupee invested	1.54	1.76		

TABLE 20.3: Treatment details of integrated farming system and conventional cropping system

Details	Area in ha.	No.
Integrated Farming System (IFS)		
Cropping Programme:		
Rice (IR-50) (May-Aug.) -Rice (IR-20) (Aug-Dec.) -Maize (Cob) (Jan-Apr.).		0.16
Rice (IR-50) -Rice (IR-20) -Groundnut (CO 2)		0.10
Maize (Co-1) -Rice (IR-20) -Sesame (TMV3)		0.10
Fisheries		
Fish pond (including mushroom shed):		

Silvercarp/Catla (Surface feeder) 0.14	120 (4)
Rohu (Column feeder)	60 (2)
Mrigal/common carp (Bottom feeder)	90 (3)
Grass carp (Grass feeder)	30 (1)
Poultry	
Bapkok layer	20
Mushroom shed	
(Production capacity 1.5 to 2.0 kg/day)	5m x 3m
Conventional Cropping System (CCS)	
Rice (IR-50) -Rice (IR-20) -Green gram (Co 3)	0.20
Rice (IR-50) -Rice (IR-20) -Green manure (Sunhemp)	0.20

This farm pond based model is intended for farmers of the coastal districts having two hectares of land with a pond. Shade loving plants such as ginger, turmeric, pine apple, black pepper and spices have been planted under the coconut trees. One poultry unit has been located at one of the corner of the pond with an idea that the droppings of the birds will directly fall into the water and help the growth of planktons. The demonstration also has a duckery unit.

Two biogas plants have been installed to meet the energy requirement of the family. Slurry from the plant will be used as manure. Fodder crops have been included for the purpose of providing greens to the cows. Banana, Mango, Ber, Lemon and Yam have also been planted around the pond and field boundaries to utilize all the available space. The plan of lay out is presented in Fig. 20.1. The resource recycling pattern is given in Fig. 20.2. Area allocation for different enterprises of FSR unit (2 ha) is presented in Table. 20.7.

The economics of some of the FSR work undertaken by the Orissa University of Agriculture and Technology, Bhubaneswar is indicated in Table. 20.8 to 20.12

The conspicuous results obtained from this unit are:

One could get a net profit of over Rs. 8,000/ ha/year with an input-output ratio of nearly 1:2 from non-crop farm enterprises.

TABLE 20.4: Statement showing the receipt, expenditure and net income for 1987-88, 1988-89, 1989-90, 1990-91 & 1991-92 in integrated farming system and conventional cropping system

Years	Receipt	(Rs.ha^{-1}Yr^{-1})	Expenditure	(Rs.ha^{-1}Yr^{-1})	Net Income	(Rs.ha^{-}Yr^{-1})
	IFS	CCS	IFS	CCS	IFS	CCS
1987-88	61232	36952	24084	15104	37148	21848
1988-89	70518	46512	29120	16688	41398	29824
1989-90	58696	28288	29683	14800	29103	13488
1990-91	82720	31190	47450	16254	35270	14936
1991-92	74928	24290	47425	16250	27503	8040
Mean	69619	33446	35552	15819	34067	17627

Net Income Rs. ha^{-1} day^{-1} (Mean of five years)

IFS: Integrated farming system; CCS: Conventional Cropping System.

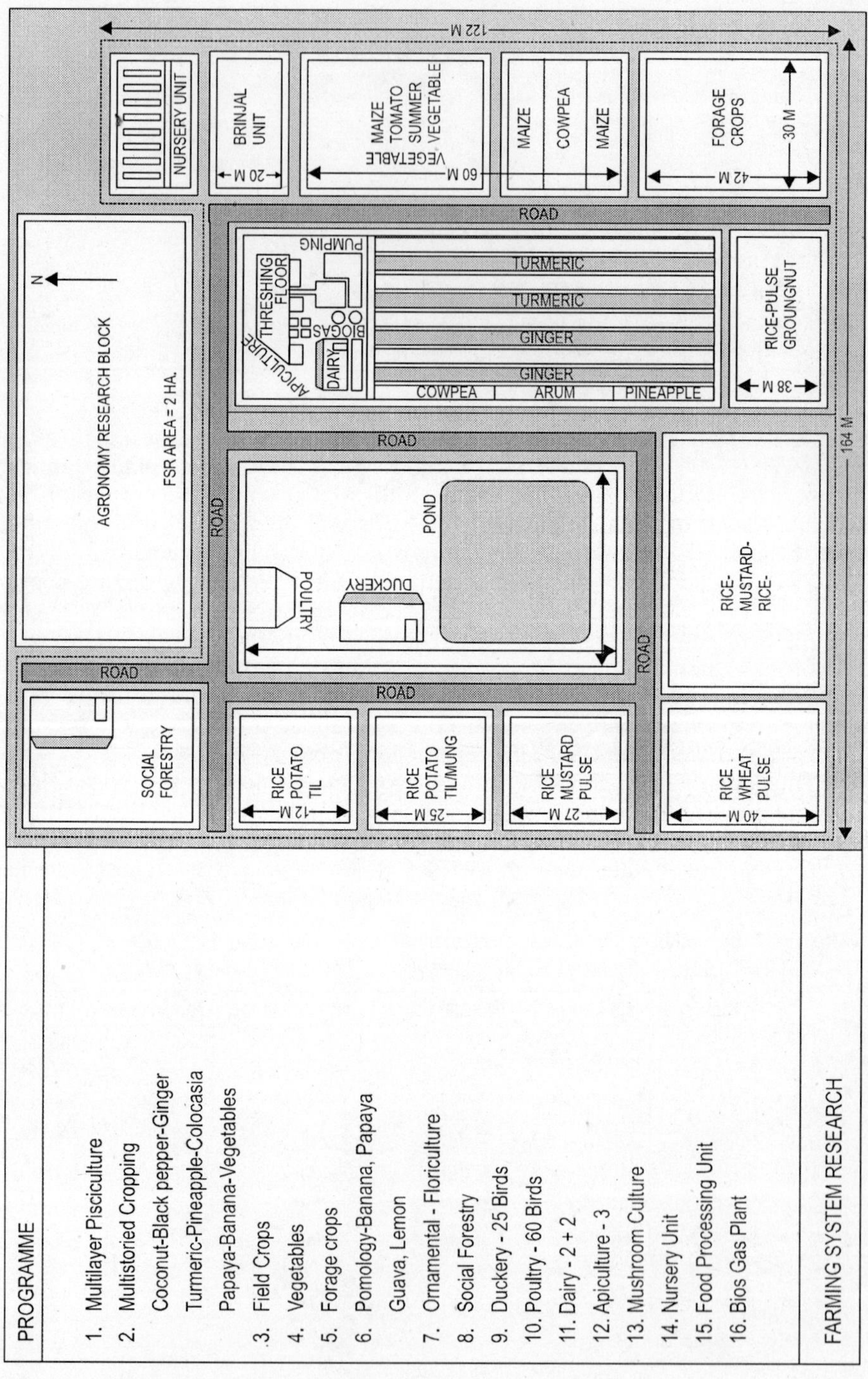

FIG. 20.1: Farming system research unit, O.U.A.T., Bhubaneswar.

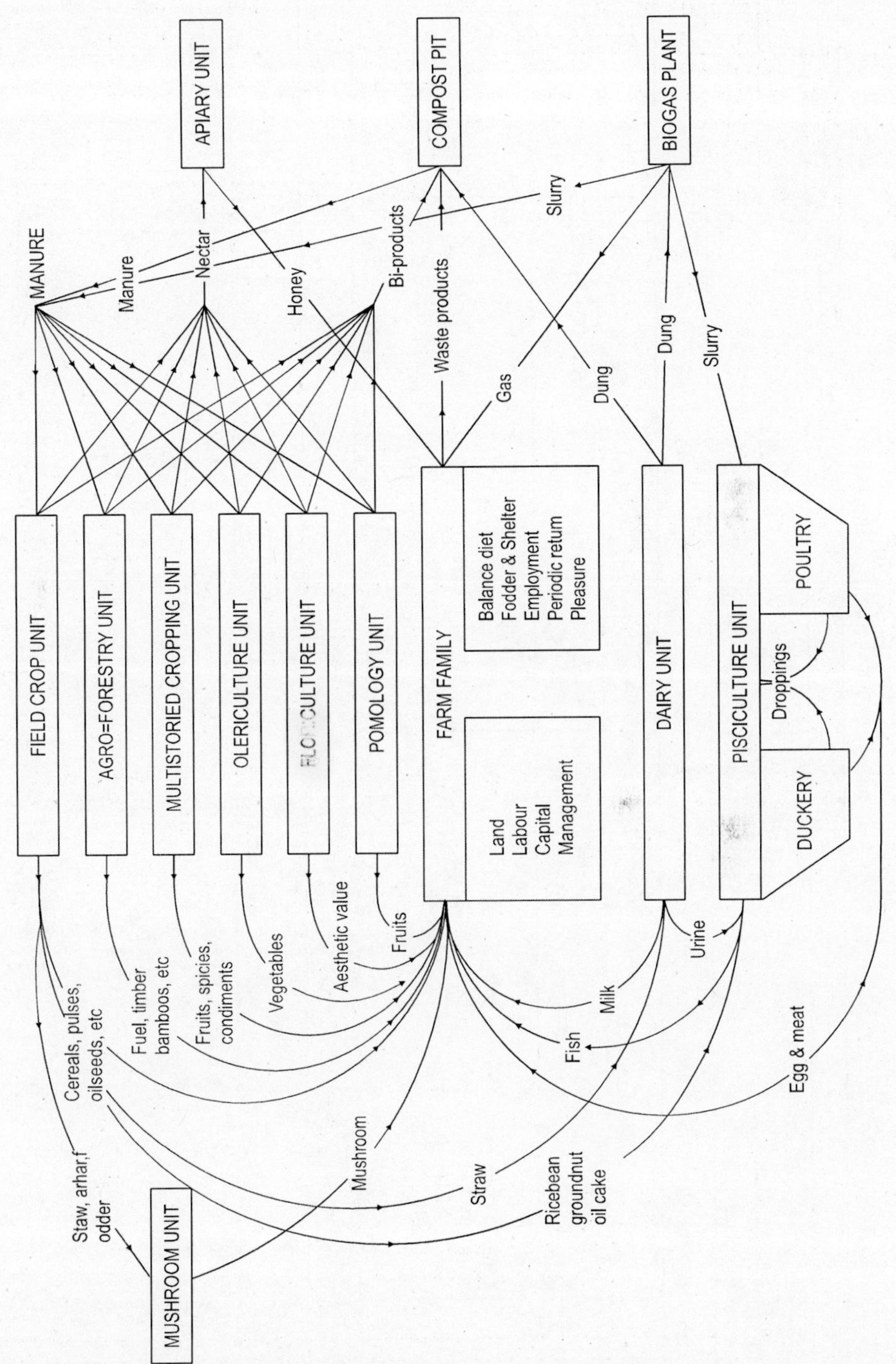

FIG. 20.2: Resource recycling in on station FSR/E unit, OUAT, Bhubaneswar.

TABLE 20.5: Employment generation in integrated farming system and conventional cropping system (mean of five years)

System	Mandays ha^{-1} Yr^{-1}
Integrated Farming System (IFS)	1338
Conventional Cropping System	930
Additional mandays generated under IFS	408

TABLE 20.6: Income-Expenditure Statement of beneficiaries of non-farm

Category	Year	Average field crops		Vegetable		Animal Hus.		Total of farmers		Total size of holding	Net	Family	Total
		Exp. (ha)	Income (Rs.)	Exp. (Rs.)	Income (Rs.)	Exp. (Rs.)	Income (Rs.)	Exp. (Rs.)	Ret (Rs.)	Ret (Rs.)	Labour (Rs.)	Income (Rs.)	
Small	90-91	1.44	4460	7040	5600	9800	570	830	10630	17670	7040	6300	13340
Marginal	90-91	0.57	1260	2630	2170	4150	750	1120	4180	7900	3720	3050	6770
Small	91-92	1.44	3960	6860	5830	10600	780	1460	10570	8920	8350	6940	15290
Marginal	91-92	0.57	1150	2470	1970	4210	640	1000	3760	7750	3990	2960	6950
Small	92-93	1.44	5160	9210	5760	10620	840	1630	11760	21460	9700	7530	17230
Marginal	92-93	0.57	1680	2890	2210	4490	1320	2670	5210	10050	5240	3465	8705

TABLE 20.7: Area allocation for FSR unit for different enterprises (2 hectares)

Particulars	Area		Percentage
	Sq. m	Cents	
Field crops (Low Land)	3,200	80.0	16.0
Olericulture (Upl and)	2,550	64.0	12.8
Multilayer pisciulture (including duckery, poultry)	3,128	78.0	15.6
Multistoried cropping	2,280	57.0	11.4
Floriculture	500	12.5	2.5
Office & Farmer's			
Cottage	100		
Road	420		

Particulars	Area		Percentage
	Sq. m	Cents	
Dairy	36	24.5	4.9
Bio-gas generation plant	120		
Threshing floor	300		
Miscellaneous (Drainage, channels, agro-forestry, area not acquired)	7,382	184	36.8
Total	**20,000**	**500**	**100.0**

TABLE 20.8: Economics of farming systems Model No. 1

Size of holding: one hectare	**No. of family members: 5 (adult – 3, children - 2)**					
Enterprise	**Input cost in Rs.**	**Labour cost in Rs.**	**Total cost in Rs.**	**Tolal return in Rs.**	**Net profit in Rs**	**Retum per rupee invested**
Fleld crop HYV paddy, Maize, sesamum, Blackgrarn, Groundnut etc.	2550	5200	7750	12650	4900	1.63
Vegetables Brinjal, tomato, Cabbage etc.	2840	4600	7440	13820	6380	1.85
Plantation crops Coconut, papaya, Banana etc.	600	1200	1800	4360	2560	2.42
Livestock						
Dairy (2 cows)	3800	2200	6000	11400	5400	1.90
Poultry (20 birds)	400	200	600	11000	500	1.83
Duckery (20 birds)	200	150	350	616	266	1.76
Pisciculture (1000 sq.mt.)	1000	750	1750	4025	2275	2030
Apiculture (one box)	500	200	700	12000	500	1.71
Mushroom (10 beds)	200	100	300	450	150	1.50
Total	**12090**	**14600**	**26690**	**49621**	**22931**	**1.85**

Total labour cost = Rs. 14600
(Family labour earning)
Total farnily income = Rs.37531 (Family labour earning + net profit)
Net benefit from 1 ha of land = Rs. 22931

TABLE 20.9: Economics of farming systems **Model No. 2**

Size of holding: Two hectares	**No. of family members: 5 (adult – 3, children – 2)**					
Enterprise	**Input cost in Rs.**	**Labour cost in Rs.**	**Total cost in Rs.**	**Total return in Rs.**	**Net profit in Rs.**	**Return per rupee invested**
Field crop *HYV paddy, Sesame, Green gram, Arhar, Groundnut etc.*	5450	11200	16650	28305	11655	1.70
Vegetables *Brinjal, Cauliflower, Tomato, etc.*	5700	10500	16200	28836	12636	1.78
Plantation crops *Coconut, Papaya, Banana, etc.*	2000	2800	4800	12048	7248	1.51
Livestock						
Dairy (2 cows)	3800	2200	6000	11400	5400	1.90
Poultry (50 brds)	1000	3000	1300	2800	1500	2.15

Size of holding: Two hectares	No. of family members: 5 (adult – 3, children – 2)					
Enterprise	Input cost in Rs.	Labour cost in Rs.	Total cost in Rs.	Total return in Rs.	Net profit in Rs.	Return per rupee invested
Duckery (-do)	300	150	450	855	405	1.90
Pisciculture (2000 sq. mt.)	3000	1600	4600	11592	6992	2.52
Apiculture (2 boxes)	1000	400	1400	2400	1000	1.71
Mushroom (30 beds)	600	300	900	1350	450	1.50
Total	22850	29450	52300	99586	47286	1.90

Total labour cost = Rs. 29450
(Family labour cost = 15500 + Hired labour cost = 13950)
Total farnily income = Rs. 62786 (Family labour earning + net profit)
Net benefit from 2 ha of land = Rs. 47286

Multistoried farming by growing cash crops like ginger and turmeric under the coconut trees would not only considerably add to the income of the farmers, but also would utilize the inter space, conserve soil and soil moisture, and provide engagement to family.

Combination of duckery, poultry and fishery has been found to be compatible, provide additional income, and reduce expenditure on fish feeds.

Inclusion of apiary in the system could generate additional income. Recently, it is obsereved that a new strain ***Apis mellifera*** (an Italian species) is able to yield 10-15 kg of honey /box/collection with four collections/year as against only 1 to 1.5 kg from the local species.

The experience with the on-farm FSR/E in different units has been quite encouraging from the beginning. The present studies, in general, reveal the following:

On-farm Trials

Trials undertaken for crop/variety substitution, nutrient, pest and disease management of crops, fish, animal, etc. in the farmers' fields have generated useful data which have benefited the farmers and provided necessary guidelines for developing suitable farming system models.

On-farm Farming Systems

Studies on the farming systems followed by different categories of farmers under rainfed and irrigated conditions have useful informations. Some of these are:

1. Under the restricted scope for horizontal expansion of farm area, the only alternate solution for increasing food production and farmer's income generation is adoption of farming system.
2. Farming systems could success fully be adopted under rainfed conditions and the farmers could get a net return of Rs. 1.37 per Rupee invested and the income is spread over the year.
3. Smaller the holding size, more is the relative efficiency of undertaking farming systems.
4. It provides alternate and sustainable avocation to marginal and sub-marginal farmers. Fruit, mushroom, apiary, animal production and poultry have been more viable with them.

5. Marginal and sub-marginal farmers landless labourers who were mostly depending on the farm avocations for maintaining their livelihood, have found farming systems a substantial source of income. The idle family labour is also suitable engaged.

TABLE 20.10: Economics of different components of the On-station FSR, Bhubaneswar (1992-93)

Component	Total cost Rs.	Gross return Rs.	Net return Rs.	Return per rupee Rs.
Field crop: *(Rice, Maize and Ricebean)*	2109.50	3233.50	1124.00	1.53
Multi-storied cropping: *(Coconut, ginger, turmeric, pneapple and colocassia)*	4831.00	11502.00	6671.00	2.38
Pomology: *(Banana, papaya)*	900.00	2866.00	1966.00	2.62
Horticulture: *(Bhindi, brinjal, poi, Kosala, cowpea, ridge gourd, tomato, bean, bottle gourd cucumber, water, melon, pumpkin and potato)*	2863.50	4650.80	1787.30	1.62
Pisciculture: *(Multi-layer)*	2401.00	6873.40	4472.40	2.86
Duckery (27 birds) *(29.10.92 to 14.1.93)*	2052.00	2591.00	539.00	1.26
Apiary (3 boxes)	1220.00	1870.00	650.00	1.53
Mushroom (15 beds)	5320	8400	3080	1.58
Total	**21697.00**	**41986.70**	**20289.70**	**1.94**

By scientific planning and use of proper technology, the income of farming system practicing farmers could be raised by 50%

In general, farming systems are found to be more viable with the small and mar ginal farmers.

Some of the activities of the on-farm farming systems in the State are:

Mr. Nath Mallick, Village: Timar of Patalda Gram Panchayat in Astaranga area of Puri district has 0.16 ha land. A pond is located in 1/5th of this area. The soil is saline in nature. Short duration rice is planted in mid July and harvested in the end of October. Crops moderately tolerant to salinity like tomato, knolkhol, wheat are grown with the pond water which has comparatively lower salinity due to dilution of soluble salts by rain water. In the pond, fresh-water fish like catla, rohu, mrigal, etc., are reared after onset of monsoon. Coconut is planted on the banks of the pond and intercropped with vegetables. The income (Rs. 15,000/- to Rs. 20,000/-) has increased by 10 times. Coconut would fruit after 2-3years.

Mr. Chema Biswal of Village: Routrapur in the Balasore district has 0.48 ha land with 385 m^2 pond area. Enterprises like crop cultivation, dairy, poultry, duckery, fish rearing, apiculture, vegetable and fruit cultivation and ornamental gardening, agro-forestry, etc., have been integrated. The upland cropping systems include early rice-tomato-mung, mesta-cole crops-groundnut, brinjal-tomato, brinjal-ratoon and turmeric/ginger-greens. He gets a net profit of Rs. 15,592/- per year from all the enterprises. The best three enterprises providing him the major income are vegetables and spices, fisheries and animal husbandry in that order. However, the gross return per Rupee invested is the highest (above 5) from plantation crops and fisheries.

TABLE 20.11: Income-expenditure statement of beneficiaries of on-farm system research at village Bhola. 1992-93

Category of farmers	Field crops		Vegetable		Animal Hus.		Total exp.	Tota return	Net return	Return per re-invested	Family labour earning	Total income
	Exp.	Income	Exp.	Income	Exp.	Income						
	(Rs.)	(Rs.)	(Rs.)	(Rs.)	(Rs.)	(Rs.)	(Rs.)	(Rs.)	(Rs.)	(Rs.)	(Rs.)	
Marginal (Six farm families)	1580	2775	1985	5565	4630	7250	8195	15590	7395	1.90	41.50	11545

It has been observed in the farmers' fields in the coastal saline block of Orissa that if one can adopt mixed farming by integrating various farm enterprises such as poultry, dairy, fishery, field crops, horticultural crops, can get a net return of Rs. 26,000/- to Rs. 28,000/- per annum from one hectare of land by investing an amount of Rs. 20,000/-. Farm-pond-technology which involves digging of a pond in a part of the field (20% area) in coastal saline low land and spreading the excavated earth in the remaining 80% of the field is now gaining popularity among the farmers in the coastal region. From the field so developed, the farmers are now able to gat a bumper crop of HYV rice in monsoon season and vegetables in winter and summer months. In addition to the above, a handsome income is obtained from stocking fish in ponds and growing banana and coconut on the pond embankment. A similar technology has also been demonstrated in West Bengal by the scientists of CSSRI, Canning, and Rama Krishna Mission in Sundarban areas.

An integrated farming system model has also been prepared and field-tested in the coastal region of Orissa in which cropping, dairy and silviculture (Plantation of Casuarina) has shown an attractive profit, and suits to the needs of the farmer within his capabilities and limitations.

TABLE 20.12: Annual expenditure and income statement of marginal farmer (Chema Biswal owner of 0.48 ha) during 1992-93

Components	Expenditure	Labour	Gross	Net	Gross
	Utilized (man-days)	Return (Rs./ha)	Return (Rs.)	Return cost ratio	
Field crops (Rice, mung, sunhemp, groundnut, maize) *(Total cropped area 2150 m²)*	795	26	1839	1044	2.31
Vegetables & spices	3445	100	9509	6064.1	2.76 IV
Fruit crops (Mango, papaya, banana, citrus)	500	5	2132	1582	3.88 III
Plantation crops (both economic product and fuel wood) *(Tamarindus indica, drumstick)*	225	4	1152	927	5.12 II
Fisheries (385 m²)	500	4	2625 III	2125 II	5.25 I
Mushroom	657	30	1707	1050	2.59 V
Poultry (20 birds) (no. of eggs 4260)	4120	23	4800	720	1.20
Duckery (10 birds) *(no. of eggs 2310 @ Rs. 11 / egg.)*	2095	23	2335	240	1.11
Apiary (one unit)	130	-	470	240	3.61
Animal husbandry (2 bullocks)	3450	45	5050	1600	1.46
Total	**15960**	**260**	**31619**	**15592**	**1.98**
Agroforestry (500 m²)	500	10	5000	4500	10.0

INCOME AND EMPLOYMENT GENERATION FOR SMALL AND MARGINAL FARMERS THROUGH INTEGRATED FARMING SYSTEMS

Model No. 3

Pond-based farming system research studies comprising enterprises like field and horticultural crops, fishery, poultry, duckery, apiary, mushroom, dairy and agro forestry were undertaken at Bhubaneswar during 1992-94, with a view to generate income and employment opportunities for small and marginal farmers. Among the above mentioned enterprises, apiculture produced the highest return of Rs. 7.94/- rupee invested. Multilayer pisciculture stood next, fetching Rs. 5.46/- rupee invested. Among the crops, the best returns were obtained with multistoried cropping involving pumpkin, ridge gourd and *poi* as ground storey; pineapple, colocasia, ginger and turmeric as first storey; and coconut as second storey. Poultry, duckery and mushroom enterprises fetched low returns. The highest level of employment (180 man days/year) was achieved in mushroom cultivation. Integrated farming system, comprising above enterprises generated net income of Rs. 58,360 and employment for 573 man days, on a small piece of land (1.25 ha), ensuring high standard of living for small and marginal farmers. The economics of different enterprises of FSR are presented in tables 20.13 to 20.16 (Behera and Mahapatra, 1999).

TABLE 20.13 Fish production and economics of the cultivated carps in the FSR pond during 1992-93

Species	Initial average weight (g)	Initial total weight (g)	No. of fingerlings released	Culture period (year)	Final average		Monthly average growth rate (g)	Estimated fish production (kg)
					Length (cm)	Weight (cm)		
Catla	0.46	253	550	1	40.6	0.900 ±0.102	75	445.5
Rohu	0.35	175	500	1	31.8	0.335 ±0.033	28	159.6
Mrigal	1.56	7.2	450	1	33.8	0.380 ±0.042	31	153.9
Grass carp	11.28	564	50	1	46.6	1.200 ±0.300	100	54.0
Total		1694	1550					813
Return from fish production @ Rs. 25.00 kg/ha								Rs. 20, 325
Expenditure on fish production								Rs. 3, 722
Net profit								Rs. 16, 603
Gross return/cost of fish production								Rs. 5.46

Model No. 4

A model was developed (Panda & Patro, 1998) for a holding of 0.6 ha. to suit the marginal farmers of Orissa and it was tested at the Department of Agronomy, Orissa University of Agriculture and Technology, Bubaneswar from 1992-93 to 1994-95. The

Components of the integrated farming system are cropping and dairy. While programming cropping system in crop component, due consideration has been given to the crops under cultivation in the locality and the crop rotations that is being followed. For comparison, conventional cropping systems practiced in the area are also tested in the area of 0.6 ha. (Table 20-17 and 20-18). The details of the different components integrated have been presented here.

TABLE 20.14: Economics of poultry and duckery units of FSR during 1992-93

Particulars	Amount (Rs.)	
	Poultry	Duckery
A. Expenditure		
(i) Cost of layer birds	1,560 (40 nos)	810 (27 nos)
(ii) Feed for 1 year	6,105	3, 422
(iii) Medicine	200	180
(iv) Labour	575	575
(v) Depreciation on cost of house	400	400
(vi) Depreciation on cage valued at Rs. 4,000/ (for poultry only)	400	-
Total expenditure	9,240	5,387
B. Income		
(i) Sale of eggs @ Rs. 1.00 / egg	8,783	5,475
(ii) Sale of empty gunny bags	100	50
(iii) Return from manure	200	100
(iv) Sale of cull birds	1,138	475
Gross income	981	713
Net income	**981**	**713**

Cropping

Four sequential cropping programmes were adopted in the area specified giving due consideration to the availability of water at different seasons and requirement of the dairy unit included in this programme.

MULTI-STORIED CROPPING

In this unit 50 coconut trees already exist. One hundred banana plants were raised in between the coconut trees along the row. In the inter-row spaces cash crops like ginger, turmeric, colocasia, radish, bassella and sweet potato were grown. This unit not only considerably added to the income of the farmer but also utilized the inter space, conserved soil and soil moisture.

TABLE 20.15: Economics of small unit of mushroom production in FSR unit during 1992-93

Particulars	Rate (Rs.)	Amount (Rs.)	
		Paddy straw (64 beds)	Oyster (100 bags)
A. Recurring cost for one month			
Paddy straw	0.5 / kg	480	100
Spawn	08.0 / bottle	512	400
Pulse powder @ 250 g/bed	10.0 / kg	160	-
Wheat grain @ 250 g/bag	04.0 / kg	-	100

Particulars	Rate (Rs.)	Amount (Rs.)	
		Paddy straw (64 beds)	Oyster (100 bags)
Labour	25 / man day	350	300
Miscellaneous including chemicals		96	50
Expenditure per month		1,598	950
B. Total expenditure			
Total expenditure for 64 beds of paddy straw mushroom for 8 months			12,784
Total expenditure for 100 bags of oyster mushroom for 4 months			3,800
Depreciation from mixed cost incurred for infrastructure development			1,600
Total expenditure			18,184
C. Total return			
Return from 768 kg of paddy straw mushroom for 8 month @ Rs. 30, 000/ kg			23,040
Return from 400 kg of oyster mushroom for 4 months @ Rs. 20, 000 / kg			8,000
Gross return			31,040
Net profit			12,856
D. Return per rupee invested			
On paddy straw mushroom			1.8
On oyster mushroom			2.1

Dairy Unit

A dairy unit with a dimension of 5 m X 4 m was constructed to accommodate two jerseys cross bred cows to provide added income to the farm family though supply of milk and cow dung for the bio-gas unit. The waste product of rice when hulled becomes a substitute of 10 per cent of dairy feed. Similarly rice bran (7%), rice straw (13%), green gram bhusa (5%) and maize flour (20%) were also utilized from the crop component. Fifty five per cent out of total constituent of the concentrate dairy feed was obtained from the crop component besides supplying green fodder round the year to the dairy unit.

TABLE 20.16: Economics of different components of FSR unit during 1992-93

Components	Total labour (man days)	Total expenditure (Rs.)	Gross return (Rs.)	Net return (Rs.)	Return per rupee invested (Rs.)
Field crops	98.2	3,315	8,954	5,638	2.70
Multistoried cropping	87.0	3,831	12,920	9,089	3.37
Pomology	18.4	900	2,366	1,466	2.63
Olericulture	96.4	3,812	12,114	8,302	3.18
Floriculture	4.0	125	225	100	1.80
Pisciculture	31.0	3,722	20,325	16,603	5.46

Poultry	23.0	9,240	10,221	981	1.11
Duckery	23.0	5,387	6,100	713	1.13
Mushroom	180.0	18,184	31,040	12,856	1.70
Apiary	1.0	170	1,350	1,180	7.94
Biogas	11.0	600	2031	1,431	3.38
Total	573.0	49,186	107,646	58,360	2.18

Bio-gas Unit

A bio-gas unit of 1 m^3 size was constructed near the dairy which provided gas for cooking and lighting and manure to crops.

Economics

The economics were worked out for the rice-dairy system. Cost of construction of dairy unit (Rs. 6000) and bio-gas unit (Rs. 1800) and purchase cost of cows (Rs. 12000) were apportioned for three years and added to the production cost. An interest rate of 14 per cent / annum was also added for the fixed capital and working capital.

TABLE 20.17: Treatment details of integrated farming system and conventional cropping system

Details	Area (ha)
Integrated Farming System (IFS) Cropping Programme	0.1
Rice (IR 36) – Mustard (M 27) – Rice (Pathara)	0.1
Maize (Ganga (101) – Mustard (M 27) – Mung (Dhauli)	0.1
Maize (Ganga 101) – Tomato (BT1) -lady finger (Parbhani kranti)	0.1
Hybrid Napier (IGFRI 6)	0.1
Multistoried Cropping	
I. Coconut (50 plants, Sakhigopal local)	
II. Banana (100 plants, Patakapura)	
III. Ginger/Turmeric/Colocasia/Bassella/Sweet potato/ Radish	0.2
Dary (2 cows)	-
Bio-gas (1 m^3)	-
Conventional Cropping System (CCS)	
Rice (IR 36) – Rice (Jajati)	0.3
Rice (IR 36) – Rice (Pathara) – Mung (Dhauli)	0.3

TABLE 20.18: Net income and employment opportunity from integrated farming system and conventional cropping system

Farming system	Net income (Rs.)				Employment opportunity /0.6 ha (Mandays)			
	92-93	93-94	94-95	Mean	92-93	93-94	94-95	Mean
Integrated farming system cropping	12545	11708	12824	12359	102	98	109	103
Multis storied cropping	4826	5428	6982	5745	126	134	121	127
Dairy	1610	1710	1970	1763	240	221	232	231
Bio gas	1094	1466	1540	1367	72	80	70	74
Total	200075	10312	23316	21234	540	533	532	535

Farming system	Net income (Rs.)				Employment opportunity /0.6 ha (Mandays)			
	92-93	93-94	94-95	Mean	92-93	93-94	94-95	Mean
Conventional cropping system (CCS)	10622	9157	9896	9892	129	143	139	137
Additional advantage in IFS over CCS	9453	11155	13420	11342	411	390	393	398
Additional advantage in IFS over CCS per day	25.56	30.56	36.76	31.07				

An average, net income of Rs. 12,359 / year was obtained from the crop component of the IFS (Table 20.18). This excludes the value of 160 kg of maize grain, 164 kg of rice bran, 900 kg of rice straw and 125 kg of green gram bhusa which was used for preparation of concentrate dairy feed and the value of 10220 kg of green fodder which was used as dairy feed. The dairy unit recorded an average net income of Rs. 1763/ year. This excludes the value of 5166 kg of raw cowdung which was utilized in the biogas. The multistoried unit and the biogas unit registered an average net income of Rs. 5,745/ year and Rs. 1,367/ year.

The total net income under rice-dairy IFS was Rs. 12359/- and the net income in CCS was only Rs. 9895/ year from 0.6 ha. / Year over CCS. Moreover, it generated an additional employment of 398 mandays per year over CCS from the specified area and thus helped in the recycling of farm wastes and by – products round the year (Fig 20.2). It also minimized the risks associated with mere crop production.

RESOURCE RECYCLING

Resource recycling clearly indicated that the interdependence of the different components of the total farming system to make the farmers self-sufficient in terms of ensuring the family members a balance diet for leading healthy life, increasing the standard of living through maximizing the total net return and providing more employment. The by-product of dairy, *i.e.* cowdung, forms a major raw material for biogas. Digested slurry of biogas forms a major part of feed of pisciculture for increasing plankton growth as well as supplying valuable manure to raise the productivity of crops. Similarly, byproduct of field crops like paddy straw forms a major ingredient of mushroom cultivation. Again straw used for mushroom production may be utilized for cattle feed and compost preparation. The by-product of poultry, *i.e.* poultry dropping, forms an important ingredient of pisciculture for increasing the plankton growth as well as increasing the fertility of land. Even an item like apiary has lot of indirect bearing on crop enterprise by increasing yield and quality though improvement in pollination, apart from giving a wholesome product like honey to the farmers. Farmers like to have flowers' which adds to the aesthetic sense of farm family and flowers also serve as foraging plants for honey bees. The entire philosophy of integrated farming system revolves around better utilization of time, money, resources and family labourers of farm families. The farm family gets scope for gainful employment round the year, thereby ensuring good income and better standards of living.

Model No. 5

Another model based on Rice-poultry-duckery-fish-apiary-mushroom Integrated

Farming System for the farmers having One hectare of land was developed. The economics of the enterprises are given in Tables 20.19-20.20.

Economics of Rice-poultry-duckery-fish-apiary-mushroom system under lowland conditions in Orissa was studied from 1992 to 1994. The results revealed that a net profit of Rs. 12,038 per year was obtained in rice-poultry-duckery-fish-apiary-mushroom Integrated Farming System (IFS) in 1.0 ha area. Conventional Cropping System (CCS) with Rice-rice-pulses/maize gave a net income of Rs. 3450 per year from the same area. By inclusion of poultry-duckery-fish-apiary-mushroom components in IFS the net income per day has been increased to Rs. 32.97 as against Rs. 9.42 in CCS.

In the irrigated lowland rice areas of Orissa, raising two crops (June-October and October-January) followed by a rice fallow pulse or oilseeds (January-May) is common after the introduction of high yielding, fertilizer responsive, short duration varieties. The yield increase came to a standstill after a linear increase initially, thus leading to deficit for the growing population. Due to escalation in the cost of inputs and lack of proper marketing of the harvested produce, the net income of the farmer has come down drastically. This apparently leads to the adoption of diverse activities, otherwise called as farming system approach. Conventional cropping system provides not only low income due to fluctuation in the market price but also leads to under employment for the family labourers of the farmer during the lean season. So, in the irrigated lowlands of Orissa, Rice-poultry-duckery-fish-apiary-mushroom system can be adopted to earn higher net return per unit area per unit time. The economics of the different enterprises are presented in Table 20.19 and 20.20

TABLE 20.19: Treatment details of Integrated Farming System (IFS) and Conventional cropping system (CCS)

S.N.	Treatment	Details	Area in ha	No.
I.	**Integrated Farming System (IFS)**			
	A. Copping programme			
	1. Rice (Lalata) (June.-Oct.)	Rice (Parijat) (Oct.-Dec.)	Groundnut (AK 12-24) (Jan–may)	0.35
	2. Rice (Lalata	mustard (M-27)	rice (Kalinga-III)	0.30
	3. Rice (CR-1009)	rice (Parijat)	sesame (Vinayak)	0.30
	B. Fisheris			
	1. Catla (Surface feeder)		0.05	150
	2. Rohu (Column feeder)			150
	3. Mrigal/ (Bottom feeder)			100
	4. Grass carp/common carp (Grass feeder)			90
	C. Poultry			
	D. Duckery			
	Indian runner layer			
	E. Apiary (Production capacity 34-48 kg/year)			
	F. Mushroom *(Production capacity 230-280 kg/year)*			

S.N.	Treatment	Details	Area in ha	No.
II.	**Conventional cropping system (CCS)**			
	Rice (CR-1009)	rice (Jajati)	greengram (Dhauli)	0.50
	Rice (CR-1009)	rice (Jajati)	Maize (Ganga-5)	0.50

TABLE 20.20: Economics of rice-poultry-duckery-fish-apiary-mushroom system

	Integrated Farming System (IFS) Area (1.0 ha)			Conventional Cropping System (CCS) Area (1.0 ha)			Additional Net Income from IFS over CCS (Rs.)
	Gross Income Rs.	Production cost Rs.	Net Income Rs.	Gross Income Rs.	Production Cost Rs.	Net Income Rs.	
	1	2	3	4	5	6	7
1992-93							
Crop	152552	10704	4548	12934	8438	4496	62
Poultry	7510	6000	1510	-	-	-	1510
Duckery	2591	2052	539	-	-	-	539
Fisheries	6873	2401	4472	-	-	-	4472
Apiary	1870	1220	650	-	-	-	650
Mushroom	8400	53220	3080	-	-	-	3080
Total	42496	27697	14799	12934	8438	4496	10303
1993-94							
Crop	17548	15197	2351	16876	15648	1226	1125
Poultry	2479	1825	654	-	-	-	654
Duckery	1293	1095	198	-	-	-	198
Fisheries	3941	2455	1486	-	-	-	1486
Apiary	2020	1364	656	-	-	-	656
Mushroom	6900	5100	1800	-	-	-	1800
Total	34181	27036	7145	16876	15648	1226	5919
1994-95							
Crop	16900	11290	5610	15525	10900	4625	985
Poultry	5420	3400	2020	-	-	-	2020
Duckery	2200	1890	310	-	-	-	310
Fisheries	8000	4600	3400	-	-	-	3400
Apiary	1990	1450	540	-	-	-	540
Mushroom	7226	4935	2291	-	-	-	2291
Total	41736	27565	14171	15525	10900	4625	9556
Mean of three years							
Crop	16567	12397	4170	15112	11662	3450	720
Poultry	5136	3282	1854	-	-	-	1854
Duckery	2028	1679	349	-	-	-	349
Fisheries	6271	3152	3119	-	-	-	3119
Apiary	1960	1345	615	-	-	-	615
Mushroom	7509	5118	2391	-	-	-	2391
Total	**39471**	**27433**	**12038**	**15112**	**11662**	**3450**	**8588**

	Integrated Farming System (IFS) Area (1.0 ha)			Conventional Cropping System (CCS) Area (1.0 ha)			Additional Net Income from IFS over CCS (Rs.)
	Gross Income Rs.	Production cost Rs.	Net Income Rs.	Gross Income Rs.	Production Cost Rs.	Net Income Rs.	
	1	2	3	4	5	6	7
Net Income per day			32.97			9.42	23.55

Model No. 6

Out of 33.5 million population of Orissa, 76% belong to small and marginal farmers and they occupy 62% of the total land holdings. Moreover, about 55% of the population mostly consisting of small and marginal farmers is below the poverty line. The average holding units for marginal farmers are less than one hectare. With the gradual decline in farm size, it would be increasingly difficult on the part of a marginal farmer to produce enough food for his family in coming years (Table 20.21).

The benefits of modern technology have, however, been restricted to favourable farming situations. Only 20 to 25 per cent of the modern technologies have reached the farmers in the ecologically favourable areas but much less (< 5 per cent) in the handicapped ecologies. Since, there is no further scope for horizontal expansion of land for cultivation and because of fragmentation and subdivision of land holdings and decline in the per capita availability of land day by day, there is no alternative left than to go in for vertical expansion and diversification of farm enterprises which have less demand on space and time particularly for small and marginal farmers with very limited resources, specially in rainfed areas (70 per cent of total cultivated land). The new farming system research strategy, should therefore, necessarily concentrate on developing technology with participatory approach within the biophysical and socio-economic environments in which the farmers operate. Keeping in view to evolve suitable farming system model for marginal farmers, a project of the Emeritus Scientist entitled **"Sustainable Farming System Development for Marginal Farmers of Orissa"** was started by Dr S. C. Panda on June 2000 at the Research Station of Orissa University of Agriculture and Technology, Bhubaneswar in an area of 0.60 ha.

TABLE 20.21: District-wise Estimated Number of Different categories of Operational Holdings in Orissa during 1985-86

S. N.	District	Number of Operational Holdings under different categories					
		Marginal	Small	Semi-Medium	Medium	Large total	
		(Below 1.0 ha)	(1.0-2.0 ha)	Medium 2.0 ha	(4.0-10.0 ha)	(10.0 & all above groups)	
1.	Balasore	2053 (60.32)	726 (21.33)	450 (13.23)	162 (4.76)	12 (0.36)	3403 (100.0)
2.	Bolangir	957 (43.79)	566 (25.90)	451 (20.65)	183 (8.38)	28 (1.28)	2185 (100.0)
3.	Cuttack	2786	1288	627	195	12	4908

S. N.	District	Number of Operational Holdings under different categories					
		Marginal	Small	Semi-Medium	Medium	Large total	
		(Below 1.0 ha)	(1.0-2.0 ha)	Medium 2.0 ha	(4.0-10.0 ha)	(10.0 & all above groups)	
		(56.76)	(26.24)	(15.78)	(3.97)	(0.25)	(100.0)
4.	Dhenkanal	1251	680	359	125	13	3565
		(52.30)	(28.43)	(15.01)	(3.80)	(0.46)	(100.0)
5.	Ganjam	2331	701	395	125	13	3565
		(65.39)	(19.66)	(11.08)	(3.51)	(0.36)	(100.0)
6.	Kalahandi	769	609	510	235	31	2154
		(35.70)	(28.27)	(23.68)	(10.91)	(1.44)	(100.0)
7.	Keonjhar	869	393	299	92	5	1758
		(49.43)	(28.04)	(17.02)	(5.23)	(0.28)	(100.0)
8.	Koraput	1382	983	780	284	24	3455
		(40.00)	(28.45)	(22.63)	(8.23)	(0.69)	(100.0)
9.	Mayurbhanj	1451	681	380	119	6	2637
		(5.02)	(25.82)	(14.41)	(4.51)	(0.23)	(100.0)
10.	Phulbani	602	309	192	64	6	1173
		(51.32)	(26.34)	(16.37)	(5.46)	(0.51)	(100.0)
11.	Puri	2232	702	339	95	8	1376
		(66.11)	(20.79)	(10.04)	(2.81)	(0.24)	(100.0)
12.	Sambalpur	1426	910	676	259	38	3309
		(36.82)	(29.22)	(24.22)	(8.77)	(0.78)	(100.0)
13.	Sundargarh	567	453	373	135	12	1540
		(36.82)	(29.22)	(24.22)	(8.77)	(0.78)	(100.0)
	Orissa	18676	9101	5823	2039	206	35855
		(52.09)	(25.8)	(16.27)	(56.9)	(0.57)	(100.0)

Figures in the brackets indicate percentage to total.

Source-Report on Number and Area of Operational Holding in Orissa 1985-86, Agricultural Census.

The net income obtained from crop husbandry was Rs. 15,480. Among the different crops taken *saga* fetched the highest profit per rupee invested (2.29) which was followed by cucumber (2.18) and tomato (1.97) (Table 20.22). Out of the crop rotations followed, Maize- radish-saga (5 times) fetched best income and was Rs 4310 from a net area of 0.04 ha.(Table 20.23). The total biomass yield of fishes was 295 kg which brought a net income of Rs 9240 with a net return per rupee Rs 3.60. Two varieties of mushroom were taken viz. straw mushroom and oyster mushroom those brought a net income of Rs. 20,745/- and recorded a net income of 1.79 rupees per rupee invested. Poultry enterprise consisted of 25 layering birds. All total 4140 eggs were obtained with a net income of Rs. 1460 /-. The profit obtained was Rs 0.31 per rupee invested. Apiary unit produced 8 kg of honey that fetched a net income of Rs 620 with a rate of Rs 3.44 per rupee invested. The net income from the integrated farming system (crop husbandry, poultry, fishery, apiary and mushroom) was found out to be Rs 68, 290/- (Table 20.24). This model can be adapted to earn higher net return per unit area per unit time.

TABLE 20.22: Gross income, cost of production and net income of different crops taken in crop husbandry (2000-2002)

	Area (ha)	Produce (Kg)	Cost/Kg of produce	Gross income (Rs)	Cost of production (Rs)	Net income (Rs)	NI per Rupee invested
Up land rice	0.04	105	5.40	567	356	211	0.59
Radish	0.06	1085	3	3255	1280	1975	1.54
Bean	0.02	65	6	390	205	185	0.90
Tomato	0.08	1252	5	6260	2106	4154	1.97
Okra	0.08	342	6	2052	857	1195	1.39
Poi	0.02	45	3	135	60	75	1.25
Ridge gourd	0.02	82	6	492	300	192	0.64
Maize	0.04	1680	1	1680	690	990	1.43
Saga	0.10	625	5	3125	950	2175	2.29
Cucumber	0.04	120	6	720	228	492	2.18
Cow pea	0.04	195	7	1365	745	620	0.83
Pumpkin	0.04	324		1134	510	624	1.22
Brinjal	0.04	512	6	3072	1600	1472	0.92
Green gram	0.08	65	13.20	858	420	438	1.04
Low land rice	0.10	280	5.40	1512	830	682	0.82

TABLE 20.23: The crop rotations, their gross income, cost of production and net income in rupees (2000-2002)

Rotations followed	Gross income (Rs)	Cost of production (Rs)	Net income (Rs) / 0.04 ha
Rice- Radish +Bean- Poi +Ridge gourd	2709	1291	1418
Maize – Radish-Amaranthus (5 times)	6800	2490	4310
Okra – Cucumber – Tomato – Amaranthus	4860	1679	3181
Okra – Tomato-Green gram	4279	1564.5	2714.5
Cow pea- Pumpkin-Brinjal- Green gram	5678	2907.5	2770.5
Rice – Green gram	2155.5	1145	1010.5

TABLE 20.24: Different enterprises with their gross income, cost of production and net income etc (2000-2002)

Enterprise	Area	Produce (kg)	Cost (Rs) / kg	Gross income	Cost of production	Net income	NI/Re invested
Crop husbandry	0.2 ha			26617	11137	15480	1.39
Mushroom	Straw	273	60	16380	7320	9060	1.23
	Oyster	399	40	15960	4275	11685	2.73
	Total			**32340**	**11595**	**20745**	**1.79**
Poultry	25 layer birds	4140 eggs	1.5/egg	6210	4750	1460	0.31
Apiary		8 kg	100	800	180	620	3.44

Pisciculture	0.28ha	295kg	40	11800	2560	9240	3.60
			Total	**110107**	**41817**	**68290**	

Socio-economic Background of the Study Area

An overview of the general characteristic of the study area lent evidence in support of the resource availability, resource constraints, and resource use pattern, socio-economic status of the inhabitants and the possibility of promoting their well being. A diagnostic survey of the area revealed that majority of the farmers were having less than one acre of land with little irrigation network. Farmers were practicing mono-cropping of only rice. Due to heavy low level of income, even the children accepted wage employment to supplement house hold income at the cost of education. The employment pattern of the adults in business, professional occupation and services was far from satisfactory constituting very insignificant proportion of the total employment. As such, the proportion of work force without any employment opportunity reflected an alarming situation

A similar analysis of the livestock position (cow, buffalo and goat) revealed that they were mostly of local breed and were reared only on sentiment ground and maintenance of entire livestock population lacked commercial consideration.

A close look at the position of water bodies depicted a sorry state of affairs. In the absence of renovation and management, the tanks, be it public or private served the purpose of bathing placing for the villagers and the village animals, rather than under taking pisciculture on commercial line.

In view of the low level cropping intensity, employment opportunity and high of poverty, constraint analysis of both social and economical aspects assumed relevance. Convincing evidence showed that Socio-economic constraints prevented farming community from taking full advantage of modern farm technology in exploiting the available land, livestock and other resources to their best advantage.

CHAPTER 21

Future Guide Line and Conclusion

India's population is expected to reach 1370 and 1660 m in 2030 and 2050 AD, respectively. The country's food production has reached an all time high of 204 m t during 2000 AD. A food production level of 289 and 349 m t is needed to satisfy the project population in 2030 and 2050 AD, respectively. Future agriculture in India, a country immersed in poverty and considered as hopeless by some forecasters and yet exporting food within five years of the deadline indicated by them is a tempting subject for speculation despite the pitfalls. It is fascinating to ponder and analyze over its performance in the past decades and to think about the future. If we see the directions in which science and population changes, economics and other forces are taking us, we may have a chance to try to steer clear of dangers. If there is the likelihood of population growth threatens to outstrip food supplies, the knowledge will help us to focus the efforts of the government and the scientists on the task of increasing food production. The prospects of Agriculture in India 2030 AD and thereafter will be immense value for the entire world, due to its size, population, pivotal position, desperate attempt to emerge from poverty and its linkage with the developed and developing countries in history, culture, industry and progress.

INDIAN AGRICULTURE

India, with 60% of its farmers and agricultural labourers as illiterates, is the third largest producer of technical manpower on the world. Though ranked as the tenth industrial country, it has one of the lowest standards of living; human and animal power is used to a largest extent for agricultural purposes along with tractors and tillers. One may witness well managed green carpet of tea plantation adjoining a highly eroded land of shifting cultivation of a small farmer.

In the agricultural technology, one may find luxuriant growth of improved rice raised with fertilizers adjacent to a pale, sparsely populated local variety manured with green leaves and sheep penning; fields irrigated with gushing water from electric pump sets adjoining to trickling water lifted by human labour. When rodents are killed by modern chemicals, temples are found wherein they are fed and worshipped. Grains are dried in modern driers while the same is sun dried using high ways as a drying yard.

The country has the largest cattle population in the world, but the yield of the milk and its consumption is one of the lowest. A country with large natural resources for raising crops throughout the year is more than 70% of the area, with assured irrigation, but suffering every year simultaneously with drought and flood in monsoon season. A country possessing one of the longest sea coasts for marine fishing, but consumption of per capita fish is one of the lowest. This, in sort, indicates the present situation wherein modern technology is trying to co-exist with primitive agriculture.

General Indian Scene Today

Availability of facilities like housing, medical care, school, drinking water supply, transport facilities, per capita consumption of food, clothing, energy, minerals and other natural resources are the principal indicators of levels of living which may have a direct bearing on the development of agriculture. The levels of living are miserably low. With more than 15% of world's population, India consumes only 1.5% of the world's energy input. Even the modes of production in agriculture are largely primitive and modern science and technology has not made much impact in deep countryside. Lack of education and adequate health care, the presence of irrational attitude towards life and work, and traditional social institutions detrimental to advancement, have constantly threatened to thwart the force of modernization.

Human resource is the most under-utilized. A sizeable part of labour force doses not engage in any form of work and most of those who do work have the job for any fraction of a day, week and year and that too not efficiently. In spite of the family planning efforts, the labour force may continue to increase by about 2% up to 2000 AD. Even if more employment is generated in the industries, they are unlikely to absorb a sizeable portion of unutilized or underutilized labour force which has to find work in agriculture. Because of this situation, the labour input per worker as well as the productivity of land is low. A series of integrated rural development with agriculture occupying the pivotal position is now in the offing suited for each region and district.

DEVELOPMENT POLICY

Development policy in India has proceeded in a much more centralized fashion than in any of the other less developed countries. A series of five year plans beginning in 1950 et the tone for Indian development policy which is retrospect have had two main thrusts. The first and primary thrust was that of heavy public investment in capital intensive industrialization including agricultural inputs like seeds, fertilizers and pesticides. Heavy industrialization was supposed to provide the focus for rapid growth generating capital for the future and eventually to provide resources with which poverty could be attacked.

A secondary thrust was the stated policy of achieving equity by a series of steps to be taken. Land laws were made more equitable and resources made available to small farmers and businessmen. But in reality, the agricultural programmes have tended to help larger farmers and education made available to appear income groups.

This old and dominant growth strategy, though intellectually attractive, had brought unnecessarily slow growth and very narrow participation in the fruits of growth as well. Thus in the context of increasing population, the extent of poverty has increased. The poor are left with little means of raising their income apart from its unequal distribution.

Desire to find means to accelerate growth, concern to reduce poverty and hunger and fear of growing food gaps have focused attention towards agriculture and rural development. In this strategy, agriculture becomes the sky for development facilitating the mobilization of otherwise idle labour. Secondly, the strategy relies on cost reducing, income raising technological change in agriculture to give an initial boost to real national income. Thirdly, increased agricultural income is used as a stimulus to other sectors, through growth in consumer demand, thereby inducing the profits necessary for capital formation in both agriculture and industry. Finally international trade necessarily plays an important role in the new strategy.

Key Role of Agriculture in the Development

Agriculture occupies one-half to two-third of the labour force, produces about half the gross national product and provides well over half the consumer goods. Mobilization of labour for productive purposes is necessary in the absence of scarce capital resources. Because the labouring classes spend the bulk of additional income on food, mobilization of labour requires increased agricultural production. If employment increases without a commensurate increase in food supplies, inflationary pressure will mount and lead to demand restraint policies that reduce employment. Thus accelerated growth in agricultural production is essential to the mobilization of labour.

Increased food consumption is essential to improved welfare for the poor. There is need for sustained growth of food production as well as consumption by the poor. Thus, if employment is to increase, if labour is to contribute to growth and if the poor are to prosper, food production must rise.

In the early decades, accelerated annual growth rate was about 2.5%, more or less keeping up with population growth mostly by increasing the area. If agriculture has to play its dynamic role in development, growth of at least 4% is required. This can be achieved only with the application of modern science is developed of high yielding crop varieties and practices. New technology not only increases yields per hectare, thereby circumventing the troublesome problem of rapidly approaching limits to expand the area, but also increases labour productivity and over-all efficiency in resource use.

A Reappraisal of Agricultural Development

Although India's major agricultural commodities are wheat, rice, millets, peanuts, sugar. tea, jute and a variety of spices and condiments, only tea, jute, spices and condiments, like pepper and cashew nuts are exported. Exports of other farm products such as coffee, sugar, scented rice etc. have started. The growing population consumes most of the primary products. In drought years import of food grains have to be resorted to.

Intensification of the new technology for increased production and growth rate needs two characteristics of importance to development planning. First, because of the the variability of agricultural conditions, new technology can not be transferred easily from one environment to another. There is a need for a vast system of agricultural experimental stations covering a wide range of conditions and trained personnel. Fortunately for India, there are research and experimental stations organized by the Indian Council of Agricultural Research, Agricultural Universities and State Departments of Agriculture throughout the country. Co-ordinated as well as *ad hoc*, time bound research Projects are carried out by thousands of well qualified research workers. The 32 agricultural universities produce thousands of graduates and postgraduates every year and most of them are employed either by the State Government or nationalized banks or firms related to agriculture. Hence, sufficient infrastructure in terms of trained manpower is available.

Secondly, modern technology requires an immense supply of purchased inputs such as fertilizers, pesticides and increased control of water, as well as ready communication with sources of technology. Besides investment in roads and electrification are necessary. The country though importing certain qualities of fertilizers and pesticides produces most of them. During the past three decades, heavy investment has been made on irrigation, electricity and communication including rural roads. Certain States like Keral and Haryana are proud of their achievement of 100% coverage of electricity. States like Tamil Nadu and Maharashtra are racing towards the goal with more than 90% coverage. Similarly States like Punjab claims the highest per hectare yield in wheat production in the world. Attempts are made in other crops and states also. Thus the second infrastructure is also available at present in the country.

But in carrying out the modern technology in agriculture, the problems of ecological regeneration of the hills and valleys have to be taken into account. Already the imbalance caused by ruthless cutting of trees and denudation of forests in the Himalayas and other hills had telling effect on erosion and floods during the monsoons. Hence, the eco-system of entire ranges is to be protected to have viable, stable agriculture down below the Indo-Gangetic plains and other valleys

Another problem needing attention in carrying out modern agricultural technology is the energy conservation and recycling. For accelerating food production fertilizers and pesticides are considered very important in the modern technology. But, both have to be obtained from oil resources, the high price of which is felt throughout the world. Hence, the need for energy conservation and recycling has to be given importance without affecting food production. Alternate sources of nutrients have to be harnessed and used, besides developing techniques by which renewable and non-renewable forms of energy have to be blended in an optimum manner.

Yet another problem peculiar to India is the dependence on aberrant weather in crop production. Growth with stability could be achieved only by developing technologies which could mitigate to the extent possible, adverse weather conditions. The Coordinated Dry land Projects are in operation for more than twenty years and establishment of International Institutions like ICRISAT in Hyderabad, technologies are available. But these have to be carried out in farmers fields through education and extension.

The new technology of crop production has to be blended with livestock production and aquaculture. Mixed farming has always been a way of life in rural India. But this system needs improvements so as to provide more income, employment and food per unit space, time, water and sunlight. Social security measures like the food for work programme initiated enables one to win his daily bread through work. This will help in the eradication of hunger without eroding human dignity. This concept of humanism where all human beings enjoy the right to work and thereby claim a share of production has to be super-imposed over science and technology.

RESEARCH AND EDUCATIONAL NEEDS AND THEIR ROLE

The research centres are meant to attract talents to solve technical problems of producing more food and to get the results of that scientific research applied in practice. The information unless applied by the farming community, would be a sterile scientific exercise. Agricultural technology, though necessary, is not the sole element in increasing food production. The process require a complex, broad scale effort which includes such other elements as incentives for farmers, availability of proper equipment, supplies of seed, fertilizers, agricultural credit and a whole series of policies which are not limited to the agricultural sector. Agricultural production does not advance by itself in any country, but only a part of the overall economic development process.

Research centres had demonstrated that they can perform several important functions: They could carry on fundamental research; they could convert this to adaptive research and they could extend the benefits of that research to areas where they could have a guaranteed performance under certain kinds of management.

Agricultural Research of the Future

Research in the country will be widened so as to cover all important food and commercial crops. Greater emphasis will be laid on research on land utilization, dry farming, integrated farming, plant protection, quick growing fuel trees, post-harvest technology, fisheries and animal production.

GENERAL FINDINGS

The present studies on the farming systems followed by different categories of farmers both under rainfed and irrigated condition generated the following information.

1. When increase on farm area is restricted, adoption of farming system is only alternative.
2. In rainfed conditions also it can be adopted with the maximum return per rupee invested up to Rs.2/- spreading over the year round.
3. For marginal farmers it can be most effective with suitable utilization of resources and family labour force instead of wage earning and services etc.
4. Smaller the farm holding site, more efficient the farming system.
5. Fruit crops, mushroom, apiary, animal production, poultry etc. have been more viable.

Future Food Needs and FSR Status

To feed the mouth of ever increasing population is a great challenge before us. The food supply in 1987-88 was 189 kg and in 1989-90 it was 210 kg per person per year (*i.e.* 575 g/day/person). Based on food supply of 1989-90, India will have to produce 210 million tonnes food in year 2000, 289 million tonnes in 2030 and 349 million tonnes in year 2050. If food supply is enhanced by 10% to eradicate mal-nutrition in lower strata of society, the demand will grow to 231 million tonnes on 2000, 318 million tonnes in 2030 and 384 million tonnes in 2050.

Food production projection, based on gross cropped area, gross irrigated area, fertilizer consumption, high yielding variety seed use, farm lands productivity, agricultural credit, fertilizer price and output price is estimated to be 211 million tonnes in 2000, 316 million tonnes in 2030 and 387 million tonnes in 2050. The share of rice and wheat will be the largest.

FUTURE PROSPECTS

Agricultural research has attained a considerable degree of sophistication by now. The farming techniques developed under the various disciplines have proven potentials of achieving the maximum production and profitability in relevant enterprises. So it is high time to ponder over an integrated system so as to cater to the needs of total farm development and the system is the farming system research.

The system should be organized and think about in a broader prospective and material approach in view of varied agro-climatic conditions ecological systems, socio- economic conditions of the farmers and the agricultural research education, extension infrastructure prevailing in the country.

Due to vagaries of monsoon, lack of irrigation facility in major areas, smaller land size etc. Farming system is most effective farm development, which will bring about.

1. Proper utilization of unutilized resources
2. Profitable use of farm wastes
3. Reduction in soil, water & air pollution.
4. Provision of suitable engagement of all farm family members.
5. Development of agro-based wastage industry.
6. Provision of avocation of landless labourers.
7. Sustained production.

Strategy to be Adopted

Following activities should be performed for effective performance of farming system with multidisciplinary team of scientists and active participation by the farmers in all stages.

a) Through undertaking of existing farming system and their components.
b) Assessment of resources availability in the farm environment and identification of bio-physical, socio-economic, institutional adminstrative and technological constraints.
c) Development of economically viable and efficient integrated farming systems suitable for various demands.
d) Diffusion of improved technologies and receiving feed back for further improvement of the system.
e) Continuous improvement in component technology to fit into a given farming system.
f) Research extension linkage through "On-Farm Adaptive Research".
g) Improvement in quality of farming system research.
h) National and International linkage.

Important Feedback and Future Lines of Work

1. Farmers of the study area were extremely poor and were unable to have two square meals in most of the days throughout the year.
2. Since they were not able to afford high cost inputs, generation of low cost technology through modification of existing practices is of great relevance.
3. Seed multiplication at village level was vital as the farmers do not get seeds at right time, nor they were able to save their seeds due to poverty. Efforts should be directed to multiply and distribute seeds of promising varieties of crops.
4. There is no organizational support and marketing of produce at the village level. As a result the farmers are subjected to distress sale.
5. Majority of the farmers are not attracted merely by the high yield of variety rather they judge the variety on the basis of consumers' preference and require low monetary inputs.
6. Appropriate farming system model is to be developed based on soil, climate and resource availability.
7. Middle men take away the profit. Prevention of exploitation through middle men is to be discouraged.
8. Input, credit and technology delivery system should be synchronized.

SPECIFIC OBJECTIVES TO BE STUDIED

1. To evolve suitable farming systems for the small and marginal farmers of under irrigated and rainfed conditions.
2. To find out optimum combination of agriculture with one or more of dairy, poultry, duckery, fishery, apiary systems as profitable and viable mixed farming systems.
3. To find out the increased employment opportunities generated by these farming systems
4. To recycle the by products of different systems and their economy in reduction of fertilizer consumption in building of soil fertility and maintaining productivity of the farm.
5. To estimate relative income from various components and total income from the system.
6. To study the short and long-term effects of these systems on over all economy of the farmer and of the system.
7. To identify constraints in increasing farm productivity.
8. To provide technological intervention options for improving farming system with a given resource base through On-farm Adaptive Research (OFAR).
9. To refine and develop appropriate farm technology to provide stability in production, employment through out the year, income at periodic intervals, improved management skill and decent living to the farmers.
10. To develop models/methods for measurement and monitoring of sustainability of farming systems.

Stategy for Farming Systems Research and Extension

The farming system approach is considered important and relevant specially for the small and marginal farmers. The technology has to be spread over the length and breadth of the country. Some strategic steps are suggested here:

1. Delineation of different parts of the country for different activities of FSRE programmes basing on similar situations.
2. The areas delineated may be
   a) Eastern Coastal Zone with plains
   b) Western Coastal Zone
   c) Northern Zone with alluvial soils
   d) Southern Zone with undulating hilly areas
   e) Central Zone with low rainfall
   f) North Eastern Frontier Zone with hilly areas and high rainfall
   g) Northern Zone with hilly areas and low temperature
   h) North Western Zone with very low rainfall
   i) Physically handicapped ecologies
3. Developing farming system models for marginal and small farmers which constitute 70% of the farm families.
4. Fixing up of a major enterprise for each situation and integrating other enterprises with it, taking into account the factors like the opportunities available, constraints, socio-economic condition, marketing, suitability for the locality, financial assistance available, transportation, storage etc.
5. Sustainability of enterprise combinations to be studied carefully before recommendations.
6. Stress 'on-farm trials' and demonstrations and component technology.

7. Suitable level of combination of enterprises.
8. Developing farming systems courses for under-graduate and post-graduate programmes in Agricultural Universities.
9. Rigorous multidisciplinary academic research on farming systems in State Agricultural Universities.
10. Engaging experienced personnel in the line departments of the Government organizations, Agricultural Universities, NGOs and other developmental agencies.
11. Providing adequate on the job training to the scientists, extension workers and the farmers on FSRE.
12. Generating adequate research information on component technology relevant to the Farming Systems Programme.
13. Establishing Lead Centres on Farming Systems and Coordination Committees in different Zones to monitor the activities in the respective zone. A Central Cell on Farming Systems may be set up that would formulate the principles of Farming Systems activities in the country.
14. International and national study tours for the farmers and extension workers to study the farming systems operations, results and economics.
15. Institution of rewards for the best activities and successful operations of farming systems.
16. Popularization of the programme through different mass media / communication systems.

CONCLUSION

In a country like India, where the population is hiking in a faster rate, all should impress much regarding stress on food grain production also with fodders, fuel, fibre requirements, whereas the land area is limiting leaving no scope for any horizontal expansion, only alternative is the intensive techniques like farming system research. One to rush in production of one or two enterprises for the imbalance monsoon; social, political and economical constraints, it is most favourable to adopt integration of various enterprises in a limited space, also in rainfed and problematic situations, F.S.R. also maintains the retain ability of environment and ecological balance giving a larger return per rupee invested with stability in income all over the year.

In this topic, it is given the objectives necessities, scope, characteristics of F.S.R./E. It also highlights on the research work done under the Universities, its area of adoption, how much it is successful in extending the technology with future strategy to be adopted for better efficiency.

There is vast scope to adopt the technology of F.S.R./E in our situation where there is limited resources and with risk-prone areas. It will be safe-guard against the food grain crisis and to meet the daily requirements.

The future of Indian agriculture depends heavily on the development of appropriate farming system as applicable to resource-poor farm families and as suited to different agro-ecological zones. The endowment of abundant sunshine, long growing season, responsive soil types, and combination of surface water, ground water and seasonal rains and above all a progressive peasantry, offer vast scope for an intensive farming system though multiple cropping and diversified farming including animal husbandry, forestry, sericulture, fisheries and the like. The challenge is to upgrade technological and social disciplines on a continuous basis and integrate these disciplines to suit the region and the farm families in a manner that may ensure increased production with stability, ecological sustainability and equitability.

Research studies carried out in the lowlands of TNAU and OUAT have shown the technical feasibility and economic viability of the integrated farming systems attempted. Besides facilitating cash income, these models generate additional employment for family labour and minimize the risks associated with mere crop production.

In view of serious limitations on horizontal expansion of land for agriculture, the only alternative left is for vertical expansion and increase in productivity of various farm enterprises requiring less space and time but giving high income. This would ensure regular and gainful employment and periodic income especially to the small and marginal farmers belonging to rainfed areas, dry lands, arid zone, hilly areas, tribal belts and problem soils. For this, several farm enterprises should be integrated viz., crop-dairy, agri-livestock, agri-livestock-poultry, agri-horticulture, agri-horticulture-sericulture, agro-forestry, silvipastoral, rice + fish rice + fish + mushroom, floriculture + apiary, fish + duck + poultry, etc. Since farming system is a new and developing concept/subject, new courses shall have to be designed and offered at the under-graduate and post-graduate levels of the Agricultural Universities. The FSRE methodology has to be taught at the post-graduate level pending finalization of such academic curriculum. Scientists belonging to various disciplines but involved in system research must be involved in the programmer without further delay. For immediate manpower requirement, farmers training on various aspects of farming systems are also equally important for successful FSRE programme implementation.

References

Anonymous 1960. Indian tobacco – A monograph, Indian Central Tobacco Committee, Ministry of Food and Agriculture, Government of India, Chenni.

Anonymous 1964. Annual Report, JARI, Bairackpur.

Anonumous 1967, 1979. Rice Prod. Man., Univ. of Philippines and IRRI, Philippines.

Anonymous 1970-78. Annual reports of All Ind. Co-ord. Maize improvement Project.

Anonymous 1970-78. Annual reports of All Ind. Co-ord. Sorghum Improvement Project.

Anonymous 1971-74. Annual reports of All Ind. Co-ord. Wheat Improvement Project.

Anonymous 1976. Annual reports of All Ind. Co-ord. Bajra Improvement Project.

Anonymous 1976. Annual reports of All India Cor-ord. Barley Improvement Project. Anonymous 1977 Trombay groundnut, BARC, Trombay, Bombay.

Anonymous 1984 Hand book of Agriculture, ICAR, New Delhi.

Abrol, P. and Palaniappan, S. P.1987. Green manure crops in irrigated and rain fed lowland rice based cropping systems in South Asia. In: Green manure in rice farming. Proc. of a Symposium on "Sustainable Agriculture", Los Banos, Philippines, pp. 71.

Abrol, I. P. and Drua Narayana, V. V. 1990. Technologies for Wastelend Development., ICAR, New Delhi.

Abrol, I. P. and Gupta, R. K. 1996. Sustaining high production systems: Research issues and approaches. Souvenir, 2nd International Crop Science Congress, Nov. 17-24, 1996, New Delhi, India

Adil, M. L. and Karte, R. K. 1992. Blue green algae in rice soil ecosystem. Proc. National seminar on "Organic Farming". Rai M. M. and LN. Verma (ed.), Ravi Printing Press, Jabalpur, India, pp. 200-210.

Agarwal, M. C. and Khanna, S. S. 1983. Efficient soil and water management in Haryana., Haryana Agri. University, Hissar.

Agarwal, R. and Tilak, K. V. B. R. 1988. Establishment of *Azospirillum brasilense* in roots of finger millet. Microbio. 143: 539-544.

Agarwal, R. R.; Jadev, J. S. P. and Gupta, R. N. 1982. Saline and alkaline soils of India., ICAR, New Delhi.

Ahmed, M. and Rab, A. B. 1992. Feasibility of adopting aquaculture to increase resource productivity in existing Bangladesh farming systems. NAGA, The ICLARM Quarterly, 15: 21-22.

Ahmed, S. 1990. Neem pest control. ILEIA Newsletter. 6: 7.

AICRPDA (All India Coordinated Research Project for Dry land Agriculture) 1982. A decade of dry land agricultural research in India - 1971-80. Hyderabad, Andhra Pradesh, India: AICRPDA.

Alan, R. Putnam and Chung Shih Tang. 1986. The Science of Allelopathy. John wiley and sons. Canada

Albert, E. H.; Norman, D. W. and Winch, F. E. 1980. Farming Systems Research. A critical appraisal. Michigan State University Rural Development-paper, 6, East Lansing Michigan, USA: Michigan State University.

Alden, S. Crafts and Wilfred W. Robbins, 1973. Weed Control – A text book and manual. Tata McGraw Hill Publishing Co. Ltd., New Delhi.

Alexander, M. 1977. Nitrogen fixation. In: Introduction to Soil Microbiology, 2nd ed. Willey, New York. pp. 305-330.

Allison, F. E. 1973. Soil organic matter and its role in crop production. Elsevier, Amsterdam, pp. 104-105.

Altieri, M. A. 1987. Agroecology: The scientific basis of alternative agriculture. Boulder/London: West view press (ITP).

Altieri, M. A. and Liebman, M. Z. 1986. Insect, weed and plant disease management in multiple cropping systems. In: Francis, CA. (ed.), Multiple cropping systems, Mac

Millan, New York, pp. 183-218.

Anandeshwari Awasthi. 1995. Indian Climatology., APH Publishing Corporation, New Delhi.

Andrews, D. J. and Kassam, A. H. 1979. The importance of multiple cropping in increasing world food supplies. *In:* Papendick, P. A., Sanchez, P. A. and Triplette, G. B. (eds) Multiple cropping. American Society of Agronomy, Pub 27, Madison, Wisconson.

Annadurai, K.; Palaniappan, S. P.; Chinnusamy, C. and Jayanthi, C. 1994a. Integrated farming system, Kisan world 21 (8): 51.

Annadurai, K.; Palaniappan, S. P.; Chinnusamy, C. and Jayanthi, C. 1994b. Integrated farming system: Need of the hour. Kisan world 21 (9): 39.

Arnon, I. 1972. Crop Production in dry regions. Vol. I. Leonard Hill London.

Arora, P. N. 1967. Seed potato should yield better in Northern India. Ind. Fmg. 16(19).

Arora, S. P. 1980. Fodder and feed resources requirement, availability and development. The Indian Dairyman. 32: 821-828.

Auld, B. A. and Kim, K. U. 1998. Weed Management in Rice. Oxford and IBH Publishing Co. Pvt. Ltd., New Delhi.

Aune, J. B. and Lal, R.1994. The tropical soil productivity calculator: A model for assessing effect of soil management on productivity. In: Lal, R. and B. A. Stewart (eds.) Sustainable management of soils. Lewis Publishers, Chelsea, Ml.

Awasthi, R. P. and Prasad, R. N. 1992. Farming Systems approach to manage resources for sustained productivity in Northeastern hill regions of India. Proceeding of the XII National symposium on "Resource Management for Sustained Crop Production" held at Rajasthan Agricultural University, Bikaner during Feb. 25-28. pp. 333-342.

Baguaraj, D. J.; Manjunath, A. and Patil, R. C. 1979. Interaction between YAM and their effects on soybean in the field. New Phytol. 82: 141-145.

Balagopal, K., 1994. Role of Chicken, Pigeon, Japanese quails and Duck rearing in sustainable IFS. In: Proceeding of Summer Institute on "Integrated Farming Systems Research and Development for Sustainable agriculture" organized by ICAR and Tamil Nadu Agric. Univ., Coimbatore, India during 6-15 June; 1994 pp. 235-240.

Balasubramanian, A., 1994. Farming system research and development in western zone of Tamil Nadu. In: Proc. of the Summer Institute on "Integrated Farming System Research and Management for Sustainable agriculture". Tamil Nadu Agric. Univ., Coimbatore. pp. 278-284.

Balasubramanian, T. N., 1988. Farming system research survey for new Moriones village in Philippines, p. 9. In: Abst. National seminar on "Farming Systems for Semiarid Tropics". Indian Council of Agricultural Research and Tamil Nadu Agric. Univ., Coimbatore.

Balasubramaniyan, P. and Palaniappan, S. P. 2001. Principles and Practices of Agronomy. Agrobios (India), Ludhiana.

Baldev, B.; Ramanujam, S. and Jain, H. K.1988. Pulse crops. Oxford and IBH, New Delhi.

Balusamy, M., 1996. Studies on nitrogen management in lowland rice-fishazolla integrated farming system. Ph.D. Thesis, Tamil Nadu Agric. Univ., Coimbatore.

Balusamy, M. and Shanugasundaram, V. S. 1994. Integrated farming system studies at Agricultural Research Station, Bhavanisagar. In: Proc. of Summer institute on "Integrated Farming System Research and Management for Sustainable Agriculture". Tamil Nadu Agric. Univ. Coimbatore. pp. 270-277.

Barik, K. C. 1996. Effect of intercropping, nitrogen, phosphorus, potassium and cutting management on sabai grass. Indian J. Agril. Sci. 66(12): 689-693.

Barik, K. C. 1997. Growth and yield characteristics of sabai grass germplasms. Indian J. Agril. Sci. 67(4): 173-174.

Barik, K. C. 1997. Effect of row x plant soacing on sabai grass. Indian J. Agril. Sci. 67 (10): 44-45.

Barik, K. C. and Panda, M. M. 1977. Sabaigrass cultivation in Mayurbhanj district of Orissa-Governs Tribal Economy. Agril. Situation India: 583-584.

Barik, K. C. 1998. Productivity and economics of sabai grass (*Eulaliopsis binata*) under different weed management methods. Indian J. Agril. Sci. 68(7) 378-9.

Barnes, A. C. 1974. The sugarcane. Leonard Hill Books.

Barry, R. G. and Chorley, R. J. 1976. Atmosphere, Weather and Cilmate. Third edition,

English language Book Society and Methuen and Co., Ltd. Suffolk.

Barwale, B. R. 1993. Hybrid Rice. Food security in India. MacMillan India Ltd., Madras, India.

Beets, W. C. 1982. Multiple cropping and tropical farming systems. Westview Press, Boulder, Colorado.

Behera, U. K. and Mahapatra, I. C. 1999. Income and Employment Generation for Small and Marginal Farmers through Integrated Farming Systems. Indian J. Agron. 44(3): 431-439.

Beldev Sahai and Ajai.1988. Application of Remote sensing techniques in Agriculture. Fert. News 33(4): 59-65.

Berry, F. A. and Bollary, E. 1945. Hand book of Meteorolgy. McGraw Hill Book Company Inc., New York.

Bhandari, S. C. and Somani, L. L.1990. Azotobacter In: Biofertilizers. Somani; L. L, S. C Bhandari, S. N. Saxena and KK Vyas (ed.), Scientific Publishers, Jodhpur, India. pp. 355-361.

Bhatnagar, K. C.; Khurana, C. L. and Leelawati, C. R.1989. Statistical assessment of crop sequences in different agroclimatic zones in the country. Bulletin of the Indian Agricultural Statistics Research Institute, New Delhi, India: ICAR.

Bhatnagar, P. S. 1976. Techniques of sugar beet Production. Ind. Fmg Digst. 9(4): 27-29.

Byerlee, D. and Collinson, M. 1980. Planning technologies appropriate to farmers: Concepts and procedures, EI Batan, Mexico: CIMMYT.

Byerlee, D.; Harrington, L. and Winkelmann, D. L.1982. Farming systems research: Issues in research strategy and technology design. Amer J. Agric. Econ. 64: 897-904.

Caldwell, B. E. 1973. Soybeans, American Soc. Of Agronomy, INC, Wisconsin, USA.

Campbell, CA.; McConkey, B. G.; Zentner, R. P.; Dyck, F. B.; Seller, F and Curtin, D. 1995. Carbon sequestration in a brown chernozen as affected by tillage and rotation. Can. J. Soil Sci. 75: 499-458.

CGIAR, 1981. Second review of the CGIAR, Review report, Nov. 1981. Washington, D. C, CGIAR, Secretariat.

Chakraverty, A. 1995. Post harvest technology of cereals, pulses and oilseeds. Oxforrd anf IBH Publishing Co. Pvt. Ltd., New Delhi.

Chambers, R. 1980. The Small Farmer is a Professional. Cares. 13(2).

Chambers, R. 1990. Micro-nutrients unobserved. Gatekeeper Series 22. London 11ED.

Chang, J. 1968. Climate and Agriculture. Aldine Publishing Company, Chicago.

Chatterjee, B. N; and Maiti, S. 1984. Cropping systems-theory and practice. Oxford and I. B. H. Publishing Co., Calcutta, India.

Chatterjee, B. N. and Bhattacharya, K. K. 1986. Principles and Practices of grain legume production. Oxford and IBH Publishing Co. Pvt. Ltd., New Delhi.

Chatterjee, B. N. and MandaI, B. K.1988. Rice based farming systems with integration of fish culture. In Abst. of National seminar on "Farming Systems for Semi-arid Tropics". Tamil Nadu Agric. Univ., Coimbatore. pp.1-7.

Chatterjee, B. N. and Das, P. K. 1989. Forage crop production- principles and practices. Oxford and IBH Publishing Co. Pvt. Ltd., New Delhi.

Chellaiah, S. and Chandrasekaran, B. 1989. Integrated farming system for Thanjavur district, Tamil Nadu. Thanjai Ulavar Mani, Dec. 88-Jan., 89 pp. 5-16.

Cheema, S. S.; Dhaliwal, B. K. and Sahota, T. S. 2001. Theory and Digest Agronomy. Kalyani Publishers, Ludhiana.

Chen T. P. and Li, Y.1980. Integrated agriculture aquaculture studies in Taiwan. In: Proc. of ICLARM-SEARCA conference on "Integrated Agriculture - Aquaculture Farming System". 4th Aug. 1979. Manila, Phillippines pp.: 239-241.

Chidda Singh, 1984. Modern techniques of raising field crops. Oxford and IBH publishing Co. Pvt. Ltd., New Delhi.

Chinnusamy, C. and Purushothantan, S. 1992. Integrated farming systems under garden land and their economics. In: Proc. National training on "Integrated Farming Systems". GOI and Tamil Nadu Agric Univ., Coimbatore. 10-17 Dec., 1992 pp. 46-48.

Choudhary, S. L. and Bains, S. S. 1966. Growing wheat – the intensive way. Ind. Fmg. 19(7): 4-7.

CIMMYT, 1974. Plan publication: Seven years of experience, 1967-1973. EL Batan, Mexico:

CIMMYT.

CIMMYT and ISNAR, 1984. Issues in organization and management of research with a farming systems: Perspective aimed at technology generation (Ed. M. Joseph) The Hague, Netherlands: CIMMYT and ISNAR.

Collinson, M. P.1982. Farming systems research in East Africa: The experience of CIMMYT and some National Agricultural Research Services, 19761981. Michigan State University International Development Paper, 7, East Lansing, Michigan, USA.

Collinson, M. P. 1983. Farm management in peasant agriculture, 2nd edition, West View Press.

Collinson, M. P. 1987. FSR procedures for technology development. ExptJ. Agric. 23: 365-388.

Collinson, M. P. 1988. The development of African farming systems – Some personal views. Agric. Admin. and Extension 29: 7-22.

Concise Handbook for field official of Non-conventional energy sources, 1993. Published by Directorate of Rural Development, Government of Tamil Nadu, Chennai.

Cornick, T. 1983. The Social organization of production in Quimas, Ecuador: A case study of small farmer production in the highlands. Ph.D. Thesis, Cornep University, Ithaca, New York.

Crafts, A. S. 1975. Modern Weed Control. University of California Press, California.

CRIDA (Central Research Institute for Dry land Agriculture), 1986. Fifteen years of dry land agricultural research in India. Hyderabad, Andhra Pradesh, India: Central Research Institute for Dry land Agriculture.

Critenfield, H. J. 1975. General Climatology. Third edition. Prentice-Hall of India, Private Ltd., New Delhi.

CSSRI, 1979. A decade of Research, Central Soil Salinity Research Institute (ICAR), Karnal, India.

Daniel Sundaraj, D. and Thulasidas, G. 1993. Botney of Field Crops. Mac Millan India Ltd., New Delhi.

Dargan, K. S.; Singh, O. P. and Gupta, I. C. 1982. Crop production in salt affected soils. Oxford and IBH Publishing Co. Pvt. Ltd., New Delhi.

DARE (Department of Agricultural Research and Education), 1990. Annual Report, New Delhi. India: Department of Agricultural Research and Education of the Ministry of Agriculture, Government of India.

Dargan, K. S. 1969. Jute Agronomy. Ind. Fmg. 18: 35-37.

Das, B. K. 1958. Jute. Ind. Agriculturist 2: 35-45.

Das, K. C.; Panda, S. C.; Sahu, S. K. and Rout, D. 1985. Effect of depth of submergence at different growth stagesof dwarf *indica* rice on the growth, yield and nutrient uptake by rice and mineral contents of soil. Oryza. 22(1): 69-83.

Das, L. K. and Panda, S. C. 2002. Economics of hybrid rice cultivation as influenced by Nitrogen and Potassium levels. Jour. Res. OUAT, 20(1):50-52.

Dash, A. N.; Mahapatra, P. K. and Panda, S. C. 1991. Yield performance of rice under different irrigation schedules and soil fertility. SUJ. Sci. & Tech. X(6-13) Silver Jubilee Volume.

Datta, R. K. 1989. Weather Forcasting – Role of medium range techniques. In Survey of Indian Agriculture 1989. The Hindu, Madras.

De Datta, S. K. 1981. Principles and Practices of Rice production. John Wiley and Sons, New York.

De Gopal Chandra. 1989. Fundamentals of Agronomy. Oxford and IBH Publishing Co. Pvt. Ltd., New Delhi

Delmando, M. N. 1980. A review of Integrated livestock, bowls and fish farming systems. In: Proc. ICLARM - SEARCA Con£. on "Integrated agriculture - aquaculture farming systems", 4th Aug. 1979, Manila, Philippines.

Demnis, A. S. 1983. Augmentation of rainfall from summer cumulus clouds. Agric. Water Manage., 7:3-14.

Dev, G. and Sander, L. L. 1989. Maximum yield Research. Proceedings of symposium on Maximum yield research. Nov. 16-18, 1988, New Delhi. Potash and Phosphate Institute of Canada, India Programme, Potash Research Institute of India, Haryana, India.

Devadas, P.; Rajammal, P. S. and Sithalakshmi, S. 1988. Appropriate home science technologies with their implications to farmwomen and farming. In: International Conference on "Appropriate Agricultural Technologies for Farm Women". Nov. 30 - Dec. 6, 1988.

Dillon, J. L. and Anderson, J. R. 1984. Concept and practice in Farming Systems, Research. In: Proceedings of AGAR Consultation in Agricultural Research in E. Africa (Ed. J. V. Martin), Canberra, AGAR.

Dillon, J. L.; Plucknett, D. L. and Vallaeys, G. J. 1978. Farming systems research in the International Agricultural Research Center, Rome Italy: Technical Advisory Committee to the Consultative Group on International Agricultural Research.

Djajadivedja, R.; Jangkaru, Z. and Mohamed Junu, 1980. Freshwater aquaculture in Indonesia with special reference small-scale agriculture. In: Proc. ICLARM - SEARCA Conf. on "Integrated Agriculture Aquaculture Farming Systems", 4th Aug., 1979, Manila, Philippines pp. 143-165.

Donn, W. L. 1965. Meteorology, McGraw Hill. New York.

Donnel Hunt. 1997. Farm power and machinery management. Iowa State University Press, Iowa, USA.

Dora, B. 1983. Traditional agriculture in India: High yields and No waste. The Ecologist 13: 84-87.

Doughlass, G. K. (ed.), 1984. Agricultural sustainability is a changing. World order Boulder: West view.

Dunham, R. S. 1957. Introduction to Agronomy. Dryden Press, New York.

Dupre, G. (ed.) 1990. Farming knowledge and development: Proeedings on the 7th section on Science, Diffusion of technical knowledge and systems of local knowledge of the 7th World congress of rural sociology, June 1988, Bologna, Italy.

Dwivedi, A. P. 1992. Agroforestry – Principle and practices. Oxford and IBH Publishing Co. Ltd., New Delhi.

Edwards, C. A. and Grove, T. C. 1991. Integrated nutrient management for crop production. Agric. Ecosystem and Environment.

Elliot, H. 1977. Farming systems research in Africa: Methods and results, Ford Foundation Farming Systems Seminar, Tunis, Tunisia, February 1 3, 1977.

Eugene, P. O. 1996. Fundamentals of Ecology. Natraj Publishers, Dehra Dun.

FAO, 1990. Comparative analysis of organic manure. Project field document. Vol. 24, FAO, Rome, Italy.

FAI. 1997. Fertilizer statistics 1995-96. The fertilizer association of India. New Delhi.

FAO, 1985. Production yearbook Vol. 37 FAa, Rome, Italy.

FAO. 1988. Sericulture manual Vol. I, II and III. Mulberry cultivation. Silk worm rearing and silk reeling. Oxford and IBH Publishing Co. Pvt. Ltd., New Delhi.

FAO. 1997. Year Book – Production Vol. 50. 1996. Food and Agricultural Organisation of the United Nations, Rome.

Francis, C. A. 1986. Multiple Cropping System. MacMillan Publishing Co., New York.

Francis, O. and Stem, W. R 1987. Cereal-legume intercropping systems. Advances in Agronomy, 41: 41-87

Franklyn, W. Cole. 1970. Introduction to Meteorology. John Wiley and Sons. Inc., New York.

Frasco. L. 1984. Comparing Anglophone and Francophone approaches to Farming systems research and extension. Paper prepared for the 4th annual conference on "Farming Systems Research" at Kansas State University, Manhattan, Kansas, USA, October, 1984.

Fresco, L. C. and Westphal, E. 1988. A hierarchical classification of farming system. Exptl. Agriculture 24: 399-419.

Gallagher, E. J. 1984. Cereal production. Butter worths and Co. Ltd., London.

Ganesan, S.; Chinnasamy, K. N.; Chandrasekaran, B.; Budhar, M. N. and Prince Jayaseelan, N. J. 1991. Duck-cum-fish culture in rice farming system in Cauvery delta region of Tamil Nadu. Indian Journal Agric. Economics 46(2): 180-185.

Gaur, A. G.; Neelakandan, S. and Dergan, K. S. 1984. Organic manures. ICAR, New Delhi.

Gaur, A. C. 1990. Principle and technology for composting urban wastes. Proc. of International Workshop on "Solid Waste Management and Resource Mobilization". Vol.

B. Paper 3-12, Kathmandu.

Gaur, A. C. and Gaind. S. 1992. Role of P solubilizing microorganims and enriched organic manure in crop productivity. In: Proc. National seminar on "Organic Farming". Ravi Printing Press, Jabalpur, India. pp. 134-143.

Glenn T. Trewartha. 1968. An introduction to Climate. McGraw Hill Book Co. New York.

Gilbert, EH.; Narman. D. and Winsh, F. E. 1980. Farming systems research. A critical appraisal. Rural development paper No.6, East Lansing Michigan State Univ., Michigan.

Gips, T. 1986. What is sustainable agriculture? In: Allen, P. and D. Van Dusen (eds.) Global perspectives on Agroecology and sustainable agricultural systems: Proc. of the 6th International scientific conference on "International Federation of Organic Agricultural Movements" (Sanfa CVUZ: Agroecology Program), University of California, 1: 63-74.

Gomez, A. A. and Gomeg, K. A. 1983. Multiple cropping in humid tropics of Asia. IDRC. Ottawa, Canada.

Gomez, A. A. 1990. Farming system research approach to identifying farmers' production problems. Farming system research approach (8): 63-70.

Gomez, K. A. and Gomez, A. A. 1984. Statistical procedures for agricultural research with emphasis on rice. John Wiley and Sons Inc., New York.

Gopalchari, N. C. 1984. Tobacco. ICAR. New Delho

Gopalswamy, N. 1994. Agricultural Meteoroly. Rawat Publications, New Delhi.

Goud, J. V. 1988. Strategies for Food Production to meet the Challenges of 21st century. Agric. Situation in India. Sept.: 505-506.

Govina, L. D. 1990. Pig-Duck-Fish-Azolla integration in La union, Philippines. NAGA the ICLARM Quarterly, Apr. 18-20.

Govindan, R. 1988. Role of Popultry-cum-Fish culture on the economics of farming systems in Thanjavur district of Cauvery delta. M.Sc. (Ag.) Thesis, Tamil Nadu Agric. Univ. Coimbatore.

Govindan, R.; Chinnasamy, K. N.; Chandrasekaran, B.; Budhar, M. N. and Price, J. 1990. Poultry - Fish culture in rice farming systems in Cauvery delta regions of Tamil Nadu, India. J. Agron. 35: 23-29.

Gulati, J. M. L.; Panda, S. C. and Misra, B. 1988. Effect of water regimes on wheat. Ind. Jour. Agro. 33(4): 364-367.

Gupta, O. P. 1999. Modern weed management. Agrobios (India), Jodhpur, India.

Gupta, O. P. 1993. Weed Management – Principles and Practices. Agrobios (India), Jodhpur, India.

Gupta, P. K. 2000. Hand book of soil, fertilizers and manures. Agrobios (India) Jodhpur, India.

Gupta, P. K. 200. Soil, plant, water and fertilizer analysis. Agrobios (India), Jodhpur, India.

Gupta, S. C. and Mehta, J. S. 1969. Sugar and sugar beet. Ind. Fmg. 19(7): 11-17.

Gupta, U. S. 1982. Physiological aspects of dryland farming. Oxford and IBH Publishing Co. Pvt. Ltd., New Delhi.

Gurmel Singh, Arora, Y. K., Pratap Narain and Grewal, S. S. 1990. Agroferestry Research, Surya Publications, New Delhi.

Gurmel Singh, Venkataramanan, C.; Sastry, G. and Joshi, B. P. 1990. Manual on soil and water conservation practices. Oxford and IBH Publishing Co. Pvt. Ltd., New Delhi.

Gururaj Hunsigi and Krish, K. R. 1998. Science of field crop production. Oxford and IBH Publishing Co. Pvt. Ltd., New Delhi.

Hake, R. E: 1982. Agroclimatic and dry-season maps of South, Southeast and East Asia Losbanos, Philippines: International Rice Research Institute.

Harpal Singh, M. 1974. Agricultural Meteorology., Punjab Agricultural University, Ludhiana.

Hart, R. D. 1982a. The ecological systems conceptual farmework for agricultural research and development. In: Readings in Farming Systems Research and Development, 44-59 (Eds. W. Shaner, P. Philip and W. Schmehl). Boulder, Colorado: West View Press.

Hart, R. D. 1982b. One farming system in Honduras: A case study in farming systems research. In: Readings in Farming Systems Research and Development, 59-73 (Eds. W. Shaner, P. Philip and W. Schemehl.) Boulder, Colorado: West View Press.

Harwood, R. 1977. Socio-economic considerations in multiple cropping systems. Invited

Paper prepared for the round table discussion on "Agricultural Production System". XVI Annual reunion of the Board of Directors, Institute Intermericano de Ciencias Agricolas (IICA), San Domingo, Dominican Republic.

Harwood, R. 1979. Small-farm development: Understanding and improving farming systems in the Humid tropics. Boulder, Coforado: West View Press.

Harwood, R. R. 1987. Agroforestry and mixed farming systems. In: O. R Clark and RD. Child code). Ecological development in the humid tropics; Guidelines for planners. Morrilton, Arkanasas: Winrock International.

Heichel, G. H. 1987. Legume Nitrogen: Symbiotic fixation and recovery by subsequent crops. In: Z. R Helsel (ed.), Energy in plant nutrition and pest control. Elsevier scientific publishers, Amsterdam. pp 63-80.

Herdt, R W. 1993. Measuring sustainability using long-term experiment. Proc. Conf. held at Rothamstead experiment station, 29-30 April, 1993. Rothamstead, U.K.

Hickling, C. F. 1960. Observations on the Growth Rate of Chinese Grass Carp, *Ctenopharingodon idellus (c & v)*. Malay Agric. J. 43(1): 43-53.

Hiebsch, C. K. and McCollum, R. E.1987. Area x Time equivalency ratio: A method for evaluating the productivity of intercrops. Agron J. 79: 15-22.

Hildebrand, P. 1981. Motivating small-farmers to accept change. In: Readings in farming systems research and development, 100-109 (Ed. W. Shaner, P. Philip and W. Schmehl). Boulder, Colorado: West View Press.

Hiran, K. S.; Jaspal Singh and Acharya, M. S. 1990. Irrigation scheduling. CBS Publishers and Distributors, New Dehi.

Horton, O. 1984. Social scientists in agricultural research: Lessons from the Mataro Valley Project, Peru, Ottawa, Canada: lORe.

Horton, P. 1984. Technology evaluation: Five case studies from West Africa. In coming full circle: Farmers' participation in the development of technology, 95-118 (Ed. P. Matlon, R. Cantrell, D. King and M Benoitcattin). Ontario: IDRC.

Howard Critchfield, J. 1962. General climatology. Prentice hall of India Pvt. Ltd., New Delhi.

Hridai Ram Yadav. 1988. Dimensions of wasteland development. Proceedings of the National Seminar on Wasteland Development, New Delhi.

IARI 1972. Recent research on multiple cropping. A bulletin. Indian Agricultural Research Institute, New Delhi.

IARI. 1977. Water requirement and irrigation management of crops in India. Water Technology Centre, IARI, New Delhi.

ICAR. 1960. Indigenous agricultural implements of India- An all India Survey. ICAR, New Delhi.

ICAR (Indian Council of Agricultural Research), 1972. A new strategy in agriculture: a collection of speeches. of C. Subramanian, New Delhi: ICAR.

ICAR. 1977. Hand book of Animal Husbandry. ICAR, New Delhi.

ICAR. 1981. Indian Farming – Special number on Pulses Vol. 31 No. 5.

ICAR. 1982. A Decade of Dryland Agricultural Research in India 1971-80. All India Coordinated Research Project for Dry land Agriculture. Hyderabad, India.

ICAR. 1982. Indian Farming – special number on Oilseeds. Vol. 32 No. 8.

ICAR (Indian Council of Agricultural Research), 1983. Shifting cultivation in Northeast IndiJi. ICAR Research Complex for NE Hill region, Shillong. New Delhi, India: ICAR.

ICAR. 1985. Rice Research in India. ICAR, New Delhi.

ICAR (Indian Council of Agricultural Research), 1989. Annual reprot 1987. 88. New Delhi, India: ICAR.

ICAR. 1996 Indian Farming Special Issue. XXth World's poultry congress 2-8 Sept. 1996, New Delhi ICAR, New Delhi.

ICAR. 1997. Handbook of Agriculture (Revised). Indian Council of Agricultural Research, New Delhi.

ICRISAT, 1980. Proceedings of the International Workshop on the Agro-Climatological Research needs of the semi-arid Tropics, 22-24 November, 1978, Hyderabad, India.

ICRISAT. 1981. Proceedings of the International workshop on Intercropping, 10-13 Jan 1979. International Crops Research Institute for Semi-Arid Tropics. Hyderabad, India.

ICTC. 1960. Indian Tobacco. Indian Central Tobacco Committee, Ministry of Food and Agriculture, Government of India, Madras.

Indian Farming, 1989. Special issue on watershed management. New Delhi, India: ICAR.

Indian Institute of Trapical Meteorology, 1976. Proceedings of the symposium pn Tropical Monsoon held at Pune-5, India, 8-10 September, 1976.

IOBC, 1993. International organization for biological and intergrated control. Technical bulletin 16(1), 1993.

IRRI. Green manuring in rice farming. International Rice Research Institute, Manila, Philippines.

ISA. 1972. Multiple Cropping. Proceedings of a symposium. Oct. 7-8, 1972, Haryana Agricultural University, Hissar and Indian Society of Agronomy and ICAR, New Delhi.

ISA. 1986. Proceedings of the National Symposium on cropping system. 3-5 April 1985. Central Soil Salinity Research Institute, Karnal and Indian Society of Agronomy, New Delhi.

Isrealsen, O. W. and Hansen, V. E. 1962. Irrigation- Principles and practices. John Wiley and Sons Inc., New York.

Jayanthi, C. 1995. Sustainable component linkage and resource recycling to lowland integrated farming system. Ph.D. Thesis, Tamil Nadu Agri. Univ., Coimbatore.

Jayanthi, C.; Rangasamy, A.; Venkitusamy, R. and Chinnusamy, C. 1994. Sustainable integrated farming systems research and management for sustainable agriculture. Summer Institute held at Tamil Nadu Agric. Univ., Coimbatore between 6-15th June, 1994, pp.: 97-109.

Jayaseelan, M. J. P.; Venugopal, M. S. and Chelliah, S. 1988. Two innovative culture methods for increased integrated fish production. Farm forum 11: 23-26.

Jena, B. K.; Patro, N. and Panda, S. C. 1994. Effect of enriched farm yard manure and fertilizer on finger millet in rainfed condition. Environment & Ecology. 15(1):136-138.

Jeremy, H., Davis C., Jonathan N, and Woolley, N. 1986. Multiple cropping with legumes and starchy roots. In: Francies, C. A. (eds.) Multiple cropping systems. Macmillan Publishing Co., New York

Jeswani, L. M. and Baldev, B. 1990, Advances in pulse production technology. ICAR., New Delhi.

Jha, L. K. 1995. Advances in Agroforestry, APH Publishing Co., New Delhi.

Jhingran, U. G. and Sharma, B. K. 1978. Operational Research Project on Aquaculture. Bull Cent. Inl. Fish Inst.27.6.

Jhingran, U. G. and Sharma, B. K. 1980. Integrated livestock fish farming in India. In: Proc. ICLARM - SEARCA Conf. on "Integrated Agricultural -Aquaculture Farming Systems", 9th Aug. 1979, Manila Philippines, pp.135-141.

John, K. C. 1991. Farming systems research in Eastern India: A new approach, New frontiers in rice research. In: Siddique, E. A. and K. Muralidharan (eds.), Directorate of Rice Research, Hyderabad.

John, Milton Poehlman, 1991. The mungbean. Oxford and IBH Publishing Co. Pvt. Ltd., New Delhi.

Johnson, H. and Clear, B. John. 1986. FSRE: Shifting the intersection between Research and Extension. Agric. Administration.21: 61-93.

Josef D. Zimmerman. 1996. Irrigation. John Wiley and Sons Inc., New York.

Joshi, N. C. 1987. Manual of weed control. Researchco Publications, New Delhi.

Kakde, J. R. 1985. Agricultural Climatology, Metropolitan Book Co., New Delhi.

Kamaraj, S. 1993. Do's and Don'ts in Biogas plants. Published by Department of Bio energy, Tamil Nadu Agric. Univ., Coimbatore.

Kamaraj, S.; Vijayaragavan, N. C. and Pugalendhi, S. 1994. Role of biogas in IFS. In: Proceedings of Summer Institute on "Integrated Farming Systems Research and Development for Sustainable Agriculture" organized by ICAR and Tamil Nadu Agri:c. Univ., Coimbatore, India during 6-15 June, 1994. pp. 203-212.

Kang, B. T. 1993. Changes in soil chemical properties and crop performance with continuous cropping on an entisol in the Humid tropics. In: Soil organic matter dynamics and sustainability of tropical agriculture. Mulongoy, K. and R. Mercisk (ed.). Wiley and Sons, U.K. pp. 297-306.

Kanitkar, N. V.; Sirur, S. and Gokhale, D. H. 1960. Dry farming in India. New Delhi, India!: ICAR.

Kanwar, J. S.; Virmani, S. M. and Das, S. K. 1992. Farming systems research in India: A Historical perspective. Expl. Agric. 28: 1-17.

Kanwar, J. S. 1972. Cropping patterns, scope and concept. In: Proceedings of Cropping patterns in India, 11-38, New Delhi, India ICAR.

Kanwar, J. S. 1978. Soil fertility – Theory and practices. ICAR, New Delhi.

Karuppiah, V. and Thangamuthu, G. S. 1986. Studies on the efficiency of phosphorus forms in combination with organic manures in lowland rice. Proc. Seminar on "Rock Phosphate in Agriculture". Tamil Nadu Agric. Univ., Coimbatore, India.

Katherkhan, S. 1994. Role of Dairy, Sheep and Goat, Piggery and Rabbit rearing in IFS development. In: Proceedings of Summer Institute on "Integrated Farming Systems Research and Development for Sustainable Agriculture" organized by ICAR and Tamil Nadu Agric. Univ., Coimbatore, India during 6-15 June, 1994. pp. 228-234.

Katyal, J. C. 1997. Sustainability:Relevance of Integrated Concept. Survey of Indian Agriculture. The Hindu, Chennai, India

Katyayan, A.2001. Fundamentals of Agriculture. Kushal Publications, Varanasi.

Kavimandan, S. K.; Subba Rao, N. S. and Mohrir, A. V. 1978. Isolation of Spirilum lipoferum from the stems of wheat and N Fixation in enrichment cultures. Current science 47: 96-98.

Khan, A. Q. 1981. Grow high yielding sugarcane varieties. Ind. Fmg. Digest. 14(2): 41-42.

Khanda, C. M.; Dixit, L. and Panda, S. C. 1997. Effect of zinc and graded levels of nitrogen on growth, yield and nutrient uptake of rice. Oryza. 34(1):43-46.

Khanna, S. S. 1989. The agro-climatic approach. In Survey of Indian Agriculture 1989. The Hindu, Madras.

Klingman, G. C. 1981. Weed Science. Principles and practices. John Wiley and Sons Inc., New York.

Kondap, S. and Upadhyaya, U. C. 1985. A practical manual of weed control. Oxford and IBH Publishing Co., New Delhi.

Korwar, G. R.; Osman, M.; Tomar, D. S. and Singh, R. P. 1991. Dry land horticulture, CRIDA, Ext. bulletin 4: 33.

Kumar, D. and Rai, M. 1992. Advances in Oilseeds Research. Vol. I. Rapeseed and Mustard. Scientific publishers, Jodhpur, India.

Kumar, K and Singh, B. N. 1984. Preliminary observation on utilization of palletized poultry litter as feed for common carp. Int. J. Acad. Ichtyol: 151-154.

Kuppusamy, M. 1989. Feasibility of introducing sericulture as a component in the farming System of Thanjavur district, Tamil Nadu. M.Sc. (Ag.) Thesis, Tamil Nadu Agric. Univ., Coimbatore, India.

LaI, R. 1993. Tillage effects on soil degradation, soil resilience, soil quality and sustainability. Soil and Tillage Res. 27: 1-8.

Lal, R. 1991. Soil structure and sustainability, J. Sustainable Agric. 1: 67-92.

LaI, R. and Miller, F. P. 1993. Soil quality and its management in humid sub-tropical and tropical environments. Proc. of XVI International Grass land Conference, Massey, New Zealand.

LaI, R.; Kimble, J.; Levine, E. and Stewart, B. A. (ed.), 1994. Soils and Global changes. Chelsea: Lewis Publishers.

LaI, R. and Miller, F. P. 1990. Sustainable farming for tropics. In: Sustainable agriculture: Issues and prospective. Vol. 1. (Ed.) R. P. Singh. pp. 69-89, Indian Society of Agronomy, IARI., New Delhi.

Lang, M. and Cantrell, R. 1984. Accenting the farmers' role: Pursue farming systems unit. In coming full circle: Farmers' participation in the development of technology, pp. 63-70 (Ed. P. Matlonn, R. Cantrell, D. King and M. Benoit-Cattin). Ottawa, Ontario: IDRC

Larson, K. J.; Cassman, K. G. and Phillips, D. A. 1989. Yield, nitrogen fixation, and above ground nitrogen balance of irrigated white lupine i", a Mediterranean climate. Agron J. 81: 538-543.

Lawrence J. King. 1966. Weeds of the world – Biology and control. Interscience Publishers, Inc. New York.

Lawton, H. W. and Wilke, P. J. 1979. Ancient agricultural systems in dry regions. In: Hall, AC, G. H. Cannell and H. W. Lawton (ed.)., Agriculture in semi-arid environments. Springer Verlag, New York. pp. 1-44.

Leewrik, D. M.; Panda, S. C. and Mahapatra, I. C. 1972. Multiple Cropping in Sambalpur. Bull. Ford Foundation. New Delhi.

Libuno, LP. 1990. Goat - fish integrated farming in the Philippines. Ambio. 19: 408-410.

Lightfoot, C.; Bimbao, M. N.; Dalsgaard, J. P. T. and Pulhin, R. S. 1992. Aquaculture and sustainability through integrated resource management. Quarterly Agric. 22: 143-150.

Lighfoot, C. and Minnick, D. R. 1990. Farmer - first qualitative methods: Farmers' diagram for improving methods of experimental design in integrated farming systems. A FSRE Symposium on "The Role of Farmers in FSRE and Sustainable Agriculture". 14-17 October 1990, Michigan State University, East Lansing, USA

Little, D. and Muir, J. 1987. A guide to integratep warm water agriculture. Stirling. Scotland. Institute of Agriculture, University of Stirling.

Lock Wood, J. G. 1979. World Climatology- An Environmental Approach. The English Language Book Society and Edward Arnold (Publishers) Ltd., Bedform Square.

Loganathan, P. 1989. Mushroom cultivation as a component in mixed farming system in Thanjavur district, Tamil Nadu. M.Sc. (Ag.) Thesis, Tamil Nadu Agric. Univ. Coimbatore.

Mahabal Ram 1980. High yielding varieties of crops. Oxford & IBH Publishing Co., New Delhi.

Mahapatra, I. C.; Bhatia, P. C.; Singh, K. N.; Dayanand and Panda, S. C. 1971. Agronomic management of upland rice in North and South Western India. Oryza 8(2) Supplement: 239-249.

Mahapatra, I. C. 1971. All India coordinated agronomic experiments scheme. In: National agricultural research systems in Asia, 226-232, (Ed. A. H. Mosenan). New York 10020, USA: Agriculture Development Council.

Mahapatra, I. C. and Panda, S. C. 1972. Uptake and utilization of nitrogen, phosphorus and potassium by dwarf indica rices. Rice News Letter. Xxi(3):1-19.

Mahapatra, I. C. and Panda, S. C. 1973. Light on fertilizing rice. Intensive Agri. XI(5): 9-11.

Mahapatra, I. C. and S. R. Bapat, 1992. Farming system research - challenges and opportunity. Proceeding of the XII National symposium on "Resource Management for Sustained Crop Production" held at Rajasthan Agricultural University, Bikaner during Feb. 25-28. pp. 382-390

Mahapatra, I. C. 1994. Farming Systems Research: A Key to Sustainable Agriculture. Fertilizer News. 39(11): 13-25.

Mahendra Pal; Singh, KA. and Ahalawat, I. P. S. 1986. Cropping systems research. 1. Concept, needs and directions. In: Proceedings of the National Symposium on "Cropping Systems", 1-20, New Delhi, India: Indian Society of Agronomy.

Mahendra Pal, Jayanta Deka and Pai, R. K. 1996. Fundamentals of cereal crop production. Tata Mc Graw Hill Publishing Co Ltd., New Delhi.

Make, R. E. 1982. Rice area by type of culture: South, Southeast, and East Asia, Losbanos, Philippines: International Rice Research Institute.

Malviya, D. D.;Sinha, N. P.; Patel, J. C. and Kalaria, K. K. 1986. Production potential and economic feasibility of different cropping sequences and its economic feasibility in wheat zone of M. P. Abstract of Symp. Efficient Cropping Syst. Zone of India. Jsan. 7-10, 1988. p 44

Maiti, S.; Hedge, M. R. and Chattopadhyay, S. B. 1988. Hand book on annual oilseed crops. Oxford and IBH publishing Co. Pvt. Ltd., New Delhi.

Mandal, R. C. 1999. Weeds, weedcides and weed control. Principles and practices. Agro Botanica publishers, Bikaner. India.

Marlon, P.; Cantrell, R.; King, D. and Benoit-Cattin, M. (Ed), 1984. Coming full circle: Farmers' participation in the development of technology. Ottawa, Ontario: IDRC.

Martin, J. H. and Leonard, W. H. 1967. Principles of field crop production. McMillan and Co., New York.

Mather, J. R. 1974. Climatology-Fundamentals and applications. McGraw Hill Book Co., New York.

Mavi, H. S. 1986. Introduction to Agro-meteorology. Oxord & IBH Publishing Co., New Delhi.

Maxwell, S. 1984. Farming systems research: Hitting a moving target. Discussion paper No. 199. SUSSEX Institute of Development Studies. Mayer, H. 1979. Land Use: Ecology and agriculture in the Mantaro Valley of Peru with special reference to potatoes, Lima, Peru: Center for International Potato Research, Social Science Unit.

McIntosh, D. H. and Thom, A. S. 1969. Essentials of Meteorology. The Lykeham Publications (London) Ltd., London.

McInfire, J. and Cryseels, G. 1987. Crop-livestock interactions in sub-Saharan Africa and their implications for farming systems research. Expt. Agrl. 23: 235-243.

Mehra, K. L. 1978. Oats, ICAR, New Delhi.

Meelu, O. P. and Morris, R. S. 1988. Green manure management in rice based cropping systems. In: "Green Manure in Rice Farming", International Rice Research Institute, Philippines. pp. 209-222.

Menon, P. A. 1989. Our weather. National Book Trust, New Delhi.

Michael Haines, 1982. An introduction to farming system. Tien Mab lither Printing Co. Pvt. Ltd., London.

Michael, A. M. 1978. Irrigation – Theory and practice. Vikas Publishing House Pvt. Ltd., New Delhi.

Mishra, R. D. and Ahmed, M. 1987. Manual on Irrigation Agronomy, Oxford and IBH Publishing Co. Pvt. Ltd., New Delhi.

Mishra, A. and Pasupalan, S. 1994. Integrated farming systems for rain fed lowland. The Hindu, 14th Sept. 1994.

Mittla, S. P. and Singh, G. 1990. Resource conservation and management in watersheds of Shiwalik foot hills. In: proceeding of International Symposium on "Water Erosion". Sedimentation and resource conservation, 338-347. Dehradun, Uttar Pradesh, India: Central Soil and Water Conservation Research and Training Institute.

M. K. Dowell, R. and Hidlebrand, P. 1980. Integrated crop and animal production: Making the most of resources available to small farms in developing countries. Working papers - The Rockfeller Foundation, New York: The Rockfeller Foundation.

Mohamed Ali, A.; Siddeswaran, K.; Jegannathan, R.; Muthusamy S. and Subramanian, S. 1998. Efficiency of mixed farming in the tribal areas of Sherveray Hills. Madras Agric. J. 71: 294-298.

Mohan, S. and Jeyarajan, R. 1994. Mushroom cultivation. In: Proceedings of Summer Institute on "Integrated Farming Systems Research and Development for Sustainable Agriculture" organized by ICAR and Tamil Nadu Agric. Univ., Coimbatore, India during 6-15 June, 1994. pp. 220227.

Mohandas, S. 1988. Field response of tomato inoculation with VAM Fungi and Rhizobium Phaseoli, Proc. Mycorrhiza Round Table. Varma, A. K. *et. al.*, (ed.) pp. 411-416.

Morachan, Y. B. 1980. Crop Production and Management, Oxford and IBH Publishing Co., New Delhi.

Munn, R. E. 1970. Bio-meteorological Methods. Academic Press, New York

Nagarajan, S.; Gyanendra Singh and Tyagi, B. S. 1998. Wheat Research needs beyond 2000 A. D. Narosa Publishing House. New Delhi.

Nagaraja, B. 1989. Effect of Crop mixture on income and Employment, A case Study of Anantapur district.

Nagarajan, K. 1980. Diseases of Tobacco and their control. Ind. Fms Digst. 13(12): 44-45.

Nair, P. K. R. 1985. Agroforestry defined. In: Agroforestry systems in the Tropics. Nair, P. K. R. (Ed.) Kluswer Academic Publishers. Dordrecht, The Netherlands. pp. 13-18.

Nanda, S. S.; Swain, K. C.; Panda, S. C.; Mohanty, A. K. and Alim, M. A. 1995. Effect of nitrogen and biofertilizers in fodder maize under rainfed condition of Orissa. Curr. Agri. Res. S(Suppl.): 45-47.

Narayana, G. V. and Seshadri, C. R. 1967. Groundnut cultivation in India. Farm Bull. 2, ICAR, New Delhi.

National Commission on Agriculture (NCA), 1976. Crop production during the pre-plan period. In Abridged report of the National Commission of Agriculture, 48-49. New Delhi, India: NCA.

National Research Council (NRC), 1993. Soil and Water quality: An agenda for agriculture. Committee on long range soil and water conservation board on agriculture. National Academy Press Washington, D. C.

Negi, M.; Sachdev, M. S. and Tilak, K. V. B. R. 1991. Nitrogen uptake by Azospirillum brasilense inoculated barley as influenced by Nand P Fertilization. J. Nuclear Agric. Bio. 20: 25-32.

Noble, R. P. and Rshide, B. 1990. Aquaculture technology transfer to small holder farmers in Malawi, Southern Africa. NAGA, The ICLARM Quartly. 16: 14-16.

Norman, D. W. 1974. Rationalizing mixed cropping under indigenous conditions; The example of Northern Nigeria. Journal of Development Studies 11: 3-21.

Norman, D. W. 1980. The farming system approach: Relevancy for the small farmers, Michigan State University Development Paper, 8, East Lansing, Michigan, USA: Michigan State University.

Norman, D. W. 1982. Farming systems approach to research. Paper prepared for the Farming Systems Research Symposium, Kansas State University, Manhattan, Kansas, November 21-22, 1982.

Norman, D. W. and Collinson, M. P. 1985. Farming Systems Research: Theory and Practice. ACIAR Proc. No. 11: 16-30.

Norman, M. J. T. 1980. Your future depends on Agronomy, J. Aust. Inst. Agri. Sci. 46:105-111.

Norman, N. Potter and Joseph H. Hotchkiss 1996. Food Science. CBS Publishers and Distributors., New Delhi.

Nyta, C. Brady. 1990. The nature and properties of soils (Tenth Edn) MacMillan Publishing Co. Inc., New York.

Obias, E. D. 1986. 'Integrated Farming System and Waste Recycling'. FFTC Book Series No. 16. Food and Fertilizer Technology Centre for the ASPAC Region, Taiwan Roc. P 7-11.

Okigbo, B. N. and Lal, R. 1980. Residue mulches, intercropping and agri-silviculture potential in Tropical Africa. In: Basic techniques in ecological farming. Hill, S. (Ed.) IFOAM Conference, Montreal, Canada. pp. 54-69.

Okigbo, B. N. 1978. Cropping system and related research in Africa. AAASA occasional Pub. Series OT-I, Addis Abeba, Ethiopia. pp. 81.

Oram, P. 1988. Building the, agroecological framework. Environment pp. 14-34.

OTA, 1988. Enhancing Agriculture in Africa: A role for development. assistance. DC Covt. Printing office, Washington, D. C

Owasu, D. Y. 1990. Experiences with agroforestry. ILEIA Newsletter 6(2): 9-10.

Palaniappan, S. P. 1985. Cropping systems in tropics-Principles and practices. Willey Eastern Ltd., New Delhi.

Palaniappan. S. P. and Annadurai, k. 1999. Organic farming: Theory and Practice. Scientific publisher, Jodhpur, p. 257.

Palaniappan, S. P. and Siddeswaran, k. 1990. Evaluation of the potential of green manures and grain legumes in rice farming system. National symposium of "Rice Farming System", Agri. College and Research Institute, Madurai, India.

Palaniappan, S. P. and Sivaraman, K. 1996. Cropping System in the Tropics. Principles and management. New Age International Pvt. Ltd. New Delhi.

Palaniappan, S. P. and Annadurai, K. 2001. Diversification in farming system Paper presented in first International agronomy congress on "Agronomy, Environment and Food Security for 21st century held at IARI, New Delhi during Nov. 23-27, 1998.

Palm, V.; Desilva, A. and Guildford, M.1987. Nutrient cycling in paddy of a traditional farming system. Tropical Agri. 87: 129-133.

Panda, S. C. 1967. The effect of number of splits in clonal multiplication on the growth and yield of rice at three levels of manuring. Ind. Jour, Agro. XII(4): 383-389.

Panda, S. C. and Leeuwrik, D. M. 1971. Effect of spacing on the growth and yield of high yielding varieties of rice. Oryza. 8(2): 30-46.

Panda, S. C. 1972. Performance of high yielding varieties of jowar under different levels of nitrogen. Ind. Jour. Agro. Xvii(1): 77-78.

Panda, S. C. 1972. Multiple Cropping in Sambalpur. Multiple Cropping: 148-151.

Panda, S. C.; Leewrik, D. M. and Mahapayra, I. C. 1972. Production Potential and Economic of Ten High Intensity one-year Crop Rotation in Sambalpur. Multiple Cropping: 102-108.

Panda, S. C. 1972. Performance of four potato varieties under Sambalpur condition. Ind. Jour. Agro. XVII(2): 110-111.

Panda, S. C. 1972. Effect of different levels of nitrogen and phosphorus on the growth and yield of Pusa Baisakhi mung. Ind. Jour. Agro. XVII(3): 239-241.

Panda, S. C.; Leeuwrik, D. M. and Mahapatra, I. C. 1972. Optimum combination of calcium ammonim nitrate and sulphate of ammonium for efficient nitrogen utilization in rice as compared by grain yield. OUAT. Jour. Res.2(1 & 2):1-4.

Panda, S. C. and Mahapatra, I. C.1972. Growth and Development of Rice Plant- I. Growth Analysis. IL RISO. XXI(4): 315-336.

Panda, S. C. and Leeuwrik, D. M. 1972. Response of high yielding varieties of rice in high levels of nitrogen. Oryza. 9(2):23-27.

Panda, S. C.; Nayak, R. C. and Mahapatra, I. C. 1973. A note on the effect of cement dust on the soil fertility and crop growth around the factory area of Rajgangpur(Orissa). Ind. Jour. Agro. 14(3): 389-391.

Panda, S. C. and Nayak, R. C. 1974. Effect of zinc on the growth and yield of rice. Ind. Jour. Agro. 19(1):9-13.

Panda, S. C.; Leeuwrik, D. M. and Mahapatra, I. C. 1974. Performance of hybrid and composite varieties of maize under Sambalpur condition. Ind. Jour. Agro. 19(1):14-17.

Panda, S. C. 1974. Performance of soybean varieties under Sambalpur condition. Ind. Jour. Agro. 19(3):163-165.

Panda, S. C. and Mahapatra, I. C. 1974. Growth and Development of Rice Plant-II. Flowering behaviour of individual tillers of different plant types. IL RISO. XXIII(3):259-262.

Panda, S. C. and Mahapatra, I. C. 1974. Growth and Development of Rice Planr- III. Analysis of Yieldcomponents. IL RISO. XXIII(3):262-276.

Panda, S. C. and Mahapatra, I. C. 1974. Growth and Development of Rice Plant-IV. Relation between grain yield and growth characters. IL RISO. XXIII(4):405-414.

Panda, S. C. and Mahapatra, I. C. 1974. Growth and Development of rice Plant- V. Correlation Studies. IL RISO. XXIII(4):414-419.

Panda, S. C. and Mahapatra, I. C. 1974. Growth and Development of Rice Plant-VI. Linear and Multiple Regression of grain yield per plant on primary, secondary and tertiary tillers for different characters. IL RISO. XXIII(4):420-438.

Panda, S. C. and Das, K. C. 1975. Relative performance of high yielding varieties of rice at different levels of nitrogen. OUAT Jour. Res. 5(1 & 2):83-89.

Panda, S. C. and Mahapatra, I. C. 1976. Relative performance of Taiwan indica, ponlai and local indica (different plant types) under varying levels of soil fertility, environmental condition and agronomic practices- I. Growth pattern of rice- Tillering behaviour and plant height. IL RISO. XXV(2):161-183.

Panda, S. C. and Mahapatra, I. C. 1976. Relative performance of Taiwan India, ponlai and local indica(different plant types) under varying levels of soil fertility, environmental condition and agronomic practices. II. Growth pattern of rice plant- Number of leaves per unit area, leaf area and leaf area index. IL RISO. XXVI (2): 157-178.

Panda, S. C. 1977. Yield performance of some maize varieties. OUAT Jour. Res. 7 (1&2): 14-17.

Panda, S. C. and Verma, S. P. 1977. Comparision of puddling implements for rice cultivation. OUAT Jour. Res. 7(1 & 2): 82-88.

Panda, S. C. 1978. Water Management of Wheat. Food Fmg. Agri. 10(4): 112-114.

Panda, S. C. and Sahu, S. K. 1978. Harvest more from Usar soil. Food Fmg. Agric. 10(1):16-19.

Panda, S. C.; Misra, B. and Singh, N. 1978. Problems of waterlogging in the Hirakud Ayacut and its management. Food Fmg Agri. 9(12): 392-395.

Panda, S. C.; Verma, S. P. and Patra, A. 1978. Maximize the production of rice through better puddling. Food Fmg. Agri. 10(2):31-34.

Panda, S. C. and Sahu, S. K. 1978. Rhizobium for legume cultivation and fixation of atmospheric nitrogen. Food Fmg. Agric. 10(5):157-160.

Panda, S. C. and Das, K. C. 1978. Effect of age of seedlings and fertilization on the high yielding varieties of rice. Nonmonitary Inputs-Field Crop Production. 83-88.

Panda, S. C. and Das, K. C. 1979. Nitrogen for Rice. Ind. Jour. Agro. 23(2): 183-188.

Panda, S. C.; Singh, N. and Misra, B. 1979. Performance of wheat varieties under different levels of soil salinity. Curr. Agri. 3 (1 & 2): 63-72.

Panda, S. C., Sahu, S. K. and Misra, B. 1979. Leaching losses of Nitrate Nitrogen under different Water Tables and Water Management Practices in Rice Field. Oryza 16(2): 107-112

Panda, S. C.; Misra, B and Singh, N. 1979. Water management of groundnut in relation to ground water tables. OUAT Jour. Res. 9(1 7 2): 110-114.

Panda, S. C. 1980. Water Management of Rice in Relation to Ground Water Table. Proc. Afro-Asian Reg. Conf. Subj. C. (9): 91-104.

Panda, S. C.; Das, K. C.; Misra, B.; Sahu, S. K. and Rout, D. 1980. Effect of depth of submergence of different growth stages of dwarf indica rice on the growth, yield and nutrient uptake of crop and mineral content of soil. I. Growth and yield of crop. Oryza. 17(2): 85-91.

Panda, S. C.; Misra. B. and Rout, D. 1980. Effect of various irrigation schedules on potato under different depths of irrigation. Curr. Agri. 4 (4-4):153-163.

Panda, S. C. 1980. Integrated nutrient supply for higher production of rice. Food Fmg. Agri. 13(1):18-20.

Panda, S. C. 1981. Research on Water Management in Hirakud Command Area and its Achievements from 1967 to 1981

Panda, S. C.; Acharya, N. and Misra, B. 1981. Effect of Irrigation Schedules on the Growth and Yield of Rice. Food Fmg. Agri. 13(9 & 10): 1282-185.

Panda, S. C. 1983. Water Management in Hirakud Command Area. 154 pp.

Panda, S. C. 1985. Research Base and Current Statues of Cropping System in Hirakud Command Area. 20 pages.

Panda, S. C. 1986. The Achievement of Research on Water Management from the inception of the Irrigation Research Centre at Chakuli/Chiplima for the period from 1967 to 1984.

Panda, S. C.; Das, K. C. and Sahu, S. K. 1986. Water table in relation to water management on rice. Proc. Sympo. Drainage of waterlogged and salt affected lands.

Panda, S. C.; Misra, B.; Taha, M., Acharya, N. Singh, N. and Dash, A. N. 1987. Effect of levels of nitrogen, spacing and planting direction on rice yield. Oryza. 24: 358-362.

Panda, S. C.; Gulati, J. M. L. and Misra, B. 1988. Effect of Water Regimes and Nitrogen on Wheat. Indian J. Agron. 33(4): 364-367.

Panda, S. C. 1995. Rice yield and physiochemical properties of soils as affected by integrated nitrogen management. National sympo. On sustainable agriculture in Sub-Humid Zone.

Panda, S. C.; Panda, P. C. and Nanda, S. S. 1995. Effect of levels of nitrogen and phosphates on yield and nutrient uptake of rice. Oryza. 32(1):146-148.

Panda, S. C.; Rath, B. S.; Tripathy, R. K. and Satpathy, L. 1996. Contingent Planting of Rice for Aberrant Weather Conditions in Coastal Plains. Environment & Ecology 14(2): 455-458.

Panda, S. C.; Pattanaik, A.; Rath, B. S.; Tripathy, R. K. and Behera, B. 1996. Response of wheat (*Triticum aestivum*) in crop geometry and weed management. Ind. Jour. Agro. 41(4):553-557.

Panda, S. C.; Rath, B. S.; Tripathy, R. K. and Das, B. 1997. Effect of water management practices on yield and nutrient uptake in the dry season rice. Oryza. 34(1): 51-53.

Panda, S. C.; Nanda, S. S. and Mishra, R. C. 1997. Response of *rabi* groundnut (*Arachis hypogaea*) in methods of sowing and sources of sulphur. Ind. Jour. Agro. 42(4):680-682.

Panda, S. C. and Patro, H. 1998. Rice-Dairy Farming System: A boon to Marginal Farmers of Orissa. Intensive Agri. March-April: 24-31.

Panda, S. C.; Mohanty, A.; Sahoo, H. K. and Sahoo, B.C. 1999. Effect of manuring in Rice-Rice cropping system. Oryza. 36(2): 139-142.

Panda, S. C.; Roul, P. K.; Tripathy, R. K. and Chaudhury, Y. C. 1999. Effect of spacing and number of seedlings per hill on transplanted rice. Environment & Ecology. 17(3): 626-

629.

Panda, S. C. 2003. Principles and Practices of Water Management. Agrobios (India), Jodhpur. India.

Panda, S. C. 2004. Cropping and Farming Systems. Agrobios (India), Jodhpur. India.

Panda, S. C. 2004. Dry land Agriculture. Agrobios (Inda), Jodhpur, India.

Panda, S. C. 2004. Agronomy. Agrobios (India), Jodhpur, India.

Pande, A. 1970. Modern Hygrometry, Somaiya Publications Private Ltd., Bombay.

Pandey, R. C. 1969. Cultivation of sugar beet for sugar in India. Farm and Parl. 4(11): 7-8.

Pandey, S. N. and Ajanta Chandha, 1996. Economic Botany. Vikas Pvt. Ltd., New Delhi.

Pandey, R. K.; Sharma, S. K. and Hegde, D. M. 1992. Approach to farming system research in ICAR: Concepts and Methodology. Project Directorate for Cropping Systems Research. Bulletin No.2. pp. 1-16.

Panse, V. G. and Sukhatme, P. V. 1967. Statistical methods for agricultural workers. ICAR., New Delhi.

Pati, M. 1992. An Economic Analysis Rice-based Farming System in Balasore District(Orissa). M.Sc(Ag) Thesis, OUAT, Bhubaneswar.

Patra, D. D. 1990. Cereal - legume associations. Effect on companion and succeeding cereals. Somani, LL, K. K. Vyas and S. N. Saxena (eds.) Biofertilizers. Scientific Publishers, Jodhpur, India, pp. 113-136.

Patterson, S. 1956. Weather Analysis and Forcasting. Sec. Edition, McGraw Hill Book Company, Inc. New York.

Paulraj, S. 1984. Composite fish culture by pig-dung. Ceobios 11: 265-267.

Perrin, R.; Winkelmann, D. L.; Moscardi, E. R. and Anderson, J. R. 1976. From Agronomic data to farmer recommendations: An economic training manual. Information training bulletin, 21. El. Batan, Mexico: CIMMYT.

Pierce, EJ.; Larson, W. E.; Dowdy, R. H. and Graham, W. A. P. 1983. Productivity of Soils: Assessing long term changes due to erosion. J. Soil and Water Conservation 38: 39-44.

Plucknett, D. L.; Dillon, J. L. and Vallaeys, G. J. 1987. Review of concepts of farming systems research: What, Why and How. In: Proc. of Workshop on "Farming Systems Research", ICRISAT, Hyderbad. pp. 2-9.

Prasad, C. 1989. Agricultural extension services. In: 40 years of Agricultural Research and Education in India, 234-267. New Delhi, India: ICAR.

Pratap Singh, Jaspal Singh and Mathur. 1995. Technology for Dryland Agriculture. Himanshu Publications, New Delhi.

Premsekhar, M. and Ranagsamy, A. 1994. Sustainable IFS research on lift irrigated garden lands. In: Proceeding of Summer Institute on "Integrated Farming Systems Research and Development for Sustainable Agriculture" organized by ICAR and Tamil Nadu Agric. Univ., Coimbatore, India during 6-15 June, 1994. pp. 106-114.

Prihar, S. S. and Sandhu, B. S. 1987. Irrigation of field crops – Principles and Practices. ICAR, New Delhi.

Pushkarnath 1964. Potato in India. ICAR, New Delhi.

Pushkarnath 1969. Potato in India, Varieties. ICAR, New Delhi.

Radha Vasiraju and Murty Krishna 1995. Practical Manual on Agricultural Meteorology. Kalyani Publishers, New Delhi.

Rajendran, P. 1994. Integrated production technology under rainfed farming system. In: Proceeding of Summer Institute on "Integrated Farming System Research and Development for Sustainable Agriculture" organized by ICAR and Tamil Nadu Agric. Univ., Coimbatore, India during 6-15 June, 1994_ pp. 65-78.

Rajkumar, S. 1988. Feasibility of introducing dairy as complementary enterprise for small farmers in Thanjavur delta, Tamil Nadu, M.Sc. (Ag.) Thesis. Tamil Nadu Agric. Univ., Coiinbatore, India.

Raju, R. A. 1996. Ecology of Aquatic weeds and their control. Kalyani Publishers, New Delhi.

Raju, R. A. 2000. Glimpses of rice technology. Agrobios (India), Jodhpur, India.

Rama Rao, M. S. V. 1962. Soil Conservation in India, ICAR, New Delhi

Ramadas, L. A. 1964. Crop and Weather in India, ICAR, New Delhi.

Raman, K. V. 1998. FSR - its implication in future research strategies in India. FSR

Newsletter 2 (-1J: 3-5.

Raman, K. V. and Balaguru, T. 1991. Farming System Research in India: Strategies for Implementation. Proc. of National Workshop, NAARM, Hyderabad. pp. 1-39.

Ramkrishnan, T. S. 1963. Diseases of millets., ICAR, New Delhi.

Randhawa_ M. S. 1979. A history of the ICAR (1929-1979), ICAR, New Delhi, India: ICAR.

Randhawa, M. S. 1980. A History of Agriculture in India, Vols. I & II, ICAR, New Delhi.

Randhawa, N. S. and Venkateswaralu, J. 1986. Farming System Research Indian Experience. ICRISAT Publication. P 111-112.

Rangasamy, A.; Shanmugasundaram, V. S.;Sankaran, S. and Subbian, P. 1990. Integrated Farming System Management: A viable approach. Research Bulletin. Tamil Nadu Agric. Univ. Coimbatore, India.

Rangasamy, A. 1992. Sustainable agriculture for rice based ecosystem. In: Proceeding of National training on "Integrated Farming System" held during 10-17th December, 1992. Gal and Tamil Nadu Agric. Univ., Coimbatore, India. pp. 5-12.

Rangasamy, A. 1994. Approaches to farming systems research. In: Proceeding of Summer institute on "Integrated Farming Systems Research and Management for Sustainable Agriculture". 6-15 June, 1994. Tamil Nadu Agric. Univ., Coimbatore, India. pp. 1-17.

Rangasamy, A. 1996. Organic farming system through integrated farming. system. Paper presented at the National Workshop on ';Organic Farming for Sustainable Agriculture" held at WALAMTARI, Hyderabad, from 1820 January.

Rangasamy, A. and Jayanthi, C. 1994. Recycling of organic wastes in integrated farming systems. In: Proceeding National training on "Organic Farming". GOI and Tamil Nadu Agric. Univ. Coimbatore, India. pp. 114-118.

Ranagasamy, A. and Premsekhar, M. 1994. Sustainable integrated farming system research of lift irrigated garden lands. Proceeding on "Integrated Farming System Research and Management for Sustainable Agriculture" held at Tamil Nadu Agric. Univ., Coimbatore, India, between 6-15, June, 1994. pp. 106-109.

Rangasamy, A.; Budhar, M. N.; Venkidusamy, R. and Palaniappan, S. P. 1988. Integrated farming system for wetlands. In: Absts. of National seminar on "Farming Systems for Semi-arid Tropics". Tamil Nadu Agric. Univ., Coimbatore, India. pp. 1-6.

Rangasamy, A.; Venkidusamy, R.; Jayanthi, C.; Purushothaman, S. and Palaniappan, S. P. 1995. Integrated farming system for rice based ecosystem. Madras Agric. J. 83: 287-290.

Rangasamy, A.; Annadurai, K.; Subbian, P. and Chinnusamy, J. 2002. Farming Systems in The Tropics. Kalyani Publishers. Ludhiana.

Rao, M. R. 1986. Cereals in multiple cropping. In: Francis, C. A. (eds.) Multiple cropping systems. Macmillan Publishing Co., New York

Rao, N. G. P. 1972. Sorghum in seventies. Oxford & IBH Publishing Co., New Delhi.

Rao, P. Y. 1984. Mushroom farms in Himachal Pradesh. Intensive Agric. 22: 19-20.

Rao, V. M. 1991. Prospect of Poultry Development in India. Agric. Situation in India. June 121-126.

Rao, V. S. 2000. Principles of weed science (setond edn.) Oxford and IBH Publishing Co. Pvt. Ltd., New Delhi.

Rath, R. 1989. Organic Fertilizer Augments Fish Production. Indian Farmers Digest. XXII(3): 31-33.

Rath, R. 1990. Integrated Fish Farming-a sound basis for the development of Integrated Rural Technology, National Seminar on Aquaculture in India-Problems and Prospects. 27-29 Nov. 1990. Thirubanantpuram Dept. of Aquatic Biology and Fisheries. Vhiv. Of Kerala p-18 Cochin.(Abstract P No. T. I. 12)

Rathore, P. S. 2000. Techniques and management of Field crop production. Agrobios (India), Jodhpur, India.

Ratikanta Maiti. 1996. Sorghum Science. Oxford and IBH Publishing Co. Pvt. Ltd., New Delhi.

Reddy, S. R. 2000. Principles of Agronomy. Kalyani Publishers, Ludhiana

Regnion, E. and Jahnke, R. 1990. Evolving strategies for managing weeds. In: Edwards, C. A. *et. al.*, (ed.) Sustainable agricultural systems. Soil and Water (::onservation Society, Ankeny, IOWA pp. 174-202.

Rennie, R. J. and Kemp, G. A. 1984. ^{15}N-Determined time course for N_2 fixation in two cultivars of field beans. Agron J. 76: 146-154.

Retner, E. T.; Label, R.; Teldhay, H. and Hartzook, A. 1979. Some characteristic of symbiotic nitrogen fixation, yield, protein and soil accumulation in irrigated peanuts. Plant Soil. 51: 373-376,

Rhoades, R. and Bootb, R. 1982. Farmer-Back to-Farmer. An model for generating acceptable technolog. International Potato Research Center, Social Sciences Department, Working Paper 1982_1. Lima, Peru: International Potato Research Center. Research 1929-1979" ICAR, New Delhi, India: ICAR.

Richards, P. 1988. Experimenting 'farmers and_gricultura_, research. paper prepared for ILEIA Workshop on "Operational, Approaches for Participatory Technology Development in Sustainable Agriculture", 11-12 April, 1988, Leusden: The Netherlands.

Riehl, H. 1965. Introduction to the atmosphere, International student Edition, McGraw Hill Book Company, New York.

Rinaudo, G.; Dreyfus, B. and Dommerges, Y. 1983. Sesbimia rostrata green manure and the nitrogen content of rice crop and soil. Soil Biochem. 15: 111-113.

Ritchie, J. T.; Singh, U.; Goodwin, D. C. and Hunt, L. 1989. A users guide to CERES. Training Manual, IFDC, Muscle Shoais, AL,

Rorenberg, N. J. 1974. Microclimate – The biological environment. John Wiley and Sons, Inc., New York.

Roy, B.; Das, D. N. and Mukahopadhayay, P. K. 1990. Rice-fish - vegetables integrated farming: Towards a sustainable ecosystem. NAG A, The ICLARM Quartly 13: 17-18.

Roy, R. N. and Seetharaman, S. 1977. Rice. The Fertilizers Association of India, New Delhi.

Royal Commission on Agriculture (RCA), 1928. Report of the Royal Commission on Agriculture in India, Bombay, India.

Sache, I. 1987. Towards a second green revolution. In: Glaeser, B. (ed.) The green revolution revised. Allen and Unwin, London, pp. 193-198.

Sahoo, B., Panda, S. C. and Mohanty, K. 1997. Effect of Rock phosphate under Rice-Green gram Cropping System in Acid Soils of Orissa. Oryza. 34(4): 358-361

Samuel, L.; Tisdale, warner. L. Nelson and Jemes D. Beaton. 1990. Soil fertility and fertilizers(fourth edition). MacMillan Publishing Co. Inc., New York.

Sankaran, S. and Subbiah Mudaliar, V. T. 1992. Principles of Agronomy. The Bangalore and Publishing Co. Ltd., Bangalore, India.

Sankaran, S. and Rangasamy,. A. 1990. Farming system research. In: Agronomic research towards sustainable agriculture (Ed.) K. N. Singh and R. P. Singh, pp. 69-80. Indian Society of Agronomy, IARI, New DIehL

Sankaran, S.; Jayakumar, R. and Kampuchetty, N. 1993. Herbicide residues. Gandhi Book House, Coimbtore, India.

Santhanagopalkrishanan, 1988. Feasibility of rearing goat under deep litter system in the farming system of Thanjavur delta, Tamil Nadu M.Sc.(Ag.) Thesis, Tamil Nadu Agric. Univ., Coimbatore, India.

Santhanam, V. 1976. Cotton, ICAR, New Delhi.

Sarkar, R. P. and Biswas, B. C. 1988. Weather forecasting and its application in Agriculture. Fert. News. 33(4):73-76.

Scroll, P. and Merrillsands, D. 1986. The Milpa and its alternatives. In:Farming systems (Eds. B. L. Turner and S. Brush).-New York: Guilford Press. Boulder, Colorado: West View Press.

Seetharam, A.; Riley, K. W. and Harinarayaana, G. 1989. Small millets in Global Agriculture. Oxford and IBH Publishing Co. Pvt. Ltd., New Delhi.

Serman, D. W.; Simmons, E. B. and Mays, H. M. 1982. Farming Systems in the Nigerian Savannah: Research Strategies for Development. Boulder, Colorado: West View Press.

Seshadri, C. R. 1961. Groundnut. The Indian Central Oilseed Committee, Hyderabad.

Shaner, W. W.; Phillips, P. E. and Schmehl, W. R. 1982. Farming Systems Research and Development: A guideline for developing countries. Boulder, Colorado: West View Press.

Shanmugasundaram, V. S. and Ravi, K. 1992. Rice-fish - azolla integration. NAGA, The ICLARM Quartly. 15: 27.

Shanmugasundaram, V. S.; Balusamy, M. and Subramaniam, K. S. 1992. Integrated farming system in wetland ecosystem. In: Proceeding of National training on "Integrated Farming System", GOI and Tamil Nadu Agric. Univ., Coimbatore, India. pp. 18-20.

Shanmugasundaram, V. S. and Balusamy, M. 1993. Rich-Fish - azolla integrated farming system in low lying wetlands. Farming Systems 9: 105-107.

Sharma, B. C. 1991. Dryland Farming- Perspectives and prospects. Daya Publishing House, New Delhi.

Sharma, B. K. and Das, M. K. 1985. Integrated Fish Livestock crop Farming Systems in M. Mohan Joshep(ed.). The first Indian Fisheries Forum, Proceedings. Asian Fisheries Society, Indian Branch, Mangalore, pp: 27-30.

Sharma, B. K.; Das, M. K. and Chakraborty, D. P. 1985. Integrated fish-cum-poultry farming. Agriculture extension manual news letter, Series. 5: 20-29.

Sharma, L. R., Bhati, J. P. and Singh, R. 1991. Emerging Farming Systems in Himachal Pradesh: Key issues in sustainability. Indian J. Agri. Econ. 46(3): 421-427.

Sharma, P. N. and Kampen. 1977. Watershed based land and water management technology – Some tentative guidelines. ICRISAT. Hyderabad, India.

Sharma, S. K.; Gangwar, K. S. and Singh, J. P. 1991. Farming Systems diagnosis, and analysis. Project report for Directorate of cropping system research on diagnostic team, Meerut, Uttar Pardesh, India. pp. 1-15.

Shende, S. T. 1987. Significance of Azotobacter inoculation for agricultural crops. Organic Manures and Fertilizers. Indian Agrl. Res. Institute, New Delhi. pp. 38-45.

Shippen, J. M.; Ellin, C. R. and Clover, C. H. 1980. Basic farm machinery. Pergamon Press, Oxford, London

Siddeswaran, K. and Palaniappan, S. P. 1990. Role of integrated farming systems in minimizing the purchase input usage in small and medium farm holdings (Unpublished).

Simmods, N. 1985. The State of the Art of Farming Systems Research, Washington, D. C USA: World Bank.

Simmonds, N. W. 1986. A short review of Farming Systems Research in the Tropics Exptl. Agric. 22: 1-13.

Simmonds, N. W. 1985. Farming Systems Research - a review. Technical Paper No. 43. World Bank, Washington, D. C

Sin A. W. C. 1980. Integrated animal-fish husbandary systems in Hong Kong with case studies on duck-fish and goat-fish systems. In: Proceeding of KLARM - SEARCA Conf. on "Integrated Agriculture - Aquaculture Farming Systems." 4th Aug, 1979, Manila, Philippines.

Sinha, V. R. P. 1989. Fisheries Development in India, Fishing Chimes. 9(3): 23-29.

Singh, B. P. 1990. Case of equity to farmers. Survey of Indian agriculture 1990. pp. 17-21. The Hindu, Chennai.

Singh, D. P. 1958. Rape seed and mustard. The Indian Central Oilseed Committee, Hyderabad.

Singh, G. B.; Singh, K. N. and Singh, R. P.1990. Sustainable land use systems in South Asia. In: Proceeding of International Workshop on "Sustainable Land use System Research". pp. 23-33. Oxford and IBH Publishing Co. (Pvt.) Ltd., New Delhi, India.

Singh, G. B. 1996. Integrated Nutrient Management: Strategy for balanced use. Survey of Indian Agriculture. The Hindu, Chennai, India.

Singh, K. N. and Singh, R. P. 1992. Farming system research and development - A holistic approach for sustainable agriculture growth. In: Proceeding of the XII National symposium on "Resource Management for Sustained Crop Production" held at Rajasthan Agricultural University, Bikaner during Feb. 25-28. pp. 391-398.

Singh, K. N. and Singh, R. P. 1990. Agronomic Research towards Sustainable Agriculture. Indian Society of Agronomy, IARI, New Delhi.

Singh, N.; Panda, S. C. and Gulati, J. M. L. 1989. Effect of moisture regimes of two stages of growth on groundnut. OUAT Jour. Agri. 2 93-4): 221-223.

Singh, S. S. 1990. Crop Management under Irrigated and Rainfed Conditions. Kalyani Publishers, Ludhiana.

Singh, P. K. 1989. Use of azolla in Asian Agriculture. Appl. Agric. Res. 4: 149-161.

Singh, P. K. and Singh, D. P. 1992. Azolla and Rice Cultivation. In: Biofertilizers. Somani, L.

L. *et. al.*, (eds.). Scientific publishers, Jodhpur, India. pp. 39-65.

Singh, R. B. 1992. Integrated farming with magur fish is profitable to small farmers. Indian Fmg. (11): 5-12.

Singh, R. B. and Paroda, R. S. 1994. Sustainability and productivity of rice-wheat systems in the Asia-pacific regions. Research and technology development needs. pp. 1-35. In: Paroda, RS., T. Woodhead and RB (eds.), Sustainability of rice-wheat production systems in Asia, FAO of United Nations Regional Office for Asia and the Pacific, Bangkok, Thailand.

Singh, R. P. 1983. Dry land Agricultural Research in India: Thrusts in the eighties, 187-191 (Eds. RP. Singh and V. Balasubramanian). Hyderabad, Andhra Pradesh, India: AICRPDA.

Singh, R. P. 1991. Farm Ponds. All India Co-ordinated Research Project for Dryland Agriculture. Hyderabad, India.

Singh, R. P.; Das, S. K.; Bhaskara Rao, U. M. and Reddy, M. N. 1990. Towards sustainable dryland agriculture practices. Central Research Institute for Dryland Agriculture, Hyderabad, India.

Singh, R. P.; Padmanabhan, M. V., Das, S. K. and Mishra, P. K. 1990. Field manual on watershed management. Central Research Institute for Dryland Agriculture, Hyderabad, India.

Singh, S. N.; Singh, K. P.; Singh, N.; Kadian, U. S. and Dahiya, S. S. 1988. Employment potentialities of different farming systems. In: Absts. of National Seminar on "Farming Systems for Semi-arid Tropics", Tamil Nadu Agric. Unlv., Coimbatore, India. p. 7.

Singh, S. P. 1987. Efficient Crop and Cropping Systems. In: Technological advances in dry land agriculture, 47-66. Hyderabad, India: CRIDA.

Singh, S. P. 1994. Hand book of agroforestry. Agrotech Publishing Academy. Udaipur. India.

Singh, S. P. and Prasad, C. 1991. Technologies for maximizing risk in rainfed agriculture. Indian Society of Extension Education and ICAR, New Delhi.

Singh, S. S. 1996. Principles and practices of Agronomy. Kalyani Publishers, New Delhi.

Singh, S. S. 1996. Soil fertility and nutrient management. Kalyani Publishers, New Delhi.

Singh, S. S. 1007. Crop management under irrigated and rainfed conditions. Kalyani Publishers, New Delhi.

Sirohiet, A. S.; Shanna, B. M.; Singh, I. and Dayal, R. 1980. Role of dairy and poultry enterprises for increasing income and employment of farm in the Union Territory of Delhi. Indian J. Agrl. Econ. 35: 112-119.

Sivaraj, S. 1989. Economics of poultry-cum-fish culture as a component in the farming system under Thanjavur district, Tamil Nadu. M.Sc.(Ag.) Thesis, Tamil Nadu Agric. Univ., Coimbatore, India.

Sivasankaran, D. 1994. Integrated farming system research under rainfed lands. In: Proc. of Summer Institute on "IFS Research and Management for Sustainable Agriculture" held at Tamil Nadu Agric. Univ., Coimbatore, India, between 6-15th June, 1994. pp. 115-121.

Sivanappan, R. K. and Karai Gowder, K. R. 1977. Irrigation and drainage. Popular Book Depot., Madras, India.

Smith, L. P. 1975. Methods in Agricultural Meteorology, Elsevier Scientific Publishing Company, Amsterdam.

Somani, L. L. and Bhandari, S. C. 1990. Rhizobium In: Biofertilizers (Somani, L. L., S. C Bhandari, S. N. Saxena and KK Uyas (eds.), Scientific Publishers, Jodhpur, India. pp. 39-65.

Somani, L. L. 1992. Dictionary of weed science. Agrotech Publishing Academy, Udaipur, India.

Somani, L. L.; Vittal, K. P. R. and Venkateswarlu, 1992. Dryland Agriculture in India – State of Art of Research in India. Scientific Publishers, Jodhpur, India.

Spencer, D. S. C. 1991. Institutionalizing the farming systems perspective in multi-commodity research institutes. The role of systems based research groups. Exptl. Agric. 27: 1-9.

Sreekumar, K. R. and Tilak, K. V. B. R. 1980. Nitrogen fixation in forage plants. Pasture and forage crops research: State of knowledge reprot. Singh P. (ed.) Proceeding of Third International Range Land Crops, New Delhi.

Sree Ramulu, V. S. 1995. Dry farming in India. Oxford and IBH Publishing Co. Pvt. Ltd., New Delhi.

Srinivasan, V. M. 1994. Role of Agroforestry for sustainable agriculture. In: Proceedings of Summer Institute on "Integrated Farming Systems Research and Development for Sustainable Agriculture" organized by ICAR and Tamil Nadu Agric. Univ., Coimbatore, India during 6-15 June, 1994. pp. 168-173.

Srivastava, H. C.; Bhaskaran, S.; Menon, K. K. G.; Ramanujam, S. and Rao, M. V. 1984. Pulse production constraints and opportunities. Oxford and IBH Publishing Co. Pvt. Ltd., New Delhi.

Srivastava, S. C.; Johari, D. P. and Gill, P. S. 1988. Manual of Sugarcane production in India. ICAR, New Delhi.

Stephen, O. Duke. 1987. Weed physiology. Vol. II Herbicide physiology. CRC Press, Inc. Boca Raton, Florida.

Steppler, H. and Lundgren, B. 1988. Agroforestry: Now and in the Future. Outlook on Agriculture 17: 146-151.

Stinner, B. and Blair, J. 1990. Biological and agronomic characteristics of innovative cropping systems. In: Edwards, CA. *et. al.*, (eds.). Sustainable Agricultural Systems. Soil and Water Conservation Society, Ankeng, Iowa. pp. 123-140.

Stocking, M. 1986. The cost of soil division in Zimbabwe in terms of the loss of three major nutrients. Consultant's working paper 3, Soil Conservation Programme, Land and Water Development division, FAa, Rome.

Stoop, W. 1984. Interdisciplinary a'11d Intercommodity aspects in the organization of a farming systems research perspective. Paper presented at the INSORMIL workshop held at CIMMYT, EI Batan, Mexico, September 17-21, 1984.

Subba Rao, N. S. 1988. In Biological N fixation - Recent trends. Subba Rao, N. S(ed). Oxford and lBH, New Delhi. pp. 1-19.

Subbaian, P.; Annadurai, K. and Palaniappan, S. P. 2000. Agriculture: Facts and Figures, Kalyani Publishers, Ludhiana, p. 359.

Subbian, P. and Palaniappan, S. P. 1992. Effect of integrated management practices on the yield and economics of crops under high inten<;ity multiple cropping systems. India, J. Agron. 37: 1-5.

Sulton, O. G. 1953. Micrometeorology, McGraw Hill Company, Inc., New York.

Sundaresan, D. 1979. The role of improved buffaloes in rural development. Indian Dairyman. 31: 73-78.

Sunith, C. P. 1975. Methods in Agricultural Meteorology. Elsevier Scientific Publishing Co., Amstardam, India.

Swaminathan, K. R. 1992. Renewable energy technologies for rural women. Paper presented during Annual Biogas Seminar of AFPRO-CHFN Network, Nagercoil, October, 28, 1992.

Swaminathan, M. S. 1975. Research advances in agriculture in relation to dairy development in India. Indian Dairyman. 27: 25-61.

Swaminathan, M. S. 1987. Inaugural address at the International symposium on "Sustainable Agriculture". The role of green manure crops in rice farming systems. 25-29 May, 1987. IRRI, Philippines.

TAC/ICGIAR, 1988. Sustainable Agricultural Production: Implications for International Agricultural Research. FAO, Rome, Italy.

TAC. 1989. Sustainable Agricultural Production: Implications for International Agricultural Research, CGIAR, Washington.

Tan, E. S. P. and Khoo, K. H. 1980. The integration of fish farming with agriculture in Malaysia. In: Proceeding of ICLARM - SEARCA Conf. on "Integrated Agriculture - Aquaculture Farming Systems". 4th Aug., 1979., Manila, Philippines. pp. 175-187.

Tejwani, K. G.; Gupta, S. K. and Mathur, H. N. 1975. Soil and Water Conservation Research 1956-71. New Delhi, India: ICAR.

Thakur, C. 1979. Scientific crop production. Vols 1 and 2, Metropolitan Book Co. Pvt. Ltd., New Delhi.

Thakur, R. C.; Bindra, A. D.; Good, R. D. and Bhargava, M. 1995. Effect of Fertilizer application on wheat crop sequence. Indian J. Agron. 40: 4-13.

Thanh Hung, LE. 1992. Integration of aquaculture with coastal rice farming in Vietnam. NAG A, The ICLARM Quartly. 15 (2): 2-29.

Thurlow, D. L and Hiltbott, A.1985. Nitrogen fixation by soybeans in Alabama. Agron. J. 77: 432-436. "

Tilak, K. V. B. R. 1985. Interaction of VAM and N Fixers. In: Proceeding of National Symposium on "Current Trends in Soil Biology". Mishra, M. M. and K. K. Kapoor (eds.). HAU. Hisar, India, pp. 219:226.

Tilak, K. V. B. R. and Dwivedi, A. 1989. Field response of barley to *Azospirillum brasilense* inoculation. Indian J. Microbiol. 29: 61-64.

Tiuy, J. 1990. Agricultural Ecology. Longman Scientific and Technical, Essex, United Kingdom.

Tiwari, P. N. 1993. Integrated Farming Research for sustaining food production. J. Nuclear Agric. BioI. 20: 1-13.

Trewartha, A. T. 1964. An Introduction to climate, McGraw Hill Book Company, Inc. New York.

Tripathi, R. P. and Singh, H. P. 1993. Soil erosion and conservation. Wiley Eastern Ltd., New Delhi.

Troll, C. 1965. Seasonal climate of the earth. In: World maps of Climatology (Eds. E. Rodenwaldt and H. Jusatz). Berlin: Springer Verlag.

Ujah, J. E. and Adeoge, K. B. 1984. Effects of shelter belts in the Sudan Savanna Zone of Nigeria on Microclimate and yield of millet. Agric Forest Meteorol., 33:99-107.

Vangriensven, L. G. 1987. The cultivation of mushroom: Its present status and future development. Outlook on Agric. 6: 131-135.

Vaidya, V. G. and Sahastrabudhe, K. R. 1970. Introduction to Agronomy and Soil and Water Management. Continental Prakasan, Poone.

Veerabadran, V. 1994. Farming systems research in southern parts of Tamil Nadu. In: Proceeding of Summer Institute on "Integrated Farming System Research and Management for Sustainable Agriculture". 6-15th June, 1994. Tamil Nadu Agric. Univ., Coimbatore. pp. 258-269.

Venkataraman, G. S. and Shanmugasundaram, S 1992. Algal biD-fertilizer technology for rice. Madurai Kamarajar University, Madurai, India. pp. 1-24.

Venkateswarlu, J. 1987. Efficient resource management system for Drylands of India. Advances in Soil Science. 7: 165-221.

Velayutham, M. 1996. Management of Natural Resources in India. Souvenir, 2nd International Crop Science Congress, Nov. 17-24, 1996. New Delhi, India.

Verma, L. N. and Bhattacharyya, P. 1992. Production, distribution and production of biofertilizer. In: Fertilizers, Organic Manures, Recyclable Wastes and Biofertilizers. Tandon, H. L. S. (ed.), FDCD, New Delhi, pp. 132-147.

Verma, S. C. 1977. Agronomy of new plant types. Tara Publication. Varanasi.

Vikraman Nair, R.; Pillai, N. G. and Shaskanaman, 1979. Mixed farming in coconut gardens. Indian Farming. 26: 24-25.

Virmani, S. M. 1997. Rainfed Farming: Watershed based Approach. Survey of Indian Agriculture. The Hindu. Chennai, India.

Vishwa Nath and Chhotey Lal. 1995. Oilseeds in India – An overview. Westvill Publishing House, New Delhi.

Wali, M. K. (ed.), 1992. Ecosystem rehabilitation: Preamable to sustainable development. SPB Academic Publishing. The Hague, Netherlands.

Walker, T. S. 1982. The perceptions of farmers and the participation in the Taddanpalle watershed project in 1982-1983. ICRISAT economics program progress report, 44. Andhra Pradesh, India: ICRISAT.

Walker, T. S. and Ryan, J. G. 1990. Village and household economics in India's Semi-arid tropics. Baltimore, Maryland, USA: Johns Hopkins University Press.

Wani,. S. P. 1990. Inoculation with associative N fixing bacteria: Role of cereal grain production improvement. Indian J. Microbiol. 30: 363-393.

Wani,. S. P.; Chandrapalaiah, I. S. and Dad, P. O. J. 1987. Nitrogen fixation potentialities of sorghum and millets. In: Biological Nitrogen Fixation: Recent Development. Subba Rao, N. S. (Ed.) Oxford and IBH, New Delhi. pp. I 25-174.

WCED, 1987. Our common future. Oxford University Press.

Weiskel, T. C. 1989. The ecological sessions of the past: An Anthropology of environment decline. The Ecologist. 19: 98-103.

Weiss, E. A. 1983. Oilseed crops. Longman Group Ltd., London

Whyte, R. O. 1964. Grassland and fodder resources of India. ICAR, New Delhi.

Whyte, W. F. 1977. Towards a new strategy for research and development in agriculture; Helping small farmers in developing countries. Desarrollo Rural En Los Americas. 9: 51-62.

Whyte, W. F. 1981. Participatory approaches to agricultural research and development: A State of the Art Paper. Cornell University, Center for International Studies. Rural Development Committee, Special series on agricultural research and extension. 1, Ithaca, New York. Cornell. University.

Wild, A. 1988. Russel's soil conditions and plant growth. Longman group, London.

Wilken G. C. 1987. Good farmers: Traditional agricultural resource management in Mexico and Central America. Berkely: University of California Press.

Willey, R. W. 1979. Intercropping-Its importance and research needs 1. Competition and yield advantage and 2. Agronomy and research approaches. Field crop Abstracts, 32: 1-10, 73-85.

Willey, R. W. and Osiru, D. S. O. 1972. Studies on mixtures of maize and beans with particular reference to plant populations. J. Agric., Sci. 79: 579-529.

Williams, R.; Zangavil, A. and Karmidi, A. 1980. A portable evaporimeter for rapid measurement of the evaporation rate of water. Agric. Forest Meteorol. 32:217-224.

Williams, J. R.; Jones, C. A. and Dyke, P. T. 1984. A modeling approach to determining the relationship between erosion and soil productivity. Trans. ASAE 27: 129-144.

Winkelmann, D. 1984. Recent views in farming systems research. In: Issues in Organization and management of research with a farming systems perspective aimed at technology generation (Ed. M. Joseph Chang). The Hague, Netherlands: CIMMYT and ISNAR.

World Resources, 1988-89. 1990-1991, 1992-1993. Reports by the World Resources Institute and International Institute for Environment and Development in Collaboration with the United Nations Environment Programme, Basic Book, New York.

Woyanarovich, E. 1980. Raising ducks on fishponds. In: Proceeding of ICLARM - SEARCA Conf. on "Integrated Agriculture-aquaculture Farming System", 4th Aug. 1979, Manila, Philippines.

Yadav, R. L. 1993. Agronomy of Sugarcane. Principles and practices. International Book Distributing Co. Lucknow, India.

Yellamada Reddi, T. and Sanka Reddi, G. H. 1992. Principles of Agronomy. Kalyani Publishers, New Delhi.

Yohida, S. 1981. Fundamentals of rice crop science. International Rice Research Institute, Manila, Philippines.

Young, A. 1989. Agroforestry for soil conservation. CAB International. Walling Ford, U.K.

Zandstra, H. G. 1982. An overview of farming systems research. Keynote address for farming systems in the field "Farming Systems Research Symposium", Kansas State University, Manhattan, Kansas, November 20-21, 1982.

Zandstra, H. G.; Pric, E. C.; Litsinger, J. A. and Morris, R. A. 1981. A methodology for on-farm cropping systems research. Losbanos, Philippines: International Rice Research Institute.

Zandstra, H. G. 1979. Cropping systems research for Asian rice farmers. Agricultural Systems, 4: 135-155.